Water Relations of Plants

Water Relations of Plants

PAUL J. KRAMER

Department of Botany
Duke University
Durham, North Carolina

ACADEMIC PRESS, INC.

(Harcourt Brace Jovanovich, Publishers)

Orlando   San Diego   New York   London

Toronto   Montreal   Sydney   Tokyo

ACADEMIC PRESS, INC.
Orlando, Florida 32887

United Kingdom Edition published by
ACADEMIC PRESS, INC. (LONDON) LTD.
24/28 Oval Road, London NW1   7DX

Library of Congress Cataloging in Publication Data

Kramer, Paul Jackson, Date
    Water relations of plants.

    Includes index.
    Bibliography: p.
    1. Plant-water relationships.   I. Title.
QK870.K7 1983   581.19'212   82-18436
ISBN 0-12-425040-8

PRINTED IN THE UNITED STATES OF AMERICA

85 86 87 88   9 8 7 6 5 4 3 2

Contents

12   Water Deficits and Plant Growth

13   Drought Tolerance and Water Use Efficiency

Preface

The importance of an understanding of the water relations of plants is indicated by the ecological and physiological importance of water. Not only is the distribution of vegetation over the earth's surface controlled chiefly by the availability of water, but crop yields are more dependent on an adequate supply of water than on any other single factor. This book attempts to explain the importance of water by describing the factors that control the plant water balance and showing how they affect the physiological processes that determine the quantity and quality of growth. It is intended for students, teachers, and investigators in both basic and applied plant science, and it should be useful as a reference to botanists, agronomists, foresters, horticulturists, soil scientists, and even laymen with an interest in plant water relations. An attempt has been made to present the information in terms intelligible to readers in all fields of plant science. If the treatment of some topics seems inadequate to specialists in those areas, they are reminded that the book was not written for specialists in soil or plant water relations, but for those plant scientists, upper-level undergraduates, and graduate students who need a general introduction to the whole field.

The need for a book summarizing modern views on plant water relations has been increased by the large volume of publications and the changes in viewpoint that have occurred in recent years. A number of books on plant water relations have appeared, but most of them are collections of papers on special topics. This book attempts to present the entire field of water relations in an organized manner, using current concepts and a consistent, simple terminology. Emphasis is placed on the interdependence of various processes. For example, the rate of water absorption is closely linked to the rate of transpiration through the sap stream in the vascular system, and it also is affected by resistance to water flow into roots and by the various soil factors affecting the availability of water. The rate of transpiration depends primarily on the energy supply, but it also is affected by stomatal opening and the leaf water supply. Proper functioning of the physiological processes involved in growth requires a favorable water balance, which is controlled by relative rates of absorption and water loss by transpiration.

These complex interrelationships are emphasized and described in modern terminology.

Although the primary objective of this book is to present a survey of modern concepts in the field of plant water relations, attention is also given to some of the older work because it constitutes the foundation on which modern concepts are based. Workers should understand that plant water relations has a long history of productive research dating back to the beginning of plant physiology as a science. Quantitative study of water relations began with the measurements of root pressure and transpiration made early in the eighteenth century by Hales. It was expanded during the nineteenth century by the research of Dutrochet, Pfeffer, Sachs, Strasburger, and others. It is interesting to note that by 1860 Sachs was aware that cold soil reduces water absorption by warm season plants more than absorption by cool season plants, and that late in the nineteenth century Francis Darwin observed that atmospheric humidity affects stomatal opening and Dixon proposed the cohesion theory of the ascent of sap. The concept of water potential was developed during the second decade of the twentieth century under the name of "suction force." In fact, most of the basic concepts in plant physiology were in existence over 50 years ago, but improvements in research methods have greatly increased our understanding of them.

The large volume of publication in recent years makes it impossible to cite all of the relevant literature and many good papers have been omitted. Nevertheless, the bibliography is extensive enough to serve as the primary source of references in almost any area of plant water relations. Today's lively research activity is producing significant changes in explanations of various phenomena, and some long-held views are being reconsidered. For example, the assumptions that stomatal closure depends chiefly on loss of bulk leaf turgor, that the leaf mesophyll is the principal evaporating surface, and that water movement outside the xylem occurs chiefly in the cell walls are all being questioned. Also, the role of growth regulators, chiefly abscisic acid and cytokinins, is receiving much attention in relation to membrane permeability and stomatal closure. Differences in opinion among various investigators are discussed, and in some instances the author has indicated his preference, but it is pointed out that in many instances more research is needed before conclusions can be reached. Readers are reminded that so-called scientific facts often are merely the most logical explanations that can be developed from the available information. As additional research provides more information, it frequently becomes necessary to revise generally accepted explanations, and those that seem logical today may become untenable next year.

There are some changes from the terminology used in earlier versions of this book. Osmotic absorption replaces active absorption of water to avoid confusion with active transport, and drought tolerance replaces drought resistance because tolerance more accurately describes the situation. The bar has been replaced by the SI unit, the MPa (1 bar = 10^5 Pa or 0.1 MPa), and the millibar by the kPa (1 mb = 0.1 kPa). There is considerable cross-referencing between chapters, but there also is some repetition of material in various chapters. For example, the osmotic properties of cells and stomatal behavior are discussed in different contexts in different chapters. This makes each chapter a fairly complete unit that can be read without excessive reference to preceding chapters and facilitates use of the book as a reference.

This book reflects my interactions with many scientists, ranging from the late E. N. Transeau, who first called my attention to the interesting problems in plant water relations, to R. O. Slatyer who broadened my viewpoint and improved my terminology, to T. T. Kozlowski who encouraged the writing of this book, and to scores of other scientists with whom I have discussed problems. I am especially indebted to the many graduate students and postdoctoral research associates for their stimulating discussions and valuable suggestions, and to the Department of Botany of Duke University for providing a good environment in which to work. I appreciate the cooperation of friends and colleagues who read parts or all of the manuscript and offered many valuable suggestions. All of the manuscript was read by M. R. Kaufmann and T. T. Kozlowski, and certain chapters were read by J. A. Bunce, E. L. Fiscus, A. W. Naylor, and J. N. Siedow, and various topics were discussed with other scientists too numerous to list. Their suggestions contributed much to the book, but they should not be held responsible for any errors that may have crept in during the several revisions. The efficient assistance of Sue Dickerson and Joanne Daniels in typing the manuscript and Shirley Thomas in preparing the bibliography also is gratefully acknowledged.

The author also wishes to acknowledge the support provided by various granting agencies. In the earliest days his research was supported by Duke University Research Council grants, later by the Atomic Energy Commission, and for the past 25 years by grants from the National Science Foundation. These grants have supported many graduate students and postdoctoral research associates and scores of publications, and have contributed both directly and indirectly to the production of this book and of the 1969 version which it replaces.

Paul J. Kramer

1

Water: Its Functions and Properties

INTRODUCTION

   Water is one of the most common and most important substances on the earth's surface. It is essential for the existence of life, and the kinds and amounts of vegetation occurring on various parts of the earth's surface depend more on the quantity of water available than on any other single environmental factor. The importance of water was recognized by early civilizations, and it occupied a prominent place in ancient cosmologies and mythologies. The early Greek phi-

1

losopher Thales asserted that water was the origin of all things, and it was one of the four basic elements (earth, air, fire, water) recognized by later Greek philosophers such as Aristotle. It was also one of the five elemental principles (water, earth, fire, wood, metal) of early Chinese philosophers. Today it is realized that the availability of water not only limits the growth of plants but can also limit the growth of cities and industries. In this chapter we will discuss the ecological and physiological importance of water, its unique properties, and the properties of aqueous solutions.

Ecological Importance of Water

Regions where rainfall is abundant and fairly evenly distributed over the growing season have lush vegetation. Examples are the rain forests of the tropics, the vegetation of the Olympic peninsula and the northwestern United States, and the luxuriant cove forests of the southern Appalachians. Where summer droughts are frequent and severe, forests are replaced by grasslands, as in the steppes of Asia and the prairies of North America. Further decreases in rainfall result in semidesert with scattered shrubs, and finally in deserts.

Even the effects of temperature on vegetation are partly produced through water relations because increasing temperature is accompanied by increasing rates of evaporation and transpiration. Thus, an amount of rainfall adequate for forests in a cool climate can only support grassland in a warmer climate where the rate of evapotranspiration is much higher. This was responsible for development of the concept of the rainfall evaporation ratio by Transeau (1905) as an indicator of the relationship between precipitation and type of plant cover. The concept was developed further by Thornthwaite (1948), and precipitation and evaporation are considered in climatic diagrams such as those constructed by Walter (1979, pp. 25–30).

Physiological Importance of Water

The ecological importance of water is the result of its physiological importance. The only way in which an environmental factor such as water can affect plant growth is by influencing physiological processes and conditions, as shown in Fig. 1.1.

Almost every plant process is affected directly or indirectly by the water supply. Many of these effects will be discussed later, but it can be emphasized here that within limits metabolic activity of cells and plants is closely related to

HEREDITARY POTENTIALITIES	ENVIRONMENTAL FACTORS
Depth and extent of root systems	SOIL. Texture, structure, depth, chemical composition and pH, aeration, temperature, waterholding capacity, and water conductivity
Size, shape and total area of leaves, and ratio of internal to external surface	
Number, location, and behavior of stomata	ATMOSPHERIC. Amount and distribution of precipitation
	Ratio of precipitation to evaporation
	Radiant energy, wind, vapor pressure, and other factors affecting evaporation and transpiration

PLANT PROCESSES AND CONDITIONS

Water absorption

Ascent of sap

Transpiration

Internal water balance as reflected in water potential, turgidity, stomatal opening, and cell enlargement

Effects on photosynthesis, carbohydrate and nitrogen metabolism, and other metabolic processes

QUANTITY AND QUALITY OF GROWTH

Size of cells, organs, and plants

Dry weight, succulence, kinds and amounts of various compounds produced and accumulated

Root-shoot ratio

Vegetative versus reproductive growth

Fig. 1.1. Diagram showing how the quantity and quality of plant growth are controlled by hereditary potentialities and environmental factors operating through the internal processes and conditions of plants. Special attention is given to factors and conditions affected by water relations.

their water content. For example, the respiration of young, maturing seeds is quite high, but it decreases steadily during maturation as water content decreases (see Fig. 1.2). The respiration rate of air-dry seeds is very low and increases slowly with increasing water content up to a critical point, at which there is a rapid increase in respiration with a further increase in water content (Fig. 1.3). Growth of plants is controlled by rates of cell division and enlargement and by the supply of organic and inorganic compounds required for the synthesis of new protoplasm and cell walls. Cell enlargement is particularly dependent on at least

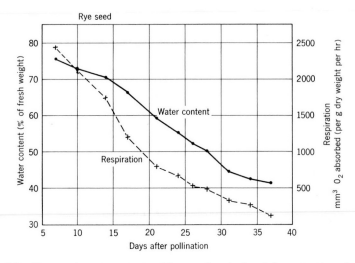

Fig. 1.2.   Decrease in water content and in rate of respiration during maturation of rye seed. (From Shirk, 1942.)

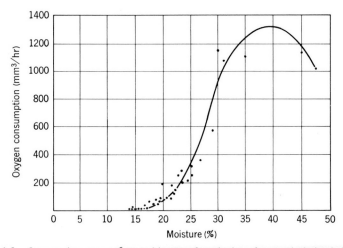

Fig. 1.3.   Increase in water content and in rate of respiration of oat seeds during imbibition of water. Note the rapid increase in respiration as the water content rises above approximately 16%. The water is probably so firmly bound at lower contents that it is unavailable for physiological processes. (From Bakke and Noecker, 1933.)

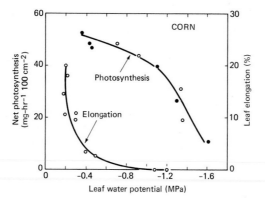

Fig. 1.4. Relationship among leaf water potential, leaf elongation, and photosynthesis of corn. Note that leaf elongation almost ceases before there is much reduction in photosynthesis. (From Boyer, 1970a.)

a minimum degree of cell turgor, and stem and leaf elongation is quickly checked or stopped by water deficits, as shown in Fig. 1.4 and in Chapter 12. Decrease in water content invariably inhibits photosynthesis (Fig. 1.4) and usually reduces the rate of respiration and other enzyme mediated processes.

In summary, decreasing water content is accompanied by loss of turgor and wilting, cessation of cell enlargement, closure of stomata, reduction in photosynthesis, and interference with many basic metabolic processes. Eventually, continued dehydration causes disorganization of the protoplasm and death of most organisms. The effects of water deficits on physiological processes are discussed in more detail in Chapter 12.

USES OF WATER IN PLANTS

The importance of water can be summarized by listing its most important functions under four general headings.

Constituent

Water is as important quantitatively as it is qualitatively, constituting 80–90% of the fresh weight of most herbaceous plant parts and over 50% of the fresh weight of woody plants. Some data on water content of various plant structures are shown in Table 1.1. Water is as important a part of the protoplasm as the protein and lipid molecules which constitute the protoplasmic framework, and reduction of water content below some critical level is accompanied by changes in structure and ultimately in death. A few plants and plant organs can be dehydrated to the air-dry condition, or even to the oven-dry condition in the case of some kinds of seeds and spores, without loss of viability, but a marked decrease in physiological activity always accompanies decrease in tissue water content. The relationship between water content and protein structure is discussed by Tanford (1963, 1980), Kuntz and Kauzmann (1974), Edsall and McKenzie (1978), and others.

TABLE 1.1   Water Content of Various Plant Tissues Expressed as Percentages of Fresh Weight[a]

	Plant parts	Water content (%)	Authority
Roots	Barley, apical portion	93.0	Kramer and Wiebe, 1952
	Pinus taeda, apical portion	90.2	Hodgson, 1953
	Pinus taeda, mycorrhizal roots	74.8	Hodgson, 1953
	Carrot, edible portion	88.2	Chatfield and Adams, 1940
	Sunflower, avg. of entire root system	71.0	Wilson *et al.*, 1953
Stems	Asparagus stem tips	88.3	Daughters and Glenn, 1946
	Sunflower, avg. of entire stems on 7-week-old plant	87.5	Wilson *et al.*, 1953
	Pinus banksiana	48–61	Raber, 1937
	Pinus echinata, phloem	66.0	Huckenpahler, 1936
	Pinus echinata, wood	50–60	Huckenpahler, 1936
	Pinus taeda, twigs	55–57	McDermott, 1941
Leaves	Lettuce, inner leaves	94.8	Chatfield and Adams, 1940
	Sunflower, avg. of all leaves on 7-week-old plant	81.0	Wilson *et al.*, 1953
	Cabbage, mature	86.0	Miller, 1938
	Corn, mature	77.0	Miller, 1938
Fruits	Tomato	94.1	Chatfield and Adams, 1940
	Watermelon	92.1	Chatfield and Adams, 1940
	Strawberry	89.1	Daughters and Glenn, 1946
	Apple	84.0	Daughters and Glenn, 1946
Seeds	Sweet corn, edible	84.8	Daughters and Glenn, 1946
	Field corn, dry	11.0	Chatfield and Adams, 1940
	Barley, hull-less	10.2	Chatfield and Adams, 1940
	Peanut, raw	5.1	Chatfield and Adams, 1940

[a] From Kramer (1969).

Solvent

A second essential function of water in plants is as the solvent in which gases, minerals, and other solutes enter plant cells and move from cell to cell and organ to organ. The relatively high permeability of most cell walls and protoplasmic membranes to water results in a continuous liquid phase, extending throughout the plant, in which translocation of solutes of all kinds occurs.

Reactant

Water is a reactant or substrate in many important processes, including photosynthesis and hydrolytic processes such as the amylase-mediated hydrolysis of starch to sugar in germinating seeds. It is just as essential in this role as carbon dioxide is in photosynthesis or nitrate is in nitrogen metabolism.

Maintenance of Turgidity

Another essential role of water is in the maintenance of the turgidity which is essential for cell enlargement and growth and for maintaining the form of herbaceous plants. Turgor is also important in the opening of stomata and the movements of leaves, flower petals, and various specialized plant structures. Inadequate water to maintain turgor results in an immediate reduction of vegetative growth, as shown in Fig. 1.4.

PROPERTIES OF WATER

The importance of water in living organisms results from its unique physical and chemical properties. These unusual properties were recognized in the nineteenth century (see Edsall and McKenzie, 1978, for references), and their importance was discussed early in the twentieth century by Henderson (1913), Bayliss (1924), and Gortner (1938). Even today there is some uncertainty about the structure of water and some of its properties, as will be seen later. However, there is no doubt that water has the largest collection of anomalous properties of any generally known substance.

Unique Physical Properties

Water has the highest specific heat of any known substance except liquid ammonia, which is about 13% higher. The high specific heat of water tends to stabilize temperatures and is reflected in the relatively uniform temperature of islands and land near large bodies of water. This is important with respect to agriculture and natural vegetation. The standard unit for measuring heat, the calorie (cal), is 4.18 joules (J), or the amount of energy required to warm 1 gram (g) of water one degree, from 14.5° to 15.5°C. The heat of vaporization is the highest known, 540 cal/g at 100°C, and the heat of fusion, 80 cal/g, is also unusually high. Because of the high heat of vaporization, evaporation of water has a pronounced cooling effect and condensation has a warming effect. Water is also an extremely good conductor of heat compared with other liquids and nonmetallic solids, although it is poor compared with metals. A substance with the molecular weight of water should exist as a gas at room temperature and have a melting point of below −100°C.

Water is transparent to visible radiation. This allows light to penetrate bodies of water and makes it possible for algae to carry on photosynthesis and grow to considerable depths. It is nearly opaque to longer wavelengths in the infrared range, so that water filters are fairly good heat absorbers (see Fig. 1.5).

Water has a much higher surface tension than most other liquids because of the high internal cohesive forces between molecules. This provides the tensile strength required by the cohesion theory of the ascent of sap. Water also has a high density and is remarkable in having its maximum density at 4°C instead of at

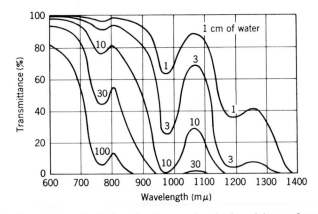

Fig. 1.5. Transmission of radiation of various wavelengths through layers of water of different thicknesses. The numbers on the curves refer to the thickness of the layers in centimeters. Transmission is much greater at short than at long wavelengths. (After Hollaender, 1956, p. 195.)

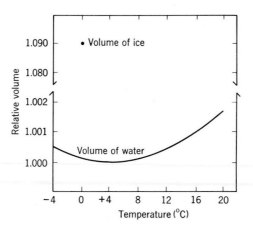

Fig. 1.6. Change in volume of water with change in temperature. The minimum volume is at 4°C, and below that temperature there is a slight increase in volume as more molecules are incorporated into the lattice structure. The volume increases suddenly when water freezes because all molecules are incorporated into a widely spaced lattice. Above 4°C there is an increase in volume caused by increasing thermal agitation of the molecules.

the freezing point. Even more remarkable is the fact that water expands on freezing, so that ice has a volume about 9% greater than the liquid water from which it was formed (Fig. 1.6). This explains why ice floats and pipes and radiators burst when the water in them freezes. Incidentally, if ice sank, bodies of water in the cooler parts of the world would all be filled permanently with ice, with disastrous effects on the climate and on aquatic organisms.

Water is very slightly ionized, only one molecule in 55.5×10^7 being dissociated. It also has a high dielectric constant which contributes to its behavior as an almost universal solvent. It is a good solvent for electrolytes because the attraction of ions to the partially positive and negative charges on water molecules results in each ion being surrounded by a shell of water molecules which keeps ions of opposite charge separated (Fig. 1.8). It is a good solvent for many nonelectrolytes because it can form hydrogen bonds with amino and carbonyl groups. It tends to be adsorbed, or bound strongly, to the surfaces of clay micelles, cellulose, protein molecules, and many other substances. This characteristic is of great importance in soil and plant water relations.

Explanation of Unique Properties

It was realized early in this century that the unusual combination of properties found in water could not exist in a system consisting of individual H_2O mole-

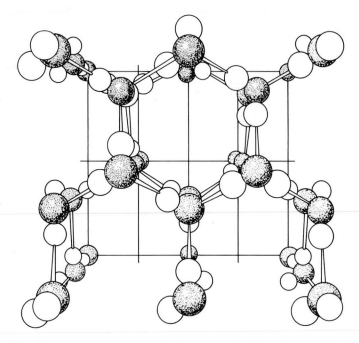

Fig. 1.7.   Diagram showing approximately how water molecules are bound together in a lattice structure in ice by hydrogen bonds. The dark spheres are oxygen atoms, and the light spheres are hydrogen atoms. (After Buswell and Rodebush, 1956.)

cules. At one time, it was proposed that water vapor is monomeric H_2O, ice is a trimer $(H_2O)_3$ consisting of three associated molecules, and liquid water is a mixture of a dimer $(H_2O)_2$ and a trimer. Now the unusual properties are explained by assuming that water molecules are associated in a more or less ordered structure by hydrogen bonding. Ice is characterized by an open crystalline lattice and liquid water by increasing disorder, and in the vapor phase the individual molecules are not associated at all. The properties and structure of water have been treated in many articles and books, including Kavanau (1964), Eisenberg and Kauzmann (1969), and a multivolume compendium edited by Franks (1975). Recent views are presented by Edsall and McKenzie (1978) and Stillinger (1980).

To explain the unusual properties of water requires a brief review of the kinds of electrostatic forces that operate among atoms and molecules. These include the strong ionic or electrovalent bonds and covalent bonds, and weaker attractive forces known as van der Waals or London forces and hydrogen bonds. Ionic bonds result from electrostatic attraction between oppositely charged partners, as between sodium and chlorine atoms in sodium chloride (NaCl). Such compounds usually ionize readily. Covalent bonds are formed by sharing electrons, as between oxygen and hydrogen atoms in water and carbon and hydrogen atoms in organic compounds. Such compounds do not ionize readily. Covalent bonds are strong, about 110 kcal/mol for the O—H bond in water, but they may be broken during chemical reactions.

If ionic or covalent bonds were the only types of bonding, there would be no liquids or solids because these do not allow individual molecules to interact with each other. However, there are intermolecular binding forces called van der Waals or London forces and hydrogen bonds that operate between adjacent molecules and affect the behavior of gases and liquids. Some molecules are polar or electrically asymmetric because they have partially positively and negatively charged areas caused by unequal sharing of electrons between atoms. These charged areas attract one molecule to another. Water shows this dipole effect rather strongly, resulting in the hydrogen bonding discussed later. Substances such as carbon tetrachloride and methane do not show permanent dipole effects because their molecules have no asymmetric distribution of electrons and consequently no charged areas.

Even electrically neutral molecules show anomalous properties, and in 1873 van der Waals suggested that the nonideal behavior of gases is caused by weak attractive forces operating between such molecules. In 1930 London developed an explanation for these attractions based on the assumption that even those molecules that on the average are electrically symmetrical or neutral develop momentary or instantaneous dipoles. These dipoles induce temporary dipoles in neighboring molecules, causing the instantaneous or momentary attraction between them known as van der Waals or London forces. This attraction is weak, about 1 kcal/mol, and effective only if molecules are very close together. In general, physical properties of liquids such as the boiling point, heat of vaporization, and surface tension depend on the strength of intermolecular bonding. For example, gases condense into liquids when cooled enough so that the van der Waals attraction between molecules exceeds the dispersive effect of their kinetic energy.

The peculiar physical properties of water result from additional intermolecular forces much stronger than van der Waals forces. These strong attractive forces are hydrogen bonds that result from the weak electrostatic attraction of the partially positively charged hydrogen atoms of one water molecule to the partially negatively charged oxygen atoms of adjacent molecules. They operate over considerable distances and have a binding force of about 1.3–4.5 kcal/mol in water. The forces produced by the asymmetric distribution of charges on water molecules bind them in the symmetrical crystalline lattice structure of ice, shown diagrammatically in Fig. 1.7. The water molecules are arranged in a lattice with unusually wide spacing, resulting in a density lower than that of liquid water.

As ice melts, 13–15% of the bonds break, and perhaps 8% of the molecules

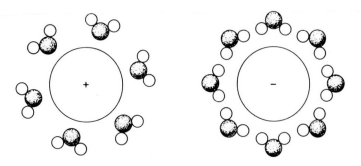

Fig. 1.8. Diagram showing approximate arrangement of water molecules in shells oriented around ions. These shells tend to separate ions of opposite charge and enable them to exist in solution. They also disrupt the normal structure of water and slightly increase the volume. (After Buswell and Rodebush, 1956.)

escape from the lattice. This results in a partial collapse of the lattice into a more disorderly but also more compact structure and an increase to maximum density at 4°C. As the temperature rises above 4°C, further increases in breakage and deformation of hydrogen bonds result in an increase in volume (see Fig. 1.6). There has been, and still is, some uncertainty about the structure of liquid water, i.e., the manner in which the molecules are oriented in relation to one another. Incidentally, the concept of structure refers only to average positions of molecules because they are continually in motion and exchanging bonds. At one time, it was believed that liquid water consisted of "flickering clusters" or "icebergs" of structured water molecules surrounded by unstructured molecules (Néméthy and Scheraga, 1962). However, Stillinger (1980) states that recent studies tend to rule out the iceberg concept. He regards liquid water as a three-dimensional network of hydrogen bonded molecules showing a tendency toward tetrahedral geometry but containing many strained or broken bonds. Only part of the structure is destroyed by heating, and about 70% of the hydrogen bonds found in ice remain intact in liquid water at 100°C. The high boiling point of water results from the large amount of energy required to break the remaining hydrogen bonds and vaporize liquid water.

The structure of water is somewhat modified by the pH, because it affects the distance of the hydrogen from the oxygen atoms, and by ions, because of their attraction for water molecules. Ions also form dipole bonds with water molecules. The result is that the ions become surrounded by firmly bound shells of water molecules (Fig. 1.8). In fact, Bernal (1965) described ions, protein molecules, and cell surfaces in general as being coated with "ice," i.e., with layers of structured water molecules. It is now considered unlikely that water can form a uniform layer over the surface of protein molecules, but it probably can form such layers on the surfaces of cellulose and other substances having a more uniform distribution of bonding sites. Solutions of alcohols, amides, or other polar liquids in water result in a more strongly structured system than occurs in the separate substances. This is seen in the high viscosity of such solutions. For example, the viscosity of an ethyl alcohol–water mixture at 0°C is four times that of water or alcohol alone. However, this structure is easily broken by high temperatures (see Fig. 1.9). The addition of nonpolar substances such as benzene or other hydrocarbons to water breaks bonds and produces "holes" or disorganized areas in the structure which are surrounded by areas with a tighter structure. The water bound to large molecules such as proteins has an important effect on their structure; and Tanford (1963, 1980) cites evidence suggesting that the relative stability of the structure of viruses, DNA, and globular proteins is affected by the water associated with them.

It was mentioned earlier that the changes in volume of water during freezing and thawing are caused by changes in the proportion of water molecules bound in an organized lattice by hydrogen bonds. The high boiling point results prin-

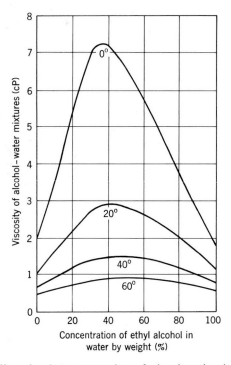

Fig. 1.9. The effect of various concentrations of ethanol on the viscosity of water at four temperatures. Mixtures of water with polar organic liquids often show large increases in viscosity at low temperatures because they have a more tightly packed structure than the component liquids alone. (After Bingham and Jackson, 1918.)

cipally from the large amount of energy required to break hydrogen bonds, two of which must be broken for each molecule evaporated. Methane (CH_4) has nearly the same molecular weight as water, but it boils at $-161°C$ because no hydrogen bonding occurs and only a small amount of energy is required to break the weak van der Waals forces holding the molecules together in the liquid.

The unusually high viscosity and surface tension of water also result from the fact that hydrogen bonds between water molecules resist rearrangement. Water wets and adheres to glass, clay micelles, cellulose, and other substances having exposed oxygen atoms at the surface with which hydrogen bonds can be formed. It does not wet paraffin and other hydrocarbons because it cannot form hydrogen bonds with them, but it wets cotton, because it forms numerous hydrogen bonds with oxygen atoms of the cellulose molecules.

Unorthodox Views Concerning Water

It is generally believed that although most of the water in cells possesses the structure and properties of bulk water, a small amount is adsorbed on the surfaces of membranes and macromolecules. This is the ice of Bernal (1965) and the bound water of Gortner (1938) and others. However, according to Ling (1969), Cope, and a few other physiologists, a significant amount of the water in living cells has a structure different from bulk water. This vicinal or associated water is said by them to affect the accumulation of ions and eliminate the classical role of cell membranes and their associated ion pumps (see Hazelwood and others, in Drost-Hansen and Clegg, 1979). This concept has not been widely accepted (Kolata, 1979), and it seems especially doubtful that it could apply to plant cells with their large volume of vacuolar water.

It has also been suggested by Drost-Hansen (1965) and others that there are anomalies in the physical properties of water at about 15°, 30°, and 45°C. For example, it is claimed that there are peaks in the disjoining pressure and viscosity of water adsorbed on surfaces, caused by phase transitions in vicinal water, i.e., water adsorbed on surfaces of macromolecules in cells (Etzler and Drost-Hansen, 1979). It was also claimed by these writers and by Nishiyama (1975) and Peschel (1976) that there are peaks in seed germination, growth of microorganisms, and other biological processes which are related to these anomalies. These claims have been received skeptically (Eisenberg and Kauzmann, 1969), and Falk and Kell (1966) concluded that the reported discontinuities in physical properties are no greater than the errors of measurement. It seems more likely that discontinuities in biological processes are related to phase transitions in membranes than to phase transitions or discontinuities in the properties of water (see Chapter 9).

Another anomaly is the polywater reported by Russian investigators in the 1960s. This was believed to be a polymeric form of water with anomalous properties, but it later turned out to be water containing a high concentration of solute (Davis *et al.*, 1971). The story of polywater was told in detail by Franks (1981). It has also been claimed by Russian investigators that water from freshly melted snow stimulates certain biological processes. Recently, other Russian investigators claimed that water boiled to remove all dissolved gas and then quickly cooled not only has greater density, viscosity, and surface tension but also stimulates plant and animal growth, and concrete prepared with it is stronger than that prepared with ordinary water (Maugh, 1978). These claims have not been verified elsewhere and should be treated with caution.

Isotopes of Water

The three isotopes of hydrogen having atomic weights of 1, 2, and 3 make it possible to differentiate tracer water from ordinary water. In the 1930s, heavy water [water containing deuterium (hydrogen of atomic weight 2)] became available and was used widely in biochemical studies. It was also used extensively in studies of permeability of animal and plant membranes (e.g., Ordin and Kramer, 1956; Ussing, 1953). However, in recent years deuterium has been largely supplanted as a tracer by water containing tritium (hydrogen of atomic weight 3), e.g., in the experiments of Raney and Vaadia (1965) and Couchat and Lasceve (1980). Tritium is radioactive and therefore more convenient as a label, being easier to detect than deuterium, which requires use of a mass spectrometer.

A stable isotope of oxygen with an atomic weight of 18 makes it possible to study the role of oxygen in water. An example is the series of experiments with $H_2^{18}O$ which demonstrated that the oxygen released during photosynthesis comes from water rather than from carbon dioxide (Ruben *et al.*, 1941).

PROPERTIES OF AQUEOUS SOLUTIONS

In plant physiology we seldom deal with pure water because the water in plants and in their root environment contains a wide range of solutes. Therefore, it is necessary to understand how the properties of water in solution differ from those of pure water. Only a brief discussion is possible here, and readers are referred to physical chemistry texts for a full development of these ideas.

The characteristics of water in solution can be shown concisely by tabulation

TABLE 1.2   Colligative Properties of a Molal Solution of a Nonelectrolyte Compared with Water

	Pure water	Molal solution
Vapor pressure	0.61 kPa at 0°C 101.3 kPa at 100°C	Decreased according to Raoult's law
Boiling point	100°C	100.518°C
Freezing point	0°C	−1.86°C
Osmotic pressure	0	2.27 MPa
Chemical potential	Set at zero	Decreased

of its colligative properties, i.e., the properties associated with the concentration of solutes dissolved in it. These are shown in Table 1.2 and include the effects of vapor pressure, boiling and freezing points, osmotic pressure, and water potential. They occur because addition of solute dilutes or lowers the concentration of the water.

Pressure Units

Vapor pressure is usually expressed in millimeters (mm) of mercury or millibars (mbar) and atmospheric pressure in bars (1 bar = 0.987 atm). However, there is a strong tendency toward the use of SI (Système International) units (Incoll *et al.*, 1977), and for the most part we will use those units in this book. The primary pressure unit is the pascal (Pa), and 1 bar = 10^5 Pa, 100 kPa, or 0.1 MPa (megapascal). In general, megapascals will be used in place of bars and kPa in place of millibars. One millibar = 0.1 kPa, and standard atmospheric pressure (760 mmHg or 1013 mbar) is 101.3 kPa.

Vapor Pressure

The decrease in vapor pressure of water in solution is essentially the result of its dilution by the addition of solutes. This is shown by Raoult's law, which states that the vapor pressure of solvent vapor in equilibrium with a dilute solution is proportional to the mole fraction of solvent in the solution:

$$e = e^o \frac{n_w}{n_w + n_s} \tag{1.1}$$

where e is the vapor pressure of the solution, e^o the vapor pressure of pure solvent, n_w the number of moles of solvent, and n_s the number of moles of solute. This is strictly applicable only to dilute molal solutions, i.e., those prepared with a mol or some fraction of a mol of solute per 1000 g of water.

Boiling and Freezing Points

The effects of solutes on the boiling and freezing points are exerted through their effects on the vapor pressure of water (see Fig. 1.10). The addition of solute lowers the freezing point because it dilutes the water and lowers the vapor pressure, thereby decreasing the temperature at which the vapor, liquid, and solid phases are in equilibrium. It can be calculated that the vapor pressure at freezing of a molal solution of a nonelectrolyte in water is decreased from 0.610 to 0.599 kPa and that a reduction in temperature of 1.86°C is required to bring about freezing. Further explanation can be found in physical chemistry texts.

Water boils when its vapor pressure is raised to that of the atmosphere. When the vapor pressure has been lowered by the addition of solute, the water in a

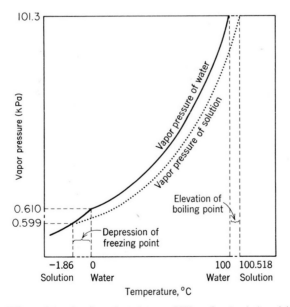

Fig. 1.10. Effects of 1 mole of nonelectrolyte per 1000 g of water (a 1 molal solution) on the freezing and boiling points and vapor pressure of the solution. Note that this diagram is not drawn to scale.

solution must be heated to a higher temperature than pure water to produce the required increase in vapor pressure.

Osmotic Pressure or Osmotic Potential

Raoult's law shows that the vapor pressure of water in a solution is lowered in proportion to the extent to which the mole fraction of water in the solution is decreased by adding solute. Therefore, if water is separated from a solution by a membrane permeable to water but impermeable to the solute, water will move across the membrane along a gradient of decreasing vapor pressure or chemical potential into the solution until the vapor pressures of solution and pure water become equal. The pressure which must be applied to the solution to prevent movement in a system, such as that shown in Fig. 1.11, is termed the osmotic pressure or osmotic potential. It is often denoted by the symbol π.

Van't Hoff developed an equation relating osmotic pressure to solute concentration in the solution. Mathematically expressed,

$$\pi V = n_s RT \qquad (1.2)$$

where π is the osmotic pressure in pascals, V the volume of solvent in liters, n_s the moles of solute, R the gas constant (0.00832 liter MPa/degree mol at 273°K),

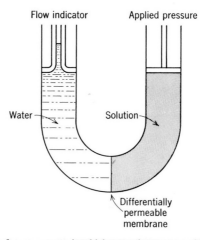

Fig. 1.11. Diagram of an osmometer in which a membrane permeable to water but impermeable to a solution separates pure water from the solution. The osmotic pressure of the solution is equal to the pressure that must be applied to prevent movement of water into it. Water movement is observed by a change in the level of water in the capillary tube on the left.

and T the temperature in °K. For 1 mol of solute in 1 liter of solvent at 273°K (0°C), this equation gives a value for π of 22.7 × 10^5 Pa or 2.27 MPa (22.4 atm or 22.7 bars).

Direct measurements have shown that this relationship is approximately correct for dilute solutions of nondissociating substances. However, there are large deviations from the theoretical value for electrolytes that ionize in solution and release more particles than nondissociating substances. Thus, the osmotic pressure of a molal solution of NaCl is approximately 4.32 MPa (43.2 bars) instead of the theoretical 2.27 MPa (22.7 bars). Assuming complete dissociation of the NaCl, the osmotic pressure should be 4.54 MPa and the discrepancy can probably be attributed chiefly to van der Waals forces operating between ions. Some nondissociating molecules become hydrated or bind water molecules. This binding of water reduces the effective concentration of water and increases the observed osmotic pressure. An example is a sucrose solution in which each sucrose molecule apparently binds six molecules of water, and the osmotic pressure of a molal solution of sucrose is approximately 2.51 MPa (25.1 bars) instead of the expected 2.27 MPa (22.7 bars).

The relationships among concentration, vapor pressure, freezing point, and osmotic pressure make it possible to calculate the osmotic pressure of a solution from the freezing point depression or from the vapor pressure. Since the theoretical depression of the freezing point of an ideal molal solution is 1.86°C and the osmotic pressure is 2.27 MPa at 0°C, the osmotic pressure of a solution can be calculated from the depression of the freezing point (T) by the following equation:

$$\pi = \frac{\Delta T}{1.86} \times 2.27 \text{ MPa} \qquad (1.3)$$

where π is the osmotic pressure of the solution in MPa and ΔT is the observed depression. The derivation of this relationship can be found in Crafts *et al.* (1949) or in a physical chemistry text. The cryoscopic method is widely used, and if suitable corrections are made for supercooling, it gives accurate results. The problems encountered in using it on plant sap are discussed in Chapter 7 of Crafts *et al.* (1949) and by Barrs (1968a).

The osmotic pressure can also be calculated from the vapor pressure or, more readily, from the relative vapor pressure or relative humidity, e/e^o × 100, according to Raoult's law [see Eq. (1.1)]. In recent years, thermocouple psychrometers have been used extensively to measure e/e^o. Their operation will be discussed in Chapter 13 in connection with the measurement of water stress in plants.

It is now customary in the field of plant water relations to substitute potential for pressure and to use the term osmotic potential in place of osmotic pressure. The two terms are numerically equal, but osmotic potential carries a negative sign. The basis for the use of the term potential is discussed in the following section.

Chemical Potential of Water

The chemical potential of a substance is a measure of the capacity of that substance to do work. It is generally considered to be equal to the partial molal Gibbs free energy (Spanner, 1964; Slatyer, 1967). In a simple solution of a nonelectrolyte in water, the chemical potential of the water depends on the mean free energy per molecule and the concentration of water molecules, i.e., on the mole fraction of the water. The degree to which the presence of solute reduces the chemical potential of the water in the solution below that of pure free water can be expressed as

$$\mu_w - \mu_w^o = RT \ln N_w \qquad (1.4)$$

where μ_w is the chemical potential of water in the solution, μ_w^o is that of pure water at the same temperature and pressure in units such as ergs per mole, R and T have the usual meaning, and N_w is the mole fraction of water. For use with ionic solutions, the mole fraction is replaced by the activity of water, a_w, and for general use, where water may not be in a simple solution, by the relative vapor pressure, e/e^o. Equation (1.4) is then written:

$$\mu_w - \mu_w^o = RT \ln \frac{e}{e^o} \qquad (1.5)$$

When the vapor pressure of the water in the system under consideration is the same as that of pure free water, $\ln e/e^o$ is zero, and the potential difference is also zero. Thus, pure free water is defined as having a potential of zero. When the vapor pressure of the system is less than that of pure water, $\ln e/e^o$ is a negative number; hence, the potential of the system is less than that of pure free water and is expressed as a negative number.

The expression of chemical potential in units of ergs per mole is inconvenient in discussions of cell water relations. It is more convenient to use units of energy per unit of volume. The measurements are compatible with pressure units which can be obtained by dividing both sides of Eq. (1.5) by the partial molal volume of water, $\bar{V}_w$ (cm^3/mol). The resultant term is called the water potential, Ψ_w:

$$\Psi_w = \frac{\mu_w - \mu_w^o}{\bar{V}_w} = \frac{RT}{\bar{V}_w} \ln \frac{e}{e^o} \qquad (1.6)$$

Energy expressed as ergs per cubic centimeter is equivalent to pressure in dynes per square centimeter or bars. However, in SI units, energy is expressed as joules per cubic meter (Jm^{-3}) and pressure is expressed in pascals, and 1 bar = 10^5 Pa or 0.1 MPa.

The water potential in any system is decreased by those factors that reduce the relative vapor pressure, including:

1. Addition of solutes which dilute the water and decrease its activity by hydration of solute molecules or ions.

2. Matric forces, which consist of surface forces, and microcapillary forces found in soils, cell walls, protoplasm, and other substances that adsorb or bind water.

3. Negative pressure or tensions such as those in the xylem of transpiring plants.

4. Reduction in temperature, T.

The water potential in any system is increased by those factors that increase the relative vapor pressure, including:

1. Pressure, such as that of the elastic cell wall on the cell contents.

2. Increase in temperature, T.

Throughout the discussion, we have explained the colligative properties of solutions as resulting from the lowering of the concentration of the solvent, water, by the addition of solute. However, Hammel (1976) and Hammel and Scholander (1976) argued that addition of a solute lowers the chemical potential of the solvent by creating a negative pressure or tension on the solvent molecules. Andrews (1976) discussed their arguments in detail and concluded that there is no mechanism by which solvent and solute molecules can sustain different pressures. The writer agrees that the classical solvent dilution theory adequately explains the behavior of solutions.

SUMMARY

Water plays essential roles in plants as a constituent, a solvent, a reactant in various chemical processes, and in the maintenance of turgidity. The physiological importance of water is reflected in its ecological importance, plant distribution on the earth's surface being controlled by the availability of water wherever temperature permits growth. Its importance is a result of its numerous unique properties, many of which arise from the fact that water molecules are organized into a definite structure held together by hydrogen bonds. Furthermore, the water bound to proteins, cell walls, and other hydrophilic surfaces has important effects on their physiological activity.

Water in plants and soils contains solutes that modify its colligative properties by diluting it. As a result of this dilution, the chemical potential, vapor pressure, osmotic potential, and freezing point are lowered in proportion to the concentration of solute present. The best measure of the energy status of water in plants and soil is the water potential (Ψ_w), which is the amount by which its chemical potential is reduced below that of pure water.

SUPPLEMENTARY READING

Alberty, R. A., and Daniels, F. (1979). "Physical Chemistry," 5th ed. Wiley, New York.

Andrews, F. C. (1976). Colligative properties of simple solutions. *Science* **194,** 567–571.

Edsall, J. T., and McKenzie, H. A. (1978). Water and proteins. I. The significance and structure of water: Its interaction with electrolytes and non-electrolytes. *Adv. Biophys.* **10,** 137–207.

Eisenberg, D., and Kauzmann, W. (1969). "The Structure and Properties of Water." Oxford Univ. Press, London and New York.

Hammel, H. T. (1976). Colligative properties of a solution. *Science* **192,** 748–756.

Pospisilova, J., and Solorova, J., eds. (1975 . . .). "Water in Plants Bibliography," Vols. 1. . . . Junk, The Hague.

Stillinger, F. H. (1980). Water revisited. *Science* **209,** 451–457.

Tanford, C. (1980). "The Hydrophobic Effect." Wiley, New York.

Tinoco, I., Sauer, K., and Wang, J. C. (1978). "Physical Chemistry: Principles and Applications in Biological Sciences." Prentice-Hall, Englewood, New Jersey.

2

Cell Water Relations

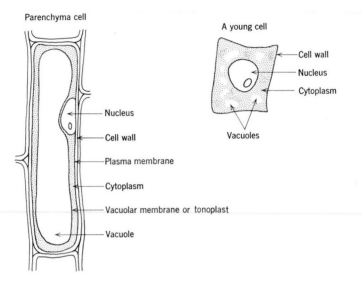

Fig. 2.1. Diagrams of a meristematic cell and a mature vacuolated parenchyma cell. The layer of cytoplasm in mature cells is usually much thinner than shown in this diagram.

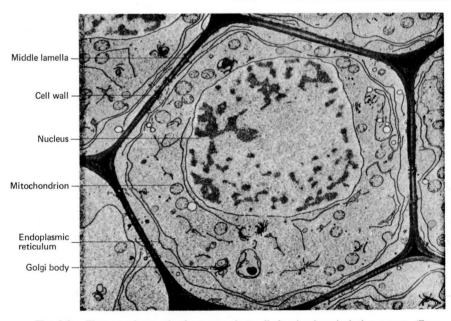

Fig. 2.2. Electron micrograph of a young plant cell showing its principal structures. (From Weisz and Fuller, 1962, by permission of McGraw-Hill Book Company.)

INTRODUCTION

Plant water relations are dominated by cell water relations because most of the water occurs in cells, chiefly in the vacuoles. Mature plant cells are characterized by their relatively rigid walls and large central vacuoles, both of which play important roles in cell water relations. A typical parenchyma cell such as that forming most of the water storage tissue in plants is shown in Fig. 2.1. In contrast to the young cell in that figure, the mature cell consists of a wall enclosing a thin layer of cytoplasm which, in turn, encloses a large central vacuole. Electron microscopy has provided much additional information concerning the structural details of cells (e.g., Robards, 1974), but considerable uncertainty persists concerning the proper interpretation of some of the observations. An electron micrograph of a plant cell is shown in Fig. 2.2. The various substructures of cells such as nuclei, chloroplasts, vacuoles, microbodies, and membranes are described in detail in Bonner and Varner (1976).

CELL STRUCTURE

Cell structure will be discussed so far as it relates to cell water relations, the distribution of water in the various structures of cells, cell membranes, and the terminology of cell water.

Cell Walls

An important characteristic of seed plants is the existence of strong cell walls with great tensile strength that limit expansion of the protoplasts, resulting in development of turgor pressure. As a result, tissue composed of turgid cells has considerable mechanical strength, and cell turgor is responsible for maintenance of the form of structures such as young herbaceous stems and leaves, flowers, and growing root tips. The framework of cell walls consists of layers of cellulose microfibrils composed of hundreds of cellulose molecules held together by hydrogen bonds. Associated with the cellulose framework are other polysaccharides, chiefly hemicelluloses and pectins, and proteins (Preston, 1979). At this stage, over one-half of the wall volume consists of water, and both plastic and elastic cell enlargement can occur. Permanent cell enlargement involves plastic extension, requiring loosening of the bonds holding the cellulose microfibrils together, a process in which auxin is probably involved, as well as slippage of the fibrils caused by turgor pressure. Thus, cell growth seems to depend on two sets of factors: the biochemical factors controlling deposition of the cell wall constituents and cell wall extensibility, and factors controlling the inward diffusion of water that produces the turgor pressure required for expansion by what Preston (1979, p. 71) terms biochemical creep.

Cell wall growth is complex, both chemically and physically, because it involves synthesis of the constituents in the cytoplasm (at least partly in the Golgi apparatus), secretion through the plasma membrane, and assemblage to form the lamellae of the wall. A further complication is the fact that there must be deposition of new lamellae and simultaneous extension of the existing lamellae in cell walls under tension.

Deposition of additional cellulose and infiltration of the cellulose framework by lignin result in secondary thickening, loss of plasticity, and cessation of enlargement. However, many mature cell walls retain enough elasticity to undergo changes in volume with change in turgor, as indicated schematically in Fig. 2.3. Measurable decreases in leaf and stem volume are often observed during wilting (see Chapter 12).

The deposition of lipids and related substances modifies the permeability of some cell walls. Everyone is familiar with the suberized walls of cork cells in bark and the cutin layer on leaves, but it is not so well known that cutin and suberin also occur in walls of cells in the interior of plants. Suberin occurs internally in the walls of specialized cells such as the endodermis. According to Scott (1950, 1964), all cell walls exposed to the air are coated with a layer of lipid material. This includes the surfaces of leaf mesophyll and root cortical cells bordering intercellular spaces, and root epidermal cells, including root hairs. However, a mucilaginous deposit of pectic substances, sometimes termed mucigel, often occurs external to the lipid layer on the outer surfaces of root epidermal cells. The effects of these modifications on water movement have never been fully evaluated.

Much emphasis has been placed on the physical properties of cell walls (Preston, 1974), but they also have a complex biochemistry (Albersheim, Chapter 9, and Karr, Chapter 13, in Bonner and Varner, 1976; see also Chrispeels, 1976). Cell walls contain numerous enzymes, some of which are involved in synthesis of wall components, but some have other functions, such as acting on extracellular metabolites (Lamport, 1970). Cell walls are also penetrated by strands of cytoplasm, the plasmodesmata, that connect the protoplasts of adjacent cells to form a continuous system of cytoplasm, called the symplast by Münch (1930) and later writers. Plasmodesmata are discussed later in this chapter.

Two properties of cell walls are important in any discussion of plant water relations. One is the amount of water occurring in the wall, the other is the degree of elasticity. The volume of water occurring in cells walls is important with respect to the possible role of the wall as a pathway for water and solute

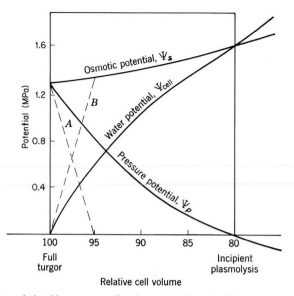

Fig. 2.3. Interrelationships among cell volume, osmotic potential, turgor pressure, and cell water potential. The solid lines represent a highly extensible cell with elastic walls. The dashed lines represent a slightly extensible cell with rigid walls, where line *A* represents pressure potential and line *B* cell water potential. Water and osmotic potentials are negative. Pressure potential is positive but decreases to zero when water potential falls as low as the osmotic potential. Note that there is a much greater decrease in pressure potential and water potential for a given change in cell volume and water content for cells with rigid walls than for cells with elastic walls. This is known as a Höfler diagram and is adapted from various sources. However, it resembles figures based on actual data, such as Fig. 15 of Hellkvist *et al.* (1974).

movement outside of the xylem, especially in roots and leaves. This will be discussed in more detail later. It has also been suggested that cell wall water contributes to drought tolerance (Gaff and Carr, 1961) and causes errors in measurement of cell water potential by dilution of the vacuolar sap (Markhart *et al.*, 1981; Tyree, 1976). Cell wall elasticity is important because it affects the relationship between cell volume and water content, turgor pressure, and water potential, as shown in Fig. 2.3. The modulus of elasticity, ϵ, of plant cells was discussed by Zimmermann (1978, pp. 129–130) and by Tyree and Jarvis (1982). Cells with rigid walls have a high modulus of elasticity and undergo a smaller decrease in volume and water content for a given decrease in water potential than cells with elastic walls. This is because wall pressure on the protoplasts decreases more rapidly and plasmolysis occurs sooner with loss of water in cells with rigid walls than in cells with elastic walls that shrink as water is lost. The situation is shown schematically in Fig. 2.3. Several writers have proposed that plants tolerant of desiccation show a smaller decrease in water content for a given decrease in water potential than plants less drought tolerant, but this is debatable (see Sanchez-Diaz and Kramer, 1971, p. 615; Zimmermann, 1978; and Tyree and Jarvis, 1982, for more details).

Cytoplasm

Within the wall, there is a layer of cytoplasm varying in consistency from a semiliquid sol to a gel. It is bounded at the outer surface by the plasma membrane, or plasmalemma, and at the inner surface by the vacuolar membrane, or tonoplast. Studies with the electron microscope indicate that cytoplasmic structure is considerably more complex than was supposed from observations made with the light microscope. Not only does it contain a nucleus and plastids but also a variety of other organelles, including mitochondria, ribosomes, glyoxysomes, peroxisomes, Golgi apparatus, and endoplasmic reticulum (see Fig. 2.2). These structures are discussed in detail by Bonner and Varner (1976) and by Beevers (1979).

Most important to us is the fact that the cytoplasm normally contains considerable water bound to its protein framework. In addition, its limiting membranes are differentially permeable and control the entrance and exit of solutes. The vacuolar membrane appears to be somewhat stronger mechanically than the plasma membrane and is probably higher in lipid content. It retains its differential permeability for some time after being separated from the cytoplasm, whereas the differential permeability of the plasma membrane is lost immediately upon separation from the cytoplasm (Frey-Wyssling and Mühlethaler, 1965). Membranes are discussed in more detail later in this chapter.

The plasmodesmata are thin tubes of cytoplasm about 0.2 μm in diameter that extend through cell walls and connect the protoplasts of adjacent cells (see Fig. 2.4). Thus, they convert a collection of cells into an organized tissue, the symplast of Münch (1930) and others, in which solutes can move considerable distances without crossing differentially permeable membranes. The role of plasmodesmata in transport has been discussed by a number of investigators, including Arisz (1956), Tyree (1970), Clarkson *et al.* (1971), and Gunning and Robards (1976).

Vacuoles

Large central vacuoles are as characteristic of mature plant cells as their rigid walls. Vacuoles range in size from the tiny spherical or rod-shaped structures

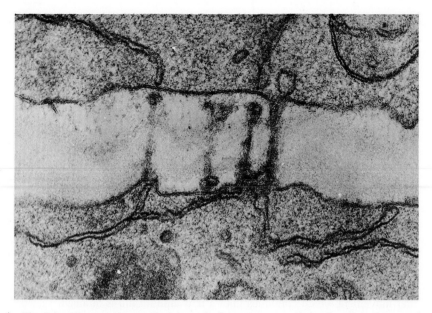

Fig. 2.4. Electron micrograph of the wall of an onion root cell showing plasmodesmata, the plasma membrane on each side of the wall, and parts of the endoplasmic reticulum. (From Laboratorium für Elektronenmikroskopie, ETH; courtesy of Professor K. Mühlethaler.)

characteristic of meristematic tissue to the large central vacuoles of mature parenchyma cells occupying more than 50% of the cell volume. Their size and shape can change; small vacuoles coalesce to form larger ones, and large ones sometimes break up to form smaller ones. For example, during seed maturation the vacuoles lose water and shrink, but they enlarge again during germination. Vacuoles contain considerable amounts of sugars and salts that largely account for the osmotic potential of the cell sap. They can also contain a wide variety of other substances, including amino acids, amides, proteins, lipids, gums, tannins, anthocyanins and other pigments, organic acids, and crystals of minerals such as calcium oxalate. Vacuoles are discussed in detail by Matile (1976, 1978). He states that whatever metabolic activity occurs in the vacuoles is controlled by the cytoplasm.

Frey-Wyssling and Mühlethaler (1965) suggested that large central vacuoles are advantageous because it is difficult for plants to produce enough protein to fill their cells with protoplasm. Wiebe (1978) also suggested that natural selection favored the evolution of large vacuoles in plant cells because they permit development of the maximum external surface area with the minimum use of energy-expensive protein. An extensive surface is advantageous for plants because essential raw materials such as carbon dioxide, nitrogen sources, and minerals occur in very low concentrations in the plant environment. Vacuoles also play an essential role in development of the turgor pressure necessary to maintain the form of herbaceous plant structures such as young stems, leaves, and flowers.

DISTRIBUTION OF WATER IN CELLS

Water is continuous in plants through the cell walls, the protoplasm, and its various organelles, but it is separated into compartments with respect to solutes by the differentially permeable membranes of the protoplasts and their organelles. Some water is bound on surfaces in cells and some is held in the microcapillaries of the walls, but most of it is free to move along gradients of difference in water potential. At times when transpiration is negligible, there may be a temporary equilibrium in water potential throughout the plant, but a change in concentration of solutes or loss of water from a tissue is followed by water movement which continues until a new equilibrium is attained.

The water in plants is sometimes described as existing in two systems: apoplastic water occurring in the cell walls and xylem elements, and symplastic water occurring in the protoplasts. The apoplastic water volume is approximately equivalent to the apparent free space (AFS) of Butler (1953) and Briggs and Robertson (1957) and the outer space of Kramer (1957). This is the part of a cell or tissue into and out of which solutes can move freely by diffusion and they accumulate in appreciable concentrations only in the symplastic water by active transport across the plasma membrane. Thus, the distinction between apoplastic and symplastic water really refers to the distribution of solutes, because the water itself forms a continuous system. The amount of water in various parts of cells and the forces holding it will be discussed briefly.

Cell Wall Water

From 5 to 40% of the water in a cell occurs in the walls, depending on age, thickness, and composition of the walls. In young cells and some mature parenchyma tissue the walls are thin, and although over 50% of the wall volume may be water, it constitutes only a small percentage of the total cell water. In thick, leathery leaves such as those of rhododendron and eucalyptus, the cells have thicker walls and a larger percentage of the cell water occurs in the walls. Some data are summarized in Table 2.1. Water is held in the walls by matric

TABLE 2.1   Distribution of Water in Cells as a Percentage of Total Water Content[a]

Plant	Apoplastic or wall	Symplastic or vacuolar	Reference
Wheat root	20–25	75–80	Butler, 1953
Eucalyptus leaves	40	(60)	Gaff and Carr, 1961
Rhododendron leaves	25–32	(68–75)	Boyer, 1967
Sunflower leaves	5–14	(86–95)	Boyer, 1967
Wheat leaves	30	(70)	Campbell *et al.,* 1979
Potato leaves	5	(95)	Campbell *et al.,* 1979
Soybean, immature leaves	(16)	84	Wenkert *et al.,* 1978
Soybean, mature leaves	(30)	70	Wenkert *et al.,* 1978

[a] The values in parentheses are estimates derived from the differences between total water and the measured values. This does not account separately for the water in the cytoplasm, at least part of which is matrically bound and not available as a solvent.

forces, including hydrogen bonding to the various cell wall constituents, and in the submicroscopic capillaries among the fibrils. This matric water is sometimes termed imbibed water (see Meyer *et al.,* 1973, pp. 53–55). The possible role of cell walls as a pathway for water and solute movement will be discussed in Chapter 7.

More than half the volume of some cell walls is occupied by water, and some types of cell walls shrink as much as 50% when dehydrated. As a result, sections of plant tissue that have been killed and dehydrated often give an erroneously low indication of the volume of cell wall available for water storage and movement. During cell wall maturation and secondary thickening, deposition of lignin, suberin, and other substances often reduces the volume available for water storage and movement and modifies permeability. The Casparian strips in the walls of endodermal cells, shown in Fig. 5.5 (see Chapter 5), are an important example.

Water in the Cytoplasm

In meristematic tissue where the vacuolar volume is small and cell walls are thin, most of the water occurs in the cytoplasm. However, in mature cells the cytoplasm usually forms only a thin layer lining the wall and may contain as little as 5 or 10% of the water in the cell. Some of this water is hydrogen-bonded to side chains of the proteins forming the framework of the protoplasm. The water in the cytoplasm is very important because changes in water content can affect protein structure (Tanford, 1963, 1980). However, this does not mean that all or even most of the water is associated in some special manner with the protoplasm, as proposed by Cope, Ling, and others (see Ling, 1969; Drost-Hansen and Clegg, 1979). Although some water is bound to protein surfaces, it probably does not form the uniform layers of structured water, or "ice," suggested by Bernal (1965) and discussed briefly in Chapter 1.

Water in Vacuoles

In plant tissue such as the mature parenchyma cells of leaves, stems, and roots, 50–80% or more of the water occurs in the vacuoles. This water usually contains considerable amounts of solutes, chiefly sugars, salts, and sometimes organic acids, and pigments such as anthocyanin, and vacuolar sap may have an osmotic potential ranging from -1.0 to -3.0 MPa, or occasionally much lower. Thus, the vacuolar water is held chiefly by osmotic forces. However, various substances, such as proteins, tannins, and mucilages, occur in colloidal dispersion, binding some water and often making the viscosity of the vacuolar sap about twice that of water. Much of the older information concerning vacuoles was summarized by Kramer (1955c), and Matile (1976, 1978) has covered the recent literature. Here we are chiefly concerned with vacuoles as osmotic systems producing the turgor pressure so important for cell enlargement and the maintenance of form in herbaceous tissues. This will be discussed further in the section on cell water terminology.

The percentage of the total cell water that occurs in the symplast, chiefly in vacuoles, depends in part on the methods used to calculate it, but even more on the type of tissue. In leaves of eucalyptus only about 50% of the cell water occurs in vacuoles (Gaff and Carr, 1961) and in rhododendron about 70% (Boyer, 1967), but in sunflower leaves (Boyer, 1967) and wheat roots (Butler, 1953), 80% or more is in the vacuoles. Roberts and Knoerr (1977) reported that by the pressure volume method 74–98% of the leaf water of several tree species contributes to the osmotic volume. A volume of 98% in mature tissue seems improbably high because the thicker walls of older tissue are likely to contain more water, as shown for soybean in Table 2.1.

CELL MEMBRANES

Any discussion of water and solute movement in plants requires consideration of the permeability of the various membranes through which these substances move. One of the most important characteristics of living cells is their ability to maintain combinations and concentrations of solutes quite different from those existing in their environment. For example, the concentration of several kinds of ions is tens or even hundreds of times higher in plants than in the soil in which they are rooted. On the other hand, the concentration of some substances is higher outside than inside cells. Such accumulation or exclusion requires the presence of membranes relatively impermeable to the ions, and active transport systems to move them across the membranes.

Definition and Structure

Membranes are boundary layers differing in their permeability from the phases or regions that they separate. Membranes that permit some substances to pass through more readily than others are termed semipermeable or, more accurately, differentially permeable. Cell membranes include the plasma membrane or plasmalemma on the outer surface of the cytoplasm, the vacuolar membrane or tonoplast on the inner cytoplasmic surface bordering the central vacuole, membranes enclosing organelles such as the nucleus, plastids, and mitochondria, and the complex system of membranes extending throughout the cytoplasm, called the endoplasmic reticulum. In addition to their role as differentially permeable boundary layers, membranes have enzymes that are involved in many important biochemical reactions bound on their surfaces.

Beginning in the nineteenth century, investigators found that increasing permeability to polar compounds is often correlated with decreasing size of their ions or molecules, whereas permeability to many nonpolar substances is correlated with their solubility in lipids. These observations led to development of the competing sieve and solubility theories of membrane permeability of the early twentieth century (Miller, 1938, pp. 107–112). In recent decades, ideas concerning the structure of protoplasmic membranes have progressed from the Davson-Danielli lipid bilayer model of the 1930s and the Robertson unit membrane of the 1960s to the lipid–globular protein or fluid mosaic model of Singer (1974). All of these models assume a double layer of amphipathic lipid molecules (molecules with hydrophobic and hydrophilic ends) oriented so that their hydrophilic ends point outward and their hydrophobic ends inward. The Singer model assumes that instead of a solid layer of protein coating the membrane surfaces, there are globular amphipathic protein molecules embedded in the lipid layer and even projecting from it, as shown diagrammatically in Fig. 2.5. An electron micrograph of a membrane surface is shown in Fig. 2.6. The protein molecules extending through the membranes presumably provide ''sites'' or pathways for movement of hydrophilic substances through the lipid layer by active transport. It has been suggested that many of the differences in membrane permeability depend on genetically controlled synthesis of the membrane proteins. Knowl-

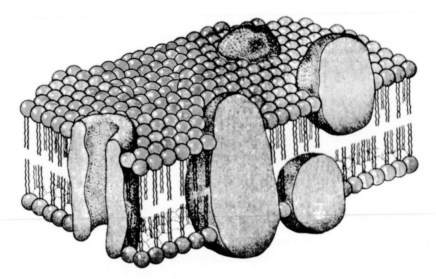

Fig. 2.5. Diagram of a fluid mosaic model of a cell membrane, consisting of a double layer of phospholipids with protein molecules embedded in it. The proteins are actually coiled polypeptide chains and are much more complex in structure than indicated in this diagram. They presumably control the passage of substances through membranes. (Modified from Keeton, 1980, and other sources.)

edge of membrane structure is increasing rapidly, and in a few years diagrams such as Fig. 2.5 probably will be modified.

In addition to their complex protoplasmic membranes, plants possess multicellular structures such as the epidermis and endodermis that in some respects function as membranes. For example, the endodermis seems to function as an ion barrier in roots. Plant organs also possess relatively impermeable layers such as the cuticle on leaves and fruits, various specialized seed coats, and the bark on woody stems and roots, all of which play important roles in limiting uptake and loss of water, as will be seen later.

Permeability of Cell Membranes

The permeability or conductance of a membrane refers to the rate of movement of a substance, usually in cm s^{-1}, across a membrane under a given driving force. For water an equation can be written as follows:

$$Lp = \frac{J_v}{\text{driving force}} \tag{2.1}$$

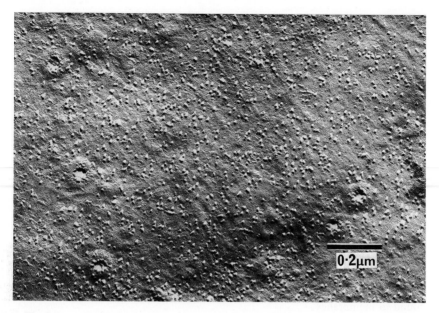

Fig. 2.6. A freeze-etched picture of the side of the plasma membrane facing the cytoplasm in a barley root cortical parenchyma cell. The numerous particles are probably protein or glycoprotein molecules embedded in the lipid matrix. The larger structures indicate the location of plasmodesmata. (Courtesy of Prof. A. W. Robards and Hilary Quine, Department of Biology, University of York, United Kingdom.)

Lp is the hydraulic conductance of the system in units such as cm^3 cm^{-2} $s^{-1}MPa^{-1}$, J_v is the volume flux in cm^3 cm^{-2} s^{-1}, and the driving force is measured in bars or MPas. Permeability to water is often measured by osmotic or plasmolytic methods in which the rate of plasmolysis or deplasmolysis is observed on individual cells. Sometimes permeability is measured by the rate of diffusion of labeled water into tissue. A few measurements of cell membrane permeability have been made by the direct application of pressure (Zimmermann, 1978).

Factors Affecting Transport across Cell Membranes

As mentioned earlier, the flux, or amount of material moving across a unit of membrane surface per unit of time, depends both on the permeability of the membrane and on the driving force. Driving forces will be discussed later. During the first half of this century, numerous papers were published describing the effects of various treatments on permeability (see Ruhland, 1956), but most of them were unable to distinguish between effects on membranes and on active transport mechanisms. The permeability of protoplasmic membranes is materially affected by factors that affect cell metabolism, such as respiration inhibitors, oxygen concentration, and temperature. Inhibitors such as sodium azide and 2,4-dinitrophenol decrease transport because they reduce the supply of energy available to maintain normal membrane structure and function. Low concentrations of lipophilic solvents such as benzene, chloroform, and ether affect permeability, presumably because they affect the lipids in membranes. Sometimes membrane permeability is reduced by low concentrations of toxic substances but increased by concentrations high enough to cause permanent injury. For example, a concentration of 2,4-dinitrophenol at 5×10^{-4} M decreased the permeability of *Vicia faba* roots to water, but at 10^{-3} M the permeability was greatly increased and the roots died (Ordin and Kramer, 1956). Other examples of this phenomenon are shown in Fig. 2.7. It should be noted that calcium seems to be necessary for normal functioning of ion transport across membranes (Epstein, 1961). In general, calcium and other multivalent ions tend to reduce permeability and monovalent ions to increase it. These effects are related to differences in the hydration of the ions and the firmness with which they are bound to membranes.

A high concentration of CO_2 reduces the permeability to water of roots (Kramer, 1940a) and sections of sunflower hypocotyl (Glinka and Reinhold, 1964), probably because it affects membrane structure. Low temperature also decreases root permeability, partly because of increased viscosity of water but also partly because of direct effects on the condition of the lipids in the cell membranes (Markhart *et al.,* 1979). This is discussed in more detail in Chapter 9. It is possible that some effects of growth regulators such as abscisic acid (ABA) are also brought about by their effects on membrane permeability.

Permeability of plant cells to water is also affected by pressure. Slight loss of turgor is reported to increase the permeability of parenchyma and algal cells (Myers, 1951; Zimmermann and Steudle, 1975), possibly because pressure of the plasmalemma against the wall in fully turgid cells reduces its permeability to water. It also appears that change in turgor pressure controls ion transport in some algae (Gutknecht and Bisson, 1977). This was discussed in several papers in the book edited by Zimmermann and Dainty (1974), but the mechanism by which pressure is transduced into control of transport is not yet fully understood. In his review article, Zimmermann (1978) proposed that pressure makes the cell membrane thinner, changing the electrical field distribution within it and thereby changing the permeability. However, Lucas and Alexander (1981) doubt if this hypothesis is applicable to land plants. Decreasing the osmotic potential of the external solution decreases the permeability of algal cells to water and urea (Dainty and Ginzburg, 1964; Zimmermann and Dainty, 1974, pp. 64–71). It also decreases the permeability of root systems to water (O'Leary, 1969), proba-

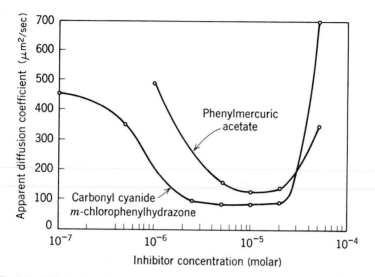

Fig. 2.7. Effects of concentration of toxic substances on permeability of corn roots to tritiated water. The first effect is to reduce permeability, but as increasing concentration causes injury, permeability suddenly begins to increase (Woolley, 1965).

bly by affecting membrane hydration and structure. According to Tepfer and Taylor (1981), dehydration of cell walls by concentrated solutions reduces their permeability to large molecules, and probably to water.

Relative Permeability of Various Cell Membranes

The relative permeability of various cell membranes is important with respect to movement of both water and solutes. The relative permeability to water of cell walls and cytoplasmic membranes is particularly important with respect to water movement outside of the xylem. It has generally been assumed that cell walls are more permeable to water than the cytoplasm, resulting in most of the extraxylary water movement occurring through the walls (Levitt *et al.*, 1936; Weatherley, 1963; Cowan and Milthorpe, 1968; Nobel, 1974, p. 395; Ferrier and Dainty, 1977). However, Newman (1976) compared data from various sources and concluded that wall permeability is probably lower than cytoplasmic permeability. Tyree and Yianoulis (1980) claimed that wall permeability is much lower than cytoplasmic permeability in leaves. However, Tyree *et al.* (1981) concluded later that water probably follows both pathways. Obviously, more reliable data are needed on the permeability of walls and cytoplasmic membranes.

There is also uncertainty concerning the relative permeabilities of the plasma membrane and the vacuolar membrane. Dainty and Ginsburg (1964) reported that the plasma membrane of *Nitella* cells was 30 times as permeable to urea as the vacuolar membrane, and they inferred from their data that also it was more permeable to water. Kiyosawa and Tazawa (1977) reported that the tonoplast or vacuolar membrane of *Chara* is more permeable to water than the plasma membrane. The plasma membrane was reported to be more permeable to ions than the vacuolar membrane in *Nitellopsis,* but not in *Chara* or *Nitella* (MacRobbie, 1962). Laties and his associates (Lüttge and Laties, 1966) proposed that at high external concentrations the plasma membrane is more permeable to ions than the vacuolar membrane. Thus, the cytoplasm would become a part of the apoplast, or free space. However, this concept has not received much support. Slatyer (1967, pp. 185–195) discussed some of the problems involved in measuring permeability of cell membranes. One of the important causes of error is the presence of "unstirred" boundary layers adjoining membranes that greatly increase their effective thickness.

MOVEMENT OF WATER AND SOLUTES IN PLANTS

The water and solutes in cells are in continual motion, moving within cells, from cell to cell, and from tissue to tissue. The distances involved vary from those across cell membranes, measured in nanometers, to those from roots to shoots, which may be measured in meters or tens of meters. Obviously, different mechanisms and factors are involved in movement over such widely different distances.

The movement of water and solutes can be classified in general terms as (1) passive movement by diffusion and mass or bulk flow and (2) active transport dependent on expenditure of metabolic energy. However, it is sometimes difficult to distinguish clearly between the two types of driving forces.

Passive Movement

Passive movement of materials obeys physical laws in the sense that movement occurs along gradients of decreasing free energy or chemical potential. If gradients of pressure potential or hydrostatic pressure constitute the driving force, the movement is generally termed mass flow. If gradients of chemical potential are involved, the movement is generally regarded as diffusion.

Mass Flow.   Movement of materials by mass or bulk flow occurs when force is exerted on the moving substance by some outside agent, such as pressure or gravity, so all of the molecules tend to move in the same direction in mass, whereas diffusion results from random movement of individual molecules (Spanner, 1956). Water flows in streams because of the hydraulic head produced by gravity and in the plumbing system of a building because of the unidirectional pressure applied to it by a pump. Water and solutes move through the xylem of plants by mass flow, caused by a tension or negative pressure potential gradient developed at the transpiring surfaces and transmitted in the xylem sap from the shoots to the roots. The pressure flow hypothesis of phloem transport assumes that mass flow occurs through the sieve tubes because of pressure developed by accumulation of sugar in the leaf phloem. Cyclosis, or the streaming of the cytoplasm in cells, can also be regarded as mass flow. Some scientists prefer the term bulk flow to mass flow.

Diffusion.   Movement by diffusion results from the random movement of molecules or colloidal particles caused by their own kinetic energy. Diffusion is a thermodynamic process operating at the molecular level, whereas mass flow is a mechanical process operating on bulk material in mass at the macroscopic level. It is often difficult to distinguish between mass flow and diffusion, and the two processes sometimes operate simultaneously, as in the roots of slowly transpiring plants. The soil water potential is often lower than the osmotic potential of the root xylem sap, so water might be expected to diffuse out along the thermodynamic water potential gradient. However, transpiration imposes a large mechanical tension on the water in the xylem, causing an inflow of water that overrides the osmotic gradient.

Familiar examples of diffusion are the evaporation of liquids, osmosis, and imbibition. The first mathematical treatment of diffusion seems to have been by Fick in 1855. Fick's law can be written as

$$\frac{dm}{dt} = -DA\,\frac{dc}{dx} \tag{2.2}$$

where *dm* refers to the amount of substance moved per unit of time *dt; D* is the diffusion coefficient, a constant that varies with the substance; *A* is the area through which diffusion is occurring; and the minus sign indicates that diffusion occurs "downhill," or from higher to lower concentration. The term *dc* refers to the difference in concentration which is the driving force and *dx* to the distance over which diffusion occurs. The equation indicates that, for a given substance, the rate of diffusion per unit area is proportional to the concentration gradient and inversely proportional to the distance over which it occurs.

Readers are reminded that all substances enter and leave cells in aqueous solution and that diffusion of gases is much slower in water than in air. The diffusion coefficient of oxygen is nearly 10,000 times greater in air than in water, and the concentration of oxygen in air at 15°C is 30 times greater than in water; hence, the actual transport of oxygen by diffusion is about 300,000 times more rapid in air than in water.

In general, the diffusion of solutes over long distances is very slow. Nobel (1974, p. 17) calculated that 8 years would be required for a small molecule with a diffusion coefficient of $10^{-5}cm^{-2} s^{-1}$ to diffuse 1 m in water but only 0.6 s to diffuse 50 μm, the distance across a typical leaf cell. Diffusion over long distances is very slow because the time required is inversely proportional to the square of the distance. Diffusion over a distance of 1 μm therefore occurs 10^8 times more rapidly than diffusion over a distance of 1 cm. Thus, movement by diffusion into and out of cells and within cells is relatively rapid, but diffusion could not possibly account for movement of water or solutes over distances measured in meters.

Diffusion is proportional to the area of the openings through which it occurs, while mass flow follows Poiseuille's law and varies with the fourth power of the radius or the square of the area. Thus, subdividing a large pore into smaller pores has a small effect on diffusion but a large effect on mass flow. This explains the fact that agar or gelatin gels and water-saturated cell walls have a low resistance to movement by diffusion but are significant barriers to mass flow through the small spaces between the micelles or fibrils.

Diffusion of gases is little affected by temperature because the rate increases in proportion to the change in absolute temperature; hence, an increase from 20° to 30°C would theoretically increase the rate by only about 3.4%. However, diffusion in aqueous solution has a temperature coefficient of about 1.2 or 1.3 because of the changing structure of water with changes in temperature. Diffusion of gases in water is also modified by their decreasing solubility in water with increasing temperature. The decrease in solubility from 20° to 30°C is about 24%% for CO_2 and 15% for O_2. Differences in solubility also affect the rate of diffusion across water-filled membranes such as cell walls, and CO_2 is about 28 times as soluble in water as O_2 at 20°C, which partly compensates for its low concentration in the atmosphere.

Osmosis.   The movement of water into and out of cells is usually described as occurring by osmosis, because it is assumed to move by diffusion across cell membranes toward regions of lower water potential. However, it has been claimed by some investigators that osmosis really involves mass flow through membrane pores (Slatyer, 1967, pp. 171–173). This claim is based on data from a number of experiments indicating that osmotic movement of water is more rapid than diffusion of water labeled with deuterium or tritium. Dainty (1965) attributed these apparent differences in rate to the effects of unstirred layers and other experimental errors that make reliable comparisons of osmotic and tracer movement difficult. Also, there is no convincing evidence for the existence of actual water-filled pores in membranes. Perhaps the problem is largely semantic, because the overall driving force is the difference in water potential across the membrane produced by differences in pressure and osmotic potential. A general transport equation for plant cells contains terms for both pressure and osmotic components:

$$J_v = Lp \, (\Delta\Psi_p + \sigma\Delta\Psi_s) \tag{2.3}$$

J_v is the volume flux of water, Lp the hydraulic conductance of the membrane, $\Delta\Psi_p$ the pressure differential, $\Delta\Psi_s$ the osmotic pressure difference across the membrane, and σ the reflection coefficient. This equation is written in various ways, but should contain the equivalent of these terms.

Lp in Eq. (2.3) is often termed the hydraulic conductivity, but it should be termed the hydraulic conductance because the dimensions of the system are not specified. Conductivity requires specification of distance and driving force, e.g., MPa cm^{-1}, and if both of these are not specified, the term conductance should be used. The same considerations apply to the reciprocals of conductivity and conductance, resistivity and resistance. Electrical resistivity is a physical property measured in ohm-centimeters or ohm-meters, but resistance does not require definition of size and is given in ohms. Usually, resistance to water movement is given in s cm^{-1} and conductance in cm s^{-1}.

A detailed understanding of osmotic movement is complicated by the interaction or coupling that exists between water and solute movement. Movement of water tends to carry solutes with it, and movement of solutes carries some water with them. The mathematical treatment of coupled transport is too complex to be dealt with here, and readers are referred to Slatyer (1967, pp. 162–171) for a discussion and references. Fiscus (1975) discussed the coupled interaction between osmotic- and pressure-induced flow of water in roots, and it is discussed further in Chapter 8.

Osmotic adjustment of cells and tissues will be discussed later in this chapter and also in Chapter 13.

Reflection Coefficient. Discussion of osmotic phenomena requires mention of the reflection coefficient σ. It is an expression of the extent to which a membrane is impermeable to a given solute and can be defined as the ratio of the observed or apparent osmotic pressure difference across a membrane to the theoretical osmotic pressure. If σ equals 1, the membrane is completely impermeable, but if it is less than 1, the membrane is permeable to the solute (Slatyer, 1967, pp. 166–169).

Disregard of the fact that plant membranes are often moderately permeable to solutes used in plasmolytic experiments may lead to various errors. For example, if cell membranes are somewhat permeable to solutes (σ less than 1), the osmotic potential of the cell sap may appear to be higher (less negative) than it really is. Likewise, if roots have σ less than 1, the external osmotic potential required to stop exudation would be higher than the true osmotic potential of the xylem sap, as reported for tomato roots by van Overbeek (1942). Observations of these situations have been interpreted as evidence for active uptake of water but may, in fact, have resulted from the leakiness of the cell membranes to solutes. Measurements of the σ of leaf and potato tuber tissue by Slatyer (1966) indicated that it is 0.6–0.7 for sucrose and 0.8–0.9 for mannitol. This may explain many of the discrepancies between osmotic potentials measured cryoscopically and plasmolytically.

Active or Nonosmotic Uptake of Water

Thus far, our discussion of cell water relations has proceeded on the assumption that all water exchange between cells and tissues and with the environment can be explained as occurring along gradients of water potential. However, over the years claims have been made repeatedly that there is active transport of water against gradients of water potential by a nonosmotic mechanism dependent on the expenditure of metabolic energy (Kramer, 1955a, pp. 211–217). These claims are based on observations that differences exist between measurements of osmotic potential made on expressed sap and by plasmolytic methods on cells of the same tissue; that respiration and anaerobiosis reduce water uptake; and that addition of auxin increases water uptake. The most recent claim that active transport of water occurs in roots was made by Russian investigators (Zholkevich *et al.*, 1979). The extensive early literature was reviewed by Kramer (1955a) and Slatyer (1967, pp. 195–197), and there will be further discussion in Chapter 8 in relation to root pressure.

In general, it is believed that these observations can be explained without invoking active transport of water. The discrepancies between cryoscopic and plasmolytic measurements probably result chiefly from errors inherent in the methods, especially failure to take into account penetration of the plasmolyzing solute into cells, as mentioned in the discussion of the reflection coefficient. Decreased water movement in the presence of respiration inhibitors and oxygen deficiency is probably the indirect result of decrease in membrane permeability and salt transport, rather than any direct inhibition of a hypothetical water transport system. Auxin-induced increase in water uptake probably results from increased cell wall extensibility (Cleland, 1971). It appears unlikely that there is any active transport of water in plants in the sense that there is active transport of solutes.

Electroosmosis.   Another mechanism sometimes proposed to explain anomalous water movement is electroosmosis (Fensom, 1957; Spanner, 1958; and others). It is well known that water can move across negatively charged cell membranes and even through plant tissue from positive to negative poles under the influence of a difference in electrical potential. However, Dainty (1963) pointed out that although the ion pumps operating in plant cell membranes might produce appreciable electroosmotic flow, cell membranes are so permeable to passive water movement that the action of an electroosmotic pump would be short-circuited by leakage outward along normal pathways. This view is supported by results of experiments such as those of Blinks and Airth (1951), who reported that applied potentials of up to 1500 mV from surface to interior of *Nitella* cells produced no significant movement of water. Brauner and Hasman (1946) attributed 10% of the water movement in potato tissue to electroosmosis. This is probably a generous estimate. The role of electroosmosis in water absorption through intact roots is discussed in Chapter 8 but is dismissed as unimportant.

Polar Movement of Water.   It has been reported that seed coats of certain species are more permeable to inward movement of water than to outward movement (Denny, 1917; Brauner, 1930, 1956). Polar water movement is also said to occur through insect cuticle (Beament, 1965) and various other animal membranes. Polar water movement in cells was discussed by Bennet-Clark (1959, pp. 164–167), who followed Brauner in attributing it to electroosmosis caused by electrical asymmetry of the membranes under study. Dainty (1963b) suggested that differences in rates of inward and outward flow might result from differences in hydration on the two sides of a complex membrane. However, he thought that many examples of supposed polar flow of water result from the effect of the unstirred layers.

Kamiya and Tazawa (1956) carried out some interesting experiments on *Nitella* cells mounted asymmetrically in a double chamber (see Fig. 2.8). The water in one side of the chamber was replaced by a sugar solution, which caused flow from the chamber containing water through the cell to the chamber containing sucrose. Observations of differences in water movement, when first the short end and then the long end of the cell were immersed in sucrose solution, led Kamiya and Tazawa (1956) to conclude that the permeability to outward movement of water is less than the permeability to inward movement. Similar results were obtained by Dainty and Hope (1959), who attributed their findings to the effect of unstirred layers inside and outside the cells. These layers reduce the driving force more when exosmosis is through the short end than when it is

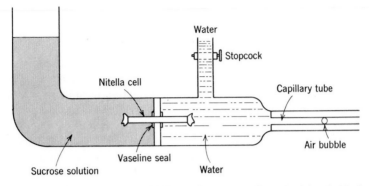

Fig. 2.8.   Apparatus for measuring the permeability or hydraulic conductivity of a *Nitella* cell by transcellular osmosis. The rate of water flow can be varied by changing the concentration of sucrose in the chamber on the left. Errors can occur from unstirred layers and from decreased permeability of the part of the cell immersed in sucrose solution. (After Dainty and Ginsburg, 1964.)

through the long end of the cell. However, further research indicated that unstirred layers are less important than the fact that the cell membrane is somewhat dehydrated by contact with the sucrose solution, and consequently its permeability is reduced (Dainty and Ginzburg, 1964). This seems to take the mystery out of polar water movement.

Water Secretion.   Any discussion of nonosmotic movement of water must include mention of examples of apparent excretion or secretion of water from various kinds of glandular structures. In his review of secretion, Stocking (1956) classified plant glands into four types: (1) glands secreting oils and resins, (2) nectaries, or glands secreting a sugary liquid, (3) glands characteristic of carnivorous plants which secrete nectar, mucilages, or digestive enzymes, and (4) water glands such as hydathodes. Hydathodes are dealt with in connection with guttation, and we are not concerned with oil- and resin-secreting glands in this book. The secretion of sugar solution from nectaries of flowers is well known, and significant amounts of sugar are also secreted from nectaries on leaves of some species. Considerable amounts of salt are secreted from glands on the leaves of some halophytes and even on leaves of some cultivated plants such as cotton (Kramer, 1969, pp. 255–256). It seems probable that the solutes are excreted by an active transport process, and the water then diffuses out along the resulting gradient in water potential. However, according to Stocking (1956), some writers claim that water and salt excretion are independent processes. More puzzling is the removal of water from the bladders of *Utricularia* and the secretion of liquid into unopened "pitchers" of various insectivorous plants. The mechanism of secretion obviously deserves more investigation. The subject was reviewed most recently by Schnepf (1974) and in a book by Fahn (1979).

Temperature and Water Movement

It was stated near the end of Chapter 1 that increasing the temperature increases the water potential. Spanner (1964, p. 207; 1972, p. 34) states that an increase of 1°C causes an increase of over 8.0 MPa in the sense that water will distill from an open solution with an osmotic potential of −8.0 MPa to pure water 1°C cooler. One therefore might conclude that since temperature differences of several degrees often occur among various parts of a plant, the effects of temperature should override the pressure, matric, and osmotic components of water potential. It is self-evident that they do not, because water moves from the cooler roots to the warmer leaves at midday. In fact, temperature differences affect movement of *liquid water* only under very special conditions (Lewis and Randall, 1961). Temperature affects water transfer across membranes only when the heat conductance of the membrane is very low and the pore size is very small. It is difficult to maintain a temperature gradient across a single cell membrane of sufficient magnitude to cause thermoosmosis, and it could not operate at all in a system such as the xylem. Even where there is a tendency for liquid water to move by diffusion from warmer to cooler regions, as in soils, the thermodynamic diffusional movement is often counterbalanced by mechanical mass flow in the reverse direction caused by gradients in hydrostatic pressure. It therefore appears that temperature has no significant direct effect on movement of water as a liquid in plants (Briggs, 1967; Wiebe and Prosser, 1977). Of course, temperature affects the viscosity of water and the permeability of membranes, and these effects will be discussed in Chapter 9 in connection with the effects of temperature on water absorption. Temperature gradients can cause measurable movement of water from the warm to the cool side of fruits and vegetables (Curtis, 1937; Veto, 1963) and from warm to cool regions in soil, but this movement occurs chiefly as vapor through the air spaces.

The direct effect of temperature on the rate of diffusion of gases is also small, as mentioned earlier. However, the effect of increasing temperature on evaporation and transpiration is much greater because it increases the vapor pressure difference between the evaporating surface and the air. For example, an increase in leaf and air temperature from 20° to 30°C, with no change in absolute humidity, and a relative humidity at 20°C of 50% would increase the vapor pressure gradient by about 260%. More data illustrating this point are given in Chapter 11.

CELL WATER TERMINOLOGY

Study of cell water relations began in the second quarter of the nineteenth century, when Dutrochet introduced the concept of osmosis. Later, Traube (1867) and Pfeffer (1877) developed the concept of differential permeability of membranes, and de Vries in 1884 studied plasmolysis and explained the conditions necessary for development of turgor in cells.

During the early twentieth century, it began to be realized that the movement of water from cell to cell cannot be explained in terms of gradients of osmotic pressure but rather in terms of gradients of what is today called water potential. This led to the introduction of a variety of terms intended to describe the ability of cells and tissues to absorb water. Among these terms were *Saugkraft* or suction force (Ursprung and Blum, 1916), water-absorbing power (Thoday, 1918), hydratur (Walter, 1931, 1965), net osmotic pressure (Shull, 1930), diffusion pressure deficit (Meyer, 1938, 1945), water potential (Owen, 1952; Slatyer and Taylor, 1960), and others. We will discuss only the most often used terms.

Diffusion Pressure Deficit

The term diffusion pressure deficit (DPD) was introduced by Meyer in 1938 and publicized in the widely used textbook by Meyer and Anderson (1939, and later editions). The DPD of water in a cell or solution was defined as the amount by which its diffusion pressure is lower than that of pure water at the same temperature and under atmospheric pressure. The DPD of a cell can also be regarded as a measure of the pressure with which water will diffuse into a cell when it is immersed in pure water. The equation for the DPD of a cell is:

$$DPD = OP - TP \qquad (2.4)$$

where *OP* is the osmotic pressure of the cell contents and *TP* is the turgor pressure in the cell.

The DPD served a useful purpose for over 2 decades, but plant science developed to a stage at which it needed and was prepared to use the basic potential terminology of thermodynamics and physical chemistry. In contrast to DPD, which was always puzzling to physical scientists, the term potential is understood by investigators in many fields. Furthermore, its use makes it easier to separate the components—osmotic, matric, and pressure potentials—that constitute the total water potential of cells and tissues.

Water Potential

The use of thermodynamic terminology is not new. In 1907 Buckingham used the term capillary potential in reference to soil moisture, and Edlefsen (1941) proposed the term specific free energy for what we now term soil water potential. The net influx free energy of Broyer (1947) is equivalent to cell water potential, and in 1952 Owen proposed the term water potential. The terms proposed by Edlefsen, Broyer, and Owen were never widely used, probably because plant scientists were not yet ready for a thermodynamic terminology. However, when Slatyer and Taylor (1960) proposed the term water potential, the intellectual climate was more favorable for its adoption and it is now generally used, in spite of a few complaints, such as those by Spanner (1972).

The concept of water potential as a measure of the free energy status of water was developed in Chapter 1 and will now be applied to plant cells. Under isothermal conditions the various factors involved in cell water relations at equilibrium can be summarized by the equation:

$$\Psi_w = \Psi_s + \Psi_p + \Psi_m \tag{2.5}$$

where Ψ_w is the potential of the water in the cell, and the other terms express the contributions to Ψ_w by solutes (Ψ_s), pressure (Ψ_p), and matric forces (Ψ_m). Ψ_s and Ψ_m are negative, Ψ_s expresses the effect of solutes in the cell solution, and Ψ_m expresses the effect of water-binding colloids and surfaces and capillary effects in cells and cell walls. Ψ_p is usually positive. In the xylem, however, Ψ_p may be either negative during transpiration or positive in guttating plants as a result of root pressure. The sum of the three terms is a negative number, except in fully turgid cells when it becomes zero. In this case, the positive pressure potential balances the sum of the negative osmotic and matric potentials, as shown below.

It is doubtful if the term for matric water should be included in Eq. (2.5). There is some uncertainty about its definition, and there is no reliable method for measuring it. Matric water is often assumed to include water bound by electrostatic forces to charged surfaces in cells and that held in the microcapillaries of cell walls. Tyree and Karamanos (1980) argue that capillary water should not be included in matric water because it can be regarded as held by negative pressure. This would reduce the matric water to a negligible fraction. There is no doubt that capillary forces in the cell walls and water binding by electrostatic forces lower the water potential, and the broader definition will be used here. However, the osmotic and matric forces are not really arithmetically additive. For example, in experiments in which sucrose solution was added to filter paper, the total potential of the system did not equal the sum of the matric potential of the filter paper and the osmotic potential of the sucrose solution (Boyer and Potter, 1973; Markhart *et al.*, 1981). If the volume of matric water is very small compared to the volume of vacuolar water, as in parenchyma cells, Ψ_m has a negligible effect on the total Ψ_w. However, in developing seeds or thick-walled cells where the vacuolar water constitutes a smaller fraction of the total water, matric potential can control the cell water potential. In both situations the two potentials come into equilibrium, but in neither are they truly additive. Weatherley (1970) and Passioura (1980) therefore argue that the term for matric potential should be dropped out of Eq. (2.5). If we disregard Ψ_m, Eq. (2.5) becomes

$$\Psi_w = \Psi_s + \Psi_p \tag{2.6}$$

Assuming no change in cell volume or Ψ_s, the relationships for cells of various turgidities can be shown in the tabulation below:

	Ψ_w	=	Ψ_s	+	Ψ_p	
Fully turgid	0	=	−2.0	+	(+2.0)	MPa
Partly turgid	−1.0	=	−2.0	+	(+1.0)	MPa
Flaccid	−2.0	=	−2.0	+	0	MPa

Another form of Eq. (2.6) is used by some writers in which hydrostatic pressure, *P*, is substituted for Ψ_p and π for Ψ_s:

$$\Psi_w = P - \pi \tag{2.7}$$

Since Ψ_w is a negative number, it becomes lower as water stress increases. The same situation occurs in reading temperatures below zero on a thermometer; negative temperatures increase in absolute magnitude but decrease in relative value. Just as a temperature of $-5°$ is higher than $-10°$, so a Ψ_w of -0.5 MPa is higher than -1.0 MPa.

The interrelationships between Ψ_w and the factors that control it can be shown by a diagram, such as that in Fig. 2.3, for a cell that undergoes considerable change in volume with change in turgor. This shows the changes in osmotic potential, Ψ_s, and in Ψ_w as the volume and the turgor pressure Ψ_p change. When $\Psi_p = \Psi_s$, the water potential of the cell is zero; conversely, when Ψ_p decreases to zero at incipient plasmolysis, $\Psi_w = \Psi_s$. The line representing Ψ_p is often drawn as though turgor pressure is linearly proportional to volume, but this is not usually correct. Thick-walled cells show little change in volume (perhaps only 2–5%), but thin-walled parenchyma cells show much larger changes in volume, and the lines for pressure and osmotic potential are assumed to be curvilinear in such cells (Bennet-Clark, 1959, pp. 171–174; Crafts *et al.*, 1949, Chapter 7; Kamiya *et al.*, 1963; Takaoki, 1969). Gardner and Ehlig (1965) show an abrupt change in the shape of the line representing turgor pressure at a pressure potential of 2 or 3 bars. Whatever the details, cell volume evidently does not change linearly with cell turgor. According to Kamiya *et al.* (1963), the curves are somewhat different in shape for enlarging and shrinking cells of *Nitella*.

Negative Turgor Pressure

Cells of plant tissue are seldom plasmolyzed in nature, but if they are, the protoplast usually pulls away from the cell wall. However, examples of adhesion were reported in the older literature (Crafts *et al.*, 1949, pp. 83–85). Occasionally, the water potential of cells is reported to be lower than the osmotic potential (Slatyer, 1957, 1960; Bennet-Clark, 1959, p. 174; Noy-Meir and Ginzburg, 1967, 1969; Richter, 1976, pp. 51–52). These discrepancies have sometimes been attributed to negative wall pressure in which the cell walls are pulled inward during plasmolysis of severely stressed cells by adhesion of protoplasts to walls, placing the water under tension. Tyree (1976) regarded values of water potential lower than osmotic potential as artifacts caused by dilution of vacuolar sap by cell wall water during extraction. Such dilution may be important in cells where 10% or more of the total cell water occurs in the walls, but it is a negligible factor in thin-walled tissue. This problem is discussed in the section on measurement of osmotic potential.

Hydrature

The term hydratur or hydrature, proposed by Walter in 1931 and used in a number of subsequent papers (Walter, 1963, 1965; Kreeb, 1967). There has been considerable uncertainty concerning the precise meaning of this term (Shmueli and Cohen, 1964; Stocker, 1960). Walter (1965) states that hydrature is the relative water vapor pressure, i.e., the vapor pressure of the water in vacuoles or cytoplasm relative to that of pure water at the same temperature and pressure. Thus, in a fully turgid cell, although the actual vapor pressure of the cell sap is equal to that of pure free water, the relative vapor pressure is lower, because, according to the definition of hydrature, the vapor pressure of the cell sap is expressed as a percentage of that of free water at the same pressure. Thus, different parts of a cell have different hydratures. As Slatyer (1967) pointed out, the hydrature of the cell as a whole is equal to its water potential, whereas the hydrature of the vacuolar sap and cytoplasm is equal to the osmotic potential. This situation suggests that hydrature is a confusing and unsatisfactory term.

Readers are referred to Weatherley (1965) and Walter (1963, 1965) for further discussion of this term.

Bound Water

The term bound water is often encountered in the older literature of plant physiology, especially that dealing with cold and drought tolerance. Gortner (1938) discussed the origin of this term and methods of measuring bound water. The subject was also reviewed by Kramer (1955b). The concept of bound water is based on the fact that a variable fraction of the water in both living and nonliving materials acts differently from free water. It has a lower vapor pressure, remains unfrozen at temperatures far below 0°C, does not function as a solvent, and seems to be unavailable for physiological processes. In practice, bound water has most often been defined as water that remains unfrozen at some low temperature, usually $-20°$ or $-25°C$. The binding forces are included in the matric potential, Ψ_m, of Eq. (2.5).

Readers should understand that there is no sharp division between unbound and bound water but a gradual transition, with decreasing water content from free water, easily frozen or removed by drying, to water so firmly bound to surfaces that it cannot be removed by temperatures far below 0°C or even by a long period in an oven at 100°C. Water bound this firmly doubtless plays an important role as a cell constituent in the tolerance of drying of some seeds, spores, microorganisms, and a few higher plants. However, efforts to explain differences in cold or drought tolerance of flowering plants by differences in amount of bound water have been disappointing (Levitt, 1980, pp. 155–156; Burke *et al.*, 1976). The amount of bound water varies widely among different kinds of tissues, with age and past history, and according to the method used to measure it. In recent years, the amount of bound water has usually been determined by nuclear magnetic resonance methods (Burke *et al.*, 1974, 1976).

COMPONENTS OF THE WATER POTENTIAL EQUATION

Equation (2.6) gives only the essential components of the water potential equation. A complete equation that includes all of the factors which might influence the free energy or potential of water (Ψ_w) might include the following (Miller, 1972; Begg and Turner, 1976):

$$\Psi_w = \Psi_s + \Psi_p + \Psi_m + \Psi_g \qquad (2.8)$$

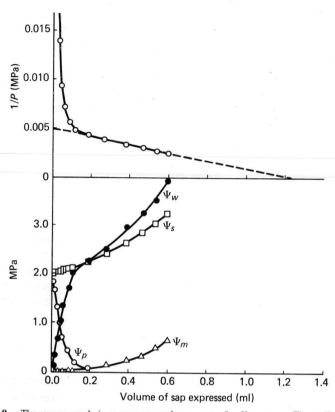

Fig. 2.9. The upper graph is a pressure–volume curve for *Ilex opaca*. The ordinate is the reciprocal of the pressure, and the abscissa is the volume of water expressed, in milliliters. The dashed line is the calculated extension of the linear part of the curve. It intersects the ordinate at 0.0051, equal to an osmotic potential of -1.96 MPa, and the abscissa at 1.23 ml. The data in the lower graph were obtained by analysis of the pressure-volume curve according to the method of Tyree and Hammel (1972). Curves are shown for turgor (Ψ_p), osmotic (Ψ_s) and matric (Ψ_m) potentials, and total water potential (Ψ_w). The turgor potential is positive, and all other potentials are negative. The abscissa of the lower graphs represents water volume in milliliters, and the ordinate is given in megapascals (1 MPa = 10 bars). (Redrawn from Roberts, 1978.)

expansion to turgor pressure even though it is really caused by inward diffusion of water.

Matric Potential. Direct measurements of the matric potential have been made on homogenized plant tissue from which the solutes have been washed or on frozen and thawed plant tissue subjected to pressure. Using the latter method, Wiebe (1966) found matric potentials of only -0.01 MPa in succulent plant tissue having an osmotic potential of -1.0 to -2.5 MPa. Boyer (1967) found only 4–6% of the water in sunflower leaves held by matric forces at -0.7 to -1.0 MPa, but in rhododendron leaves 25% of the water was held by matric forces at -0.1 to -0.3 MPa water potential. Shayo-Ngowi and Campbell (1980) measured matric potential by a simplified method which has been used successfully in the author's laboratory by R. G. Musser (unpublished). The method is discussed in Chapter 12.

In general, the amount of matrically bound water depends on the amount of cell wall tissue, being much greater in cells with thick walls than in cells with thin walls. Also, the proportion of matric water increases as the total water content decreases. Thus, matric forces are likely to be more important in sclerophyllous than in mesomorphic leaves (Miller, 1972, p. 216).

Interactions among Components

There has been an interesting change over the years in the emphasis placed on various components of the total cell water potential. During the period from 1910 to 1940, the only value measured extensively was the osmotic pressure or osmotic potential, because it was the only value for which a satisfactory method was available. During the 1930s, it was realized by a few physiologists that water movement is not controlled by the osmotic potential, but by what is now termed the water potential. However, there was no satisfactory method available to measure it, and measurement of the osmotic potential was passing out of favor. As a result, during the 1940s and 1950s, plant water status was seldom measured at all, decreasing the value of the research on plant water relations during that period. About 1958, usable thermocouple psychrometers were introduced (Monteith and Owen, 1958; Richards and Ogata, 1958), and a few years later Scholander and his colleagues (1964, 1965) introduced the pressure chamber method. The availability of relatively easy methods of measuring water potential led to overemphasis on this value and neglect of the other terms in the water potential equation.

Water potential is an important and sensitive measure of plant water status, varying from near zero in unstressed plants to a value as low as or lower than the osmotic potential in severely stressed plants (see Fig. 2.3). Furthermore, a given water potential indicates the same degree of water stress in all kinds of tissue, whereas the osmotic potential at full turgor may be -1.5 MPa in a well-watered mesophyte and -3.0 MPa or less in a halophyte. Also, soil water potential can be measured and compared with plant water potentials. This is important because water movement from soil to plants, as well as within plants, occurs from regions of higher to regions of lower water potential. Nevertheless, water potential by itself is not an adequate description of the water status of plants.

Osmotic and Turgor Potentials

During the 1970s, renewed interest developed in the osmotic potential and in osmotic adjustment, probably because of increased appreciation of the importance of turgor pressure. It has already been pointed out that turgor pressure plays an important role in cell enlargement, stomatal opening, and maintenance of form in herbaceous plants. It also affects or controls cell membrane permeability in some algae and possibly in other cells (Gutknecht and Bisson, 1977; Zimmermann, 1978). Hsiao (1973) suggested that the most likely mechanism through which water stress can affect metabolic processes is by changes in turgor pressure. This process might operate by changing spatial relationships of enzymes on cell membranes and thereby affecting enzyme-mediated processes.

Osmotic Adjustment

It has been known for many years that water stress results in lowering of the osmotic potential of plant tissue in some plants. This contributes to the maintenance of turgor and presumably to maintenance of normal metabolic activity. Part of the decrease in osmotic potential is caused by concentration of cell solutes resulting from the loss of water. Osmotic adjustment refers to a net increase in solutes, as distinguished from the passive increase in concentration caused by loss of water (Turner and Jones, 1980, p. 89). It is important because it results in maintenance of turgor at a lower Ψ_w than would otherwise be possible. The interest in osmotic adjustment as a factor in drought tolerance is reviewed in detail by Turner and Jones (1980), and it will be discussed in connection with drought tolerance in Chapter 13.

SUMMARY

The water relations of plants are dominated by cell water relations because most of the water occurs in cells, chiefly in the vacuoles. Experimentally, the volume of water in the walls is important because it can dilute the vacuolar sap, producing incorrect values for the osmotic potential. The elasticity of cell walls is also important because it affects the relationship between cell volume and the water content, turgor pressure, and water potential, as shown on Höfler diagrams. Water in plant tissues is conveniently separated into the symplastic water occurring in the vacuoles and cytoplasm and the apoplastic water in the walls and xylem elements. This distinction is based on the fact that the content of the symplast is enclosed in membranes relatively impermeable to most solutes, and movement across them usually requires expenditure of metabolic energy to drive an active transport process. Metabolic energy is also required to maintain cell membranes in a physiologically normal condition.

Movement of water into and out of cells occurs by diffusion along gradients of decreasing water potential, and there is no evidence that active transport of water, dependent on metabolic energy, occurs in plants. Mass flow occurs over long distances in the xylem, and in the phloem. The terminology of cell and tissue water relations is somewhat confusing, but most investigators use the potential terminology proposed by Slatyer and Taylor (1960). Because of the ease of measurement and because it controls water movement, the cell water potential has received major attention. However, it is now realized that the turgor pressure is important both at the whole plant level and in cell water relations. There is also increasing interest in the osmotic potential and the role that osmotic adjustment can play in increasing the drought tolerance of plants.

SUPPLEMENTARY READING

Bonner, J., and Varner, J. E., eds. (1976). "Plant Biochemistry," 3rd ed. Academic Press, New York.
Esau, K. (1965). "Plant Anatomy," 2nd ed. Wiley, New York.
Fahn, A. (1979). "Secretory Tissue in Plants." Academic Press, New York.
Robards, A. W., ed. (1974). "Dynamic Aspects of Plant Ultrastructure." McGraw-Hill, New York.
Slatyer, R. O. (1967). "Plant Water Relationships." Academic Press, New York.
Slavik, B. (1974). "Methods of Studying Plant Water Relations." Springer-Verlag, Berlin and New York.
Smith, H. W. (1962). The plasma membrane, with notes on the history of botany. *Circulation* **26,** 987–1012.
Tyree, M. T., and Jarvis, P. G. (1982). "Water in Tissues and Cells." Encyclopedia of Plant Physiology N.S.12B. Springer-Verlag, Berlin and New York.
Zimmermann, U., and Dainty, J., eds. (1974). "Membrane Transport in Plants." Springer-Verlag, Berlin and New York.

Soil and Water

INTRODUCTION

For readers of this book, interest in soil is centered chiefly on its role as a medium for root growth and storage place for water, but it is also the source of mineral nutrients and provides anchorage for plants. Furthermore, it contains an active population of microorganisms, earthworms, and small insects which have important effects on its chemical and physical characteristics and on root growth. Soil imposes various physical, chemical, and biological limitations on root growth, which in turn affect absorption and plant growth (Russell, 1973; Cassell, 1982). Most of these will be discussed in later chapters in connection with root growth and the absorption of water and minerals. In this chapter, we will deal chiefly with the characteristics affecting the suitability of soil as a medium for root growth and as a source of water. Readers are referred to books such as those by Baver *et al.* (1972), Black (1968), Brady (1974), Donahue *et al.* (1977), Hillel (1980a,b), and Russell (1973) for more details.

IMPORTANT CHARACTERISTICS OF SOIL

Composition

Soil is a complex system consisting of varying proportions of four principal components. These are the mineral or rock particles and the nonliving organic matter that form the solid matrix, and the soil solution and air that occupy the pore space within the matrix. In addition to these four components, soil usually contains numerous living organisms such as bacteria, fungi, algae, protozoa, insects, and small animals that directly or indirectly affect soil structure and plant growth.

Mineral particles are the chief components of most soils on a volumetric basis, except in organic soils such as peat, and they are also the most stable. They consist of rock particles developed *in situ* by weathering or deposited in bulk by wind or water. Nonliving organic matter usually constitutes less than 5% of the volume, except in the surface layer and in peat soils, but it can vary considerably in a given soil with cultural practices. Organic residues consist of a wide variety of materials ranging in size from large root fragments through rootlets and litter to colloidal products of decomposition.

The most conspicuous property of the solid matrix is its particulate nature. In contrast, the pore space forms a continuous but geometrically complex system which usually constitutes 30–60% of the total volume (see Fig. 3.1). It may be filled entirely with water, as in saturated soils, or largely with air, as in dry soils. In agricultural soils at field capacity, the water fraction usually ranges from 40 to 60% of the pore space. The high degree of continuity in the water phase is of great importance in respect to water and salt movement in the soil and to roots.

In addition to these nonliving components, the soil is permeated by living roots and contains large populations of microorganisms, especially in the surface layers and in the vicinity of roots, the rhizosphere. Soil organisms play an important role in decay of organic matter and release of nitrogen and mineral nutrients, which then become available for reabsorption. They also tend to deplete the oxygen content and increase the carbon dioxide content of soil, thus modifying the soil atmosphere in which roots grow. Organic products of decomposition are said to play a significant role in cementing soil granules together, thereby improving soil structure. The roots, together with small animals such as earthworms (see Barley, 1962), insects, and tunneling vertebrates, exert important effects on soil water relationships, particularly the infiltration and distribution of water in the soil. The role of insects and the soil microflora was discussed in detail by Russell (1973) and Donahue *et al.* (1977), and the role of soil organisms in plant disease was covered in a symposium edited by Baker and Snyder (1965).

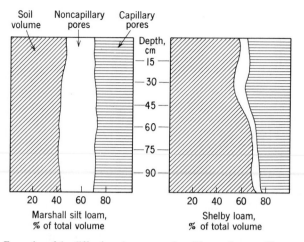

Fig. 3.1. Examples of the differences in amount of capillary and noncapillary pore space in two dissimilar soils. A large proportion of noncapillary pore space is desirable because it promotes drainage and improves aeration. (Adapted from Baver, 1948.)

The physical characteristics of a soil depend chiefly on the texture or size distribution of mineral particles, on the structure or the manner in which these particles are arranged, on the kind of clay minerals present and the kind and amount of exchangeable ions adsorbed upon them, and on the amount of organic matter incorporated with the mineral matter.

Characteristics of the Clay Fraction. The clay fraction provides most of the internal soil surface; therefore, it controls the important soil properties. For this reason, it will be discussed in more detail.

There are three major types of clay minerals: kaolinite, which is most common in mature, weathered soils, and montmorillonite and illite, which are the chief constituents of young soils. Unit crystals or micelles of kaolinite consist of silica and alumina platelets in a 1:1 ratio. These micelles form rigid lattices, so that soils composed chiefly of kaolinite show little swelling and shrinking with changes in hydration. Montmorillonite and illite micelles are composed of silica and alumina platelets in a 2:1 ratio. In illite, potassium ions occurring between silica platelets of adjacent micelles form chemical bonds strong enough to prevent separation and swelling. No such bonds exist between micelles in montmorillonite; hence, soils containing a large proportion of montomorillonite swell and shrink markedly with changes in hydration. Such soils often develop broad, deep cracks during prolonged droughts. Two types of clay minerals are shown in Fig. 3.2. A form of montmorillonite, called bentonite, is sometimes used as a sealing compound in irrigation canals, on dams, and in drilling oil wells, because it swells when wet and becomes highly impermeable to water.

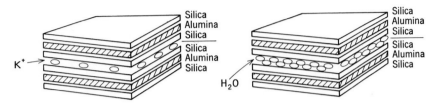

Fig. 3.2. Diagram showing arrangement of silica and alumina sheets in illite crystals (left) and in montmorillonite crystals (right). Penetration of water between silica layers causes the swelling characteristics of soils containing a large proportion of montmorillonite. (From Thompson, 1952.)

Clay micelles are negatively charged, chiefly by replacement of silicon and aluminum ions with other cations within the crystal lattice. Another source of charge is the incomplete compensation of charge where bonds are broken at lattice crystal edges. The intensity of the negative charge determines the cation exchange capacity, or capacity to hold cations. The amount of water and ions bound on soil colloids has important effects on soil properties and plant growth. Organic matter also has a large ionic exchange capacity (see Table 3.1), and consequently, ionic exchange reactions are influenced by organic matter content as well as the amount and kind of clay. Soils low in clay, but high in silt and sand, have a low cation exchange capacity because silt and sand have fewer binding sites than clay particles.

Soil Texture

Textural classifications of soil are based on the relative amounts of sand, silt, and clay predominating in the solid fraction. Soils are classified as sand, loam, silt, or clay, with various intermediate classes such as sandy loam, silt loam, or clay loam, as shown diagrammatically in Fig. 3.3. Table 3.2 gives examples of soils classified according to particle size. The least complex soil is a sand, which

TABLE 3.1 Cation Exchange Capacity of Humus and Clay Minerals in Milliequivalents per 100 g Dry Soil[a]

Vermiculite	160
Humus	100–300
Montmorillonite	100
Illite	30
Kaolinite	10

[a] From data of Thompson (1952).

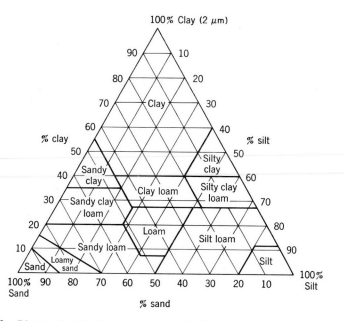

Fig. 3.3. Diagram showing the percentage of sand, silt, and clay in various soil classes. (From *Soil Sci. Soc. Am. Proc.* **29**, 347, 1965.)

by definition contains less than 15% of silt and clay. Such soils form relatively simple capillary systems with a large volume of noncapillary pore space that ensures good drainage and aeration. Sandy soils are relatively inert chemically, are loose and noncohesive, and have a low water-holding capacity and usually a low cation exchange capacity.

Clay soils are at the other extreme with reference to size of particles and complexity because they contain more than 40% of clay particles and less than 45% of sand or silt. The clay particles are usually aggregated together in complex

TABLE 3.2 Classification of Soil Particles According to the System of the International Society of Soil Science, and Mechanical Analysis of Three Soils[a]

Fraction	Diameter (mm)	Sandy loam (%)	Loam (%)	Heavy clay (%)
Coarse sand	2.00–0.20	66.6	27.1	0.9
Fine sand	0.20–0.02	17.8	30.3	7.1
Silt	0.02–0.002	5.6	20.2	21.4
Clay	Below 0.002	8.5	19.3	65.8

[a] From Lyon and Buckman (1943, p. 43).

granules. Because of their plate-like shape, clay particles have a much greater surface area than cubes or spheres of similar volume. Their extensive surface enables clay particles to hold more water and minerals than sandy soils. The surface possessed by even a small volume of particles of colloidal dimensions is tremendous. A cubical bit of rock 1 cm on the edge has a surface of only 6 cm^2, but if it is divided into particles the size of colloidal clay, 0.1 μm on the edge, the total resulting surface would be 600,000 cm^2. Day *et al.* (1967) state that the surface available to bind water ranges from less than 1000 cm^2/g in coarse sands to over 1,000,000 cm^2/g in clays. The size of the clay fraction largely controls the chemical and physical properties of mineral soils.

Loam soils contain more or less equal amounts of sand, silt, and clay, and therefore have properties intermediate between those of clay and those of sand. Such soils are considered most favorable for plant growth because they hold more available water and cations than sand and are better aerated and easier to work than clay.

Textural classification has only an approximate relationship to the behavior of a soil as a medium for plant growth because external properties may be modified appreciably by organic matter content, the kinds of clay minerals present, and the amounts and kinds of ions associated with them. Aggregation effects of organic matter tend to give a fine-textured soil high in clay some of the pore space properties of a coarser-textured soil, and colloidal effects of organic matter additions to a coarse-textured, sandy soil give it some of the moisture and cation-retention characteristics of a finer-textured soil. A clay soil composed chiefly of swelling clay minerals of the montmorillonite type has more of the properties of a finer-textured soil than one composed of kaolinitic clay minerals. In addition, when the exchange complex is dominated by sodium ions, clay micelles are dispersed and the soil appears to have a finer texture and less pore space than when calcium or hydrogen are the dominant ions.

Structure and Pore Space

Combination of the smallest soil particles into aggregates (crumbs, granules, or even clods) produces soil structure. The degree of structure existing in a soil affects the amount and size of pores and thereby greatly affects water movement, soil aeration, and root growth. The numerous factors affecting soil structure are discussed in detail by Baver *et al.* (1972, Chapters 4 and 5) and by Hillel (1980a, Chapter 6).

The pore space is that fraction of the soil volume occupied by air and water, and usually amounts to about one-half of the soil volume, but varies with soil type (Fig. 3.1) and past treatment (Fig. 3.4). However, total volume of pore

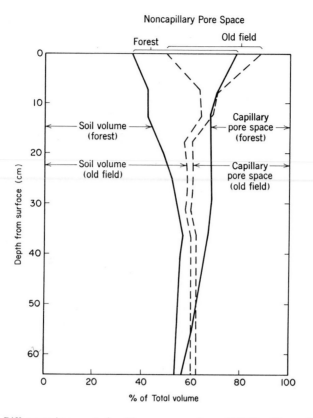

Fig. 3.4. Differences in amount of capillary pore space in an old field and in an adjacent forest on the same type of soil. The large percentage by volume occupied by noncapillary pore space in the forest soil provides better aeration for roots. It also increases the rate of infiltration, as shown in Fig. 3.13, and decreases runoff during heavy rains. (From Hoover, 1949.)

space is less important than the proportions of capillary and noncapillary pore space. Noncapillary pore space refers to that fraction from which water drains by gravity, providing the air space so important for root aeration. Capillary pore space refers to the fraction of pore space composed of pores 30–60 μm or less in diameter that hold water against gravity. This determines the amount of water available to plants that is retained in a soil after a rain or irrigation. Pores small enough to hold water at a potential of -1.5 MPa are also too small for easy penetration by roots which are usually much more than 60 μm in diameter. The large noncapillary capacity of sandy soils results in good drainage and aeration, but it also results in a much lower storage capacity for available water than exists in clay soil, with its larger proportion of capillary pore space. In a soil good for root growth and water retention, the pore space is about equally divided between large noncapillary and small capillary pores. An example is the Marshall silt loam shown in Fig. 3.1.

Soil Structure.   Soil structure and pore space depend both on particle size and on the manner in which the particles are assembled together in stable crumbs, or aggregates. Maintenance of stable aggregates is very important in agricultural soils, in which the structure is often degraded by compaction and cultivation when too wet, and sometimes by saturation with sodium ions from salty irrigation water. The nature of the mechanism responsible for stability of soil structure is not fully understood, but it seems to be related to the presence of organic colloids produced by microorganisms, probably chiefly those in the rhizosphere, which produce substances that aid in cementing particles together. Soil structure is generally much more stable under forests and pastures than in similar soils that are cultivated (see Fig. 3.5). Freezing and thawing, and even alternately wetting and drying soil, seem to aid in aggregation of clay micelles into the stable crumbs necessary for a good structure. Soil structure is discussed in detail in most soil texts.

Preservation of soil structure is especially important in heavy soils to permit the infiltration of water as well as for maintenance of the aeration necessary for good root growth. Sometimes pelting rain causes dispersion of the surface of bare soil, resulting in the formation of a surface crust that hinders infiltration of

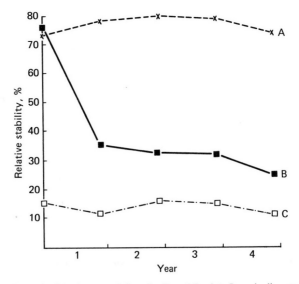

Fig. 3.5.   Effect of cultivation on stability of soil particles 3 to 5 mm in diameter in a calcareous soil. (A) Permanent grassland for about 100 years, (B) grassland ploughed at beginning of the experiment and cultivated annually thereafter, (C) grassland cultivated for about 100 years. (After R. S. Russell, 1977.)

water, gas exchange, and emergence of seedlings. Presence of a mulch prevents crust formation. Compaction by farm machinery or salt accumulation in a particular soil horizon often causes formation of dense clay pans or hardpans that are relatively impermeable to water and roots (see Fig. 6.9). Under some conditions, impermeable layers develop even in sandy soils. The modification of soil structure is discussed by over 50 authors in the book edited by Emerson *et al.* (1978), and the presence of impermeable layers is discussed by Cassell (1983).

Soil Profiles

Although some soils, such as recently deposited alluvium, may have a uniform texture to a depth of several feet, generally there are important changes in texture

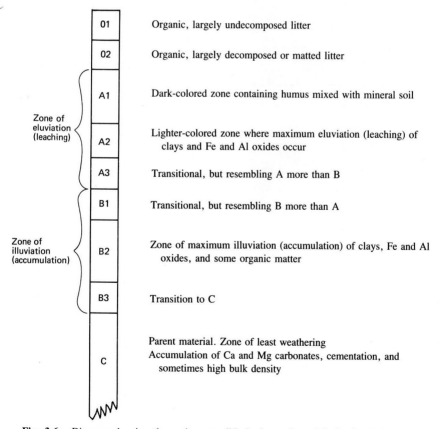

Fig. 3.6. Diagram showing the various possible horizons that might be found in a highly developed soil. (Adapted from various sources.)

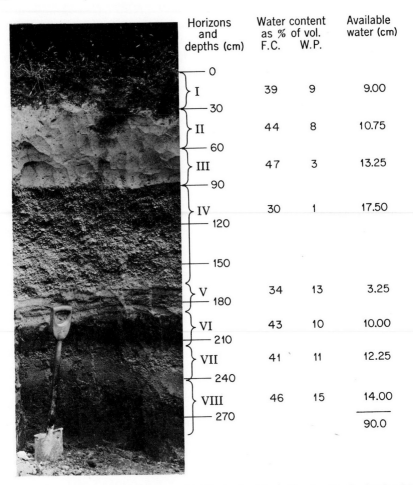

Horizons and depths (cm)	Water content as % of vol. F.C.	W.P.	Available water (cm)
0			
I	39	9	9.00
30			
II	44	8	10.75
60			
III	47	3	13.25
90			
IV	30	1	17.50
120			
150			
V 180	34	13	3.25
VI	43	10	10.00
210			
VII	41	11	12.25
240			
VIII	46	15	14.00
270			90.0

Fig. 3.7. Profile of a deep pumice soil in New Zealand formed by deposits of volcanic origin. Water content at field capacity (FC) and at permanent wilting (WP), expressed as percentages by volume, are shown for each horizon. Also shown are the readily available water in each horizon in centimeters and the total amount of water to a depth of 270 cm. This soil has an extraordinarily high storage capacity for readily available water, and forests growing on it are very productive. Root distribution of a Monterey pine tree growing in this soil is shown in Fig. 6.11. (Photograph of the soil profile by H. G. Hemmin of the New Zealand Forest Service; moisture data from Will and Stone, 1967, for a similar soil.)

and even in structure at different depths in the soil. This is particularly true in older soils, where much downward leaching has occurred. As a result, undisturbed soils usually possess a characteristic profile consisting of definite horizons, or layers of soil, that differ in their properties. A hypothetical soil profile is shown in Fig. 3.6. The upper or "A" horizon usually differs appreciably in texture and hence in water retention characteristics such as field capacity and permanent wilting percentage from the "B" horizon which lies below it. It is therefore necessary to sample at least as deeply as roots penetrate and to determine the water-retaining properties of all horizons penetrated by roots in order to understand the water supply of plants. An example of an actual soil profile and its water-holding capacity at various depths is shown in Fig. 3.7. Tree roots penetrated to at least 180 cm in this soil (see Fig. 6.11), and Patric *et al.* (1965) reported absorption of water from a depth of 6 m under forest trees in a deep soil. This topic is discussed in more detail in Chapter 6 in connection with factors limiting root growth.

HOW WATER OCCURS IN SOIL

The water content of a soil is usually stated as the amount of water lost when the soil is dried at 105°C, expressed either as the weight of water per unit weight of soil or as the volume of water per unit volume of soil. The volume per unit volume of soil indicates the amount of water available to plant roots more clearly than the weight per unit weight of soil. However, water content on a percentage basis tells little about the amount of water available for plants, because a sand may be saturated at a water content that would be too dry for plant growth in a clay soil (see Table 3.3). Sometimes the water content is expressed as centime-

TABLE 3.3   **Water Content of Soils of Various Textures at Matric Potentials of 0.03 and 1.5 MPa and at First Permanent Wilting**

Name of soil	Water content as percentage of dry weight[a]		
	0.03 MPa	1.5 MPa	First permanent wilting
Hanford sand	4.5	2.2	2.9
Indio loam	4.6	1.6	2.6
Yolo loam	12.6	7.1	8.4
Yolo fine sandy loam	12.6	5.5	8.3
Chino loam	19.7	8.0	10.2
Chino silty clay	40.8	21.9	23.2
Chino silty clay loam	48.9	15.0	23.3
Yolo clay	45.1	26.2	29.6

[a] From Kramer (1969).

ters of water per unit of depth, as in the first 25 or 100 cm, or in various horizons, as in Fig. 3.7.

When water is applied to the soil surface, it infiltrates and drains downward through the larger soil pores by gravitational flow. However, some is retained by capillary forces in the smaller pores (diameter less than 30–60 μm) and by adsorption on the surfaces of the soil particles. These constitute the matric forces mentioned earlier, and they hold that part of the soil water available to plants. A variable fraction of the matric water is held so firmly and moves so slowly that it is treated as unavailable to plants.

Soil water is discussed in detail in a book edited by Nielsen *et al.* (1972), in Hillel (1980a,b), Slatyer (1967), and various soil texts, such as the one by Baver *et al.* (1972).

Terminology of Soil Water

The only satisfactory measure of the availability of soil water to plants is the water potential. However, other terms, such as field capacity, permanent wilting percentage, and readily available water content, are widely used and will be discussed. The relationships among some of these values are shown graphically in Fig. 3.8 for a loam and a clay soil and in Fig. 3.9 for various soil textures.

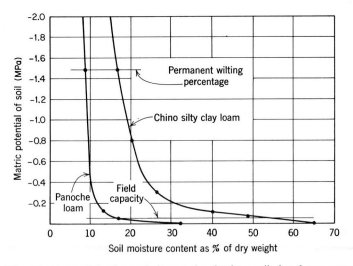

Fig. 3.8. Matric potentials of a sandy loam and a clay loam soil plotted over water content. (Curve for Panoche loam is from Wadleigh *et al.*, 1946; curve for Chino loam from data of Richards and Weaver, 1944.)

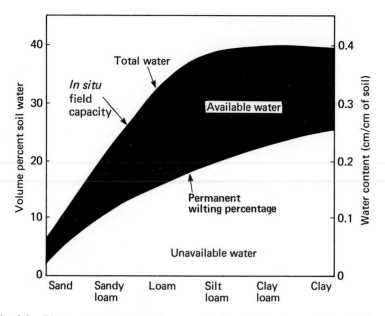

Fig. 3.9. Diagram showing the relative amounts of available and unavailable water in soils ranging from sand to clay. Amounts are expressed as percentages of soil volume and as centimeters of water per centimeter of soil. (From Cassell, 1983.)

Water Potential. The importance of the chemical potential of water with respect to cell and plant water relations was discussed in Chapter 2, and it is equally important with respect to soil water. The soil water potential includes four components:

$$\Psi_{soil} = \Psi_m + \Psi_s + \Psi_g + \Psi_p \tag{3.1}$$

In this equation Ψ_m represents the matric potential produced by capillary and surface forces, Ψ_s the osmotic potential produced by solutes in the soil water, and Ψ_g the gravitational forces operating on the soil water. Ψ_p refers chiefly to pressure in the gas phase, but it is sometimes used with reference to tensiometer pressures and then refers to matric forces. The troublesome problem of terminology is presented in detail in Bulletin 48 of the International Society of Soil Science and is discussed by Corey *et al.* (1967).

Field Capacity. The field capacity, or *in situ* field capacity, of a soil is usually described as the water content after downward drainage of gravitational water has become very slow and water content has become relatively stable. This situation usually exists several days after the soil has been thoroughly wetted by rain or irrigation. In some soils slow drainage continues for many days, as shown in Fig. 3.10. The field capacity has also been termed the field-carrying capacity, normal moisture capacity, and capillary capacity. It is not a true equilibrium value, but only a condition of such slow movement of water that the moisture content does not change appreciably between measurements. Although deep soils and sandy soils reach field capacity rather quickly, the presence of a water table or an impermeable layer near the surface will prolong the time required for drainage. Also, if a deep soil is initially saturated to a depth of several meters, drainage of the surface layer to field capacity will be much slower than if only the surface meter is wetted. Lack of homogeneity in the soil also affects the water

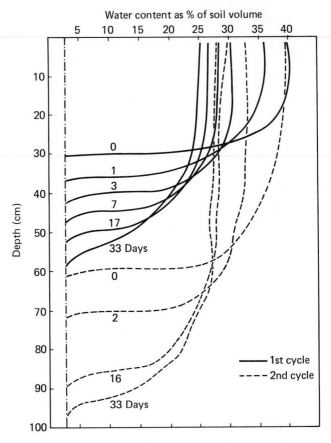

Fig. 3.10.   Rate of drainage of water applied to a column of air-dry soil. The vertical dashed line is the original soil water content. The upper set of curves shows progressive drainage after applying 10 cm of water, and the second set of curves after applying another 10 cm of water. Measurable downward movement occurred for at least 35 days, and no distinct field capacity could be distinguished in this soil. (From Gardner *et al.*, 1970.)

content at field capacity. For example, a fine-textured soil overlying a coarse-textured soil will have a higher water content than a uniformly fine-textured soil. Thus, the field capacity of a soil is related to the conditions under which it is measured as well as to the characteristics of the soil itself. For example, the water content of soil allowed to drain in the field might be quite different from the water content of a cylinder of the same soil drained in the greenhouse on a layer of sand. Such factors should be considered in the interpretation of field capacity data.

Field capacity is determined most simply by pouring water on the soil surface and permitting it to drain for 1–3 days (depending on soil type) with surface evaporation prevented. Soil samples are then collected by auger for gravimetric measurement, or soil water content is measured by one of the techniques discussed in Chapter 4 and the results are expressed on a gravimetric or, preferably for water storage purposes, a volumetric basis. Field capacity is a relatively reproducible value when determined in this manner, so long as precautions are taken to avoid sampling in the transition zone or at the wetting front. The wide variation in water content at various depths in the soil is shown in Fig. 3.11.

Because field capacity is affected by soil profile and soil structure, laboratory determinations are not always reliable indicators of the value in the field. Nevertheless, it is often desirable to make laboratory determinations of field capacity. Most laboratory measurements are made by simulating the tension that develops during drainage in the field by use of pressure membranes or tension tables (see Chapter 4). There is some difference of opinion about the proper tension or pressure to apply, but laboratory measurements are usually based on the water content of soil subjected to 0.033 MPa or 0.3 bar. Hillel (1980b, pp. 381–384) discussed the problem of determining field capacity. The amount of water retained at field capacity decreases as the soil temperature increases (Richards and Weaver, 1944). This results in increased runoff from watersheds as the soil warms. As field capacity has no fixed relationship to soil water potential, it cannot be regarded as a soil moisture constant, and some soil scientists have suggested that it be abandoned. However, it serves a useful purpose if users are aware of its limitations.

Permanent Wilting Percentage. This is the soil water content at which plants remain wilted unless water is added to the soil. Briggs and Shantz (1911, 1912) first emphasized the importance of this soil water constant, terming it the wilting coefficient. They made measurements on a variety of plants and found that all wilted at about the same water content in a given soil. Richards and Wadleigh (1952) found that the soil water potential at wilting ranged from −1.0 to −2.0 MPa, with the average at about 1.5 MPa; the value of 1.5 MPa (15 bars) is generally used as an approximation of soil water at permanent wilting. Slatyer (1957) strongly criticized the concept of the permanent wilting percentage as a

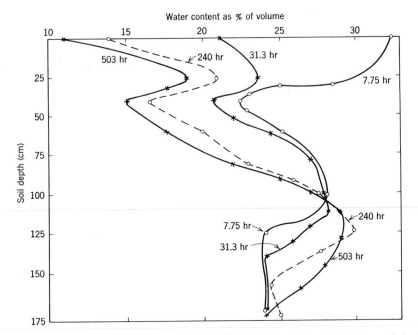

Fig. 3.11. Profiles of water content of a nonuniform soil after drainage for various periods of time following irrigation. The surface horizon is a sandy loam, changing to a fine sandy loam at about 25 cm and to a clay which holds much more water at 75 to 100 cm. The original water content at a depth of 125 to 175 cm was about 24%. The profiles show progressive decrease in water content near the surface and increase in the 100- to 175-cm horizon with the passage of time. Slow drainage and change in water content occurred for 20 days after irrigation. (From Rose *et al.*, 1965.)

soil constant because wilting really depends on the osmotic properties of plants. It occurs when water loss causes leaf cells to lose their turgor. This depends on the osmotic potential of the leaf tissue, on meteorological factors affecting the rate of transpiration, and on soil factors affecting the rate of absorption. However, because of the shape of the water potential–water content curve of most soils (Fig. 3.8), large changes in soil water potential accompany small changes in water content near the wilting point. Also, most crop plants have osmotic potentials in the range of −1.5 to −2.0 MPa, so for practical purposes the water content at −1.5 MPa is near the point at which soil moisture usually becomes severely limiting. Thus, although the permanent wilting percentage is useful for practical purposes, there is no sharply defined lower limit of availability for various physiological processes (Gardner and Nieman, 1964).

   Readily Available Water.   This expression, referring to the availability of soil water for plant growth, is the amount of water retained in a soil between field capacity and the permanent wilting percentage. Since field capacity represents the upper limit of soil water availability and the permanent wilting percentage the lower limit, this range has considerable significance in determining the agricultural value of soils. The available water capacity of different soils varies widely, as shown in Figs. 3.8 and 3.9 and Table 3.3.

In general, fine-textured soils have a wider range of water between field capacity and permanent wilting than coarse-textured soils. Also, the slope of the curve for water potential over water content in fine-textured soils indicates a more gradual release of the water with decreasing soil water potential. In contrast, sandy soils, with their larger proportion of noncapillary pore space, release most of their water within a narrow range of potential because of the predominance of large pores, and release of additional water requires very low water potentials.

Data on readily available water must be used cautiously because the availability of water depends on several variables. For example, in any given soil, increased rooting depth in the profile as a whole can compensate for a narrow range of available water in one or more horizons. Conversely, restricted root distribution combined with a narrow range of available water results in considerable hazard for plant growth from an inadequate water supply, especially in climates where summer droughts are frequent. Also, it should be remembered that in many soils the range of water available for survival is substantially greater than that available for good growth. Furthermore, within the range of available water, the degree of availability usually tends to decline as soil water content and Ψ_{soil} decline (Richards and Wadleigh, 1952; Hagan *et al.*, 1959). It should be clear that there is no sharp limit between available and unavailable water and that the permanent wilting percentage is only a convenient point on a curve of decreasing water potential and decreasing availability. A more detailed discussion of soil water availability is given in Chapter 9 and in a paper by Cassell (1983). For the present, it need only be noted that the range of soil water between field capacity and the permanent wilting percentage constitutes an important field characteristic of soil when interpreted properly.

MOVEMENT OF WATER IN SOIL

Several aspects of water movement in soils are important. These include the infiltration of water and saturated and unsaturated flow through the soil. Unsaturated flow is particularly important in terms of water supply to roots. Under some conditions, movement in the form of vapor is of importance. We will discuss each of these.

Infiltration

The rate of infiltration of water into soil is important with respect to recharge of soil moisture by rain and irrigation. Also, if infiltration is slow, surface runoff causes erosion. Infiltration and rewetting of a dry soil is a somewhat complicated process, as shown in Fig. 3.12, after Bodman and Colman (1944). They found that the wetted portion of a column of uniform soil into which water was entering at the top and moving downward appeared to comprise a stable gradient through which water was transmitted, ranging from a saturated zone at the top to a wetting zone at the lower end. Five zones in series were described as (1) a saturation zone, i.e., a zone presumed saturated which reached a maximum depth in their soil of 1.5 cm; (2) a transition zone, a region of rapid decrease of water content extending to a depth of about 5 cm from the surface; (3) the main transmission zone, a region in which only small changes in water content occurred; (4) a wetting zone, a region of fairly rapid change in water content; and (5) the wetting front, a region of very steep gradient in water content which represents the visible limit of water penetration.

It can be seen from Fig. 3.12 that the transmission zone is a continually lengthening, unsaturated zone of fairly uniform water content and potential. According to Marshall (1959), in this zone the matric potential, Ψ_m, is close to zero and the pore space is about 80% saturated. The gradient of Ψ_m in the transmission zone is usually very small after the water has penetrated to a reasonable depth, so that movement within that zone is caused primarily by gravity. The rate of advance of the wetting front depends on the rate at which

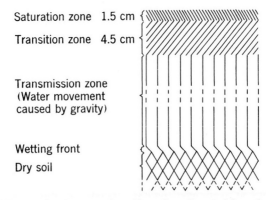

Fig. 3.12. Diagram showing zones in uniform dry soil wetted from the top. There is little difference in water content from top to bottom of the progressively lengthening transmission zone, through which water passes by gravity, but a steep decrease in water content and matric potential occurs at the wetting front. (After Bodman and Colman, 1944.)

water is supplied to it through the transmission zone and is little affected by the size of the potential gradient between it and the dry soil beyond it. This explains the sharp boundary between wet and dry soil observed when only the surface soil is wetted by a shower following a period of dry weather. It also explains why soil cannot be rewetted partway to field capacity by adding water to the surface, but is either wetted to field capacity or not at all. Following application of surface water, the infiltration rate usually decreases with time to a fairly stable minimum value, as shown in Fig. 3.10. Infiltration is discussed in more detail in various soils texts and in an article by Miller and Klute, in Hagan *et al.* (1967).

Factors Affecting Infiltration.   Rapid infiltration of water into the soil is important in agriculture because if it is impeded, recharge of soil water is hindered and rapid runoff occurs, often accompanied by erosion and flooding. The major factors affecting infiltration of water into soil are the initial water content; surface permeability; internal characteristics of the soil affecting hydraulic conductivity, such as pore space; degree of swelling of soil colloids and organic matter content; duration of rainfall; and temperature of the soil and water. Infiltration is more rapid at higher temperatures.

The rate of infiltration into clay soils decreases as initial soil water content increases because hydration and swelling of clay particles reduce the cross-sectional area available for the entrance of water. Infiltration is greatly decreased by zones of low soil permeability such as surface crusts, claypan and hardpan layers, surface compaction caused by farm implements and human and animal traffic, and clay structure dispersal caused by an excess of alkali. Puddling of the surface of bare soil often reduces infiltration, but puddling can be reduced by use of surface mulches and the incorporation of organic matter. Differences in rate of infiltration into a forest soil and into the soil of an adjacent cultivated, eroded old field are shown in Fig. 3.13. This difference is to be expected in view of the differences in noncapillary pore space in the two soils, shown in Fig. 3.4.

Occasionally, rainfall and irrigation water fail to penetrate properly because of puddling of clays that swell when wetted and close the pores, or because the surface does not wet. Jamison (1946) reported very uneven penetration of water under citrus trees on sandy soils in central Florida because the sand particles were resistant to wetting, and Adams *et al.* (1970) found that annual vegetation was excluded from beneath shrubs and trees in a desert scrub because the soil was water-repellent. They cite several other papers dealing with soil wettability. Attempts have been made to improve infiltration of water on burned slopes and on watersheds with hydrophobic soils by applying wetting agents (Valoras *et al.*, 1974; Mustafa and Letey, 1970). Excessive incorporation of peat into greenhouse bench and potting soil also decreases infiltration of water (Lunt *et al.*, 1963). Various aspects of the infiltration problem are discussed in the book edited by Emerson *et al.* (1978).

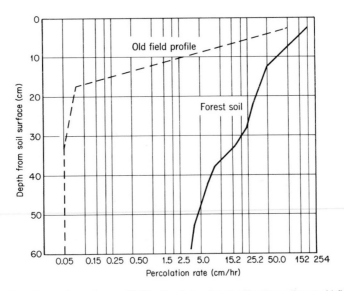

Fig. 3.13.   Comparison of rates of infiltration into a forest soil and an adjacent old field on the same soil type. Fig. 3.4 shows the differences in capillary pore spaces in these two soils. (From Hoover, 1949.)

Movement of Water within Soils

Movement of water within soils controls the rates of infiltration, of supply to roots, and of underground flow to springs and streams. Soil water is not pure, but consists of a complex solution of minerals; hence, a distinction must be made between the bulk or mass movement of the soil solution and the movement of pure water as vapor. The soil solution percolates downward through the non-capillary pore space by bulk or mass flow under the influence of gravity, and moves through the capillary pore spaces and in films surrounding soil particles under the influence of surface tension or matric forces. Pure water diffuses as vapor through air-filled pore spaces along gradients of decreasing vapor pressure.

It is often stated that water *always* moves toward regions of lower total water potential, but this is not always true. Corey and Kemper (1961) pointed out that no single component of the soil solution, such as the water potential, can serve as an indicator of the net transfer of water in soil, and this has been reemphasized by Arnold Klute in a private communication to the writer. Total water potential, or thermodynamic potential, includes all forces affecting the energy status of water [see Eq. (3.1)], but these forces are not equally important with respect to move-ment of water within the soil. For example, although the osmotic potential is an important component of the total soil water potential in saline soils, it has little effect on water movement through the soil, that being controlled by matric and gravitational forces. However, the osmotic potential does affect water movement from soil to roots of plants because the soil solution is separated from the root cell solution by differentially permeable membranes. Thus, it appears that water movement through soil is not always along gradients in total water potential, but movement from soil to roots is.

An example of movement of water vapor through soil in one direction along a vapor pressure gradient and movement as liquid in the opposite direction along a gradient of matric potential is shown later in Fig. 3.15. This illustrates the fact that the two phases of water, liquid and vapor, can move independently of one another, each along its own gradient in potential.

It has been customary to differentiate between saturated flow in saturated soils and unsaturated flow, or capillary conductivity, in unsaturated soils. However, the term hydraulic conductivity or conductance is now being used for both saturated and unsaturated flow. The chief difference is that in saturated soils gravity is the principal force causing water movement through the larger pores, whereas in drained soils water movement occurs through small pores and in films surrounding the soil particles, and it is controlled largely by the matric potential. However, there is considerable interaction of the two kinds of forces, as in unsaturated flow down a slope, in which both gravity and matric forces are involved (Hewlett, 1961; Beasley, 1976; Patric and Lyford, 1980).

Saturated Flow.　The downward movement of water in saturated soils was discussed in the section on infiltration. It is caused by gravitational forces, but there may be some horizontal movement caused by matric forces when water is applied to a point on the surface of a dry soil, as in an irrigation ditch. An example is shown in Fig. 4.16 (Chapter 4). The theory of movement of liquid water is based on Darcy's law, which states that the quantity of water passing a unit cross section of soil is proportional to the difference in hydraulic head. Usually, a term for the hydraulic conductivity of the soil is also introduced. Downward flow of water in saturated soils seems to be a simple process, but Hillel (1980a) devotes a chapter to it, and discussions can be found in other soil texts cited at the end of this chapter.

Unsaturated Flow.　As drainage proceeds and the larger pores are emptied of water, the contribution of the hydraulic head or gravitational component to total potential becomes progressively less important and the contribution of the matric potential, Ψ_m, becomes more important. The effect of pressure is generally negligible because of the continuous air spaces, and the solute potential, Ψ_s, does not affect the potential gradient unless there is an unusual concentration of salt at some point in the soil. Such concentrations sometimes develop at the surface of saline soils. Darcy's law is applicable to unsaturated flow if the hydraulic conductivity is regarded as a function of water content. As the soil water content and soil water potential decrease, the hydraulic conductivity (K) decreases very rapidly, as shown in Fig. 3.14, so that when $\Psi_{soil} = -1.5$ MPa, K is only about 10^{-3} of the value at saturation. According to Slatyer (1967, pp. 105–109), the rapid decrease in conductivity occurs because the larger pores are emptied first, greatly decreasing the cross section available for liquid flow. When the continuity of the films is broken, liquid flow no longer occurs. The low conductivity in drying soil limits the water supply to roots.

Differences in Soil Conductivity. The values for hydraulic conductivity vary widely, ranging from 0.0025 cm/hr in the least permeable to 25 cm/hr in the most permeable soil (Smith and Browning, 1946). Soils with a hydraulic conductivity of less than 0.25 cm/hr are poorly drained, whereas those with conductivities greater than 25 cm/hr do not hold enough water for good plant growth. Permeability and conductivity decrease with decreasing pore space and are sensitive to changes in cation content, which affect the degree of swelling of clay

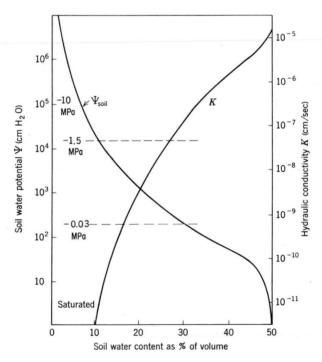

Fig. 3.14. Diagram showing approximate decrease in hydraulic conductivity (K) and soil water potential (Ψ) with decreasing soil water content. (After Philip, 1957.)

colloids. Entrapment of air greatly reduces permeability by blocking soil pores, whereas decrease in temperature decreases it by increasing the viscosity of water.

Water Vapor Movement

As soil water content decreases, the continuity of the liquid films is broken; finally, water moves chiefly in the form of vapor. Although vapor movement in soil is very slow, it occurs more rapidly than would be expected from calculations of diffusion rates. Philip and de Vries (1957) attributed this discrepancy to the fact that part of the path through the soil is liquid, and water condenses on one surface and evaporates from the other surface of isolated wedges of water. They also found that the temperature gradients across air-filled pores might be twice as steep as the average temperature gradient in the soil mass. These two factors result in a rate of vapor movement much more rapid than expected from theory.

The reason for negligible vapor transfer under isothermal conditions, at moderate to high water contents, is primarily the fact that even quite steep gradients of Ψ_w are associated with relatively small gradients of vapor pressure because of the nature of the relationship of Ψ_w to $e/e°$. For example, at Ψ_{soil} of -1.5 MPa and temperature of 20°C, the vapor pressure is reduced by only 25.8 Pa. However, steep vapor pressure gradients develop if large differences exist in pressure, solute concentration, or temperature. Although pressure is seldom of importance and salts only occasionally cause large gradients, steep temperature gradients are frequently established, particularly near the surface.

The experiments of Gurr *et al.* (1952) demonstrate the relative importance of liquid and vapor movement in soils of varying water content. These workers used 10 horizontal columns of soils, each of initially uniform water and salt content, the water content in the columns ranging from practically dry to practically saturated levels. A uniform temperature difference was then applied across all columns, and the net transfer of water and salt after a period of 5 days was used as an indication of the net water movement and total water movement, respectively. The results showed that at low and intermediate water contents there was a net transfer of water toward the cold side, but the salt moved in the reverse direction. Moreover, the greatest net transfer of water was in the low to medium water content treatments, and the greatest transfer of salt occurred in the medium water contents. There was little change of salt or water distribution in the very dry and very wet treatments (see Fig. 3.15).

These apparently contradictory results can be explained as follows. At soil water contents so low that there was no continuity in the liquid phase and transfer of water occurred solely in the form of vapor, there was no salt movement. Once

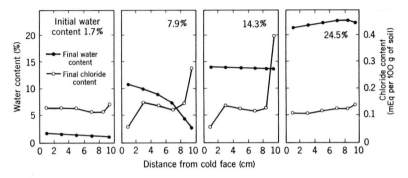

Fig. 3.15. Distribution of water and chloride ions in columns of loam soil of various water contents after being subjected to a temperature gradient ranging from 10° (on left) to 25°C (on right) for 5 days. There was marked transfer of water to the cold side and salt to the warm side at a moisture content of 7.9%. Salt moved to the warm side at a water content of 14.3%, but there was no net movement of water. In moist soil there was negligible net movement of either salt or water. (From Gurr *et al.*, 1952.)

liquid continuity started to develop, however, the net transfer was affected by both vapor and liquid movement, and the temperature-induced movement of vapor toward the cold face tended to be partly compensated for by a return flow of liquid water along water potential gradients, carrying salt to the warm side. The net transfer of water toward the cold face could therefore be expected to decrease rapidly as water content increased, even though for a time there would still be a substantial flow of vapor in one direction and liquid in the other, leading to salt accumulation at the warm face but no net change in water content. Once complete liquid continuity developed and the soil approached saturation, vapor flow would become negligible, and any movement of liquid water containing salt associated with temperature differences would be compensated for by a return flow associated with water potential differences. Hence, neither salt nor water content appeared to change in moist soil.

In nature there are large seasonal fluctuations in soil temperature, the surface warming up in the summer and cooling down in the winter, relative to the deeper horizons. As a result, there can be significant upward movement of water in the winter and downward movement in the summer. Edlefsen and Bodman (1941) observed such movements in experiments at Davis, California, and Lebedeff (1928) reported this upward movement to be an important source of groundwater in southern Russia, where cooling of the surface at night results in condensation in the surface layer. Such movement is usually assumed to occur as vapor, but Smith (1943) and others have pointed out that water movement along thermal gradients can occur both as vapor and as liquid. Readers are referred to Slatyer (1967, pp. 109–118) for a detailed discussion of this problem. Movement of water as vapor is also discussed in Nielsen *et al.* (1972, pp. 112–117) and in Hillel (1980b, Chapter 5).

Upward and Horizontal Movement and Evaporation

Thus far, we have been concerned chiefly with the downward movement of water, but horizontal movement toward roots and upward movement toward the soil surface are also important. The movement of water toward absorbing roots is discussed in Chapter 9. Upward movement occurs because evaporation and absorption by plants decrease the water content of the surface soil, thereby decreasing its water potential and causing upward movement against gravity. Assuming a constant rate of evaporation, the rate of upward movement to an exposed soil surface depends on the depth of the water table and the soil texture. Gardner (1958) estimated that the upward flow to an evaporating surface from a water table at 60 cm would be only half as rapid in a coarse-textured soil as in a fine-textured soil. Under conditions favorable for rapid evaporation, water loss can exceed the rate at which water can move to the surface and the surface soil becomes dry. This is most likely to occur in sandy soils because they contain less water per unit volume of soil and have a smaller pathway for water flow per unit of cross section than clays. Evaporation is drastically reduced after the soil surface becomes dry because movement through the surface layer as vapor is much slower than movement as liquid.

Gardner and Fireman (1958) reported that lowering the water table of Chino clay or Pachappa fine sandy loam from 90 to 180 cm reduced evaporation from the surface to about 12% of the value at 90 cm because of slower upward movement to the evaporating surface (Fig. 3.16). However, further lowering to 3 or 4 m had little additional effect on evaporation, and significant upward movement of water occurred from water tables as deep as 8 or 9 m. Patric *et al.* (1965) reported recharge of soil water in covered plots during the winter, when the surrounding soil was moist, by vertical and horizontal movement over distances of several meters. Gardner and Ehlig (1962) also observed measurable upward movement of water in soil below the field capacity. Such movement of water must be taken into account in considering the amount of water available to plants rooted in the surface soil. For example, if significant upward movement of water into the root zone occurs (see Fig. 3.16), water use based on successive measurements of soil water content will be underestimated. Conversely, if there is significant downward percolation out of the root zone (see Fig. 3.11), the amount of water used will be overestimated. Evaporation often results in excessive concentration of salt at the surface of saline soils.

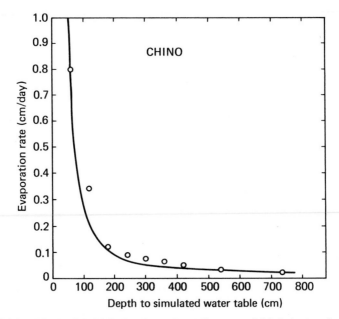

Fig. 3.16. Calculated (solid line) and experimentally measured (circles) rates of evaporation from a column of Chino clay with the simulated water table at various depths below the soil surface. (From Gardner and Fireman, 1958.)

Considerable attention has been given to the possibility of reducing evaporative losses by shallow tillage or mulching. Shallow tillage ideally results in a surface layer of small clods of soil through which water must diffuse as vapor. Layers of crop residues, pine needles, chipped bark, and even pebbles are often used, especially around ornamental plants. They not only reduce water loss but usually improve filtration. A soil covering of plastic film is sometimes used on valuable crops such as pineapples. However, mulches can have undesirable secondary effects on plant growth by modifying the soil temperature. They usually cool the soil, but black plastic results in warmer soil. Hillel (1980b) devotes a chapter to evaporation from soil and the effects of mulching and cultivation. This subject is also discussed in some of the other books cited at the end of this chapter.

SUMMARY

Soil consists of four fractions: the mineral particles and nonliving organic matter forming the matrix, and the soil solution and air occupying the pore spaces within the matrix. It provides the anchorage which enables roots to maintain plants in an erect position, and it acts as a reservoir for water and mineral nutrients. Much of the success of plants in any given habitat depends on the suitability of the soil as a medium for root growth and functioning.

Soil water is held largely by matric forces which bind water on the soil particles and hold it in the smaller pore spaces. The availability of soil water for plants depends on its potential and on the hydraulic conductivity of the soil. The water readily available for plants occurs in the range between field capacity and permanent wilting percentage. Field capacity is the water content after drainage of gravitational water has become very slow and represents a water potential of -0.3 bar or less. Permanent wilting percentage is the water content at which plants become permanently wilted and corresponds to a water potential of -10 to -20 bars. The exact value varies with the kind of plant and the conditions under which wilting occurs. Because field capacity and permanent wilting percentage are not consistently related to the soil water potential, they cannot be regarded as physical constants.

The rate of infiltration of water into soil is important in connection with recharging the soil with water by rain or irrigation. Movement of water through the soil is controlled chiefly by the gravitational potential in soils above field capacity and by the matric potential in soils drier than field capacity. Hydraulic conductivity decreases very rapidly with decreasing water potential, so movement of liquid water becomes very slow in dry soil and practically ceases. In dry soils some water moves as vapor.

There appears to be more movement of water in soil below field capacity than was formerly supposed. Significant upward movement of water occurs from depths of several meters, and if the water table is within 1 m of the surface, the upward movement should be sufficient to supply a crop. Upward movement of water also results in considerable losses by evaporation. These can be partially controlled by proper tillage or by mulching.

SUPPLEMENTARY READING

Baver, L. D., Gardner, W. H., and Gardner, W. R. (1972). "Soil Physics," 4th ed. Wiley, New York.

Black, C. A. (1981). "Soil-Plant Relationships." Wiley, New York.

Donahue, R. L., Miller, R. W., and Shickluna, J. C. (1977). "Soils: An Introduction to Soils and Plant Growth," 4th ed. Prentice-Hall, Englewood Cliffs, New Jersey.

Emerson, W. W., Bond, R. D., and Dexter, A. R., eds. (1978). "Modification of Soil Structure." Wiley, New York.

Hillel, D. (1980a). "Fundamentals of Soil Physics." Academic Press, New York.

Hillel, D. (1980b). "Applications of Soil Physics." Academic Press, New York.

Nielsen, D. R., Jackson, R. D., Cory, J. W., and Evans, D. D., eds. (1972). "Soil Water." Am. Soc. Agron. and Soil Sci. Soc. Am., Madison, Wisconsin.

4

Measurement and Control of Soil Water

INTRODUCTION

This chapter deals with the measurement and control of soil water, including irrigation and drainage, which involve control of soil water on a large scale. Recognition of the importance of the available water content of the soil in relation to plant growth and irrigation scheduling resulted in development of a number of methods of measuring or estimating it. Control of soil water content or potential on a small scale is necessary for research dealing with the effect of soil water supply on plant growth and on a large scale for timing irrigations. Control of excess soil water requires drainage to prevent saturation of soil and damage to root systems from inadequate aeration.

FIELD MEASUREMENTS OF SOIL WATER

Soil Water Balance

The hydrologic cycle is shown in Fig. 4.1. Measurements of change in soil water storage are commonly used to estimate evapotranspiration. A water balance equation can be written as follows:

$$\Delta W = P - (O + U + E) \tag{4.1}$$

where ΔW is the change in soil water storage (initial content minus final content) during the period of measurement and P, O, and U are the precipitation, runoff, and deep drainage, respectively. U is defined as the amount of water passing beyond the root zone, or, for experimental purposes, as the amount passing below the lowest point of measurement. E is evaporation (including transpiration) from the plant and soil surfaces. This expression can be used on any scale, ranging from continental land masses and hydrologic catchments down to indi-

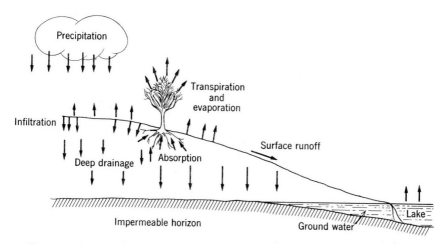

Fig. 4.1. The hydrologic cycle, showing disposition of precipitation by surface runoff, infiltration, and deep drainage, and its removal from the soil by evaporation and transpiration.

vidual plants. During periods of dry weather between rains or irrigations, and neglecting or otherwise measuring U, $E = \Delta W$ and can consequently be measured by determining changes in soil water storage under the plant community being studied.

Measurements of ΔW are most accurately conducted by the use of weighing or floating lysimeters (Fig. 4.2), provided they are properly designed and located (Ritchie and Burnett, 1968). However, they cannot be used if the mixed species composition, spatial distribution of the vegetation, depth and ramification of the root system, plant size, or other factors make it impossible to simulate the natural root environment in the lysimeter. Examples of such limitations are found in arid plant communities and in natural mixed forests. The ultimate in size of lysimeters was one built to accommodate a Douglas-fir tree 28 m high (Fritschen et al., 1973), but it caused considerable damage to the root system. In many situations,

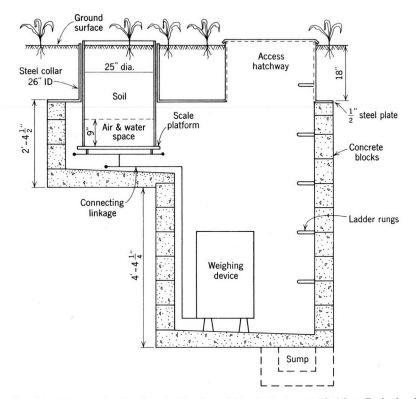

Fig. 4.2. Diagram showing the principle of a weighing lysimeter, modified from England and Lesesne (1962). It consists of a large container filled with soil, mounted on a weighing device. Electronic weighing mechanisms are often used. The lysimeter must be surrounded by a border of similar vegetation if the results are to be applicable to crops or stands of plants. Some lysimetry problems are discussed by Hagan et al. (1967, pp. 536–544).

determinations of soil water storage at different points in the community provide the only technique for evaluating ΔW. Even so, interpretation of observations is difficult, and the presence of a water table near the surface can limit application of the method unless fluctuations of water table depth themselves can be used (Holmes, 1960). Examples of water storage determination by soil water measurement are cited by Slatyer (1967). The quality of such measurements can be improved considerably if adjustments are made for the frequently significant loss of water to, or gains of water from, the soil zones below the depth of measurement (Patric *et al.*, 1965; Rose and Stern, 1965).

Sampling Problems

It is difficult to obtain reliable estimates of changes in soil water content in the entire root zone because of the great vertical and horizontal variability in soil water distribution in the field. Water content often varies drastically over short distances. This is partly because of irregularities in root distribution which cause some areas to be depleted of water sooner than others, and partly because of variations in physical characteristics of the soil, particularly its clay content, which influences its capacity to retain water. Furthermore, the available water content of samples obtained only short distances apart or at different depths may vary widely. Variation in water-holding capacity is shown in Figs. 3.7 and 3.11.

Because of this variability, considerable replication of sampling is required to give valid estimates of the bulk soil water content. For example, Aitchison *et al.* (1951) found that in a typical loam soil more than 10 samples were required to show that differences of 1.0% in soil water content (1 g water/100 g dry soil) were significant at $p = 0.05$. More than 40 samples were required to show that differences of 0.5% were significant at this level. Staple and Lehane (1962) found a similar order of variability in clay loam soils which were regarded as reasonably uniform. Hewlett and Douglass (1961) also present a discussion of sampling problems.

Intensive direct sampling often disturbs the vegetation, requires care in refilling the holes from which samples are removed, is very difficult in stony soil, and requires large amounts of labor. As a consequence, field workers have an increasing tendency to use indirect methods which permit installation of a number of sensing elements in a study area so that repeated measurements can be made at the same point. Although the equipment for such installations usually costs more initially than that for direct sampling, the saving in labor costs and the advantages of repeated measurements at the same point compensate for the greater initial expenditure.

The choice of a method to study soil water content depends partly on the objectives of the investigator. The usual objective in measuring soil water content is to determine how much water is present and available for plant growth at a particular time and place. The availability of soil water depends primarily on its potential (Ψ_{soil}), and the most useful description of the moisture characteristics of a soil is a graph showing water potential plotted against water content (Fig. 3.8).

Direct Measurements of Soil Water Content

The basic measurements of soil moisture are made on soil samples of known weight or volume; the water content is expressed as grams of water per gram of oven-dry soil or milliliters of water per cubic centimeter of oven-dry soil. Samples for gravimetric determinations are usually collected with a soil auger or sampling tube. Samples for volume determinations are usually collected in special containers of known volume, with a tube sampler, or from the side of an exposed soil profile (see, for example, Coile, 1936, and Lutz, 1944). The bulk density of the soil can be determined separately (Vomocil, 1954). The water is usually removed from the soil by oven-drying at 105°C to constant weight. The time required can be shortened somewhat by using an oven with forced ventilation. Several methods requiring less time than oven-drying were discussed by Rawlins (1976).

Expression of water content as a percentage of dry weight is not useful to the plant scientist unless the water potential curve or the field capacity and permanent wilting percentage are known, because a percentage water content representing saturation in a sandy soil might be below the permanent wilting percentage of a clay soil (see Table 3.3). For some purposes, it is more useful to convert water content per unit of weight into water content per unit of volume, because the soil occupied by a root system is measured by volume rather than by weight. Also, additions to and losses of water from soil are often measured in inches or millimeters, which on an area basis become the volume. The conversion from weight to volume units can be made by multiplying the percentage by weight times the bulk density of the soil under study. The water content can then be expressed in convenient units such as inches per foot or centimeters per meter of soil depth.

Indirect Measurements of Soil Water Content

Most of the methods of measuring soil water content are indirect. This means that the property measured must be related to water content by some kind of calibration procedure. Methods of measuring soil water were discussed by Holmes *et al.* (1967) and Rawlins (1976).

Neutron Scattering.   The most commonly used indirect method for measuring soil water content is probably the neutron probe. This method is based on the fact that hydrogen atoms have a much greater ability to slow down and scatter fast neutrons than most other atoms, so that counting slow neutrons in the vicinity of a source of fast neutrons provides a means of estimating hydrogen content. Since the only significant source of hydrogen in mineral soils is the water, the technique offers a convenient means of estimating soil water content. In soils with high root density or high levels of organic residues, the amount of hydrogen in organic matter may affect the estimates. However, the amount is generally small enough, compared with the hydrogen in the soil water, to be neglected. Portable commercial instruments are available and are reasonably convenient to use in the field. An instrument consists of a probe containing a source of fast neutrons and a counter tube for detecting slow neutrons connected through an amplifier to a portable scaler (Fig. 4.3). In use, the probe is lowered into a plastic- or aluminum-lined access tube, and counting rates are determined at the desired depths. The count rates, adjusted for background and standardized, are then calibrated against direct volumetric determinations of soil water content. Theoretical calibrations have also been developed (Holmes and Jenkinson, 1959), and Hewlett *et al.* (1964) discuss sampling problems. The development and operation of neutron probes are described by Rawlins (1976).

The method has several important advantages over most other techniques, including the absence of a lag period while the soil water equilibrates with a sensing instrument and the fact that a large volume of soil (about 20 cm in radius) is monitored, thus smoothing out local variability. However, it does involve the disturbing influence of an access tube, and the large volume of the sample prevents sampling at a point on or near the surface unless special precautions are taken. Also, the results are influenced by other sources of hydrogen atoms, such as inorganic matter, and by other elements, notably chlorine, iron, and boron (Holmes, 1960). Late models of neutron meters can be self-calibrated and corrected for neutron-absorbing elements and organic hydrogen.

Gamma Ray Absorption or Attenuation.   Ashton (1956) described measurement of changes in soil water content by change in amount of radiation absorbed. The amount of radiation passing through soil depends on soil density, which varies chiefly with change in water content. If the soil does not shrink and swell appreciably, change in water content can be measured from change in amount of radiation passing through the soil. Ferguson and Gardner (1962), Gurr (1962), Gardner *et al.* (1970), and others have found this method useful for measuring movement of water in soil columns. However, it is reliable only in

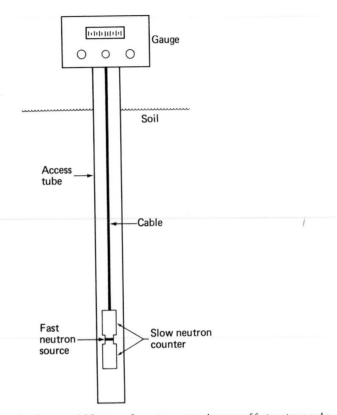

Fig. 4.3. Diagram showing essential features of a neutron meter. A source of fast neutrons and a counter for slow neutrons are lowered to any desired depth in the access tube installed in the soil. The slow neutrons reflected by the hydrogen in soil water are counted and the results indicated on the attached gauge. The water content of a spherical mass of soil surrounding the counter is measured, the size of the mass increasing with decreasing soil water content. A special model is available to measure water in the surface soil.

soils where the change in bulk density is very small compared with the change in water content.

The method requires a source of gamma radiation such as cesium, a detector such as a geiger tube or scintillation probe, and a scaler. Ashton placed two plastic tubes on opposite sides of large pots and dropped the radiation source down one tube, the detector down the other. In work with small pots or columns of soil, the source and detector can be positioned in a jig on opposite sides of the container so that the radiation can be carefully collimated. Rawlins (1976) discussed both gamma ray absorption and neutron probe measurements in detail.

Other Methods. Attempts have been made to measure changes in soil water content by measuring changes in capacitance, a method widely used to measure the water content of grain, flour, dehydrated foods, and various industrial products. It depends on the fact that water has a much higher dielectric constant (insulating value) than air or dry materials (water, 80; dry soil, 5); hence, changes in water content are reflected in changes in capacitance (Wallihan, 1946). Although the method is theoretically attractive, it has not proved practical in the field, chiefly because of difficulties in obtaining uniform contact with the soil.

Heat conduction decreases with decreasing water content, and attempts have been made to measure changes in water content from changes in thermal conductivity. A heating element is buried in the soil and warmed by passing an electric current through it, and the rate of heat dissipation, which varies with soil water content, is measured. De Vries and Peck (1958) and Bloodworth and Page (1957) produced improved models of thermal conductivity equipment, but it has not been widely used because it does not work well in dry soils.

Measurements of Matric Potential

Several techniques used for estimating soil water content really measure matric potential, or pressure potential, directly or indirectly. Two important methods are electrical resistance units and tensiometers; another is change in electrical resistance.

Electrical Resistance Blocks. Near the end of the nineteenth century, attempts were made to measure changes in soil moisture by means of change in electrical resistance. The early attempts were unsuccessful because variations in electrode contact with the soil, variations in salt content of the soil, and temperature-induced changes in resistance obscured the changes produced by variations in soil moisture. The most serious of these difficulties was eliminated by Bouyoucos and Mick (1940), who embedded the electrodes in small blocks of plaster of paris. The blocks are buried in the soil and connected by well-insulated leads to a resistance bridge. The water content of the blocks changes with that of the soil, producing measurable changes in the electrical conductivity of the solution between the electrodes. The blocks can be left in the soil for months or possibly for years, although ordinary plaster of paris or gypsum blocks tend to disintegrate rather rapidly in wet, acid soils and often last for only one season. The life of commercially available blocks has been prolonged by impregnating them with resin (Bouyoucos, 1953a,b). Gypsum blocks are sensitive over a range of about -0.05 to -1.5 MPa of matric potential; hence, they are more satisfactory in dry than in very moist soil. A diagram of a gypsum block is shown in Fig. 4.4a, and a graph relating resistance to available water content is seen in Fig. 4.4b.

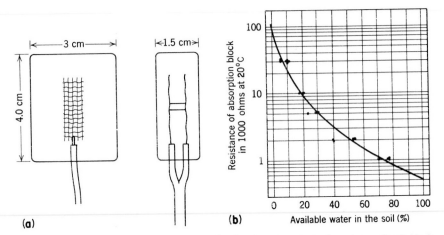

Fig. 4.4. (a) Surface and edge views showing location of electrodes in a plaster of paris block designed to measured changes in soil water content by changes in resistance. The electrodes are pieces of stainless steel screen separated by a plastic spacer and enclosed in plaster of paris. (b) Resistance in ohms of a plaster of paris resistance block plotted over available soil water content of a silt loam soil. (After Bouyoucos, 1954.)

Units consisting of electrodes wrapped in nylon (Bouyoucos, 1949, 1954) or Fiberglas (Bouyoucos and Mick, 1948; Colman and Hendrix, 1949) are considered to be more durable, to respond more rapidly, and to be more sensitive to soil water potentials higher than −0.1 MPa than gypsum blocks. However, they are much more sensitive to changes in salt content of the soil than gypsum blocks and are less sensitive in drier soils. The gypsum blocks are less sensitive to salt because they are buffered by the dissolved calcium sulfate, and gypsum block readings are said to be unaffected by up to about 2.2 metric tons of fertilizer per hectare (Bouyoucos, 1951). The importance of the salt error can be expected to vary in different types of soil (Richards and Campbell, 1950). The nylon and Fiberglas units tend to give different readings during successive wetting and drying cycles (H. A. Weaver and Jamison, 1951). England (1965) reported that field calibration was necessary every year during the first 5 or 6 years because entrance of colloidal clay and other materials caused the relation of resistance to moisture content to change. He estimated that the life of Fiberglas units would probably be less than 15 years. Resistance measurements change somewhat with temperature, and thermistors are sometimes installed with the blocks to measure the temperature. All porous-block methods show hysteresis effects, and they are more sensitive on drying cycles than on wetting cycles because drying is slower and gives time for a better equilibration between soil and blocks. Fortunately, the drying cycle is also more important in relation to plant growth.

Resistance blocks are sometimes calibrated by placing them in a pressure membrane apparatus and measuring the resistance under various pressures (Haise and Kelley, 1946). Such calibration permits estimation of the matric potential of the soil from resistance readings of the blocks. They can also be calibrated against soil water content in the field by taking samples for gravimetric determinations from the vicinity of the blocks. Occasionally, they are calibrated in the laboratory in samples of the soil in which they are to be used (Kelley, 1944). Details of installation and calibration can be found in papers by Aitchison *et al.* (1951), Kelley *et al.* (1946), Knapp *et al.* (1952), and Slatyer and McIlroy (1961). In spite of certain shortcomings, resistance blocks have been used very extensively, especially in dry soil, where tensiometers do not function. They are particularly useful in monitoring gross changes in soil water content between irrigations. Also, the progress of a wetting front through the soil can be followed by the sudden reduction in block resistance through a series of blocks buried at various depths in the soil.

Tensiometers. Direct field measurements of the matric or capillary potential can be made only with tensiometers. These devices can also be used to estimate soil water content. They consist essentially of a porous ceramic cup filled with water which is buried in the soil at any desired depth and connected by a water-filled tube to a manometer or vacuum gauge (see Fig. 4.5). The manometer indicates the pressure drop on the water in the porous cup, which is in equilibrium with the matric potential of the water in the soil. Pressure transducers are replacing vacuum gauges in transistors used for research (Rawlins, 1976). Tensiometers are excellent measuring instruments in moist soils; but when the matric potential drops to about -0.08 MPa, bubbles of air and water vapor form, and they become useless. Although most rapid plant growth occurs within the sensitive range, lower potentials are of great interest to agriculturists and ecologists. The other limitations are relatively minor, including the necessity for refilling the cup with water after gas bubbles appear, the tendency for roots to become concentrated around the porous cup, and the occasional diurnal fluctuations in reading resulting from heat conduction along the water-filled tube.

Although tensiometers read in negative pressure units, they can be calibrated against soil moisture content so that readings can be converted to percentage water content. Richards (1949, 1954) and Scofield (1945) published extensive discussions of the use and calibration of tensiometers. They are widely used as guides to irrigation (Henry, 1978).

Hysteresis. A minor problem in measuring soil water content results from the occurrence of hysteresis. This term refers to the fact that the water content at a given water potential is higher during a drying or desorption cycle than during a wetting cycle, as shown in Fig. 4.6. The reasons for this are complex, involving differences in curvature of menisci in advancing (wetting) and receding (drying)

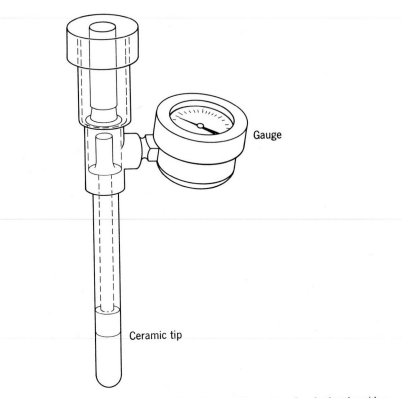

Gauge

Ceramic tip

Fig. 4.5. A commerciably available type of tensiometer. It consists of a plastic tube with a ceramic cup on the lower end, a screw cap for refilling with water on the upper end, and a Bourdon-type vacuum gauge attached at the side.

soil, swelling and air entrapment in wetting soil, and differences in rate of emptying and refilling of pore spaces of various sizes (Baver *et al.,* 1972; Hillel, 1980b). Thus, a given tensiometer reading is associated with a higher water content in a drying soil than in a wetting soil. The porous blocks used for conductivity measurements also show hysteresis.

Measurements of Soil Water Potential

Field methods of measuring soil water that use porous blocks and tensiometers can be used to estimate the matric component but not the osmotic component of the total soil water potential. The pressure and gravitational potentials can usually be neglected, but the osmotic potential may be important in heavily fertilized

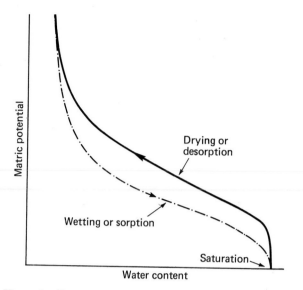

Fig. 4.6. Hysteresis effect, resulting in differences between matric potential at a given water content in wetting and drying soil. (Modified from Hillel, 1980a.)

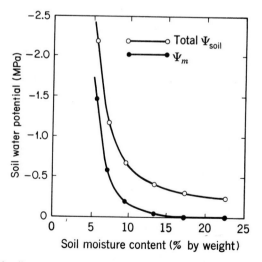

Fig. 4.7. Total soil water potential of a greenhouse soil measured with a thermocouple psy-chrometer and matric potential measured with a pressure plate, plotted over soil water content for a greenhouse soil consisting of sand, loam, and peat. The difference between matric and total potential is the osmotic potential, in this instance produced chiefly by heavy fertilization. (From Newman, 1966.)

soil (see Fig. 4.7) and in arid regions where salt accumulates. The best method of measuring total soil water potential seems to be by the use of thermocouple psychrometers. The thermocouples are enclosed in porous porcelain bulbs, buried in the soil, and operated as Peltier-type psychrometers (Brown and Van Haveren, 1972; Rawlins, 1976). Thermocouple psychrometers are discussed in Chapter 12.

LABORATORY MEASUREMENTS OF SOIL WATER

Although the techniques already described can be used in laboratory or greenhouse work, they are intended primarily for field measurements. Considerable attention has been given to techniques for evaluating the soil water characteristics of isolated samples and soil cores under controlled laboratory conditions. These techniques are also used to develop calibrations between water content and the components of soil water potential.

Measurements of Soil Water Potential

In soils low in salt, the water potential is essentially equal to the matric or capillary potential, and field measurements with tensiometers give useful indications of the water potential in the range higher than -0.08 MPa. Likewise, estimates of water potential can be made from measurements of water content if a graph relating matric potential to water content is available. However, if enough salt is present to produce a measurable osmotic potential, the matric potential is not an adequate measure of the total water potential. This is often true of heavily fertilized soils (see Fig. 4.7) as well as saline soils.

Vapor Equilibration.   One of the earliest attempts to measure soil water potential was made by Shull (1916), who measured the uptake of water by dry cocklebur (*Xanthium pennsylvanicum*) seeds from soils at various moisture contents. Since he had determined how much water these seeds would absorb from solutions of various osmotic pressures, he could calculate the approximate force with which water was held by a given soil. Gradmann (1928), Hansen (1926), and other European workers measured the water potential of soil with strips of paper saturated with solutions of various concentrations. The strips were weighed before and after exposure to the soil, and the soil water potential (*Saugkraft*) was regarded as equal to the osmotic pressure of the strip which neither gained nor lost weight. This method can be used in the field or laboratory, but it requires careful temperature control. The vapor equilibration method described by Slatyer (1958) can also be used for soil.

Thermocouple Psychrometers. Most laboratory measurements of water potential are based on measurements of relative vapor pressure, although some have been based on depression of the freezing point. However, it is difficult to obtain a sharp freezing point in dry soil, and the cryoscopic method is not recommended. The most useful technique for measuring total soil water potential is the thermocouple psychrometer, which measures the relative vapor pressure. This depends on the relationship between the chemical potential of water and depression of its vapor pressure, according to Eq. (1.7), developed in Chapter 1. Use of this method for field measurements was mentioned earlier, but in the laboratory a small sample of soil is placed in a chamber containing the thermocouple junctions and immersed in a water bath at constant temperature during equilibration. The operation of thermocouple psychrometers is discussed in Chapter 12, in Brown and Van Haveren (1972), in Rawlins (1976), in Slavik (1974), and in Wiebe *et al.* (1971).

Measurements of Matric Potential

Laboratory measurements of capillary or matric potential are almost always made with the pressure plate or pressure membrane equipment developed primarily by Richards (1949, 1954). In this procedure, a previously wetted sample of soil is placed on a membrane permeable to water and solutes, enclosed in a container, and subjected to a pressure difference across the membrane. The pressure difference is produced either by suction beneath the membrane or by gas pressure applied above, usually from a cylinder of compressed air. When water outflow ceases, indicating that equilibrium has been reached between the capillary potential and the imposed pressure, the sample is removed and the water content is determined gravimetrically. Sometimes repeated determinations are made on the sample by measuring the water outflow associated with successive increases in pressure.

A pressure membrane apparatus is shown in Fig. 4.8. Ceramic plates are used for suction or low pressure. For higher pressures, cellulose membranes supported on metal screens have been used, but ceramic plates are now available for pressures up to 1.5 MPa.

Measurements are usually made on small soil samples contained in rings about 5.0 cm in diameter and 1.5 cm in height. Approximately 20 samples can be measured at one time. The samples may be naturally occurring soil or may have

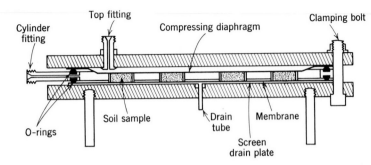

Fig. 4.8. A pressure membrane apparatus for measuring the matric potential of soil at various water contents. The soil samples are usually contained in metal rings 5.0 cm in diameter and 1.2 cm deep, placed on a cellulose acetate membrane. Compressed air is supplied through the cylinder fitting, and air at a slightly higher pressure is supplied through the top fitting to keep the soil samples pressed firmly against the membrane. In some models, the cellulose acetate membrane is replaced by a porous ceramic plate. Data obtained with this type of equipment are shown in Fig. 3.8.

been pulverized and screened to remove rock fragments. Elrick and Tanner (1955) found that the sieving procedure can result in overestimating water retention by up to 30% at tensions less than 0.04 MPa and by about 10% at tensions greater than 0.1 MPa. Young and Dixon (1966) also report overestimates of field capacity from sieved samples. The possible effect of sample pretreatments should always be kept in mind in work with soil samples. Difficulty in maintaining good contact between samples and membrane is sometimes experienced with soils that shrink while drying. The water content of soil at 1.5 MPa, as measured by this method, is a satisfactory approximation of the permanent wilting percentage.

Measurements of Osmotic or Solute Potential

Measurement of the osmotic pressure, or the osmotic or solute potential, can be made on soil solution only after it has been removed from the soil. Extraction is often done by adding water to the soil to a specified supersaturated level and filtering off a "saturation extract" (Richards, 1954). The osmotic potential can then be determined cryoscopically by the same method used for plant sap or in the vapor pressure psychrometers described later for measurement of total water potential. The electrical conductivity of the soil solution is often measured because there is close agreement between electrical conductivity and osmotic potential of the saturation extract (Campbell *et al.,* 1949; Richards, 1954). This relationship is shown in Fig. 4.9.

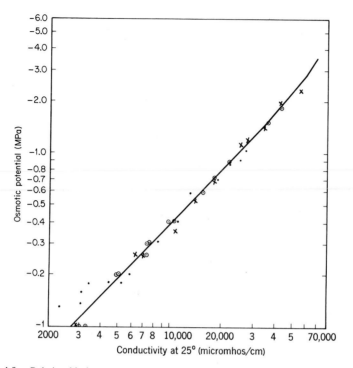

Fig. 4.9. Relationship between osmotic potential and conductivity of soil solution. The points include measurements of soil solution extracted by pressure and of nutrient solutions. (Adapted from Richards, 1954, Table 6.)

The osmotic pressure of the saturation extract is adjusted to that of the originally more concentrated field soil solution by a simple proportional correction. Although this procedure is satisfactory for most purposes, it should be recognized that it gives only an approximation of the real values because the degree of dissociation and the osmotic coefficients of various salts vary with their concentration and the concentration of other salts in the solution, as well as with temperature. Rawlins (1976) discussed the use of several types of salinity sensors.

EXPERIMENTAL CONTROL OF SOIL WATER CONTENT

There are two problems in supplying water to potted plants grown for study of the effects of soil water potential on plant processes and growth. One is maintenance of a uniform supply of water to plants to be kept under minimum stress. The other problem is that of subjecting plants to various known and controlled levels of soil water stress.

Maintenance of a Uniform Water Supply

One of the most conspicuous deficiencies in experimental control of plant environment is failure to maintain the water supply at a uniform level. In greenhouses, phytotrons, and plant growth chambers, relatively large plants are often grown in small pots filled with vermiculite, gravel, or other media containing limited water reserves. As a result, the water supply frequently fluctuates widely. Large fluctuations in soil moisture also often occur in greenhouses, where irregular watering of benches results in marked variations in the supply of water to individual pots in an experiment, and the regular schedule of watering is seldom adjusted to compensate for differences in the rate of evapotranspiration.

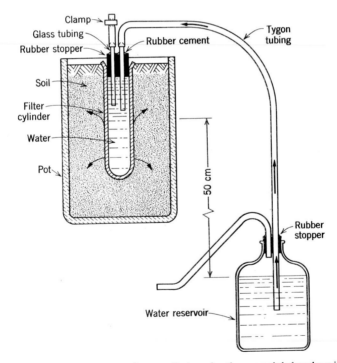

Fig. 4.10. An autoirrigator system for pots. By lowering the reservoir below the soil mass, the matric potential of the soil can be lowered slightly. (Redrawn from Read *et al.*, 1962.)

Early attempts to deal with this problem involved the use of autoirrigated plants in which water was supplied to the soil from a porous porcelain cone or cylinder buried in the center of the pot and attached to a reservoir which supplied water as it was used (Livingston, 1918). An example is shown in Fig. 4.10. Unfortunately, the soil a few centimeters from the cone tended to dry out, and roots became massed around the cone. Water distribution was improved by constructing double-walled pots with a space for water between the glazed outer wall and the porous inner wall. Even with this improvement, the water supply is sometimes inadequate for large, rapidly transpiring plants (Richards and Loomis, 1942).

Water Control in Greenhouses and Nurseries

Greenhouse and nursery operators have given considerable attention to development of methods for watering potted plants that reduce labor and increase uniformity. Many of the earlier methods were described by Post and Seeley (1943). One simple method is to pull one end of a piece of glass rope through the hole in the bottom of the pot and spread it, thus making good contact with the soil. The other end dips in a reservoir of water (see Fig. 4.11). This arrangement maintains the soil at approximately the field capacity and is particularly useful for house plants, which are notoriously either under- or overwatered.

Post and Seeley (1943) described several arrangements for subirrigation of greenhouse benches. The most common one uses waterproof benches with a V-shaped bottom about 5 cm deep with half-tiles or pieces of eave trough inverted over the bottom to provide a channel for water movement. The V is filled with sand or fine gravel, and the pots standing on it are supplied with water by

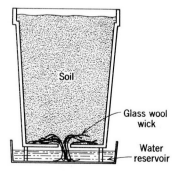

Soil

Glass wool wick

Water reservoir

Fig. 4.11. Section through an autoirrigated pot designed for small house plants. A wick of glass wool conducts water from the reservoir into the pot. The pot and water reservoir are of plastic.

capillarity. More recently, it has been found that ordinary flat-bottom benches can be lined with plastic and used just as satisfactorily as the special V-bottom benches. Water can be distributed the length of the bench through a piece of plastic tubing pierced with holes at 15-cm intervals. A diagram of such an arrangement is shown in Fig. 4.12. In some instances, water is supplied manually or is turned on by a solenoid valve controlled by a tensiometer or a time clock. In other installations, a constant water level is maintained in the bottom of the bench by the use of a float valve in a tank connected to the supply tube in the bench. The amount of water supplied can be controlled somewhat by the depth of sand above the water table. Modifications of this method were developed by the National Institute of Agricultural Engineering in England, and it can even be adapted for use in plant growth chambers.

Trickle or drip irrigation systems are used extensively to water potted plants. Usually, a header tube is run along a bench, and small plastic tubes extend from this to each pot. These tubes usually end in weighted nozzles which hold them in place in the pots. The amount of water or nutrient solution supplied depends on the length of time it is supplied to the system. Some systems use tensiometers or resistance blocks to monitor soil water and open solenoid valves at a predetermined soil water deficit. An example of a simple trickle irrigation system is shown in Fig. 4.13. If plants are grown in sand, or in gravel and vermiculite, and watered with nutrient solution, they should be watered until liquid drips from the bottoms of the pots to avoid salt accumulation.

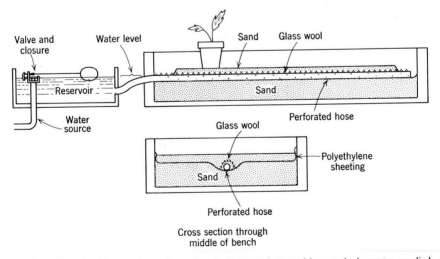

Fig. 4.12. A subirrigated greenhouse bench. Pots stand on sand kept moist by water supplied through a perforated hose from a constant-level reservoir. Various modifications of this general principle are in use.

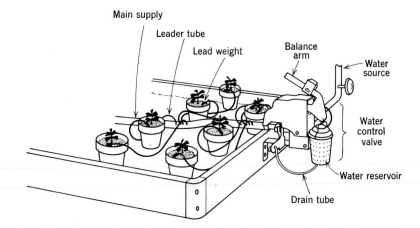

Fig. 4.13. Apparatus for surface or trickle irrigation of pots. In this diagram the water supply is controlled by a weighted arm and valve, but it can be controlled by a time clock or tensiometer operating a valve.

Maintenance of Definite Levels of Soil Water Stress

One of the most troublesome problems in plant water relations research is that of maintaining plants growing in soil at uniform levels of water potential lower than field capacity. In the older literature, experiments were frequently described in which plants were said to have been grown in soil maintained at arbitrary water contents such as 10, 20, or 30% of the dry weight of the soil, or at some percentage of capillary or field capacity.

The impossibility of doing this should have been realized by all who have observed the distribution of moisture in dry soil after a rain or who have considered the physical forces acting on soil moisture (see Chapter 3). Long ago, both Shantz (1925) and Veihmeyer (1927) called attention to the fact that if a small quantity of water is applied to a mass of dry soil, the upper layer is wetted to field capacity and the remainder of the soil mass remains unwetted. Addition of more water results in wetting the soil to a greater depth, but there will always be a definite line of demarcation between the wetted and the unwetted soil, as indicated in Fig. 3.12. This situation has been observed by everyone who has dug in soil after a summer shower and observed the well-defined boundary between the wet soil above and the dry soil beneath it. Since the field capacity is the amount of moisture held against gravity by a soil, it is obviously impossible to wet any soil mass to a moisture content less than its field capacity. If a container is filled with dry soil having a field capacity of 30% and enough water is added to the surface to wet the whole mass to 15%, the upper half of the soil mass will be wetted to field capacity and the lower half will be dry. Obviously, the earlier investigators did not really maintain their plants at the specified soil moisture contents but merely gave them various amounts of water distributed in different proportions of the soil mass used in their experiments.

Attempts have been made to control the soil water tension or matric potential by placing autoirrigated pots at various heights above the water supply. Read *et al.* (1962) were able to maintain good control at a tension of 50 cm of water, but at 200 cm the variation in soil water tension was ±50%. Livingston (1918) attempted to control the tension by inserting mercury columns of various heights between the irrigator cones and the pots in the apparatus shown in Fig. 4.10, but this also resulted in uneven wetting of soil in the pots. Moinat (1943) and others placed pots on top of sand columns of various heights standing in pans of water. The tension on the water in the pots increases with their height above the water table. Plants are also sometimes grown in soil in an inclined bench. However, it is difficult to produce tensions of more than 100 or 200 cm of water by these methods.

Another method is to divide the soil mass into several layers separated by layers of paraffin, asphalt, or some other material impermeable to water but capable of being penetrated by roots (Emmert and Ball, 1933; Hunter and Kelley, 1946; Vaclavik, 1966). Water is then added to each layer separately through tubes extending above the surface. Difficulties are encountered with soil aeration, leakage of seals, and failure to obtain uniform root distribution in each layer of soil. A somewhat better method is to inject small amounts of water into various areas of the soil mass instead of applying it all to the surface. Vaclavik (1966) used a long needle attached to a large syringe to inject small amounts of water at various levels and at various points around the soil mass. By replacing the water lost each day in this manner, he obtained relatively uniform distribution of water and of roots in the soil. However, this procedure does not bring the entire soil mass to a uniform water potential; it only brings masses near the point of injection approximately to field capacity.

Because of the difficulties inherent in other methods, most investigators limit the water supply to plants in containers by varying the intervals between irrigations. The plants are allowed to reduce the soil water content from field capacity to some predetermined water content or water potential, and enough water is then added to bring the entire soil mass back to field capacity. This is essentially what occurs in nature when rain or irrigation periodically wets the soil to field capacity after it has been dried to various levels of water content by evapotranspiration. However, this method requires frequent weighing and daily replacement of water in those containers kept near field capacity, and there are wide variations in soil water potential between replacements of water if the intervals are longer than a day or two (see Fig. 4.14). Wadleigh (1946) discussed methods of calculating what he termed the integrated soil moisture stress over cycles of drying and rewetting.

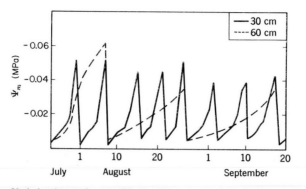

Fig. 4.14. Variation in matric potential of soil in an avocado orchard, as measured with ten-siometers at 30 and 60 cm. The peaks in Ψ_m result from irrigation, either with enough water to wet the soil to a depth of 30 cm (solid line) or every third irrigation with enough water to wet it to 60 cm. (From Richards and Marsh, 1961.)

There is continuing interest in maintaining the soil at a constant water potential lower than field capacity. A promising possibility is to grow plants in thin layers of soil in containers with side walls made of a differentially permeable membrane such as cellulose acetate. A diagram of such a device is shown in Fig. 4.15. This type of apparatus was described by Gardner (1964) and his co-workers, by Kemper *et al.* (1961), by Painter (1966), and by Zur (1967). Zur reported that he could maintain a relatively uniform water potential in a small chamber containing a sunflower seedling. The chief problems with this method are the limitation on size of the soil mass and rapid deterioration of the membrane. The soil layer must be relatively thin if water is to move to the center as rapidly as it is absorbed. More work is needed to establish the critical thickness, which will certainly depend on the water potential used, the capillary conductivity of the soil, and the rate of transpiration of the plants. Cellulose acetate membranes have been used in most experiments, but they are often destroyed by microorganisms in 1–2 weeks. Some of the new membrane materials are probably more resistant to decay.

In some kinds of experiments, it is possible to subject plants to water stress by growing them in nutrient solution plus additional solute to lower further the water potential. Unfortunately, solutes such as sodium chloride, potassium nitrate, and even sucrose are absorbed by the roots, producing a decrease in osmotic potential of the plant somewhat in proportion to the decrease in osmotic potential of the root medium (Eaton, 1942; Slatyer, 1961). For example, Boyer (1965) reported that the osmotic potential of cotton leaves decreased 0.12–0.15 MPa for each 0.1 MPa of reduction in the osmotic potential of the root medium brought about by addition of sodium chloride. Mannitol is not entirely satisfactory because it is absorbed to a limited extent (Groenewegen and Mills, 1960), and it is also attacked by microorganisms. Polyethylene glycol is used because it is less sub-

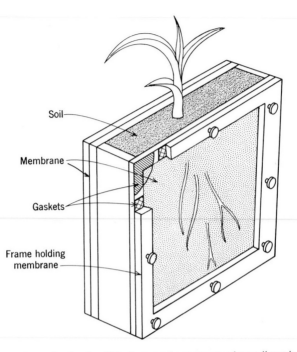

Soil

Membrane

Gaskets

Frame holding
membrane

Fig. 4.15.   A plastic chamber in which plants can be grown at various soil matric potentials. The flat sides are made from cellulose acetate or some other substance permeable to water but not to solutes, and are usually supported by plastic screen. The chamber can be immersed in polyethylene glycol of the proper concentration to provide the desired decrease in water potential. The soil mass can only be a few centimeters thick because of slow water conduction at low potentials. The membranes are often destroyed by microorganisms after a few days. (Designed by Dr. David Lawlor.)

ject to attack by microorganisms. There have been some complaints of toxic effects (Jackson, 1962), perhaps partly because of the impurities found in some lots of polyethylene glycol. These can be removed by dialysis (Lagerwerff and Eagle 1961). Other investigators have observed no visible toxic effects after long periods of exposure, provided that no entry occurs through damaged roots (Lawlor, 1970). The use of polyethylene glycol was discussed more recently by Thill *et al.* (1979), who regard it as the best solute available for producing stress.

IRRIGATION

Irrigation, as stated in the introduction, is an example of soil water control on the broadest possible scale. Efficient irrigation requires the use of information from many areas of plant and soil water relations. It is important to know the readily available water content of the soil at various bulk water contents in order to determine when to irrigate. It is desirable to know the water stress at which plant growth is inhibited during various stages of development from seed germination to maturation of the crop, and in humid regions it would be helpful if rainfall in the near future could be predicted. Information about the drainage characteristics of a soil are also desirable to prevent waterlogging and losses in yield from inadequate aeration. Finally, there are questions concerning the best irrigation system to use and the best method of timing the application of water. Although the answers to these questions seem to be available for some crops in some soils and climatic conditions, much more research is needed. As a result, many decisions about irrigation still depend on the judgment and experience of the operator. Nearly every aspect of irrigation was discussed in the monograph edited by Hagan, Haise, and Edminster (1967).

Irrigation is probably simpler to manage in arid regions, where crop production is completely dependent on irrigation, than in humid regions, where it is used as a supplement to rainfall. Timing of irrigation is more difficult in humid regions because it must take into account not only past but also possible future rainfall, and an unexpected downpour shortly after an irrigation can waterlog the soil. Thus, soil drainage is important. The intermittent and uncertain scheduling of irrigation in humid areas also makes it more difficult to apply fertilizer with the irrigation water than in arid regions, where the number and frequency of irrigations are predictable. Some of the problems of irrigation in humid climates are discussed by Sneed and Patterson (1983), and irrigation in arid regions and saline soils are discussed in books cited at the end of this chapter.

In this chapter we will discuss briefly the history of irrigation, methods, timing, and some of the problems encountered.

History

Prehistoric man seems to have realized that rainfall was often inadequate for his crops because irrigation was being practiced at the beginning of recorded history. It was in use in Egypt about 5000 BC, was well developed in Babylonia and China by 2000 BC or earlier, and was widely practiced in the Middle East and along the eastern Mediterranean. The Egyptians and Babylonians built dams to store water, the Persians (Iran) built extensive tunnels to bring water down from the hills, many of which are in use today, and large tanks or reservoirs were built in India and Ceylon. Irrigation was also practiced in pre-Columbian America. The Spanish invaders found elaborate irrigation projects in Peru and Mexico, and the Indians of southwestern Arizona had extensive canal systems (Masse, 1981).

The history of early agriculture and irrigation was reviewed by Hagan *et al.* (1967), Hall *et al.* (1979), and Parks (1951).

Irrigation within the present boundaries of the United States in modern times seems to have begun on the farms operated by the Franciscan missionaries in California before 1800, and on a much larger scale by the Mormons in Utah about the middle of the last century. In 1860 the United States census reported 16,250 ha of irrigated land, and in 1979 the irrigated area was said to be 24,000,000 ha (Sneed and Patterson, 1982). There have also been extensive expansions of irrigation in other countries, including Australia, China, India, and Israel. Perhaps the most notable change has been the increase in supplementary irrigation in the relatively humid climates of England, Western Europe, and the eastern United States. For example, in the decade ending in 1979, the irrigated area increased 77% in the eastern United States, compared with an increase of 17% for the western states. Originally, irrigation in humid areas was restricted to vegetable crops, but now it is being used on all kinds of field crops. This expansion results from increased realization among farmers that drought often reduces crop yields even in humid climates, and from development of sprinkler irrigation.

Methods

A wide variety of irrigation systems are in use, depending on local custom, soil, water quality, crops, and terrain.

Basin and Furrow Irrigation.   The earliest method of surface irrigation was probably basin irrigation, in which the entire soil surface was flooded. Low dikes are often built around individual trees or plots of ground to control spread of water, and elaborate systems of dikes and terraces are sometimes used on hillsides. Row crops are often furrow-irrigated, water being allowed to flow in furrows between the rows. Successful furrow irrigation requires careful leveling of the land and considerable labor to control water flow into the furrows from the canals supplying the water. There are differences in water supply along the furrow and loss of water by seepage and evaporation. Wetted zones in irrigated soils are shown in Fig. 4.16.

Sprinkler Irrigation.   In recent decades, there has been a great increase in the use of sprinklers and a remarkable increase in their size and the area which they cover. In the 1950s, most sprinkler systems were small and required much labor to move them from place to place in fields. Various types of movable, self-propelled sprinklers were introduced that require less labor and work well on

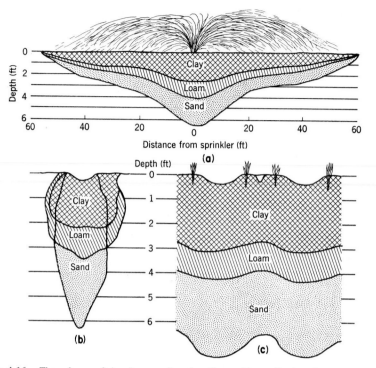

Fig. 4.16. The volumes of clay, loam, and sandy soil wetted by application of a given volume of water. Since downward movement of water greatly exceeds lateral movement, furrows and sprinklers must be close enough together to wet the soil between them. (From Doneen and MacGillivray, 1946.)

level, rectangular fields. Recently, there has been a trend toward large center-pivot, rotating sprinklers that cover large areas with a minimum of labor. In addition to saving labor and water, sprinkler irrigation can be used on rolling or sloping land, reducing the cost of soil preparation. Sprinkler irrigation also cools leaves and reduces midday closure of stomata. Sometimes leaves are injured if the water is too high in salts, leaching of minerals sometimes occurs, and the wet leaf surfaces may increase spread of some pathogenic organisms. Excessive irrigation also produces soil conditions favorable for the spread of root pathogens such as *Phytophthora*.

Trickle or Drip Irrigation. Another method is drip or trickle irrigation. This involves distributing water through tubing and allowing it to run out in the soil through small nozzles. It was first used for plants growing in containers (see Fig. 4.13), but it is also used extensively for trees and some row crops in the field. It permits precise placement of known volumes of water in the immediate vicinity of plants, thereby reducing losses from seepage and evaporation. This is important in areas where water is scarce or high in salt. It also reduces weed growth because most of the soil surface usually remains dry. Trickle irrigation may be economical in young orchards where only the soil in the immediate vicinity of recently planted trees needs to be irrigated.

In nurseries irrigators and sprinklers often are connected to tensiometers to turn on the water when the soil water tension decreases to a certain level. Containers in which plants are grown with drip or trickle irrigation, using water high in salt, should be flushed out occasionally to prevent accumulation of salt to toxic concentrations.

Subsurface Irrigation.   This method also places water precisely in the root zone and avoids losses from evaporation. It involves installing an underground distribution system provided with outlets for water which allow it to trickle out in the root zone. This method uses the minimum amount of water, but there is always concern lest some of the outlets become plugged, causing a water deficiency in certain areas.

Subsurface irrigation systems use the least possible amount of water because it is placed very precisely and the losses from evaporation are low. They also provide an excellent method of combining fertilization with irrigation. However, there is a possibility of salt accumulation at the soil surface since no surplus water is applied to leach out salt. The cost of installation is also high, and subsurface irrigation is likely to be used chiefly where the water supply is limited.

High-Frequency and Deficit Irrigation.   During the 1970s, interest developed in the use of high-frequency irrigation, meaning the application of water at least once a day. It has been suggested that misting plants with water all day on a very short cycle, such as 15 s every 2 min, would benefit growth more than the same amount of water applied to the soil in one application. The improved growth was attributed to elimination of midday closure of stomata and to cooling of leaves (Lawlor and Milford, 1975). Wright *et al.* (1981) found that 5 min of sprinkling every half hour significantly reduced the temperature of onion umbels, and they predicted that this would increase the yield of seed.

The increased use of sprinklers makes it relatively easy to supply water on any desired time schedule. To more frequent application has been added the concept of deficit irrigation, in which less water is supplied than is lost by evapotranspiration. The deficit is made up by absorption from the soil. This system seems to operate most successfully where the soil is fully charged with water at the beginning of the growing season, and it is probably more successful for some crops than others. For example, in the state of Washington, daily irrigation of sugar beets at 50–100% of the pan evaporation rate produced the same yield of sugar, although the yield of roots was somewhat reduced by the larger deficits (Miller and Aarstad (1976). However, the sugar yield remained high because sugar content of the beets increased with decreasing frequency of irrigation. Miller and Hang (1980) found that irrigation of sugar beets could not be reduced much below the loss by evapotranspiration on a sandy soil, but it could be reduced to 35–50% of evapotranspiration on a loam soil. In contrast, at Davis, California, where the soil was not fully charged with water at the beginning of the season, deficit irrigation reduced the yield of several crops (Fereres *et al.*, 1981). Rawlins and Raats (1975) have an extensive discussion of high-frequency irrigation.

Where soil water is available at the beginning of the growing season or is occasionally recharged by summer rains, deficit irrigation can sometimes save considerable water and energy. However, to be successful, it requires more careful monitoring of evapotranspiration and available soil water than is likely to be practicable on most farms.

Timing

Assuming that water is available, the most important problem is to determine when and how much to apply. This depends on several factors, including the kind of crop, the stage of development of the crop, the rate of evapotranspiration, the water-holding capacity of the soil, and the extent of the root system. Plants are usually most susceptible to injury by water stress during the reproductive stage. Clear, hot weather produces high rates of evapotranspiration and exhausts soil water more rapidly than cloudy, cool weather. In fact, plants in moist soil near field capacity and in water cultures often wilt temporarily on hot, dry days. The larger the volume of soil occupied by roots, the larger the volume of water available, so that plants with large, deep root systems can survive longer between rains or irrigations than plants with shallow root systems. Deep tillage often increases' the drought tolerance of crops because it usually results in deeper rooting.

In general, successful irrigation depends on supplying enough water to prevent reduction of yield by plant water stress without saturating the soil or incurring the cost of supplying more water than is required. In arid regions where water is scarce and expensive, it is sometimes economically preferable to permit a small reduction in growth and yield in order to increase the area irrigated. This seems to be true in Israel, where the irrigated area was expanded by about 30% from 1958 to 1969, with no increase in water supply, by increasing the efficiency of water use (Shmueli, 1971).

In spite of modern technology, much irrigation is based on rule-of-thumb methods instead of on knowledge of soil–plant–atmosphere relations. There are three approaches to the problem of determining when to irrigate. These are based on measurements of soil moisture, estimation of water usage from evaporation data, and measurement of plant water stress. Each of these methods will be discussed briefly.

Soil Water Measurements.   The traditional method of determining the need for irrigation is to dig into the soil and decide from its appearance and "feel" whether or not water should be applied. Determination of water content by gravimetric sampling or by use of neutron meters is not much better unless the water potential has been determined for various water contents, because the availability of water depends on its potential rather than on the content as percentage of weight or volume. A sand might be at field capacity ($\Psi_m = -0.03$ MPa) at a water content which is below the permanent wilting percentage ($\Psi_m = -1.5$ MPa) in a clay (see Table 3.3). Also, the available water content differs in various soil horizons, as shown in Fig. 3.7 and Table 6.2. Thus, the only reliable indication of the soil water status in terms of plant growth is its water potential. The matric potential of the soil can be measured directly with tensiometers down to about -0.08 MPa, which covers the range for best growth of most vegetable crops. Use of the calibrated electrical resistance blocks mentioned earlier in this chapter permits measurements to much lower soil water potentials. Resistance blocks have been used successfully in a variety of situations. Tensiometers are widely used, but are most effective in sandy soils because in clay soils they become inoperative while much of the available water remains. Also, if the soil shrinks as it dries, contact between tensiometer cups and the soil may be reduced. The successful use of tensiometers and resistance blocks depends on installation in the zone of maximum root concentration, where water is being removed most rapidly.

Use of Evaporation Data.   Because of practical difficulties in measuring soil water over large areas, there is increasing use of evaporation data, combined with knowledge of the water storage capacity of the soil. This is called the water budget method. It was used by van Bavel and his colleagues to estimate the frequency of drought in various regions of the United States (van Bavel and Verlinden, 1956; Blake *et al.*, 1960), and its use in California was discussed by Fereres *et al.* (1981). If the depth of rooting, the water-holding capacity of the soil, the allowable percentage of depletion, and the rate of water removal by evapotranspiration are known, the date at which irrigation will be required can be predicted quite precisely. Sometimes the evaporation data are obtained from evaporation pans or Livingston atmometers, and sometimes they are calculated from meteorological data. A widely used method is that of Jensen and Haise (1963), and Tanner (1967) discussed several methods of measuring and calculating evapotranspiration. Sometimes the evaporation data are corrected by a "crop coefficient" which is based on experimental determination of the relationship between pan evaporation and actual evapotranspiration of a crop in a particular region. An example of the procedure used in Minnesota for corn is given by Dylla *et al.* (1980). Methods for estimating the frequency of irrigation are given for cabbage in Arizona by Bucks *et al.* (1974) and for sugar beets in Washington by Miller and Arstaad (1976). In some areas of the United States, weather forecasts and newspapers tell homeowners how much water must be applied to replace losses by evapotranspiration.

Plant Water Stress. At least in theory, plants should be the best indicators of the need for irrigation because they integrate the soil and atmospheric factors controlling the plant water balance. Among the early visual indicators are changes in leaf color, change in angle of leaves, and rolling of leaf blades, and some of these have been used successfully as indicators of the need for irrigation (Haise and Hagan, 1967). Nurserymen sometimes use a sensitive plant such as hydrangea as an early warning signal of water stress. It must be remembered that on hot, sunny days, even plants in moist soil may show stress. However, if they show stress early in the morning, the soil water supply is becoming limiting to plant growth.

A more sensitive indicator of plant water stress than wilting is needed because by the time plants have wilted, growth has already been reduced. One of the earliest attempts to use quantitative measurement of plant water status as an indicator for irrigation was Clements and Kubota's (1942) use of the water content of selected leaf sheaths of sugarcane. Leaf water potential is a more reliable indicator of plant water stress than leaf water content, but measurements of water potential are seldom possible on farms. However, some farmers are beginning to use pressure chambers to measure plant water stress. Premature closure of stomata is an early indicator of water stress and has been used successfully on several kinds of plants as an indicator of need for irrigation (Oppenheimer and Elze, 1941; Shmueli, 1953; Alvim, 1961). Reductions in rate of growth of leaves, fruits, and stems are also indicators of developing water stress and are sometimes used as indicators.

Hiler and his colleagues (Hiler and Clark, 1971; Hiler *et al.*, 1972) developed the stress day index, which accounts for differences in stress tolerance among species and at various stages of development. This is the product of the stress day factor, which is a measure of plant water stress, usually the leaf water potential, and the crop susceptibility factor, which is obtained experimentally by subjecting the crop to a specified stress at various stages of development and determining the yield as a fraction of the yield of unstressed plants at the same stage of development. A paper by Hiler and Clark (1971) shows the wide differences in sensitivity to stress at various stages of development among various crops. In many respects, this appears to be a good approach to irrigation timing, but unfortunately it is too complex for everyday use on a farm.

Remote Sensing.   Improvements in aerial photography and related technology make it possible to determine the need for irrigation by various kinds of remote sensing. For example, Blum (1979) used infared photography on a small scale to determine differences in drought tolerance among sorghum lines in a breeding nursery. This was possible because stressed sorghum canopies with low water potentials are lighter in color than unstressed canopies, and differences in color can be used on a much larger scale to detect the beginning of water stress over large units. It is possible to use radiation in three different regions of the spectrum—the visible or short-wave region, the thermal or long-wave region, and the microwave or radar region—to measure soil moisture. Idso and his colleagues (1975) described a number of experiments in which microwave techniques were used to monitor changes in moisture content of bare soil, and they regarded such measurements as useful in predicting the incidence of various insect pests and diseases as well as yields. Later, these investigators (Idso *et al.*, 1980) described other experiments indicating that, in dry climates such as those of Davis, California, and Phoenix, Arizona, wheat yields can be predicted from solar radiation over the period of vegetative growth and the sum of stress degree days during the reproductive period. Stress degree days are identified on the basis of differences between crop canopy and air temperatures at about 2 PM. This difference increases as leaf water potential decreases. There are opportunities for increased use of remote sensing in agriculture as the techniques and capabilities improve. Clawson and Blad (1982) made usable measurements of differences in average leaf temperature between plants in a well-watered reference plot and those in the field, using a hand-held infrared thermometer from the top of a 4-m ladder.

Comparison of Methods of Timing Irrigation.   Bordovsky *et al.* (1974) reported that water use efficiency was higher when irrigation of cotton in Texas was based on leaf water potential and the stress day index than when it was based on soil water status. Bucks *et al.* (1974) found that the irrigation could be timed equally well from gravimetric sampling of soil water before and after irrigation and from evapotranspiration data calculated by the Jensen-Haise method, which is based on radiation. Dylla *et al.* (1980) found that estimates of evapotranspiration based on pan evaporation were only slightly more accurate than those based on meteorological data, and both were improved by correction from soil water data. Prihar *et al.* (1976) reported that scheduling the irrigation of wheat in the Punjab by a system based on pan evaporation was effective and simple. According to Sammis (1980), comparable yields of lettuce were obtained with trickle, sprinkler, furrow, and subsurface irrigation at a soil water tension of -20 and

−60 kPa at a depth of 15 cm, but higher water use efficiency was obtained with trickle and subsurface irrigation of potatoes. It appears that there is no best method of timing irrigation, and choice of the method depends on experience and the amount of information available to the manager.

Various mathematical models for predicting the timing of irrigation are being tested and eventually will become useful. However, in too many instances the data needed to improve their reliability is unavailable. For example, more information is needed concerning the degree of stress that can be tolerated by various crops without permanent injury and the effects of water stress at various stages of growth. It is also necessary to have reliable information concerning the available water storage capacity of the soil and the rate of water extraction from the soil. The first often varies from field to field and even within fields; the second varies widely depending on variation in the rate of evapotranspiration, which in turn depends both on weather and the crop canopy. Examples of the problems encountered are given by Curry and Eshel and Palmer *et al.* in Raper and Kramer (1983).

Irrigation Problems

In addition to the problems of methodology and timing, various other problems accompany the use of irrigation, the most important of which are soil aeration and salt accumulation. Hillel and Rawitz (1972) and Hagan *et al.* (1967) discuss causes of inefficient water use in irrigation.

Soil Aeration and Drainage. The importance of adequate soil aeration for good growth and functioning of roots is mentioned repeatedly in this book. Irrigation by flooding saturates the surface soil and produces at least temporary decrease in oxygen supply to roots. Basin irrigation, often used on trees, presumably is worst in this respect, but furrow irrigation also reduces the oxygen supply, as indicated in Fig. 4.17. Sojka and Stolzy (1980) found that low soil oxygen causes stomatal closure, which is undesirable. Sprinkling at moderate rates ought to cause the least interference with gas exchange, unless it is continued long enough to saturate the soil. Perhaps worse than the temporary saturation of the surface layer is the prolonged saturation of deeper soil horizons, which sometimes occurs as a result of too frequent or too heavy irrigation combined with poor internal drainage. Letey *et al.* (1967) reported a case of chlorosis in a citrus orchard where sufficient water was applied to wet the entire root zone whenever tensiometer readings in the surface soil indicated need for water. As a result, the subsoil was kept too wet for adequate aeration of the roots.

Soil saturation or waterlogging is a serious problem in both arid and humid

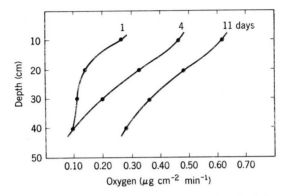

Fig. 4.17.   Effect of irrigation on oxygen diffusion rate at various depths in the soil 1, 4, and 11 days after irrigation, measured with a platinum electrode. Diffusion rates of less than 0.20 μg cm^{-2} min^{-1} are regarded as inadequate for root growth. (Redrawn from Hagan *et al.*, 1967, p. 946.)

regions. In arid regions, in addition to causing deficient aeration of root systems, it causes salt accumulation at the soil surface and injury from excess salinity. It is believed that the decline of several early civilizations in semiarid regions was hastened by salinity and waterlogging problems in their cultivated land. Proper drainage of irrigated land is as important as an adequate supply of water. It is a troublesome problem because use of an excess of water to remove surplus salt also leaches out fertilizer and aggravates salinity problems downstream. In humid regions, wet soil delays planting operations in the spring, and heavy rain soon after an irrigation can cause injurious waterlogging. Installation of effective drainage systems depends on a knowledge of the physics of water movement in various types of soils as well as practical engineering experience. The movement of water in soils in relation to drainage is discussed in soils texts such as those of Baver *et al.* (1972, Chapter 12) and Hillel (1980b, Chapter 4). Some of the problems of drainage in relation to irrigation are discussed by Donnan and Houston (1967).

Salt Accumulation.   One of the most serious problems in arid regions is the accumulation of salt in the soil. Large areas of land all over the world have been rendered unproductive by the accumulation of salt either deposited by irrigation water or derived from natural sources. In humid areas where rainfall exceeds evaporation, salt is leached out, but in arid regions there is insufficient rainfall to remove it. Irrigation water usually contains appreciable amounts of salt. For example, water used in the southwestern United States is said to contain 0.1–5.0 tons of salt per acre-foot (0.25–12.5 metric tons/ha). Thus, depending on the quantity and quality of water used, 1–10 tons of salt per acre (2.5–24.0 metric tons/ha) might be added to the soil in a year, and most of it remains when the water is removed by evapotranspiration. The only way to prevent salt accumulation is to supply enough water to leach it out, but often this is impossible because of either lack of water or lack of drainage to remove it. Salt accumulation is therefore a major problem in almost every irrigated area in the world (Reeve and Fireman, 1967).

There are two principal effects of salt accumulation, namely, osmotic effects and alkali effects. Mere accumulation of salt reduces the soil water potential and decreases the availability of water. Osmotic potentials equivalent to -0.4 MPa at the permanent wilting percentage are said to reduce the growth of even salt-tolerant crops such as alfalfa, cotton, and sugar beets in soil kept near field capacity (Magistad and Reitemeier, 1943). The complex problem of how high salt concentration reduces plant growth is discussed in Chapter 9 and by Slatyer (1967). In some areas there is an accumulation of excessive amounts of exchangeable sodium which replaces calcium and magnesium. This leads to decreased permeability, poor drainage, and development of a black surface film which causes such land to be called black alkali. These areas support little or no plant growth. The problem of saline and alkaline soils in relation to irrigation was reviewed by Reeve and Fireman (1967) and in a monograph edited by Kŏvda *et al.* (1967). Methods of diagnosing and improving such soils were discussed by Richards (1954) and in Hagan *et al.* (1967).

Irrigation with Saline Water. In many arid and semiarid regions, the only water available contains too much salt for optimum growth of many crop plants. High initial salt content also hastens salt accumulation by evaporation. This can be minimized by trickle and subsurface irrigation, but the salt can be removed only by sufficient rainfall or irrigation water to leach it out. This requires a permeable soil such as sand and good drainage. Boyko (1966) seems to have achieved some success in watering plants with very saline water, perhaps because he used deep sand where the salt was leached out. A monograph on irrigation and drainage in relation to salinity, edited by Kŏvda *et al.* (1967), was published by UNESCO, and Poljakoff-Mayber and Gale (1975) also discuss salinity problems.

During the 1970s considerable interest developed in the possibility of using saline water for crop production either by using wild halophytes more extensively or by finding salt-tolerant strains of cultivated crops. The first approach is described in an article by Somers (1979), who pointed out that dry matter production is high in intertidal zones and salt marshes. He suggests more extensive use of such halophytic species as *Spartina alterniflora, Atriplex patula, Chenopodium album,* and *Kosteletzkya virginica.* The other approach is exemplified by Epstein and his colleagues (Epstein and Norlyn, 1977), who are attempting, by breeding and selection, to produce strains of crop plants more tolerant of saline water. It was formerly claimed that there is little genetic diversity among cultivated plants with respect to salt tolerance, but Epstein and his co-workers were able to isolate strains of barley that could be grown in deep sand irrigated with seawater. The possibilities for selecting and breeding salt-tolerant plants were discussed in Section V of Rains *et al.* (1980) and in ''The Biosaline Concept,'' edited by Hollaender *et al.* (1979).

Cold Water.   Occasionally, the low temperature of the water used for irrigation becomes a limiting factor, especially for rice. Cold water has been reported to be a limiting factor for rice production in northern Italy, Hokkaido, and the Sacramento Valley of California. In those areas, much of the irrigation water comes from melting snow. In some instances, cold water is held in warming basins before being applied to crops. Occasionally, use of cold water on greenhouse crops such as cucumbers becomes a limiting factor in the winter (Schroeder, 1939). The effects of low temperature on water absorption are discussed in Chapter 9.

SUMMARY

Ability to measure and control the soil water supply to plants is basic to effective research on plant water relations and essential for efficient irrigation of crops. The best measure of the availability of soil water to plants is the water potential, the chief components of which are the matric and osmotic potentials. The matric potential can be measured in the field directly by tensiometers down to -0.08 MPa and estimated to much lower values from readings of calibrated electrical resistance blocks; it can also be measured in the laboratory on a pressure plate apparatus. The osmotic potential can be measured on soil extracts, and the total soil water potential can be measured with a thermocouple psychrometer. Changes in soil water content can be monitored with resistance blocks or neutron meters. Field capacity (Ψ_w -1.5 MPa) and permanent wilting percentage (0.03 MPa) are useful values because they indicate approximately the upper and lower limits of soil water readily available to plants. However, they are not physical constants but merely somewhat arbitrary points on the water potential–water content curve.

Maintenance of uniform soil water potential at any value lower than field capacity for prolonged periods of time is impossible in large-scale experiments. In such experiments, water stress is ordinarily produced by withholding water until the desired level of stress is reached and then rewatering to field capacity. This produces a gradually increasing stress during each cycle. Small plants can be subjected to more uniform stress by growing them in nutrient solution to which a solute such as polyethylene glycol is added to produce a desired level of water potential.

Irrigation involves control of water supply on a large scale. Its success depends on proper timing, which in turn depends on knowledge of the water storage capacity of the soil, depth of rooting, and rate of water extraction. Timing of irrigation can be based on measurment of leaf water potential, measurement of soil water, or rate of evapotranspiration calculated from meteorological data or pan evaporation. Irrigation sometimes produces drainage and soil aeration problems, and in arid regions it often results in salt accumulation and salinity problems.

SUPPLEMENTARY READING

Epstein, E., and Norlyn, J. D. (1977). Seawater-based crop production: A feasibility study. *Science* **197**, 249–251.

Hagan, R. M., Haise, H. R., and Edminster, T. W., eds. (1967). "Irrigation of Agricultural Lands." Am. Soc. Agron., Madison, Wisconsin.

Hall, A. E., Cannell, G. H., and Lawrence, H. W., eds. (1979). "Agriculture in Semi-Arid Environments." Springer-Verlag, Berlin and New York.

Hillel, D. (1980). "Applications of Soil Physics." Academic Press, New York.

Hollaender, A., Aller, J. C., Epstein, E., San Pietro, A., and Zaborsky, O. R., eds. (1979). "The Biosaline Concept." Plenum, New York.

Kŏvda, V. A. (1980). "Land Aridization and Drought Control." Westview Press, Boulder, Colorado.

Poljakoff-Mayber, A., and Gale, J., eds. (1975). "Plants in Saline Environments." Springer-Verlag, Berlin and New York.

Rains, D. W., Valentine, R. C., and Hollaender, A., eds. (1980). "Genetic Engineering of Osmoregulation." Plenum, New York.

Richards, L. A., ed. (1954). "Diagnosis and Improvement of Saline and Alkaline Soils," U.S. Dep. Agric. Handb. 60. USDA, Washington, D.C.

Slavik, B. (1974). Methods of Studying Plant Water Relations. Springer-Verlag, New York.

Van Schilfgaarde, J., ed. (1974). "Drainage for Agriculture." Am. Soc. Agron., Madison, Wisconsin.

5

Root Growth and Functions

INTRODUCTION

This chapter deals with the growth and functions of roots. Physiologists generally concentrate on the activities of shoots and neglect the roots of plants, partly because they are out of sight and are more difficult to study than shoots. However, even casual consideration of their functions indicates that vigorous root systems are as essential as vigorous shoots for growth of healthy plants. In fact, the growth of roots and shoots is completely interdependent, and one cannot succeed if the other is failing. Factors affecting root growth will be discussed in Chapter 6.

FUNCTIONS OF ROOTS

Although the chief concern in this chapter is with roots as absorbing organs, some of their other functions, such as anchorage and synthesis of growth regulators, are equally important for successful growth.

The role of roots in anchorage is generally taken for granted, but it is important because the success of most terrestrial plants depends on their ability to stand upright. There are wide differences among plants in their ability to withstand wind, depending on differences in the extent and mechanical strength of their roots as well as on the strength of their stems. Mechanical strength of roots is also important in preventing injurious frost heaving of wheat and some other plants, which results in winter injury. Considerable food, chiefly carbohydrate, is stored in roots of some kinds of plants. This is economically important in crops such as beets, carrots, sugar beets, cassava, and yams.

The importance of roots as sites of synthetic activities has been appreciated only in recent decades. In many kinds of plants, inorganic nitrogen is converted to organic compounds in the roots before being translocated to the shoots. Nicotine is synthesized in the roots of tobacco and translocated to the shoots, and other important compounds are produced in the roots of other species. It was suggested long ago by Went (1938, 1943) that roots probably produce hormones, termed caulocalines, which are essential to shoots. Chibnall (1939) proposed that roots produce a substance essential for normal nitrogen metabolism in leaves, and later this was found to be cytokinin (Kende, 1965). Skene (1975) described the history of research on cytokinins which indicates that they are synthesized in roots and translocated to shoots in the transpiration stream. Cytokinins are also said to play a role in developmental activities in root meristems (Feldman, 1975). Roots synthesize significant amounts of gibberellins that are translocated to the shoots in the xylem sap (Skene, 1967). Abscisic acid (ABA) is said to be synthesized in the root or root cap (Audus, in Torrey and Clarkson, 1975; Walton *et al.*, 1976), although leaves are regarded as the principal site of synthesis by most writers (Hoad, 1975; Loveys, 1977; Milborrow, 1974).

It has been suggested that the reduced shoot growth observed when roots are subjected to unfavorable conditions, such as water stress, flooding, or low temperature, is caused in part by reduced synthesis of cytokinins and gibberellins (Itai and Vaadia, 1965; O'Leary and Prisco, 1970; Skene, 1975). Drew *et al.* (1979) consider disturbance of nitrogen metabolism to be an important factor in flooding injury, but this might be related to a reduced supply of cytokinin. This problem deserves further study.

Root cells possess many of the synthetic functions of shoots and can even produce functional chloroplasts when exposed to light. However, they are dependent on shoots for thiamin, and some require niacin and pyridoxine. Roots also receive auxin from the shoots, and its role in root growth is discussed by Batra *et al.* (1975). The synthetic activities of the bacteria and other microorganisms associated with roots in the rhizosphere must also be taken into account. The role of *Rhizobium* bacteria in fixing nitrogen is well known, and the role of mycorrhizal-forming fungi is very important [see papers in Torrey and Clarkson (1975) and the discussion later in this chapter].

ROOT GROWTH

The effectiveness of roots as absorbing organs depends on the anatomy of individual roots and on the extent and degree of branching of the root system.

Primary Growth

During growth and maturation, roots undergo extensive anatomical changes that greatly affect their permeability to water and solutes. Elongating roots usually possess four regions—the root cap, the meristematic region, the region of cell elongation, and the region of differentiation and maturation (see Fig. 5.1)—but these regions are not always clearly delimited. Although the root cap is composed of loosely arranged cells, it is usually well defined. However, it is absent from certain roots, such as the short roots of pine. Since it has no direct connection with the vascular system, it probably has no role in absorption. It is said to be the site of perception of the gravitational stimulus (Juniper, 1976), but this is debatable (Jackson and Barlow, 1981). Root cap development and function are discussed in detail by Barlow, in Torrey and Clarkson (1975).

The meristematic region typically consists of numerous small, compactly arranged, thin-walled cells almost completely filled with cytoplasm. Relatively little water or salt is absorbed through this region, largely because of the high resistance to movement through the cytoplasm and the lack of a conducting system. Growth in the apical portion of the meristematic region is probably limited by food supply because the phloem is not differentiated to the apex and food must move to it by diffusion through a thick layer of cells.

Near the root apex growth is by cell division, but usually there is a zone of rapid cell elongation and expansion a few tenths of a millimeter behind the apex. It is difficult to indicate a definite zone of differentiation because various types of cells and tissues are differentiated at different distances behind the root apex.

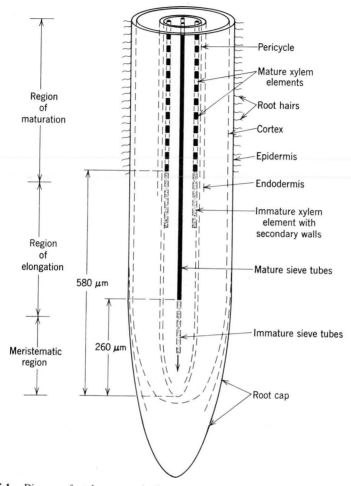

Fig. 5.1. Diagram of a tobacco root tip showing relative order of maturation of various tissue. The distance from the tip at which the various tissues differentiate and mature depends on the kind of root and the rate of growth. (Modified from Esau, 1941.)

Sieve tubes of *Phleum* mature within 230 μm of the tip, well within the zone of cell division, but xylem elements first become differentiated about 1000 μm behind the apex (Goodwin and Stepka, 1945). Phloem also becomes differentiated closer to the tip than xylem in roots of Valencia orange (Hayward and Long, 1942) and pear (Esau, 1943), and this is probably the typical situation (see also Esau, 1965, p. 503).

The control of growth in roots is obviously complex because cell division, enlargement, and differentiation occur somewhat independently. For example, tissues and cells often continue to differentiate after root elongation ceases, especially when cessation is caused by unfavorable environmental conditions. Branch root formation also often continues down to the apex, indicating that requirements for growth of branch roots are somewhat different from those for growth of the central axis. As a result, it is impossible to identify the zones of elongation and differentiation in terms of specific distances behind the root tips because their location depends to a considerable extent on the species and the rate of root elongation. In slowly growing roots, differentiation of tissues extends much closer to the root tips than in rapidly growing roots; if elongation ceases, differentiation extends almost to the apex, leaving a very short meristematic region. In *Abies procera,* mature protoxylem occurs only 50 μm from the tips in dormant roots, but as much as 5 mm behind the tips in rapidly elongating roots, where differentiation lags behind elongation (Wilcox, 1954).

Supported behind by the older, more rigid tissue and on the sides by soil particles, root tips are pushed forward through the soil by cell elongation at rates of a few millimeters to a few centimeters per day. Wilcox (1954, 1962) and others have observed rhythmic elongation of roots, and it has been reported that roots often grow more rapidly at night than during the day (Reed, 1939; Head, 1965). Apple root tips also show rhythmic nutational movements (Head, 1965), an occurrence observed by Darwin and Sachs in the nineteenth century and reviewed recently by Ney and Pilet (1981). The course of roots through the soil is often tortuous, as they follow the path of least resistance through crevices between soil particles and around pebbles and other obstacles. In spite of these temporary deflections, roots of at least some plants tend to grow outward in a generally straight line because the tips tend to return to their original direction of growth after being turned aside by obstacles. This tendency, termed exotropy by Noll in 1894, is very strong in the roots of some trees that grow straight outward for up to 25 m (Wilson, 1967). The cause of this behavior is unknown, but it is important because it results in widely spreading root systems.

As the newly enlarged, thin-walled cells at the base of the zone of enlargement cease to elongate, they become differentiated into the epidermis, cortex, and stele, which constitute the primary structures of a root. The arrangement of the principal tissues in a dicot root is shown in Fig. 5.2. A scanning electron micrograph of a cross section of a young barley root is shown in Fig. 5.3. The conductive tissues of roots usually form a solid mass in the center, instead of being dispersed in bundles around the periphery of the pith, as in stems of most herbaceous, dicotyledonous plants. This arrangement contributes to flexibility without loss of tensile strength. The primary xylem usually consists of two to several strands extending radially outward from the center, with the primary phloem located between them.

The narrow layer of parenchyma cells separating the xylem from the phloem later becomes the cambium and produces secondary xylem and phloem. The

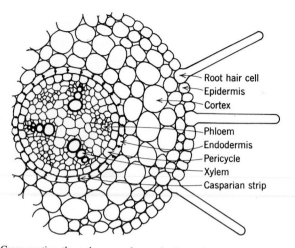

Root hair cell
Epidermis
Cortex
Phloem
Endodermis
Pericycle
Xylem
Casparian strip

Fig. 5.2. Cross section through a squash root in the region where water and salt absorption occurs. Later, the endodermal cell walls usually become much thickened, except for the passage cells opposite the xylem, which often remain unthickened. (After Crafts and Broyer, 1938.)

outermost layer of the stele is the pericycle. Its cells retain their ability to divide, and they give rise not only to branch roots but also to the cork cambium found in many older roots. The endodermis usually consists of a single layer of cells and forms the inner layer of the cortex. Because of its unique structure, it has received much attention from anatomists and physiologists (Esau, 1965, pp. 489–493; Van Fleet, 1961). Its role in water and solute absorption will be discussed later.

Epidermis and Root Hairs.   Considerable attention has been given to the epidermis, and to root hairs because of their presumed importance as absorbing surfaces. The epidermis is usually composed of relatively thin-walled, elongated cells that form a compact layer covering the exterior of young roots. Sometimes a second compact layer, the hypodermis, lies beneath the epidermis. The permeability of the walls of these two layers seems to vary considerably, depending on the amount of suberization (Clarkson *et al.,* 1978). The most distinctive feature of epidermal cells is the production of root hairs. These usually arise as protrusions from the external walls (Fig. 5.4).

In plants of some species, root hairs can arise from any or all of the epidermal cells, but in others they arise only from specialized cells known as trichoblasts (Cormack, 1944, 1945). In other plants, including citrus and conifers, they arise from the epidermis, the layer of cells beneath the epidermis, or even from deeper in the cortex (Kramer and Kozlowski, 1979, pp. 100–101). Development of root hairs is inhibited in very dry or very wet soil, and they are often absent on roots

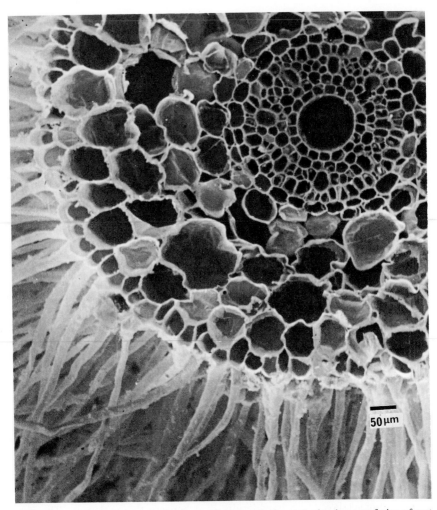

50 μm

Fig. 5.3. Scanning electron micrograph of a young barley root, showing a profusion of root hairs, a large central xylem vessel, and a few smaller xylem vessels. (Courtesy of Prof. A. W. Robards and A. Wilson, Department of Biology, University of York, England.)

growing in water culture. The number of root hairs varies widely among different kinds of plants, and they are said to be rare or absent on some kinds of roots, including avocado and pecan, but they are easily overlooked. Kozlowski and Scholtes (1948) reported 520 root hairs cm^{-2} on black locust roots and only 217 on loblolly pine roots, but Dittmer (1937) reported 2500 root hairs cm^{-2} on roots of winter rye. The abundance of root hairs on barley roots is shown in Fig. 5.3

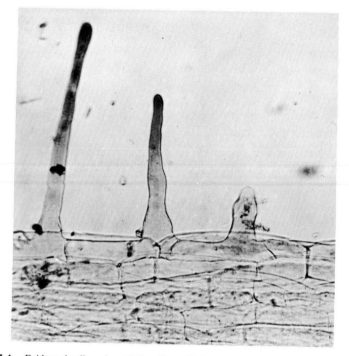

Fig. 5.4. Epidermal cells and root hairs. (From Weisz and Fuller, 1962. By permission of the McGraw-Hill Book Company, New York.)

Root hairs are generally short-lived, either dying or being destroyed by destruction of the epidermis and cortex during secondary growth. However, they are reported to persist for 10 to 12 weeks on some herbaceous plants (Weaver, 1925; Dittmer, 1937), and on some woody plants they become suberized or lignified and persist for months or years (Hayward and Long, 1942). On growing roots, the root hair zone usually migrates outward, new hairs being formed on the new epidermal cells and older ones being destroyed by suberization of older epidermal cells as the roots elongate.

It appears that root and root hair surfaces are covered with a hydrophobic lipid layer, or cuticle, and Scott (1963, 1964) reported that cell walls bordering intercellular spaces in roots are also covered with a lipid layer. The importance of this layer with respect to water and salt uptake has not been fully evaluated, but it is thin and contains so many pores that it is probably not a serious barrier to water and solute entrance. There are also said to be pits and plasmodesmata in the outer walls of epidermal cells (Scott, 1964). The root cap and root hairs are usually covered with a layer of mucilaginous material, the mucigel, apparently secreted chiefly by root cap cells. The function of this layer has not been determined, although it presumably improves the contact between roots and root hairs and soil particles (Jenny and Grossenbacher, 1963; Russell, 1977), and it certainly provides a medium favorable for growth of rhizosphere microorganisms.

Endodermis. A prominent feature of the primary structure of most roots is the endodermis, the inner layer of cells of the cortex which separates it from the stele. It becomes conspicuous as it matures because of extensive thickening of the inner tangential walls. However, this thickening occurs very unevenly in various parts of roots (Esau, 1965, p. 492). Early in its development, suberin is deposited in bands on the radial walls, forming the Casparian strips shown in Fig. 5.5, rendering them relatively impermeable to water and presenting a barrier to inward movement of water and solutes in the apoplast. However, there are numerous pits in the inner walls of the endodermis and plasmodesmata extend through these, connecting the protoplasts of the endodermal cells to those of the pericycle cells, facilitating symplastic transport (Clarkson and Robards, 1975). Also, in the roots of some plants, the endodermal cell walls opposite the xylem remain unthickened, facilitating transport across the endodermis.

In many kinds of roots, especially those of grasses but including some woody plants, the endodermis is pierced by numerous branch roots that provide temporary openings for lateral movement (Dumbroff and Peirson, 1971; Karas and McCully, 1973; Queen, 1967). The branch roots themselves provide possible pathways for lateral movement of water and solutes before suberization reduces their permeability. Whatever the pathway, water enters some kinds of roots in

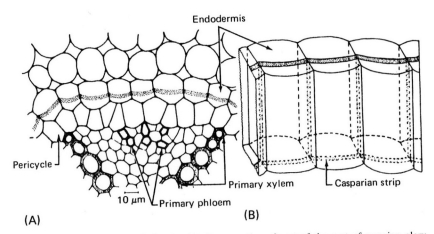

Fig. 5.5. Structure of endodermis. (A) Cross section of part of the root of morning glory (*Convolvulus arvensis*) showing location of the endodermis. (B) Diagram of endodermal cells, showing position of the Casparian strip on the radial walls. (Adapted from Esau 1977, by permission of John Wiley and Sons, New York.)

regions where the endodermis is suberized, as shown later in Fig. 5.8. Ferguson (1973) reported uptake and translocation of calcium and phosphorus 20–30 cm behind the tips of corn roots, where suberization had occurred. In contrast, Nagashi *et al*. (1974) reported that lanthanum, which does not enter the symplast, cannot pass the endodermis in young corn roots. However, calcium, which is also thought to move in the apoplast, entered over the entire length of corn roots in Ferguson's experiments.

Development and functioning of the endodermis was discussed in detail by Clarkson and Robards, in Torrey and Clarkson (1975, Chapter 19). The growth and development of roots were discussed in monographs edited by Kolek (1974) and Torrey and Clarkson (1975). Fayle (1968) made a detailed study of the growth and anatomy of tree roots, and there is much interesting information in the chapter by Head in Kozlowski (1974).

Secondary Growth

Secondary growth usually destroys the epidermis and root hairs, as shown in Fig. 5.6. Sometimes hypodermal cells become suberized, and a cork cambium or phellogen develops in the outer part of the cortex. In roots of Valencia orange and some other woody species, the hypodermal cells produce secondary root hairs. Lenticels and areas of thin-walled, radially elongated cells also develop, through which water is absorbed (Hayward and Long, 1942). Usually, as cambial activity increases the diameter of the stele, the cortex and epidermis collapse and disappear, the dying cells being destroyed by decay or by feeding of soil animals. Frequently, a cork cambium arises from the pericycle, and the roots become covered with a layer of corky tissue containing lenticels. In woody roots, successive cork cambia develop until the cortex disappears, and the arrangement of tissues in older woody roots becomes similar to that in woody stems. However, the bark of roots appears to be considerably more permeable to water than the bark of stems.

Root Contraction

A seldom mentioned aspect of root growth is the contraction which occurs in the roots of some species. This is brought about by change in shape of cortical cells which contract longitudinally while expanding radially, causing roots to become shorter and thicker. The vascular tissues are usually considerably distorted during contraction, and shoots are drawn closer to the ground and bulbs

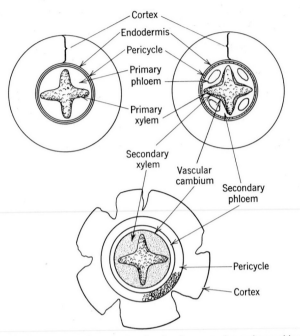

Fig. 5.6. Secondary growth of a root, showing development of vascular cambium and secondary phloem and xylem. Enlargement by secondary growth crushes the primary phloem and results in loss of the cortex and endodermis. (Adapted from Esau, 1965.)

deeper into the soil. It occurs in some bulbous plants and in rosette-forming plants such as dandelion. Root contraction is discussed by Esau (1977, p. 216) and by Wilson and Honey (1966).

THE ABSORBING ZONE OF ROOTS

In this chapter we are chiefly concerned with roots as absorbing organs, and it is important to determine how much of the root surface is available for the entrance of water and minerals. Water obviously will enter most rapidly through those regions offering the lowest resistance to its movement. However, the location of the region of lowest resistance varies with the species, the age and rate of growth, and sometimes with the magnitude of the tension developed in the water-conducting system.

Young Roots

Consideration of the anatomy of primary roots suggests that the absorption of water and minerals occurs chiefly in a region a few centimeters behind the root tip, roughly equivalent to the zone where root hairs are present. A number of studies have been made of the absorbing zone by attaching potometers at various distances behind the root tip. Little water enters through the meristematic region because of the high resistance offered by the dense protoplasm and lack of xylem elements to carry it away. Farther back the xylem is functional, but suberization and lignification of the hypodermis and especially of the endodermis seriously reduce the entrance of water and minerals (Fig. 5.7).

Potometer studies indicate that water absorption through growing corn roots more than 10 cm long increases to a maximum about 10 cm behind the tip and then decreases toward the base (Hayward *et al.*, 1942). In onion roots maximum absorption occurs 4–6 cm behind the root tips in roots over 7 cm long and decreases toward both the tip and the base. In roots less than 5 cm in length, there

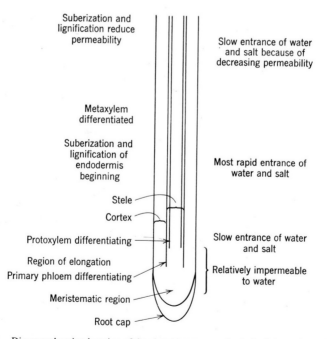

Fig. 5.7. Diagram showing location of the absorbing zone and principal tissues in an elongating primary root. The distance behind the root tip at which various degrees of maturation occur depends on the species and the rate of growth.

is greater absorption toward the base than toward the tip (Rosene, 1937). Sierp and Brewig (1935) found that maximum intake of water through roots of *Vicia faba* L. over 10 cm long occurred 1.5–8.0 cm behind the tip. When the rate of transpiration was increased, the absorbing zone was extended, and the region where most rapid absorption occurred shifted toward the base of the root. Brouwer (1953) (Fig. 7.8) and Soran and Cosma (1962) also observed a shift in the region of most rapid absorption toward the base of the root as stress was increased. Fiscus (1977) suggested that this shift can be explained in terms of coupled solute and water flow (Fiscus, 1975). There is a decrease in ion concentration in the root xylem sap and in osmotic movement with increasing distance behind the tip, accompanied by increasing pressure flow through the basal region. Figure 5.8 shows rates of water uptake at various distances behind the tips of barley and *Cucurbita pepo* roots. It appears that in at least some plants, suberization of the endodermis is not a serious barrier in the terminal 8–10 cm of rapidly elongating roots, and measurable absorption occurs after it is complete. In slowly elongating roots, maturation of tissues occurs much nearer the tip, shortening the region of low resistance to water movement.

Until recently, textbooks indicated that mineral absorption is restricted to a region near the root apex. This view was based chiefly on data for ion accumulation, such as those of Steward *et al.* (1942). However, experiments in which radioactive tracers are supplied to roots attached to plants at various points

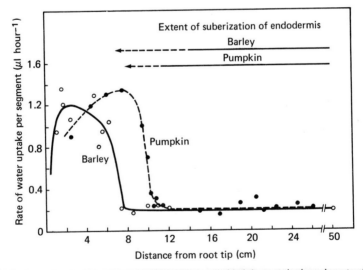

Fig. 5.8. Rate of water uptake at various distances behind the root tip through roots of barley and pumpkin (*Cucurbita pepo*). Note that there was measurable uptake in the region where suberization of the endodermis was complete. (From Agricultural Research Council Letcombe Laboratory Annual Report, 1973, p. 10.)

behind the tip indicate that uptake of ions occurs far behind the apex (Wiebe and Kramer, 1954; Clarkson *et al.*, 1975; Richter and Marschner, 1973). The difference between the zones of mineral accumulation in roots and mineral absorption through roots is shown in Fig. 5.9.

Root Hairs. It is often stated that root hairs are important or even essential for absorption because they increase the root surface area in contact with the soil. For example, the root hairs on the winter rye plants studied by Dittmer (1937) more than doubled the root–soil contact. This might be more important for uptake of relatively immobile elements, such as phosphate, than for uptake of mobile substances such as nitrate or water. Tinker (1976) suggested that root

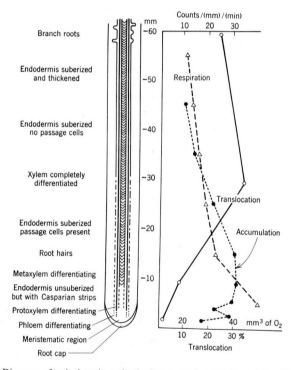

Fig. 5.9. Diagram of apical region of a barley root, showing the relationship between root structure and the regions where salt is accumulated and where it is absorbed and translocated to the shoot. Compare with region of maximum water absorption by barley roots, shown in Fig. 5.8. The curve for respiration is based on data of Machlis (1944). The curve for translocation represents the percentage of ^{32}P absorbed that was translocated away from the region where it was absorbed (Wiebe and Kramer, 1954). The relative positions of these regions are probably similar in most or all growing roots, but the actual distances behind the root tip will vary with the species (see Fig. 5.8) and with the rate of elongation. (From Kramer, 1969.)

hairs might be particularly useful in maintaining root–soil contact when drying soil and roots are shrinking, but it is likely that most root hairs collapse in drying soil. Newman (1974) concluded that resistance to longitudinal movement of water through root hairs may be higher than through the soil surrounding them, and R. S. Russell (1977, pp. 233–235) stated that there is not enough evidence to make a decision concerning the importance of root hairs in increasing the absorbing surface.

The Role of Older Roots

Most discussions of absorption deal with young roots and leave the impression that the older suberized roots do not function as absorbing surfaces. However, it is probable that a large part of the absorption of water and solutes by many perennial plants occurs through roots that have undergone secondary growth and are covered with layers of suberized tissue.

It has been noted by several observers that when the soil is cold or dry, or the soil solution becomes too concentrated, few or no unsuberized roots can be found. McQuilkin (1935) and Reed (1939) found few growing tips on various species of pine in dry or cold weather, and few or no white unsuberized root tips were found on citrus trees during the winter (Chapman and Parker, 1942; Reed and MacDougal, 1937). Pine and citrus trees lose large amounts of water by transpiration on sunny winter days and obviously, over a period of time, must be absorbing equivalent quantities of water through their suberized roots. Roberts (1948) found only a small percentage of unsuberized roots under a stand of *Pinus taeda*. In another study, Kramer and Bullock (1966) found in midsummer that an average of less than 1% of the root surface under stands of *P. taeda* and *Liriodendron tulipifera* was unsuberized. Head (1967) reported a large reduction in root growth of apple and plum in midsummer and concluded that a considerable proportion of the water and minerals must be absorbed through the older suberized roots.

Unsuberized roots are more permeable to water and solutes than suberized roots, but they constitute too small a percentage of the root surface to account for absorption of all the water and minerals required by trees. Long ago, Crider (1933) and Nightingale (1935) reported instances in which young trees possessing no actively growing roots were able to absorb sufficient water and minerals for growth. More recently, Chung and Kramer (1975), Kramer and Bullock (1966), and Queen (1967) measured significant uptake of water and phosphorus through suberized roots of woody plants. Table 5.1 shows that removal of unsuberized roots did not prevent absorption of water or minerals. It seems probable that the unsuberized root surface of many perennial plants is too limited

TABLE 5.1   Effects of Removal of Unsuberized Roots on Uptake of Water and ^{32}P through 1-Year-Old Loblolly Pine Seedlings under a Pressure of 31 cm Hg[a,b]

Description	Total surface area (cm^2)	Rate of H$_2$O uptake (cm^{-3} cm^{-2} s^{-1})	Rate of ^{32}P uptake (cpm cm^{-2} -hr)	Concentration factor
Unpruned root systems	147.3	4.69	333	0.478
Part of unsuberized root	112.0	4.28	239	0.365
surface removed	(24%)	(9%)	(28%)	(24%)
All unsuberized roots	86.0	3.61	178	0.324
removed	(42%)	(23%)	(47%)	(32%)

[a] From Chung and Kramer (1975).

[b] Numbers in parentheses are percentage reductions caused by pruning. Removal of all unsuberized roots reduces root surface by 42%, rate of water uptake by 23%, and ^{32}P uptake by 47%, indicating a high rate of uptake of water and salt through the suberized roots. The low concentration factor indicates existence of an effective ion barrier in suberized roots.

in extent and occupies too small a volume of soil to supply the water and minerals required by them (Chung and Kramer, 1975).

Pathway of Radial Water Movement

Having discussed the location of the absorbing zone, it seems appropriate to consider the path followed by water in moving from the surface of the root into the xylem elements. The pathway in primary roots will be discussed first, then the pathway in roots that have undergone secondary growth.

Primary Roots.   A somewhat diagrammatic cross section of a young root of an herbaceous plant is shown in Fig. 5.2 and a scanning electron micrograph in Fig. 5.3. The epidermis, and the hypodermis if present, are important because they are composed of very compact layers of cells containing no intercellular spaces. As mentioned earlier, their walls sometimes become more or less suberized, increasing resistance to the entrance of water and solutes. The cortical parenchyma usually consists of loosely arranged cells containing numerous intercellular spaces and even large lacunae, and its walls can be regarded as free or appoplastic space, accessible to solutes by diffusion and mass flow. The intercellular spaces are usually filled with gas, even in water culture, and infiltration by liquid results in abnormal growth (Burström, 1959, 1965).

As mentioned earlier, the endodermis seems to present a major barrier to water and solute movement because the radial walls are usually made impermeable by the Casparian strips. Thus, where the endodermis is intact, water and solutes must pass through its protoplasts to reach the stele. Flow through the endodermis must be several times more rapid than through the epidermis because of its much smaller circumference. As pointed out earlier in the section on endodermis, it often contains gaps that might permit passage of water, but on the other hand, it seems to be an effective barrier to mass movement of ions. Evidently, we need to know more about the structure and function of the endodermis in order to understand the inward movement of water and solutes in roots.

It seems that by this time we should know the exact pathway followed by water and solutes moving from the epidermis across the cortex and into the stele. Unfortunately, we are less certain about the pathway today than we were in 1970. There is general agreement that at some point most of the water must pass through living membranes because the rate of movement is greatly reduced by respiration inhibitors, high concentration of CO_2, low concentration of O_2, and low temperature. It is also generally agreed that at the endodermis much or all of the water must pass through protoplasts because the Casparian strips on the radial walls of the endodermal cells prevent movement of water through the walls. However, it is not certain how water moves across the 6 to 12 layers of cortical parenchyma cells lying between the epidermis and the cortex (Fig. 5.2).

There are three possible pathways for water and solute movement across the cortex: (1) through the vacuoles, (2) through the cytoplasm, and (3) through the walls (apoplast) as far as the endodermis (Fig. 5.10). During the nineteenth and

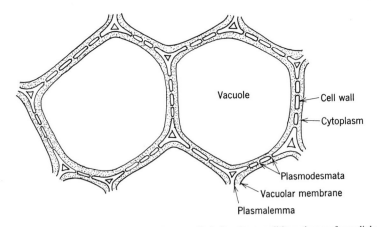

Fig. 5.10. Diagram of cortical parenchyma cells indicating possible pathways for radial water and salt movement into roots through the walls, through the symplast formed by plasmodesmata connecting the protoplasts of adjacent cells, or across the vacuoles. Movement may occur by all three pathways. The size of the plasmodesmata is greatly exaggerated (see Fig. 2.4).

early twentieth centuries, it was assumed that water moved from vacuole to vacuole across parenchyma tissue. However, Scott and Priestley (1928) pointed out that the soil solution can move as far as the endodermis by diffusion and mass flow in the microcapillaries of the cell walls and in films on their outer surfaces. This constitutes the free space or apoplast mentioned in Chapter 2. Later, Strugger's (1943, 1949) experiments with fluorescent dyes led him to conclude that water movement occurs in the walls of parenchyma cells of both roots and leaves. This conclusion was supported by the results of other experiments, such as those of Tanton and Crowdy (1972).

Attempts have been made to calculate the most likely pathway from estimates of the cross-sectional areas of the cell walls and the permeability of the walls and the protoplasmic membranes. Examples are those of Russell and Woolley (1961), who concluded that the wall is the pathway of least resistance, and of Tyree (1969), who also concluded that a significant amount of water moves through the walls. Later, Tyree and Yianoulis (1980) abandoned the cell wall pathway and concluded that most water movement from cell to cell occurs through the vacuoles, at least in leaves. Still later, Tyree *et al.* (1981) compared actual rehydration data with models using two compartments in series (the symplast pathway) or in parallel (apoplast and symplast). Data from one-half of the experiments fitted either model, but those of the other half fitted neither, so no conclusion could be drawn concerning the pathway. Molz and Ikenberry (1974) developed two diffusion equations, one for flow in the wall and the other for a cell-to-cell pathway, using a resistance–capacitance analogy to describe water flow. Their theory predicts comparable water flow in the wall and in the cell-to-cell pathway. Newman (1974, 1976) examined the data available and concluded that the resistance to movement in the wall is too high, movement across the vacuoles is questionable because of high membrane resistance, and movement in the symplast is most likely. However, this depends on the assumption that water can move freely through the plasmodesmata. Unfortunately, as Newman (1976) and Dainty (1976) agree, there are not enough reliable data on wall and membrane permeability of root cells to permit valid calculations concerning the pathway.

The experimental data published thus far do not provide convincing evidence concerning the pathway of water movement. The use of dyes is unsatisfactory because their presence in a structure does not prove that it is a major pathway for water movement. The dye may have diffused out into the structure from the major pathway. Diffusion of water labeled with deuterium into roots is relatively rapid, the half-time for living roots of *Vicia faba* being 36 s and for dead roots 6–12 s (Ordin and Kramer, 1956). The large increase in permeability of dead roots suggests that the protoplasmic membranes offer a much larger barrier to diffusion than the walls. Woolley (1965) also reported rapid diffusion of water labeled with tritium into roots of maize and a threefold increase in permeability after the roots were killed. He estimated that 70–80% of the movement was in the protoplasm. Raney and Vaadia (1965) found that the half-time for equilibration with THO of detached sunflower roots and attached roots on plants in darkness was only 30 s, but it was about 15 min for roots on transpiring plants. These data suggest that the transpiration stream bypasses the vacuoles of the root cells, and transpiration even decreases the rate of diffusion into them. Perhaps this occurs because when transpiration is rapid, water is removed from the root cortical cells, slowing down diffusion of THO into them. Biddulph *et al.* (1961) reported that there is slow equilibration of labeled water in roots and leaves and regarded this as evidence that a portion of the root and leaf tissue is not involved in movement of the transpiration stream. They also cited other investigators who reported slow equilibration of roots and stems with tracer water. Another factor bearing on the pathway of water movement is the volume of free space or apoplastic water in roots. According to what appear to be reliable estimates, it amounts to about 10% of the root volume (Levitt, 1957; Ingelsten and Hylmö, 1961). This free space seems to provide a significant pathway for apoplastic water flow. The writer suspects that the cross-sectional area of cell walls is generally underestimated because it shrinks during killing and fixation of material. However, it may also shrink somewhat when plants are water-stressed.

It seems impossible to rule out any of the pathways, and it is probable that some movement occurs by all three (Molz and Ikenberry, 1974). The writer favors the cell wall or apoplast as the major pathway. The resistance involved in crossing four membranes per cell when moving through the vacuoles seems excessive, and the small cross-sectional area of the plasmodesmata, about 1% of the area in common between adjacent cells (Tyree, 1970), makes the resistance high for mass flow of water through the symplasm in roots of transpiring plants. Nevertheless, Robards and Clarkson (1976) concluded that the plasmodesmata can carry the flux of water and solute that must occur across the endodermis. One hopes that in another decade sufficient information will be available to decide which pathway is most important.

Water Movement after Secondary Thickening.   The preceding discussion dealt with roots possessing an intact primary structure. As mentioned previously, during secondary growth the cortex, including the endodermis, usually disappears and the xylem becomes enclosed in a cylinder of phloem with layers of suberized cells on the outside, as shown in Fig. 5.11. Also, as mentioned earlier, these are often the only absorbing surfaces available on the root systems of perennial plants. No serious attempts have been made to study the path of radial movement in such roots, but it presumably occurs chiefly through parenchyma cells. However, the important ion barrier formed by the endodermis of primary roots has disappeared. It is reasonably certain that older roots contain an ion barrier, as shown by the low concentration factor in Table 5.1. However, no

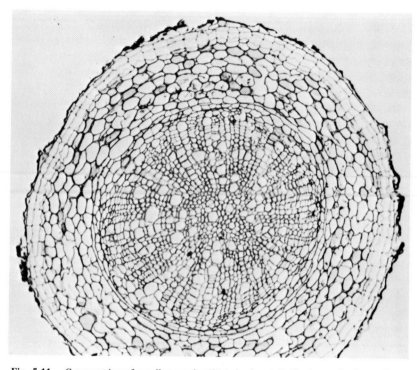

Fig. 5.11. Cross section of a yellow-poplar (*Liriodendron tulipifera*) root that has undergone secondary thickening. Much of the outer parenchyma tissue has sloughed off, and a layer of cork tissue has begun to develop. (Courtesy of R. A. Popham. From Kramer and Kozlowski, 1979, by permission of Academic Press.)

effort has been made to locate it. Chung and Kramer (1975) suggested that it might be the cambium, because the cells in the cambial region are thin-walled and provide the minimum free space for apoplastic movement of water and solutes. An example of a completely suberized but functional root system is shown in Fig. 5.12.

Mycorrhizae

The root systems and absorbing surfaces of most trees and herbaceous plants are invaded by fungi, forming associations called mycorrhizae. Two general types of mycorrhizae occur, ectotrophic and endotrophic or vesicular arbuscular. Combinations of the two also occur. Ectotrophic mycorrhizal development results in hypertrophy and dichotomous branching of roots (Figs. 5.13 and 5.14),

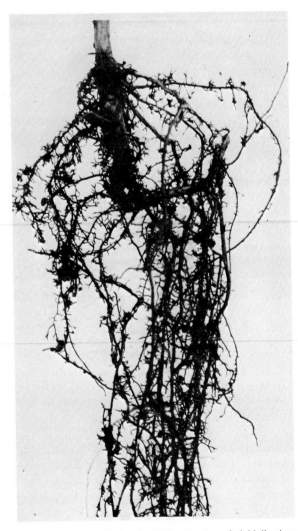

Fig. 5.12. Photograph of a completely suberized root system of a loblolly pine seedling. Numerous clusters of mycorrhizal roots are also shown. (From Kramer and Kozlowski, 1979, by permission of Academic Press.)

but endotrophic types develop within the root and have little effect on the external appearance. The mycorrhizal association is usually regarded as symbiotic, or beneficial to both the host and the fungus, but occasionally the fungus may become parasitic.

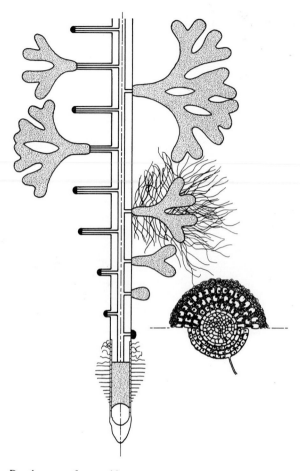

Fig. 5.13. Development of mycorrhizae on a pine root. The long root bears a normal root cap and root hairs, but mycorrhizal branches, similar to those shown in Fig. 5.14 are developing from the branches (short roots). The upper part of the cross section represents a mycorrhizal root, the lower part an uninfected root. (From Hatch, 1937.)

Ectotrophic Mycorrhizae. Most important tree species develop ectotrophic mycorrhizae, usually formed by *Basidiomycetes* (Kramer and Kozlowski, 1979, pp. 48–50; Marks and Kozlowski, 1973). Exceptions are *Acer, Liquidambar,* and *Liriodendron,* which bear endotrophic mycorrhizae. Trees most likely to develop ectotrophic mycorrhizae produce both long and short roots. Long roots elongate rapidly, seldom produce mycorrhizae, usually produce root hairs from the hypodermal layer, branch racemosely, are relatively long-lived, and undergo

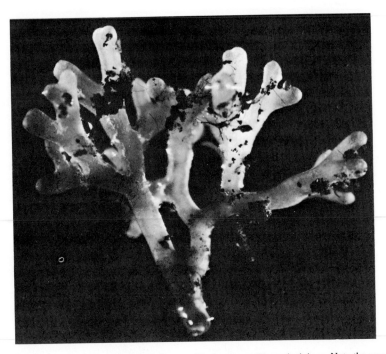

Fig. 5.14. Photograph of a cluster of mycorrhizal roots on *Pinus virginiana*. Note the mycelial strands and particles of soil adhering to the surfaces. (Courtesy of Edward Hacskaylo, U.S. Department of Agriculture.)

secondary thickening. Short roots elongate slowly, bear few root hairs, do not undergo secondary thickening, usually live one season or less, and often are invaded by fungi and become mycorrhizal. Such roots are usually thickened and blunt, branch dichotomously, and form the coralloid clusters shown in Fig. 5.14. The fungal hyphae grow between the cortical cells but do not usually penetrate into them. A feltlike covering or sheath of mycelium is often produced over the root surface, and individual hyphae and rhizomorphs (bundles of hyphae) extend out into the soil. Slankis (1973) suggested that the profuse branching of mycorrhizal roots is caused by auxin produced by the fungus, but other factors may be involved.

Björkman (1942) suggested that mycorrhizal development depends on a surplus of carbohydrates in the roots, and Wenger (1955) found that loblolly pine seedlings growing in the sun bore many more mycorrhizal roots than those growing in the shade. Hacskaylo discussed the role of carbohydrates in mycorrhizal development in Marks and Kozlowski (1973). They seem to develop most profusely where moderate deficiencies of minerals inhibit growth and allow carbohydrates to accumulate (Marx *et al.*, 1977). In very fertile soil, carbohydrates are used in vegetative growth and no surplus occurs in the roots. If there is a severe deficiency of minerals, or heavy shading that reduces photosynthesis, the carbohydrate deficiency in the roots suppresses mycorrhizal formation. However, it appears that mycorrhizal formation is a complex process depending on a variety of internal conditions (Slankis and Hacskaylo, in Marks and Kozlowski, 1973).

Endotrophic Mycorrhizae. These are also known as vesicular-arbuscular mycorrhizae, sometimes abbreviated as VA mycorrhizae, because the fungal hyphae penetrate the host cells and form vesicles and clusters of hyphae within them. The fungi belong to the Endogonaceae, a family of the order Mucorales. Although less well known, endotrophic mycorrhizae are much more common than ectotrophic mycorrhizae, occurring in mosses, ferns, gymnosperms, and angiosperms, including crop plants. They produce little or no visible modification of root morphology and form no fungal mantle on the root surface. Endotrophic mycorrhizae are often overlooked because the fragile hyphae extending out into the soil are easily broken, leaving no trace of their presence unless the roots are sectioned and stained.

Endotrophic mycorrhizae occur in some woody plants, including *Acer, Liquidambar, Liriodendron,* citrus, and some members of the Ericaceae, and they occur in most herbaceous plants, including crop plants (Gerdemann, 1975). Endotrophic and ectotrophic mycorrhizae sometimes, perhaps often, occur on the same kinds of plants.

Effects of Mycorrhizae. It is generally assumed that the beneficial effects of mycorrhizal roots result from the increased absorbing surface provided by the hyphae and rhizomorphus extending into the soil from the roots. It has been reported that mycorrhizal roots accumulate more minerals per unit of dry weight than nonmycorrhizal roots (Harley, 1956; Hodgson, 1954; Kramer and Wilbur, 1949). Bowen and Theodorou (1967) found that the rapidly elongating apical regions of pine roots accumulated phosphate as rapidly as mycorrhizal roots. However, they regarded the uninfected roots as functional for only a few days, whereas the mycorrhizal roots are functional for many weeks. Where extensive mycelial growth into the soil occurs, the absorbing surface is materially increased. It was demonstrated by Melin and his co-workers that various substances, including minerals, supplied to fungal hyphae attached to conifer roots were absorbed and translocated to the shoots (Melin, 1953; Melin *et al.*, 1958). Increased absorption of minerals, especially phosphorus, has also been demonstrated for crop plants (Russell, 1977, pp. 130–131), but the importance of this under field conditions needs further study.

Although mycorrhizae seem necessary for good growth of trees in phosphorus-deficient soils, they are not essential if an adequate supply of nutrients is available. Vigorous seedlings can be grown in sand cultures (Addoms, 1937; Hatch, 1937) and in nursery beds (Mitchell *et al.*, 1937; Hacskaylo and Palmer, 1957) without mycorrhizae if abundant nutrients are supplied. However, according to Gerdemann (1975), mycorrhizae are essential for good growth of citrus, and seedlings develop very poorly in soil where mycorrhizae-forming fungi have been eliminated by soil fumigation. Readers interested in more detailed discussions of the role of mycorrhizae in woody plants are referred to books by Harley (1969) and Marks and Kozlowski (1973). Maronek *et al.* (1981) reviewed the role of mycorrhizae for horticultural crops.

The presence of mycorrhizae decreased resistance to movement of water into soybeans (Safir *et al.*, 1972), but Sands *et al.* (1982) found no effect on resistance to flow into pine roots. Mycorrhizae are said to increase drought tolerance of tree seedlings (Marks and Kozlowski, 1973, p. 182), but the data are limited. Mycorrhizal roots are also said to be more resistant to root diseases and parasites because the fungi secrete fungistatic substances, forming a protective barrier, and their presence favors the development of protective organisms in the rhizosphere (Zak, 1964; Marx, 1969; Gerdemann, 1975, pp. 584–585; Marks and Kozlowski, 1973, Chapter 9).

It has been demonstrated that fungal hyphae associated with VA mycorrhizae can facilitate transfer of ^{32}P from one plant to another (Heap and Newman, 1980). Woods and Brock (1970) suggested that mycorrhizal fungi were involved in transfer of radioactive tracers from the roots of donor stumps to other vegetation in the vicinity. It has also been suggested that the mass of mycorrhizal fungi in the litter beneath tropical rain forests plays an important role in intercepting minerals released by decaying vegetation and returning them to the trees (Went and Stark, 1968; Stark and Jordan, 1978).

Gerdemann (1975) and Russell (1977) discuss the general benefits of mycorrhizal infection to cultivated plants. It has been shown that the growth of both annual and perennial plants in sterilized or fumigated soil can be improved by inoculating the soil with mycorrhizae-forming fungi.

SUMMARY

The role of roots in the absorption of water and minerals is well known, but they have other important functions. They are essential for the anchorage of plants in the upright position, and some roots store considerable amounts of food. They are also important synthetic organs. Roots convert inorganic nitrogen into organic nitrogen compounds, and synthesize growth regulators such as cytokinins and gibberellins, and other compounds such as nicotine. In turn, roots are dependent on the shoots for carbohydrates, auxin, and certain vitamins.

During growth and maturation, roots undergo extensive changes in anatomy that greatly modify their permeability to water and solutes. The radial walls of the endodermal cells develop bands of suberized tissue, and the epidermal cells and root hairs often collapse and die. During secondary growth the cortex, including the endodermis, sloughs off, and roots become enclosed in secondary phloem covered with a layer of suberized tissue. There is some uncertainty about whether water moves across roots chiefly from vacuole to vacuole, in the symplast, or in cell walls, and more information is needed concerning the relative conductivity of the three possible pathways. It is generally assumed that most water and solutes are absorbed near the tips of roots, but there is increasing evidence that considerable absorption occurs through the older regions, including completely suberized roots.

The roots of many plants are invaded by fungi, forming mycorrhizal associations that modify root structure. The ectotrophic mycorrhizae common in conifers and some other woody plants produce hypertrophy and extensive branching. The endotrophic types produce no important changes in root anatomy, but numerous hyphae extend from the roots into the surrounding soil. The importance of endotrophic mycorrhizae in crop plants is just beginning to be realized. They increase absorption of minerals, especially phosphorus, decrease resistance to the entrance of water into roots, and increase resistance to organisms that attack and injure roots.

SUPPLEMENTARY READING

Brouwer, R., Gasparikova, O., Kolek, J., and Loughman, B. G., eds. (1982). "Structure and Function of Plant Roots." Junk, The Hague.

Carson, E. W., ed. (1974). "The Plant Root and Its Environment." University Press of Virginia, Charlottesville.

Harley, J. L. (1969). "The Biology of Mycorrhiza." Leonard Hill, London.

Kolek, J., ed. (1974). "Structure and Function of Primary Root Tissues." Slovak Acad. Sci., Bratislava.

Marks, G. C., and Kozlowski, T. T., eds. (1973). "Ectomycorrhizae: Their Ecology and Physiology." Academic Press, New York.

Scott, G. D. (1969). "Plant Symbiosis." St. Martin's Press, New York.

Torrey, J. G., and Clarkson, D. T., eds. (1975). "The Development and Function of Roots." Academic Press, New York.

Whittington, W. J., ed. (1969). "Root Growth." Butterworth, London.

Development of Root Systems

INTRODUCTION

Chapter 5 dealt with the growth of individual roots and their function as absorbing surfaces. However, the amount of water and mineral nutrients available to plants depends first of all on the volume of soil with which their roots are in contact. Thus, the development of large root systems by vertical and horizontal extension and branching is important to the success of plants, especially when the soil water content falls much below field capacity. Development of root systems involves complex interactions between roots and shoots and between roots and their environment. For example, it is stated in textbooks that roots are positively geotropic, i.e., they grow downward in response to gravity. However, many roots grow outward horizontally, and occasionally roots even grow upward. Thus, various roots on the same plant respond quite differently to gravity, and no fully satisfactory explanation is available for these diverse responses. This chapter discusses the development of root systems and the plant and environmental factors that control their growth.

ROOT SYSTEMS

This section discusses different types of root systems, their rate of development and longevity, and modifications such as root grafts.

Depth and Spread

Among the best-known studies of development of root systems are those made by Weaver and his colleagues (Weaver, 1919, 1926; Weaver and Bruner, 1927; Weaver and Clements, 1938; Weaver *et al.*, 1922). Most of these studies were made in deep, well-aerated prairie soils, where roots penetrate to great depths. Corn and sorghum roots regularly penetrate to a depth of 2 m (Fig. 6.1a), alfalfa roots have been found at a depth of 10 m, and Wiggans (1936) reported that roots of 18-year-old apple trees had penetrated to a depth of at least 10 m and had fully occupied the space between the rows, which were about 10 m apart. Proebsting (1943) found that roots of various kinds of fruit trees growing in a deep soil in California penetrated at least 5.0 m, and the greatest number of roots occurred at a depth of 0.6–1.5 m. Hough *et al.* (1965) studied root extension by placing ^{131}I in the soil in a forest stand and measuring radioactivity in the surrounding trees. It was absorbed in detectable amounts from as far away as 17.0 m in longleaf pine and 16.5 m in turkey oak. Hall *et al.* (1953) used uptake of radioactive phosphorus to measure root extension of various species of crop plants, as shown in Fig. 6.1b. Recently neutron radiography has been used as a nondestructive method of studying root growth in thin layers of soil (Couchat, *et al.*, 1980). Böhm (1979) reviewed various methods used to study root systems.

The situation is very different with plants growing on heavy soils. Pears growing on a heavy adobe soil in Oregon had about 90% of their roots in the upper meter of soil (Aldrich *et al.*, 1935). Coile (1937) found that over 90% of the roots less than 2.5 mm in diameter occurred in the top 12.5 cm under pine and oak stands in the heavy soils of the North Carolina Piedmont. Even in sandy soils, trees often form mats of roots near the surface, probably because the surface soil contains more nutrients and is wetted by summer showers (Woods, 1957). An example of restriction of root penetration by a hardpan layer is shown later in Fig. 6.10.

The branching and rebranching of root systems often produce phenomenal numbers of roots, especially on grasses. Pavlychenko (1937) estimated that a 2-year-old plant of crested wheat grass possessed over 500,000 m of roots occupying about 2.5 m^3 of soil. Nutman (1934) estimated that 3-year-old coffee trees growing in the open possessed about 28,000 m of roots, 80% of which occurred in a cylinder 1.5 m deep and 2.1 m in diameter. Kalela (1954) reported that a 100-year-old Scotch pine possessed about 50,000 m of roots with about 5,000,000 root tips. A 6-month-old dogwood seedling grown in greenhouse soil bore over 5 m of roots; a loblolly pine seedling possessed only 0.38 m of roots (Kozlowski and Scholtes, 1948). Much information on the extent of root systems was summarized by Miller (1938, pp. 137–148), by Weaver (1926), and by

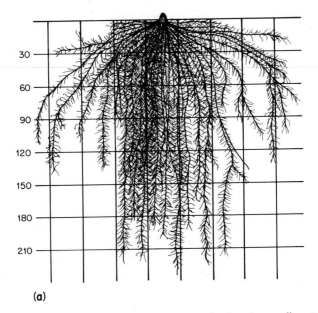

(a)

Fig. 6.1. (a) Root system of a mature corn plant growing in a deep, well-aerated soil. Each square is about 30 cm across. Where soil is too compact or too poorly aerated, roots do not penetrate so deeply. Refer to Figs. 6.9, 6.10, and 6.11 for other examples of depth of rooting. (b) Root extension of corn root systems growing in a clay loam soil, based on uptake of ^{32}P placed in the soil at various distances and depths from the seedling. The figures at the right are percentages of total ^{32}P absorbed from various depths; those across the bottom are the percentages absorbed at various distances from the corn plants. [(a) From Weaver *et al.*, 1922; (b) from Hall *et al.*, 1953.]

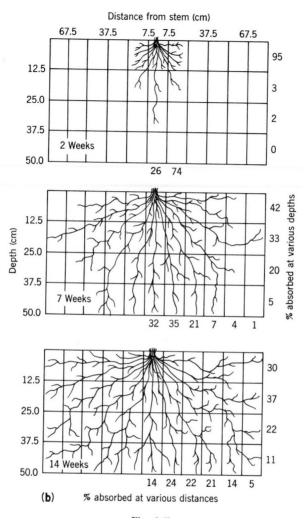

Fig. 6.1b

Taylor and Terrell (1982); Böhm (1979) published a book on methods of studying root systems.

The depth of rooting and extent of branching are important in choosing plants for soil stabilization and watershed cover. Deep-rooted species are preferable for stabilizing soil, but they also remove water from greater depths, so shallow-rooted species are preferable where maximum water yield is important (Hellmers *et al.*, 1955). Further discussion of depth of rooting occurs later in this chapter in the section on hereditary characteristics of roots.

Absorptive Capacity

Readers are reminded that the capacity of roots to absorb water and minerals does not increase in direct proportion to the increase in length or area of roots, because while new roots are being added, older roots are maturing and becoming less permeable. Thus, the actual absorptive capacity of root systems is the product of their surface area and their permeability or conductance. This can be written as

$$L_R = Lp\ A_R \qquad\qquad (6.1.)$$

where L_R in cm^3 s^{-1} bar^{-1} is the absorptive capacity or root system conductance, Lp is permeability or hydraulic conductance in cm^3 cm^{-2} s^{-1} bar^{-1}, and A_R is the root surface in cm^2 (Fiscus and Markhart, 1979).

Fiscus and Markhart (1979) made a quantitative study of changes in absorptive capcity (L_R) of a growing bean root system. The hydraulic conductance (Lp) of the root system changed in a complex manner with age and increasing root surface area (Fig. 6.2). There was a rapid increase in Lp in young root systems, followed by a decrease as the permeability (Lp) of older roots began to decrease. A small increase in Lp occurred later, probably caused by an increase in new roots. In another study, Fiscus (1981a) found that the conductance of entire bean root systems peaked at an area of 1000 cm^2 and then declined, probably because of increasing suberization of the older roots. He also found that the ratio of root

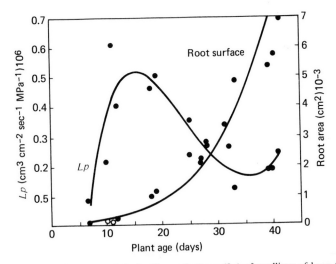

Fig. 6.2. Root surface area and hydraulic conductance (Lp) of seedlings of bean (*Phaseolus vulgaris*). Root surface area increased steadily with age, but the hydraulic conductance changed with increasing age of the root system. (From Fiscus and Markhart, 1979.)

to leaf surface area was relatively constant for growing plants after the root surface area exceeded 1000 cm².

By midseason, the roots of crop plants have usually completely occupied the surface soil and are invading deeper soil horizons. Whether further root branching in the surface soil will result in increased absorption of water and minerals depends on whether it will increase the absorptive capacity in that region. Apprently, under at least some crops the soil already is so fully occupied by roots that loss of some absorbing capacity can occur without affecting yield (Fig. 6.3).

The situation is different under young forest stands, especially those on shallow soils, where competition for water and minerals is particularly intense. Here reoccupation of soil already containing roots may be beneficial because rooting is usually less dense under forest trees than under crop plants. However, the benefits may depend in part on the effectiveness of mycorrhizal fungal hyphae as absorbing surfaces, a topic discussed in Chapter 5.

The effectiveness of a deep, wide spreading root system depends in part on its longitudinal or axial conductance. This is usually assumed to be high (Stone and Stone, 1975a), and Veihmeyer and Hendrickson (1938) stated that water 6 m

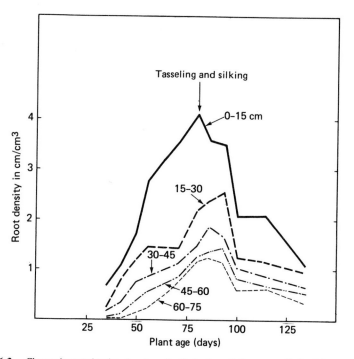

Fig. 6.3. Change in root density at various depths in the soil, in cm cm⁻³ of soil for corn during an entire growing season. (From Mengel and Barber, 1974.)

from a tree was absorbed just as readily as water 2 m from the tree. Also, it has been reported that corn (Reimann *et al.*, 1946) and some grasses (McWilliam and Kramer, 1968) can absorb sufficient water for growth from a depth of more than 1 m after the surface soil is dried to the permanent wilting percentage. Sands *et al.* (1982) reported that longitudinal (axial) resistance to water movement in bean and pine roots varies with the species and with root development but is negligible compared with radial resistance. On the other hand, Wind (1955) and Passioura (1972) found high resistance to longitudinal or axial water movement in certain grasses and in wheat. Some diverse values for longitudinal or axial resistance are given by Taylor (1980, p. 80), and the problem is discussed further in Chapter 7.

Rate and Periodicity of Root Growth

The rate of extension of root systems varies widely among species and with soil conditions. The principal vertical roots of corn have been observed to grow downward at a rate of 5–6 cm/day for 3 or 4 weeks (Weaver, 1925). The shoots were probably growing twice as rapidly at this time. A rate of 12 mm/day is common in roots of grasses, but Reed (1939) found that the rate for pine roots was usually less than 2.5 mm/day. Barney (1951) found that the maximum rate of elongation of roots of loblolly pine seedlings was 3.4–5.2 mm/day at the optimum temperature, which was 20°–25°C, and Rogers (1939) reported a rate of about 3 mm/day for apple roots. According to Wilcox (1962), roots of incense cedar grow 1–2 mm/day in the autumn and 3–5 mm/day in the spring. The most rapid rate observed was 7 mm/day.

It has been reported in a few studies that root elongation is slower during the day than at night (Reed, 1939; Lyr and Hoffmann, 1967). This behavior is most likely when the rate of transpiration is high enough to produce daytime water stress.

There has been some uncertainty concerning the existence of autonomic seasonal rhythms in root growth because it is not always clear whether the cycles often observed are caused by internal or external factors. Romberger (1963) and Lyr and Hoffmann (1967) reviewed the early literature in this field. Turner (1936) observed root growth of loblolly and shortleaf pine in Arkansas every month in the year. Least growth occurred in the winter, most in spring and autumn, and little during dry periods in the summer. These species also grew every month in the year at Durham, North Carolina, as shown later in Fig. 6.12, but their growth was greatest in April and May and least in January and February (Reed, 1939). The periods of slowest growth in summer coincided with periods of lowest soil moisture. In colder climates there is complete cessation of root growth during the winter, although root growth sometimes continues on seedlings brought into the greenhouse (Stevens, 1931). On the other hand, seedlings of incense cedar (*Callocedrus decurrens*) kept under favorable temperature and photoperiod made little root growth from December to April (Wilcox, 1962). This suggests the operation of an internal control.

Another indication of some kind of internal control over root growth is the fact that seedlings transplanted at one season develop new roots better than when transplanted at another. For instance, some hardwoods produce new rootlets more rapidly when transplanted in the spring than when transplanted in the autumn. It is reported that seedlings of several western conifers produce few or no roots when transplanted during the summer, and root regeneration increases during the autumn and is greatest about the time of bud opening in the spring (Kozlowski, 1971, Chapter 5). There also seem to be environmental effects on the capacity of seedlings to regenerate roots. For example, in California, fir seedlings (*Abies concolor* and *A. magnifica*) transplanted to the field in 1976 and 1977 showed good survival in spite of droughts, but in 1978 only 30% of the seedlings survived. Controlled environment studies indicate that the unusually warm weather of late autumn and early winter of 1977 reduced the capacity of the seedlings to produce new roots (Stone and Norberg, 1979). We need to learn what kind of growing conditions will produce seedlings with the greatest capacity for root regeneration, because seedling survival usually depends on rapid development of new roots after transplanting.

Root Longevity and Replacement

Roots differ widely in their life span, and there seems to be little relationship between the longevity of roots and the plants on which they grow. Some fine lateral roots on trees live only a week or two, whereas other roots live as long as the plant. In general, a large amount of photosynthate is used in replacement of short-lived roots.

Longevity of Roots. Although trees have the largest and oldest roots, many tree roots are short-lived. Many of the short roots on apple and other fruit trees are ephemeral and die after a week or two, and there is increasing evidence that many or most of the fine roots in forest stands are replaced each year. There are various causes of this short life, including heredity, attacks by nematodes, small insects, and fungi, and unfavorable physical conditions such as droughts or flooding. Woods (1980) concluded that all of the fine roots in a 60- to 65-year-old deciduous forest in New England die each winter, and Grier *et al.* (1981) reported a very high rate of turnover in fine roots in a mature forest of *A. amabilis* in the Washington Cascades. High rates of root replacement are also reported for shrubs in cool deserts and mixed deciduous forests (Caldwell, 1976).

The growth and death of roots of woody plants is discussed in more detail in Chapter 5 of Kozlowski (1971, Vol. II), by Head, in Kozlowski (1974), and by Lyford, in Torrey and Clarkson (1975).

Among grasses and cereals, the primary or seminal roots that are produced during germination and seedling growth are usually supplemented or even supplanted by an extensive system of adventitious secondary roots that arise from the lower nodes of the stem. The size of the root system of annual plants usually increases rapidly during the period of vegetative growth, then ceases or even decreases in dry weight during seed filling, and some of the roots often die before the plants mature (R. S. Russell, 1977). An example is shown in Fig. 6.3. It is sometimes stated that primary or seminal roots of grasses are short-lived, but this is not always true. Weaver and Zink (1946) found that at least a part of the seminal roots of 14 species of perennial grasses survived for one or two seasons, and some were alive after three seasons. Stuckey (1941) observed that whereas some kinds of grasses produce new root systems each year, the root systems of others are perennial.

Cost of Root Replacement.   Research in recent years indicates that a surprising amount of photosynthate goes into root growth. Caldwell (1976) cited data indicating that over 50% of the annual net primary production of a deciduous forest and a fescue meadow and about 75% of the annual production of shortgrass prairie and shrub steppe communities are used in the growth of new roots. Persson (1979) stated that over 50% of the carbohydrate produced by a Scotch pine stand in Sweden is used to develop new fine roots. Woods (1980) concluded that the annual turnover in root biomass of a 60- to 65-year-old deciduous forest in New England exceeds that of leaves. Grier *et al.* (1981) also reported a very high turnover of fine roots in a mature forest of *A. amabilis* in the Cascade Mountains of Washington. Harris *et al.* (1977) reported that root production is 2.8 times greater than above-ground wood production in both pine and hardwood forests of the southeastern United States.

The data on size of root systems and on longevity and rate of replacement raise two questions. First, why do many plants possess more roots than are necessary for survival or even for good growth, and second, why are so many roots replaced so often?

There is little doubt that many plants produce more roots than are necessary for survival. Trees often survive the loss of one-half or more of their root system during excavation for roads, sidewalks, and buildings, and trees and shrubs survive the loss of many roots during transplanting. Weaver and Zink (1946) reported that one-half of the roots could be removed from some native grasses without apparent injury. More important is the fact that the total root mass of some crop plants is reduced during the reproductive phase without any injury (R. S. Russell, 1977). Figure 6.3 shows that the root density of maize was reduced by 40–50% during the period from tasseling to maturity, yet increase in dry weight of shoots continued, suggesting that reduction in root density had not been injurious. However these plants were growing in a deep, fertile soil, with adequate soil moisture, and root density never fell below 2 cm/cm³ in the top 30 cm of soil. Taylor (1980) stated that a root density of 0.2–0.3 cm/cm³ was adequate for cotton and soybeans, but Jordan and Miller (1980) claimed that increasing the root density of sorghum at a soil depth below 50 cm to 1–2 cm/cm³ would result in increased water absorption from deep in drying soil and postpone dehydration. This is probably most important in areas where crops must be matured entirely or chiefly on water present at the beginning of the growing season. Thus, what constitutes an adequate root system depends at least in part on the rainfall pattern.

There is probably an unnecessary amount of food diverted into the roots of many plants when they are grown with an abundance of water. Deep, much-branched root systems are important for good growth and even for survival in years when severe droughts occur and in soils where minerals are deficient. However, where crops are grown with irrigation and fertilization, they can thrive with smaller root systems than are needed in the absence of irrigation. This fact needs to be taken into account in plant-breeding programs.

Caldwell (1976) discussed the advantages and disadvantages of root extension in detail and was unable to find a fully satisfactory explanation for the amount of energy used in root growth. Reynolds, in Torrey and Clarkson (1975), argued that short-lived roots are efficient because their disappearance during periods of drought reduces use of food in respiration at times when they cannot absorb water, but this neglects the large amount of food required to replace them. Part of the concern over this apparent waste of energy arises from the current emphasis on optimization. There seems to be a belief that successful plants must have attained a condition of maximum efficiency in respect to structure and functions. All that is really necessary is that plants possess a combination of characteristics in which the advantageous ones outweigh the disadvantageous ones (Bradshaw, 1965). Most plants have a higher potential rate of photosynthesis than is generally used (Kramer, 1981, p. 32), and diversion of a large amount of photosynthate into root growth does not necessarily constitute a serious drain on their carbohydrate supply. On the other hand, an excess of roots and frequent replacement of absorbing roots are occasionally advantageous in respect to water and mineral absorption. To expect a perfect balance in processes and functions in plants is as unreasonable as to expect a perfect balance in the economic and social functions of a state.

Root Grafts

The extent of the root system of an individual tree is sometimes increased by natural grafting to the roots of adjacent trees. Bormann and Graham (1959) found so many root grafts in stands of white pine that they regarded the entire stand as a physiological unit, and Kozlowski and Cooley (1961) found a similar situation in stands of both angiosperms and gymnosperms. An example of extensive root grafting is shown diagrammatically in Fig. 6.4. Root grafting is also common in tropical trees (LaRue, 1952). The grafts provide pathways for translocation of water, solutes, and even fungus spores from one root system to another (Kuntz and Riker, 1955). Bormann and Graham (1960) found that 43% of the untreated trees in a 30-year-old white pine plantation were killed by "backflash" through root grafts when the plantation was thinned with ammonium sulfamate. Root grafts are also common in plantations of Monterey pine in New Zealand (Will, 1966) and in slash pine in Florida (Schultz, 1972). Root systems and stumps of

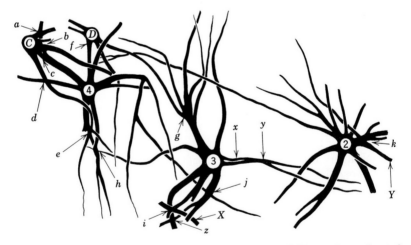

Fig. 6.4. Root grafting among roots of three 18-year-old trees of *Pinus radiata* and roots from living stumps of two trees removed 9 years previously. Grafts *a* through *g* were between trees 3 and 4 and roots of stumps C and D. Grafts *h* through *k* were between roots of trees 3 and 4 or between trees 3 and 4 and roots *X* and *Y* of trees removed during root excavation. Grafts *x, y,* and *z* are between two roots of the same tree. (From Will, 1966.)

cut trees sometimes survive on carbohydrates from intact trees to which they are grafted, and apparently such root systems function to at least a limited degree as absorbing surfaces for the attached intact trees. Root grafting in trees was discussed in Kramer and Kozlowski (1979, pp. 50–51) and in Kozlowski (1971, Vol. II, Chapter 6). Apparently, it occurs only rarely in herbaceous plants, probably because the roots are too short-lived for grafts to develop. However, Bormann (1957) found that when roots of tomato plants became firmly intertwined, water moved from one plant to the other even when no grafting occurred.

It appears that materials can be transferred between adjacent roots by grafting, through connecting fungal hyphae, and by diffusion of materials from root to root through the soil. Woods and Brock (1970) reported that radioactive calcium and phosphorus supplied to stumps of red maple were found in the foliage of 19 other species occurring at distances of up to 8 m from the donor stumps. They also cited several papers on transfer of herbicides and other organic compounds from plant to plant through roots. This led them to suggest that the root mass of an ecosystem should be regarded as a functional unit rather than an association of independent entities.

INTERNAL FACTORS AFFECTING THE DEVELOPMENT OF ROOT SYSTEMS

The manner in which a root system develops depends both on internal factors such as its hereditary potentialities and shoot growth and on environmental factors such as soil texture, depth, moisture content, aeration, kind and concentration of solutes and competition with other roots. These factors will be discussed briefly here and are reviewed in more detail in Carson (1974) and in R. S. Russell (1977).

Hereditary Characteristics

The type of growth of root systems of seedlings of many species is determined by heredity, although it may be modified later by environmental factors. There are also important interactions between shoots and roots which affect the size of root systems. An example of the interaction of hereditary and environmental factors on root development is shown in Fig. 6.5. The importance of hereditary factors in controlling root development is seen where a number of species grow in the same soil (Fig. 6.6). Some species always develop tap root systems, and others always develop fibrous root systems. Some species are always deep-

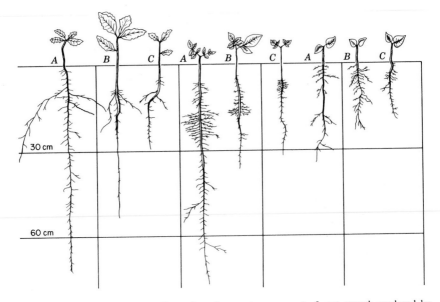

Fig. 6.5. Interaction of heredity and environment on amount of root growth produced by seedlings of three tree species grown in three different environments. Seedlings of *Quercus rubra* (left), *Hicoria ovata* (center), and *Tilia americana* (right). Seedlings *A* were grown in an open prairie habitat, *B* in an oak forest, and *C* in the deep shade of a moist linden forest. The linden seedlings in the prairie habitat were watered to prevent death from desiccation. Oak developed the deepest and largest root system in all three habitats, linden the shallowest. The shade of the linden forest greatly reduced the size of the oak root systems. (From Holch, 1931.)

rooted and others always have shallow root systems, but there are still other species (Fig. 6.7) which develop different types of root systems in different kinds of soils (Toumey, 1929; Weaver and Clements, 1938).

Differences in type of seedling root systems can significantly affect establishment and survival of seedlings. Baldcypress [*Taxodium distichum* (L.) Rich] and yellow birch (*Betula alleghaniensis* Britton) can become established only in moist soil because their shallow root systems do not enable them to survive droughts. In contrast, upland species such as oaks and hickories typically develop tap roots that penetrate deeply into the soil and provide water for the shoots even after the surface soil has become dry during summer droughts (Toumey, 1929). Roots of bur oak seedlings (*Quercus macrocarpa* Michx.) which grow on dry ridges penetrated 1.7 m in the first season, but roots of linden (*Tilia americana* L.) penetrated only about 0.3 m in the same soil, although they spread out laterally (Holch, 1931). As a result of this difference, most linden seedlings in open prairie soil died during summer droughts, whereas the deeper-rooted bur oak seedlings survived. Albertsen and Weaver (1945) concluded that survival of

Fig. 6.6. Differences in spread and depth of root systems of various species of prairie plants growing in a deep, well-aerated soil: *h, Hieracium scouleri; k, Koeleria cristata; b, Balsamina sagittata; f, Festuca ovina ingrata; g, Geranium viscosissimum; p, Poa sandbergii, ho, Hoorebekia racemosa; po, Potentilla blaschkeana.* (From Weaver, 1919.)

trees in the prairie region of the United States during the prolonged drought of the 1930s was largely dependent on depth of rooting. In the southeastern United States, shallow-rooted dogwood suffers much more from summer droughts than its deeper-rooted neighbors. In contrast is the view of Kummerow (1980) that in the shallow soils of the coastal region of California, root adaptations are less important than leaf adaptations for survival of the woody chaparral vegetation.

The success of cultivated plants subjected to drought is often dependent on development of deep, profusely branched root systems that absorb water from a large volume of soil. This was emphasized by Hurd (1974) for wheat, by Taylor (1980) for cotton and soybeans, and by Jordan and Miller (1980) for sorghum. General observations indicate that shallow-rooted crops such as potatoes, onions, and lettuce suffer more from droughts than deeper-rooted species such as alfalfa, maize, sorghum, and tomato. According to Burton, deep rooting is an important factor in drought tolerance of grasses. For example, the greater drought tolerance of coastal Bermuda as compared with common Bermuda grass (*Cynodon dactylon* Pers.) results from the much deeper rooting of the former (Burton *et al.,* 1954). A great deal of information concerning the root systems of vegetable crops was summarized by Weaver and Bruner (1927), and little work on those crops seems to have been done since.

Cassell (1983) suggested that improved capacity of roots to penetrate plow pans and soils of high bulk density would contribute to improvement of the absorbing systems of crops. It seems likely that one of the best possibilities for

Fig. 6.7.   Effects of soil conditions on development of the adaptable root systems of red maple (*Acer rubrum*) seedlings. (Left) A typical seedling from a swamp. (Right) A seedling from a well-drained upland site. (After Toumey, 1929.)

plant-breeding programs to increase drought tolerance is the development of root systems which penetrate more deeply into unfavorable soil. Unfortunately, very little is known about the reasons for hereditary differences in depth of rooting. Perhaps there are differences in tolerance of the decreasing oxygen supply with increasing soil depth, but this has never been investigated. Nor is much known about the reason why roots of some kinds of plants overcome mechanical resistance and penetrate dense soils better than others. More information about the physiological processes limiting root penetration in the soil seems necessary before much progress can be made in a breeding program for deeper-rooting plants to be used in regions subject to drought. Mechanical hindrance of root growth by compact soils is discussed by R. S. Russell (1977, Chapter 8) and examples are shown later in Figs. 6.9 and 6.10.

Root–Shoot Relationships

Roots are dependent on shoots for carbohydrates, growth regulators, and organic substances such as thiamin and niacin; shoots are dependent on roots for water, minerals, and growth regulators. Successful growth of plants therefore depends on maintenance of a balance in growth and function between roots and shoots such that neither organ suffers serious deficiencies in the contribution of essential substances from the other. Thus, considerable emphasis is often placed on the root–shoot ratio as an indicator of a satisfactory balance between the two.

Some data on root–shoot ratios are shown in Table 6.1, but such data are not very reliable because various investigators use different methods for recovering roots. It should be noted that *Medicago* (alfalfa) and *Zea mays,* which have large root–shoot ratios, are also relatively drought-tolerant. The root–shoot ratio is sometimes regarded simply as the result of a source–sink relationship in which lack of carbohydrate limits root growth and lack of water and minerals limits shoot growth. This explanation seems reasonably adequate during vegetative growth of many species, but it does not explain why so much food is diverted into roots of sugar beet and other root crops. This must be controlled by heredity, but the mechanism by which heredity controls partitioning of food remains unknown. Nor does a simple source–sink system explain why, during the reproductive stage, so much food formerly translocated to roots is translocated to fruits and seeds that root growth slows and parts of root systems sometimes die

TABLE 6.1 Amounts of Dry Matter in Metric Tons per Hectare Incorporated Annually into Roots and Shoots of Various Plant Species[a]

Species	Roots	Shoots	Root–shoot ratio
Zizania aquatica	0.6	4.0	0.15
Hordeum	3.0	12.0	0.25
Andropogon scoparium (1st year)	3.5	14.2	0.25
Triticum (average)	2.0	6.8	0.29
Medicago sativa (average)	3.2	7.4	0.43
Zea mays (average)	4.5	8.7	0.52
Solanum tuberosum (average)	4.0	2.6	1.54
Beta (average)	9.5	3.1	3.06
Pinus sylvestris (average)	1.6	8.9	0.17
Picea abies (average)	2.1	11.9	0.18
Fagus sylvatica (average)	1.6	8.2	0.19
Ghana rain forest	2.6	21.7	0.12

[a] From Bray (1963).

(Fig. 6.3). Perhaps growth regulators synthesized by the new sinks gain control of partitioning and translocation and divert food from the roots to the developing seeds and fruits. A better understanding of how the partitioning of photosynthate is controlled is badly needed.

Effects of Shoots on Roots.   Because root growth depends on a supply of carbohydrate from the shoots, shading and reduction in leaf area usually reduce root growth. Overgrazing of pastures and too frequent cutting of hay crops can reduce the dry weight of roots to as low as 10% of that of control plants (Weaver and Darland, 1947). Shading usually reduces both the absolute size of root systems and the root–shoot ratio, but there are exceptions. Kozlowski (1949) found that the root system of *Quercus lyrata* Watt. was reduced only slightly by an amount of shading that reduced the weight of roots of *Pinus taeda* L. to 25% of the weight of the roots in unshaded seedlings and the root–shoot ratio by about 30%. Fiscus (1981a) reported that a 25% reduction in light intensity did not reduce the ratio of root surface to leaf surface in bean, although plant size was reduced.

Development of fruits and seeds often reduces root growth. For instance, a heavy crop of coffee sometimes reduces the carbohydrate reserves in the roots so much that many die (Nutman, 1933), and root growth of tomato is reduced during fruiting (Hudson, 1960). Eaton (1931) reported that both root dry weight and root–shoot ratio of cotton were nearly tripled by preventing boll and branch formation. Growth of corn roots is reduced by ear formation, but if the ears are removed, root growth continues until frost occurs (Loomis, 1935). In a more recent study, it was found that the weights of corn root systems actually decrease after silking and tasseling (Mengel and Barber, 1974). Some data from their paper are shown in Fig. 6.3. Similar results have been reported for oats (R. S. Russell, 1977, p. 49), and the depressing effect of vigorous shoot growth on apple root growth is shown in Fig. 6.8. According to Van der Post (1968) appearance of flowers on cucumber stops root growth. In his experiments, root growth stopped before fruits were large enough to become significant sinks, suggesting that cessation of root growth was not caused merely by competition for food.

Effects of Roots on Shoots.   Some effects of roots on shoots, such as their role in absorption and anchorage, seem obvious, but others are less obvious. It is not surprising to learn that damage to root systems serious enough to reduce water and mineral absorption inhibits shoot growth. However, it is surprising to learn that the mineral content of the leaves and the quality of citrus fruit are affected by the rootstock on which the trees are growing, yet this effect has been observed all over the world (Sinclair and Bartholomew, 1944; Haas, 1948). For example, greater amounts of soluble sugars and total acids are found in oranges

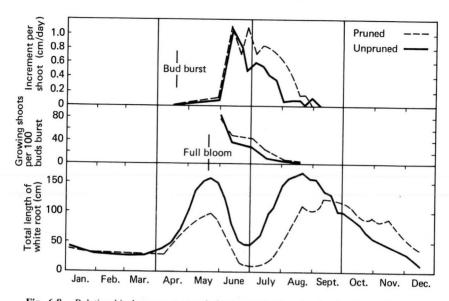

Fig. 6.8. Relationship between root and shoot growth of apple. As shoot growth increased (upper curves), the production of new roots decreased (lower curves). Pruning stimulated shoot growth and reduced midsummer root growth. (From Head, 1967.)

from trees grown on citrange and trifoliate orange root stocks than in fruit from trees grown on rough lemon roots. In other experiments, it was found that juice of the Washington navel orange is much less bitter in fruit grown on trifoliate orange roots than in fruit grown on other rootstocks, such as sour orange, sweet orange, rough lemon, or its own root system. The reasons for these effects are unknown, although Skene and Antcliff (1972) argued that growth regulators formed in the roots play a role in grapes.

It is well known that root stocks show differences in disease resistance and in tolerance of salinity, flooding, and cold. Horticulturists have known for centuries that the kind of rootstock on which scions are grafted or budded affects the size and vigor of trees. The dwarfing effect of M9 and certain other rootstocks on apples is well known, and pears are often grafted on quince roots to dwarf them. Dwarfed trees usually bear fruit at an earlier age and often produce a larger crop in proportion to the size of the top than do trees of normal size. For example, apple trees on M9 rootstocks usually grow to only one-third of normal size and begin to produce fruit in 2 or 3 years. On the M9 rootstock, about 70% of the photosynthate is directed to the fruit, whereas only 40 or 50% goes to the fruit in normal trees. More information about dwarfing rootstocks and the reciprocal effects between roots and shoots can be found in Tubbs (1973) and in a review by Lockard and Schneider (1981). The latter proposed that auxin moving downward through the phloem is degraded to different extents in different cultivars, causing differences in root metabolism that affects shoot growth. However, the physiology and biochemistry of root-shoot interactions are poorly understood.

ENVIRONMENTAL FACTORS AFFECTING ROOT GROWTH

The ability of soil to support vegetation depends chiefly on its suitability as a medium for root growth. This depends on soil texture and structure, which in turn affect physical properties such as water-holding capacity and aeration, and on chemical properties such as pH, available minerals, and toxic substances such as an excess of aluminum. Sometimes biological factors become important because there are important interactions between roots and soil organisms. Substances formed by roots and microorganisms affect aggregation of soil particles and soil structure, and substances released by living roots or during the decay of dead roots can affect the root growth of other plants and subsequent crops (see the section on allelopathy).

Physical Properties of Soil

The success of plants depends to a considerable extent on the volume of soil occupied by their root systems. Soil texture and structure play important roles in root growth, both directly by restricting root penetration and indirectly through effects on available water content and aeration.

Soil Texture. It was stated in Chapter 3 that the texture of soil depends on the relative proportions of sand, silt, and clay, which in turn affect the amount of available water and the volume of air space in the soil. As shown in Figs. 3.8 and 3.9, sandy soils have a limited water-storage capacity, whereas clay soils contain more available water but often limit root growth by deficient aeration and high physical resistance to root penetration. The texture and water-holding capacity often change with depth, as shown in Table 6.2 and in Figs. 3.7 and 3.11.

Increasing clay content is generally accompanied by decrease in average size of the soil pores. As a result, more of the pore space is occupied by water (noncapillary pore space) and less by air, causing poor aeration. Also, roots are usually unable to penetrate pores smaller than their own diameter, which is usually in excess of 60 μm for the smallest roots and may be 10 times that for some absorbing roots. However, a small percentage of large pores will suffice to

TABLE 6.2 Bulk Density, *in Situ* Field Capacity, 15-bar Water Content, Available Water, and Cumulative Days of Water Storage in a 150-cm-Deep Norfolk Soil Profile[a]

Horizon	Depth (cm)	D_b (g/cm³)	In situ field capacity (cm)	15-bar water content (cm)	Avail. water (cm)	Cum. avail. water (cm)	Cum. water supply[b] (day)	Cum. water supply with pan[c] (day)
Ap	0–25	1.62	4.2	1.1	3.1	3.1	6.2	6.2
A21	25–35	1.75	1.8	0.5	1.3	4.4	8.8	—
A22	35–45	1.70	1.6	0.6	1.0	5.4	10.8	—
B1	45–75	1.66	8.7	4.5	4.2	9.6	19.2	—
B2	75–150	1.51	19.9	11.3	8.6	18.2	36.4	—

[a] From Cassell (1983).

[b] An evapotranspiration rate of 0.5 cm/day is assumed.

[c] Plow plan at 25 cm.

accommodate the roots (R. S. Russell, 1977, p. 152), and root hairs and fungal mycelium can occupy smaller pores.

Soil Structure.   The structure of a soil depends on the manner in which the primary soil particles are aggregated together (Chapter 3). In some soils the structure is easily broken down when wet, and traffic over it leads to a rearrangement of the particles and compaction into soil with a greater density, which is less favorable for root growth. Figure 6.9 shows the effects of surface compaction by farm machinery on depth of rooting in a clay soil. Table 6.2 shows the effect of a plow pan developed at a depth of 25 cm by several years of moldboard plowing. Such impermeable layers are common in cultivated soils. They often result from compaction of the wet soil in the furrow bottom by tractor wheels. Figure 6.10 shows the effect of such a plow pan on root penetration and the improvement resulting from breaking it up by deep tillage. There are areas of shallow soil underlain by naturally formed horizons of very high density called frangipans. They are often extremely acid and restrict air and water movement and root penetration. In the Great Plains, a calcareous hardpan layer usually marks the lower limit of water penetration during periods of average rainfall (Weaver and Crist, 1922). Cassell (1983) discussed in detail the effects of plow pans on crop growth and mapped the areas of the eastern United States in which physical and chemical barriers to root growth exist.

Root penetration is limited in some areas by rock, and vegetation on such shallow soils is more subject to injury from drought than on deeper soils. In contrast Fig. 6.11 shows uniform root penetration to a depth of over 180 cm in a deep volcanic soil in New Zealand. As mentioned in the section on depth of root

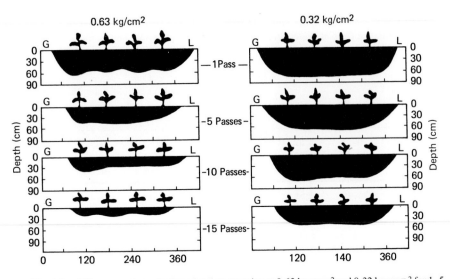

Fig. 6.9. Effect on root penetration of soil compaction at 0.63 kg cm^{-2} and 0.32 kg cm^{-2} for 1, 5, 10, or 15 passes over the soil prior to seeding. (From Cassell, 1983).

systems, root penetration to a depth of several meters is common in deep loess soils, such as those studied by Weaver and his colleagues in Nebraska. In contrast, in the clay soils of the North Carolina Piedmont and many soils in Western Europe, root penetration is severely limited by mechanical resistance and deficient aeration. This is illustrated by the difference in growth of corn on a well-aerated silt loam soil in Illinois and on a Cecil clay loam in North Carolina. Corn flourished on the Illinois soil even though by early August no available water remained in the upper meter and absorption was occurring to a depth of 2 m (Reimann *et al.*, 1946). However, corn was so shallow-rooted on a heavy clay loam in North Carolina that it was unable to absorb all the water from even the upper meter of soil (James, 1945).

Soil Moisture

Either an excess or a deficiency of soil water limits root growth and functioning. Water is not directly injurious to roots, as shown by their vigorous growth in well-aerated nutrient solutions. However, an excess of water in the soil displaces air from the noncapillary pore space and produces an oxygen deficiency, causing the death of many roots. This will be discussed in the section on soil aeration. A deficiency of water brings about cessation of root growth, and there is usually

Fig. 6.10. Effect of a compacted layer of soil on root penetration by 11-week-old oat plants. (Left) Undisturbed soil with dense mass of roots above the compacted layer, but little below it. (Right) Uniform penetration of roots into soil loosened by tillage to a depth of 50 cm. Restriction of root penetration was caused by mechanical resistance, as aeration was not limiting below the compacted layer. (Courtesy of H. C. DeRoo, Connecticut Agricultural Experiment Station.)

Fig. 6.11.   Partially excavated root system of a *Pinus radiata* tree growing in a deep pumice soil in Kangaroa Forest, New Zealand. Note the 6-ft (180-cm) measuring stick. The water-holding characteristics of this soil are shown in Fig. 3.7. (From Will, 1966.)

little or no root growth in soils with a water content near the permanent wilting percentage.

In dry regions, root penetration is generally limited by the depth to which the soil is wetted by rain. Weaver (1920) stated that root penetration is deepest in well-watered prairie soil, less in the mixed prairie-plains vegetation, and least in the shortgrass plains, where the soil is usually wetted no deeper than 45–120 cm. E. W. Russell (1973, pp. 532–537) also discussed depth of rooting in relation to soil water.

There are few quantitative studies of the effect on root growth of soil water stress. Newman (1966) found a reduction in growth of flax roots at a soil water potential of -0.7 MPa. The growth rate was only 80% of the control rate at -1.5 MPa, but some growth occurred in soil drier than -2.0 MPa. It also appeared that root growth at any depth was independent of the water potential at other depths because at a stage when root growth was much reduced in the upper, drier layer, it was not yet reduced in the deepest, wettest layer of soil. Kaufmann (1967) found that growth of loblolly pine roots was reduced to 25% of the control rate at a soil water potential of −0.6 to −0.7 MPa. Also, when pine roots were subjected to repeated soil-drying cycles, less root growth occurred during the second or third cycle than prior to drying. In general, root growth is decreased or stopped by soil water stress and roots tend to become suberized to their tips, reducing their capacity to absorb water. As a result, plants subjected to severe droughts usually do not regain their full capacity to absorb water until several days after the soil is wetted (Brix, 1962; Kramer, 1950; Leshem, 1965; Loustalot, 1945). The relationship between soil moisture and root growth of loblolly pine in the field is shown in Fig. 6.12.

Hydrotropism. It has often been claimed that roots grow toward water or moist soil, and the average layman has a firm belief that roots can detect water at a distance and grow to it. This is based chiefly on observations of root development in leaky drain pipes and other moist areas. For example, large masses of roots sometimes develop under leaky water taps and where roots penetrate into old-style dug wells. This probably results from a combination of abundant water

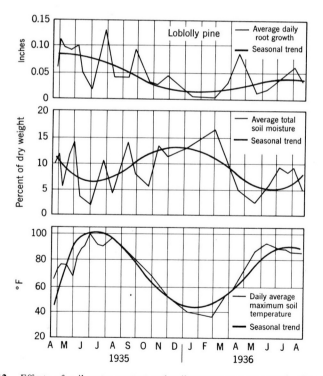

Fig. 6.12. Effects of soil water content and soil temperature on growth of roots of *Pinus echinata* in the forest at Durham, North Carolina. (From Reed, 1939.)

and good aeration (Oppenheimer, 1941). Roots continually grow outward randomly in all directions to distances of many meters. Where these randomly growing roots encounter moisture and/or mineral nutrients, they tend to branch profusely.

There is some difference of opinion concerning the ability of roots to penetrate dry soil. In some experiments, roots have been observed to grow short distances into dry soil (Hunter and Kelley, 1946; Volk, 1947), but it seems unlikely that under field conditions any important amount of root growth occurs into soil at or below the permanent wilting percentage. Hydrotropism, if it exists at all, is weaker than geotropism because roots frequently grow downward out of moist soil into dry soil, and then cease to elongate.

Soil Aeration

There are two aspects of the aeration problem: inadequate aeration caused by lack of sufficient noncapillary pore space for adequate gas exchange, and inadequate aeration caused by high water tables or actual flooding. Aeration is often a limiting factor for root growth and functioning in soils that are not saturated. Respiration of roots and soil organisms uses large quantities of oxygen and produces equally large quantities of carbon dioxide, especially during the summer. The data in Table 6.3 indicate that under a crop approximately one-half of the oxygen is used by soil organisms and one-half is used by roots and organisms in the rhizosphere. The summer rate of oxygen use shown in Table 6.3 would require a complete turnover in oxygen every day to prevent depletion. Woolley (1966) estimated that diffusion might supply the oxygen requirement to a depth of 1 m if at least 4% of the soil volume consists of interconnected, gas-filled pores. Diffusion may be aided to some extent by mass flow caused by temperature changes and wind.

TABLE 6.3 Oxygen Consumption and Carbon Dioxide Production from a Bare Soil and a Soil under Kale in Summer and Winter at Rothamsted[a]

| | Soil temperature at 10 cm | | | |
| | 17°C | | 3°C | |
	Cropped	Bare	Cropped	Bare
Oxygen consumption	24	12	2.0	0.7
Carbon dioxide production	35	16	3.0	1.2

[a] Rates are in g m^{-2} day^{-1}. From E. W. Russell (1973).

The rate of gas exchange depends largely on soil texture and structure. Aeration is seldom limiting in sandy soils but is often inadequate in fine-textured soils, where less than 10% of their volume consists of noncapillary pore space, such as the soils shown in Figs. 3.1 and 3.4 (Robinson, 1964; Vomocil and Flocker, 1961). As mentioned earlier, compaction of soil by trampling and filling, and paving of the surface greatly reduce gas exchange and results in decreased oxygen and increased carbon dioxide concentrations (Yelenosky, 1964). It is often difficult to separate the mechanical effects of compaction on root extension from the aeration effects, and often both are involved (Rickman *et al.*, 1966). The soil atmosphere and gas exchange are discussed by E. W. Russell (1973, Chapter 6).

It seems likely that growth of both cultivated crops and native vegetation is often reduced by undetected deficiencies in root aeration. It has been suggested that root respiration and decomposition of organic matter produce conditions in prairie soils sufficiently anaerobic to hinder tree root growth (McComb and Loomis, 1944). Howard (1925) claimed that in India several species of trees were killed whenever a dense growth of grass developed over their roots, producing anaerobic conditions. Other factors may have been involved, but it was demonstrated in pot experiments that addition of starch to soil high in nitrogen resulted in anaerobic conditions severe enough to kill wheat seedlings (Karsten, 1939).

Limiting Concentrations of Oxygen and Carbon Dioxide. There is uncertainty concerning the relative importance of low oxygen versus high carbon dioxide under field conditions. In laboratory experiments, replacement of air in the root medium by carbon dioxide decreased root permeability more and sooner than replacement by nitrogen. However, such high concentrations are not found in soils. It is impossible to determine the effective oxygen concentration from measurements of oxygen in bulk soil air because they indicate the concentration in large pores rather than the oxygen concentration at root surfaces. Attempts have been made to measure the oxygen diffusion rate to roots by using a platinum electrode to simulate a root (Letey and Stolzy, 1964). In general, it appears that oxygen diffusion rates of 0.2 μg cm^{-2} min^{-1} or less are definitely limiting for most roots of most plants, and rates greater than 0.4 μg are adequate, but the adequacy of values in the range between 0.2 and 0.4 μg depend on factors such as plant species and temperature. Stolzy *et al.* (1981) discuss methods of measuring soil aeration.

It seems probable that bulk air concentrations of oxygen above 10% are adequate for growth, and it is said that high concentrations, maintained by energetic aeration with air containing 20% oxygen, can even be inhibitory (Loehwing, 1934; J. Letey, Jr., private communication). There is uncertainty concerning the concentration at which carbon dioxide limits growth. It was reported that cotton (Leonard and Pinckard, 1946), corn, and soybean (Grable and Danielson, 1965) all tolerate 20% and tomato at least 6.8% (Erickson, 1946) without injury, and carbon dioxide concentrations in soil rarely rise above 5–10% (Fig. 6.13). Geisler (1963) reported that carbon dioxide in concentrations usually found in soil increases lateral branching and root dry weight of pea roots, and he cites several other reports of carbon dioxide stimulating root growth. In general, it appears that under field conditions root growth is more likely to be limited by a low concentration of oxygen than by a high concentration of carbon dioxide.

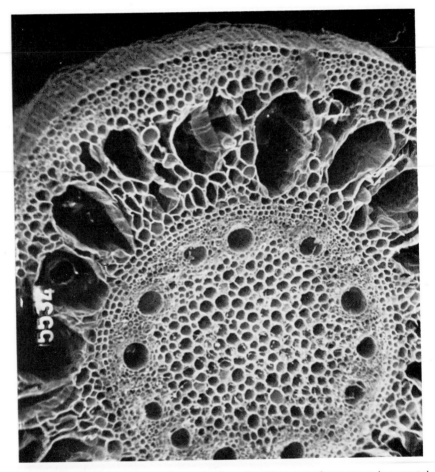

Fig. 6.13. Scanning electron micrograph of an adventitious root of corn grown in unaerated nutrient solution. The section was made 8 to 10 cm behind the root tip. The air spaces in the cortex are formed lysigenously by the breakdown of cells. (From Agricultural Research Council Letcombe Laboratory Annual Report, 1978, p. 42; courtesy of M. C. Drew.)

Flooding Injury

The most severe examples of deficient aeration result from saturation of the soil with water because this displaces the soil air, and the low solubility and slow diffusion of oxygen in water drastically reduce the supply to the roots. Nearly everyone is familiar with the wilting, epinasty, yellowing of leaves, reduction in growth, and eventual death of plants in flooded soil, and so much has been published that it is impossible to cite all of the important papers. A great deal of information can be found in a book edited by Hook and Crawford (1978), "Plant Life in Anaerobic Environments," and in reviews by Armstrong (1979) and Kozlowski (1982b). Kozlowski made a thorough review of the effects of flooding on woody plants, including causes of injury and adaptations to flooding.

Effects of Flooding. Flooding drastically reduces the oxygen supply because of the low solubility and slow rate of diffusion of oxygen in water. Prolonged flooding often occurs in the winter and spring, but temporary flooding sometimes occurs as a result of heavy summer rains. Injury from flooding usually develops sequentially over a period of several days and becomes more severe as flooding continues. Wilting is likely to be the first symptom if atmospheric conditions are favorable for transpiration. For example, if low areas in a tobacco field are flooded by a heavy rain, followed immediately by bright sunshine, the leaves often wilt so suddenly and severely that farmers term the occurrence flopping. Wilting is less severe in most other species and seldom occurs in cloudy weather. The wilting is generally assumed to result from decreased water absorption caused by a sudden increase in root resistance in the water-saturated soil (Kramer, 1940a). Kramer attributed the increased root resistance to the toxic effect of high carbon dioxide, but Hunt *et al.* (1981) attribute it to ethylene produced in the soil and the plant. After a short period of water stress, the stomata usually close, partially reducing water loss, and plants regain their turgor. Recently, Sojka and Stolzy (1980) reported that oxygen deficiency in the soil causes closure of stomata even when no plant water stress exists.

If the soil drains within 24 hr, the plants usually recover with little or no obvious injury, although it is reported that only 24 hr of flooding reduces the yield of some crops (Cannell *et al.*, 1979; Erickson, 1965). If the soil remains saturated for several days, other symptoms appear. Epinasty is often noticeable within 24–48 hr, the leaves turning downward so that they appear to be badly wilted. Epinasty has been observed in a number of woody and herbaceous species, and several decades ago investigators noted that the leaves appeared as though they had been exposed to ethylene (Furkova, 1944; Williamson, 1950). Development of convenient methods for determining the concentration of ethylene led to increased attention to its role in flooding injury. Kawase (1974, 1976) proposed that accumulation of ethylene in flooded plants is at least partially responsible for epinasty and leaf injury. It was suggested by M. B. Jackson and Campbell (1976) and Jackson et al. (1978) that a substance produced in flooded roots moves to the shoots and causes the production of ethylene. Later, Bradford and Yang (1980) reported that an ethylene precursor, 1-aminocyclopropane-1-carboxylic acid, is synthesized in the roots under anaerobic conditions and translocated to the shoots, where it is converted into ethylene. The synthesis of ethylene in roots is said to be inhibited by the absence of oxygen (Jackson et al., 1978).

Within a few days to a week, the lower leaves of flooded plants begin to turn yellow and die. Kramer and W. T. Jackson (1954) attributed this to the upward movement of toxic substances from the dying roots, but later W. T. Jackson (1956) concluded that leaf chlorosis results from lack of some essential substance ordinarily formed in the roots and translocated to the shoots. It now seems probable that the substance is cytokinin, ordinarily synthesized in the roots and translocated to the shoots (Burrows and Carr, 1969). This is supported by the observations of Railton and Reid (1973) and Drew et al. (1979) that application of benzyladenine to the foliage delays or prevents chlorosis in flooded plants. Reid and Crozier (1971) also reported that there is a reduction in gibberellic acid in shoots of flooded plants. However, Drew and Siswora (1979) claimed that although water stress and hormone imbalance play a role, the major factor in injury to the shoots is disturbance of nitrogen metabolism, including premature translocation out of older leaves. Obviously, more research is needed on this problem.

Adventitious roots often begin to develop near the water line on stems of flooded plants, and sometimes the stems enlarge in that region. It seems possible that hypertrophy and adventitious root formation occur because flooding inhibits downward translocation near the water line, and Phillips (1964) reported accumulation of auxin in stems of flooded plants. However, Drew et al. (1979) concluded that formation of adventitious roots is stimulated by ethylene. Apparently, the adventitious roots take over the functions of the dying root system because plants that quickly develop adventitious roots recover from flooding better than plants that do not (W. T. Jackson, 1955). However, some kinds of plants tolerate prolonged flooding even though they develop no adventitious roots.

There is extensive evidence that products of anaerobic respiration, such as ethanol, aldehydes, and lactic acid, accumulate in the roots of flooded plants and presumably cause injury (Bolton and Erickson, 1970; Crawford, 1967; Francis et al., 1974; Hook et al., 1972). However, Jackson et al. (1982) found that the concentrations of ethanol occurring in flooded plants are not toxic and concluded that it probably does not cause injury to flooded plants. Compounds such as methane, sulfides, reduced iron, and other minerals also accumulate in flooded soil and cause injury to roots. The complex chemistry of flooded soils is discussed in Chapter 25 of E. W. Russell (1973).

Depth and Season of Flooding. The depth of flooding may be quite important, especially to seedlings. Cypress (*T. distichum*) seedlings can become established only on exposed soil, but after establishment they are uninjured by flooding as long as the tops remain above water. In bogs and swamps many plants become established on hummocks above the water surface, although eventually most of their roots occupy saturated soil. Land (1974) found that increasing the depth of flooding by only a few centimeters increased the mortality of pine seedlings. It has also been demonstrated that establishment of pine seedlings on wet soil is generally improved by planting them on ridges so that at least part of the root system is well aerated (McKee and Shoulders, 1974). Increased depth of flooding was also reported to increase injury of a number of different kinds of grasses (Rhoades, 1964).

Fluctuation in the level of the water table often increases injury to plants. Plants tend to develop root systems just above the water table, and if it is lowered abruptly, the root systems are often left in drying soil, whereas if the water table is raised abruptly, they are injured by flooding. In the Netherlands, crops and fruit trees are often grown successfully with a water table only 60 or 100 cm below the surface because it is carefully maintained at a fixed level.

In general, flooding during the dormant season is much less injurious than flooding during the growing season (Hall and Smith, 1955; Rhoades, 1964). This is not surprising, as the oxygen requirement of dormant tissue is much lower than that of growing tissue, and the rate of respiration is also much lower in cool weather. Also, flooding by flowing water is less injurious than flooding by standing water.

It was reported by Cannell *et al.* (1979) that flooding for only 24 hr at flowering reduces the yield of peas, and Erickson (1965) reported that 1 day of oxygen deficiency early in the life of tomatoes reduces the yield. It is probable that crop yields are sometimes reduced by a few days of saturated soil caused by heavy rain or irrigation (see Fig. 4.17, Meek and Stolzy, 1978), but growers seldom realize what causes the reduction.

Differences in Flooding Tolerance. It is well known that there are wide differences in tolerance of flooding. Cattails, *Spartina, Phragmites,* rice, and woody plants such as cypress, willow, and mangrove tolerate long periods in saturated soil, but corn, wheat, barley, tobacco, and trees such as dogwood and *Liriodendron* are quickly injured or killed. Pond pine (*Pinus serotina*) is more tolerant of wet soil than shortleaf (*P. echinata*) or longleaf pine (*P. palustris*), and among shade trees, American elm (*Ulmus americana*) and honey locust (*Gleditsia triacanthas*) are more tolerant than most other species (Yelenosky, 1964).

There are two principal reasons why some plants tolerate flooding better than others. Either their stem structure permits relatively rapid diffusion of oxygen from shoots to roots, or the roots tolerate anaerobic respiration. In some species, such as swamp tupelo (*Nyssa sylvatica* v. *biflora*), both mechanisms seem to operate. (Hook and Brown, 1973).

Aeration and Root Structure. Plants native to wet habitats often contain extensive air spaces that form continuous passageways from shoots to rhizomes and roots. This tissue, called aerenchyma, is usually regarded as an important adaptation that improves the oxygen supply to roots and rhizomes growing in poorly aerated soil. Incidentally, Williams and Barber (1961) argued that the chief function of aerenchyma is provision of anchorage with the smallest possible amount of living, oxygen-consuming tissue. Roots of mesophytes such as corn, sunflower, tomato, and many other plants also develop large air spaces when grown in poorly aerated media. Such a corn root is shown in Fig. 6.13. Usually, these air spaces are formed lysigenously by breakdown of masses of cortical cells, leaving spokelike strands of living cells extending from the outer cortex to the endodermis. These strands of cells seems healthy and function effectively in ion and water transport (Drew *et al.*, 1980).

The aerenchyma is generally regarded as a useful acclimation, permitting growth in a poorly aerated medium. However, it is really caused by injury from inadequate aeration, although it doubtless facilitates downward movement of oxygen. It has been suggested that the high ethylene concentration found in flooded plants stimulates cellulase activity which in turn causes cell breakdown (Kawase, 1979), and it is reported that application of ethylene to stem and root tissue causes formation of aerenchyma. A puzzling feature is the difference in resistance to injury among adjacent cells of the root. Aerenchyma seldom occurs nearer than 2 or 3 cm behind the root tips, the epidermis, endodermis, and stele are usually unaffected, and the strands of cortical cells between air spaces remain functional. There is no obvious reason for the degeneration of patches of cells within the cortex while adjacent cells are uninjured.

It has been shown in a number of experiments that measurable amounts of oxygen move from shoots to roots in rice (van Raalte, 1940; Armstrong, 1969), *Nuphar* (Laing, 1940; Dacey, 1980), mangroves (Scholander *et al.*, 1955) conifers (Armstrong and Read, 1972), swamp tupelo (Hook *et al.*, 1972), and other species which survive or even thrive in saturated soil. In some instances, so much oxygen reaches the roots that it diffuses out into the soil surrounding them (Armstrong, 1968, 1978; Barber *et al.*, 1962; Greenwood, 1971). However, it seems unlikely that significant amounts of oxygen can diffuse over distances of more than a few centimeters or tens of centimeters, and it is probable that the deeper roots of plants in saturated soil are subjected to anaerobic conditions. Crawford (1976) suggested that root meristems usually suffer from lack of oxygen, and Berry (1949) found that cells in the meristematic region of onion roots receive less than the optimum amount of oxygen, even in air containing 21% of oxygen. Fiscus and Kramer (1970) also found the oxygen concentration in the interior of roots to be low and concluded that the internal tissues usually operate under an oxygen deficit. However, Bowling (1973) doubted if the deficit is large enough to affect ion transport significantly.

It is generally assumed that oxygen movement from shoots to roots occurs by diffusion along concentration gradients resulting from photosynthetic production in the leaves and its use in root respiration (Laing, 1940). However, in some woody plants studied by Armstrong (1968), air entered the stems through lenticels in the region a few centimeters above the water line, and the leafy shoots contributed no oxygen. Dacey (1980) claimed that there is a temperature-induced mass flow of air from the leaves to the rhizomes in *Nuphar*.

Development of aerenchyma in the absorbing region of roots greatly reduces the pathway for translocation across the cortex to scattered strands of parenchyma cells, and this would be expected to decrease uptake and inward movement of minerals. However, Drew *et al.* (1980) reported that in the presence of oxygen, there was little reduction in transport of rubidium across aerenchyma as compared with normal roots. This might be construed as indicating that ion transport is limited at the endodermis or in the stele rather than in the cortex.

Biochemical Adaptations.   There is increasing evidence for biochemical adaptations as a factor in tolerance of flooding by roots. In the presence of oxygen, the end product of glycolysis, pyruvic acid, is converted to carbon dioxide and water, but in its absence the end products are incompletely oxidized compounds such as ethanol, aldehydes, and organic acids, and only a small fraction of the energy in the substrate is released. Accumulation of products of anaerobic respiration is usually regarded as injurious, and the work of Crawford and others has shown that plants intolerant of flooding show large increases in ethanol and alcohol dehydrogenase, whereas tolerant species show little increase. For example, there was a 30-fold increase in alcohol dehydrogenase in roots of flooded *Trifolium subterraneum* (Francis *et al.*, 1974). However, Jackson *et al.* (1982) question the toxicity of ethanol at the concentration found in flooded roots. Perhaps the toxicity of other products of anaerobic respiration ought to be investigated more critically. Various aspects of biochemical effects of flooding are discussed in the monograph edited by Hook and Crawford (1978), and by Crawford (1976).

Miscellaneous Effects of Deficient Aeration.   In addition to the effects already discussed, deficient aeration reduces mineral uptake and produces deficiency symptoms. In fact, Drew (1977) and Drew and Siswora (1979) suggest that disturbance of nitrogen metabolism is a major cause of the symptoms of flooding injury.

Attacks on roots by fungi are often increased by deficient aeration because a number of species of pathogenic fungi thrive in poorly aerated soils, and reduced

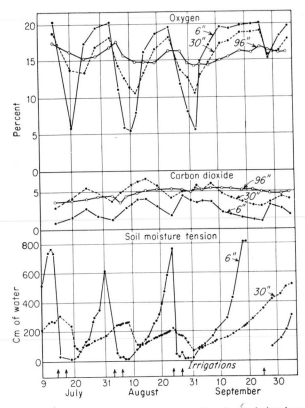

Fig. 6.14.   Soil aeration and soil moisture tension at various depths during the growing season. Increasing soil moisture tension (decreasing water content) was accompanied by an increase in oxygen content. (From Furr and Aldrich, 1943.)

root vigor makes them more susceptible to infection. Examples are the "decline" of avocado and citrus, and littleleaf disease of pine. In contrast, the Panama wilt of banana is controlled by flooding the soil to reduce growth of the causal organism. Zentmyer (1966) summarized considerable literature on soil aeration and plant diseases, and other information is summarized in books edited by Baker and Snyder (1965) and Kozlowski (1978). The effect of irrigation on soil oxygen content is shown in Fig. 6.14.

Soil Temperature

Root growth is often limited or stopped by low temperatures, and occasionally the surface soil becomes hot enough to stop root growth. The optimum temperature varies with species, stages of development, and oxygen supply, but it is probably about 20°–25°C (Batjer *et al.*, 1939). Pecan roots are reported to grow very slowly below 10°C (Woodroof and Woodroof, 1934) and are killed at −2°C. Barney (1951) found that roots of loblolly pine seedlings grew most rapidly in the laboratory at 20°–25°C and that the rate of elongation at 5° and 35°C was less than 10% of the maximum rate. Data on growth of loblolly pine roots in the forest are given in Fig. 6.12 and laboratory data in Fig. 6.15.

Roots of species native to warm climates cease growth at higher temperatures than those from cool climates. The minimum temperature for growth of roots of grapefruit, sweet orange, and sour orange in solution culture was found by Girton (1927) to be 12°C, the optimum 26°C, and the maximum 37°C. Soil temperatures in the Los Angeles area are never sufficiently high for optimum growth of citrus, and in the winter they are low enough to limit seriously root growth (North and Wallace, 1955). According to Arndt (1945), the minimum temperature for elongation of roots of cotton is 16° to 17°C, and the optimum decreases from 33°–36° to 27°C as the roots grow downward into cooler soil. In experiments carried out by Brown (1939), roots of Bermuda grass made no growth at 4.5°C and little at 10°C, but the optimum was only 10°C for Canada bluegrass and 15°C for Kentucky bluegrass. Roots of both bluegrass species were severely injured by the high temperature favorable for Bermuda grass.

High temperatures can severely limit root growth, and temperatures in exposed soil can become high enough to injure or kill roots and bases of stems (Bates, 1924; Korstian and Fetherolf, 1921; Pearson, 1931; Shirley, 1936). It is claimed that the root surface of strawberries is sometimes reduced so much by high soil temperatures that the tops suffer from lack of water. The small number of roots found in the surface 30 cm of soil in many California orchards is

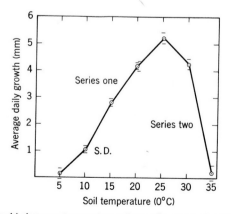

Fig. 6.15. Relationship between temperature and rate of root growth of *Pinus taeda* seedlings in the greenhouse. (From Barney, 1951.)

attributed by Proebsting (1943) to high summer soil temperatures. Gur *et al.* (1972) found differences among apple rootstocks in tolerance of high temperature, and temperatures above 30°C inhibited both root and shoot growth. Less tolerant types accumulated more ethanol and aldehydes, and the roots and leaves contained less cytokinin than tolerant types.

Injury to roots from frost heaving is an indirect effect of low temperature. Much damage is caused to alfalfa, winter cereals, and sometimes even tree seedlings in winter because repeated freezing and thawing lift plants and break off their roots below the soil surface. Mechanical strength of roots may be an important factor in winter survival of some plants (Lamb, 1936).

Brouwer and his colleague at Wageningen studied intensively the effects of soil temperature on plant growth. The best root and shoot growth of bean (*Phaseolus vulgaris*) occurred between 20° and 30°C, and very little occurred at 5°, 10°, and 35°C. The reduction in shoot growth was attributed to water stress caused by reduced absorption of water at both low and high temperatures (Brouwer, 1964). On the other hand, Davis and Lingle (1961) claimed that the reduction in shoot growth of tomatoes at low soil temperatures was not caused by reduced absorption of water or mineral nutrients. It seems likely that low root temperatures decrease synthetic activities and nitrate reduction (Duke *et al.*, 1979), in addition to decreasing water uptake. It was also reported that low temperature causes irreversible injury to the meristematic region of corn roots (Kleinendorst and Brouwer, 1972).

Unfavorable temperatures also affect root differentiation and anatomy. In bean roots, elongation is limited and roots become differentiated up to the apex. Branching often continues so that branches occur almost to the root apex. The walls of endodermal cells and of the cells forming the external surfaces of roots become heavily suberized and less permeable to water and salt. Leaves of plants grown at unfavorable root temperatures tend to develop xeromorphic characters, presumably because they are subjected to high water stress (Brouwer and Hoogland, 1964). Lyr and Hoffmann (1967) report a number of instances in which temperature differences produced morphological differences in roots. They also found that heating the soil 5°C above normal increased height growth of seedlings of *Robinia pseudoacacia*. Their experiments indicated that the optimum temperature for root growth in short-term experiments (20°C) was much higher than the usual soil temperatures in the region near Berlin where their study was made. They suggest that soil temperatures in that region are often too low for maximum root growth. Hellmers (1963) noted that the optimum temperature for root growth of Jeffrey pine is lower than that for shoot growth.

Low soil temperatures in the spring often delay planting of warm season crops such as beans, cotton, and maize. Heavy mulches that reduce erosion and prevent deep freezing during the winter have the disadvantage of slowing soil warming in the spring. Wet soil also warms up more slowly than well-drained soil. In the tropics, mulches may be beneficial because they lower soil temperatures (R. S. Russell, 1977, pp. 264–267). Soil temperature is discussed in Chapter 17 of E. W. Russell (1973), and the early literature was summarized by Richards *et al.,* in Shaw (1952).

Minerals, Salt Concentration, and pH

It is well known that soil pH and the kind and concentration of ions have important effects on root growth. An abundance of certain essential elements, particularly phosphorus and nitrogen, stimulates root growth; but shoot growth is stimulated even more, so the root–shoot ratio is usually lower in fertile than in infertile soil. Not enough is known about the effects of specific ions, but it is recognized that phosphorus stimulates root growth and that deficiencies of boron and calcium in the root environment result in short, stubby branch roots and death of root tips. Failure of roots to penetrate deeply in certain soils is related more to lack of nutrients than to mechanical resistance or deficient aeration. Thus, it often happens that loosening the subsoil does not increase depth of rooting unless nutrients are added to it (Bushnell, 1941; DeRoo, 1961; Pohlman, 1946).

Calcium and boron seem to have direct effects on root growth and must exist in the immediate root environment if growth is to occur. Haynes and Robbins (1948) found that when part of a root system was grown in a complete nutrient solution and the other part in a solution lacking calcium, the roots in the solution lacking calcium died. If boron was supplied to only part of the root system, the other part survived but did not grow. Apparently, neither boron nor calcium is readily translocated from one part of the root system to the other. In addition to specific ion effects on root growth, there is a general osmotic effect. Growth is reduced due to a decrease in cell division and elongation, as osmotic potential of the substrate increases (González-Bernáldez *et al.,* 1968). The reduction is probably caused by water stress.

In arid regions all over the world where irrigation is practiced, salt tends to accumulate in the soil until the concentration becomes too high for satisfactory plant growth (see also Chapter 4). As a result, large areas of land have become unproductive or have even reverted to desert. The direct effects of excess salt on water absorption are discussed in Chapter 9, but noticeable effects on root growth also occur. High concentrations of salt tend to slow down or stop root elongation and hasten maturation. This results in roots that are suberized to the tips and appear dormant (Hayward and Blair, 1942). Considerable differences exist among roots of different species in tolerance of salt. Wadleigh *et al.* (1947) grew crop plants in soil containing various amounts of salt. Few bean roots penetrated soil containing 0.1% sodium chloride, and few corn roots penetrated soil containing 0.2%, but some alfalfa roots penetrated soil containing as much as 0.25% sodium chloride, and cotton roots were abundant in that soil.

Soil pH has little direct effect on root growth over a wide range and becomes important only when it renders elements such as iron and manganese unavailable (high pH) or increases the solubility of others, especially aluminum, to toxic concentrations in very acid soils (low pH). Aluminum toxicity is a serious problem in acid soils. Plants have been grown successfully over the range from pH 4 to 9 if precautions were taken to keep all of the essential elements in solution. Soil pH is discussed by E. W. Russell (1973, pp. 121–128).

Root Interactions and Allelopathy

The growth of roots is influenced by a variety of environmental factors in addition to those already discussed. These include competition with roots of other plants for water and minerals, root exudates, the activities of organisms in the rhizosphere, and the products of decay or organic matter in the soil. There are also toxic effects of specific substances released by roots or leached out of leaves and crop residues, usually described as allelopathic effects.

Root Competition.   The size of root systems is usually much reduced when they are grown in competition with other systems. For example, Pavlychenko (1937) reported that root systems of barley and wheat were nearly 100 times larger when grown without competition than when grown in rows 15 cm apart. It has also been claimed that grass inhibits root growth of trees by depleting oxygen and increasing the carbon dioxide concentration in the soil (Howard, 1925; McComb and Loomis, 1944). Although roots often seem to be intertwined in the soil, there is evidence of a mechanism that prevents roots of some kinds of plants from growing very close to one another. This has been reported for tree roots by Lyford and Wilson (1964) and for soybean by Raper and Barber (1970). An example of the difference in root development between an isolated plant and plants growing in rows is shown in Fig. 6.16. The detrimental effects of one kind of plant or crop on another growing with it or following it can be attributed to (1) depletion of water or nutrients, especially nitrogen, by the first crop, (2) release of toxic substances from its roots or leaves, or (3) the production of toxic substances during its decomposition.

Allelopathy.   The term allelopathy refers to substances, sometimes termed allelochems, released by plants that affect the growth of other plants either favorably or unfavorably. In the broadest sense, these include (1) volatile sub-

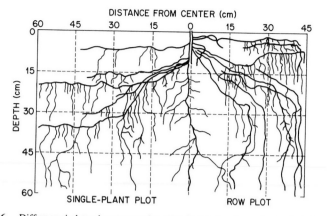

Fig. 6.16. Difference in lateral root extension of an isolated soybean plant and of plants growing in rows. (From Raper and Barber, 1970.)

stances such as the terpenes released by *Artemisia* and *Eucalyptus,* (2) substances leached from living or dead tissues, (3) root exudates, and (4) substances released during the decay of plant materials (Rice, 1974). Presumably, the antibiotics produced by certain fungi can also be termed allelochems. It has been known for at least 2000 years that some plants seem to stimulate or inhibit their neighbors. The elder Pliny reported that walnut trees are injurious to other vegetation, and the injury is now attributed to a compound called juglone produced in the roots (Gries, 1943). Colton and Einhellig (1980) reported that aqueous extracts from the leaves of velvetleaf (*Abutilon theophrasti*) are injurious to soybean, and Lodhi and Killingbeck (1980) found extracts from the needles and bark of ponderosa pine to be toxic to nitrifying bacteria. However, conclusions based on laboratory or greenhouse tests of plant extracts should be treated cautiously because in nature most organic compounds are either leached out of the soil or decomposed by microorganisms. Norby and Kozlowski (1980) summarized the information on allelopathy among woody plants, and Rice (1974, 1979) reviewed the literature on this interesting but controversial subject.

Attempts have been made to explain success or failure of plants in natural succession in terms of effects of substances released directly or indirectly by the competing vegetation. Muller (1969, and other papers) claimed that biochemical products of plants are often limiting factors for species in both natural and agricultural plant communities. In contrast, Loehwing (1937) argued that plant injury ascribed to toxic root secretions could be explained in terms of disturbed nitrogen nutrition. The truth probably lies somewhere between these two views.

Early in this century there was wide interest in the effects of crop residues on subsequent crops, and some of the early work was summarized by Miller (1938, pp. 164–174). There is renewed interest among crop scientists in this complex and frustrating problem, and examples of recent work are papers by Elliott et al. (1981) and Patterson (1981). The former reported that the inhibitory effect of wheat residues on the following crop is caused by immobilization of nitrogen rather than by the release of phytotoxic substances. However, Patterson found that several compounds produced by weeds and decaying crop residues are toxic at a concentration of 10^{-3} M. Putnam and Duke (1978) reviewed some of the literature dealing with crop residues, crop rotation, and the replant problem, but it is difficult to arrive at any definite generalizations. Closely related to the foregoing discussion are the "replant problem" and the relative merits of various kinds of crop rotation.

The Replant Problem.   When old orchards and vineyards are replanted, the new trees often grow poorly or even die. This problem has been observed for many kinds of fruits, including apple, peach, and other stone fruits, citrus, and grapes. For example, hundreds of thousands of peach replants have died prematurely in the southeastern United States. Although often attributed to toxic products of peach root decay or to substances released from injured roots (Proebsting and Gilmore, 1941; Israel et al., 1973), poor growth probably has several causes in addition to allelochems. These include nematodes, fungi such as *Phytophthora, Clitocybe,* and *Armillaria,* poor drainage, and inadequate aeration (Ducharme, 1977). Tree survival is often increased by deep plowing or subsoiling and by soil fumigation. The success of soil fumigation suggests that nematodes and fungi are more important than allelochems in causing replant difficulties. The replant problem seems not to have arisen as yet in forestry, but it may become important with shorter rotations. Yadava and Doud (1980) reviewed the replant problem in horticulture.

Crop Rotation.   The problem of crop rotation is related to the replant problem, but on a shorter time scale. Earlier in this century, rotation of crops was considered to be an essential feature of a good agricultural program, and many studies were made to determine if one sequence of crops was better than another. Unfortunately, the results were often inconclusive or even contradictory (Miller, 1938, pp. 165–174). Although some sequences appeared to be better than others, the reasons were never very clear. Rotation was supposed to improve control of weeds, diseases, and insect pests, decrease injury from toxic products of root decay, and increase the nitrogen supply by growing legumes in the rotation. However, increased use of fertilizers, and the introduction of herbicides and more disease-resistant varieties have resulted in the abandonment of traditional crop rotations, and the same crop is being grown year after year for long periods of time. It is too soon to evaluate the results in terms of soil structure, fertility, and yield, but the practice deserves careful study.

Soil Organisms

Numerous references have been made to the activity of soil organisms, but it should again be emphasized that they play an important role in root growth. The soil teems with bacteria and is usually permeated with fungal hyphae, some attached to mycorrhizal roots. It also contains an extensive fauna, including nematodes, earthworms, and small insects. Some of these organisms are beneficial, whereas others are harmful. The roles of bacteria in nitrogen fixation and denitrification are well known, and it is believed that products of bacterial activity play a role in the development of soil structure by the aggregation of primary soil particles into larger masses.

The development of bacteria and other soil organisms in the immediate vicinity of roots, the rhizosphere, is stimulated by the release of root exudates. These include sugars, amino acids, vitamins, enzymes, and other substances that escape from healthy roots (Rovira and Davey, 1974). The quantity of exudate from the surfaces of growing roots is surprisingly large. In one series of experiments, over a period of 12 days it amounted to 50% of the root dry weight (Prikryl and Vancura, 1980). In addition, considerable material is released by the death and decay of root caps and epidermal and cortical cells sloughed off during secondary growth. These substances stimulate multiplication of microorganisms in the rhizosphere to the point where they may even compete with roots for essential mineral nutrients. It has also been suggested that bacteria in the rhizosphere synthesize significant amounts of growth regulators or stimulate plant growth in some other manner (R. S. Russell, 1977, Chapter 6).

Direct injury to root systems by fungi and nematodes is well known, and various small insects feed on roots. The beneficial activity of earthworms, first publicized by Darwin (1881), is also well known. The soil fauna is discussed by W. E. Russell (1973) in Chapter 11 and the flora in Chapters 9 and 10.

Soil fumigation is usually followed by increased plant growth, but the reasons for this are complex. It can sometimes be attributed to destruction of nematodes and pathogenic fungi, but it might also decrease competition for minerals from soil organisms. Fumigation is occasionally injurious, chiefly when it kills fungi needed to form the mycorrhizal association essential for some plants.

SUMMARY

The depth and spread of root systems and the density of root branching affect the success and survival of plants subjected to drought. The type and size of root systems are controlled by heredity and soil conditions. There are also important interactions between roots and shoots; roots are dependent on shoots for carbohydrates, growth regulators, and certain vitamins, and shoots are dependent on roots for water, minerals, and certain growth regulators.

Roots vary widely in size and longevity, and many of the smaller absorbing roots live from only a few weeks to a few months. Root systems of many plants are modified by root grafts and the presence of mycorrhizae, and in general they present very large surfaces to the soil. The absorptive capacity of root systems is a function of root area and root permeability.

The types of root systems, tap or fibrous, deep or shallow, are determined first of all by the hereditary potentiality of the species, but they are often modified by environmental constraints. The principal environmental factors affecting root growth are the physical properties of the soil, aeration, soil moisture, and temperature. Most vigorous root growth of perennial plants in temperate climates occurs in the spring and early summer, when temperatures are favorable, and in the early autumn after rain occurs. Growth is often reduced or stopped in the summer by soil water deficit and in the winter by low temperature. There is also evidence of some degree of internal control of root growth by shoot growth and by seed production, and there seems to be some seasonal periodicity in regeneration of roots and seedlings.

There is evidence that roots of plants of one species sometimes inhibit growth of roots of another species by the release of toxic substances known as allelochems. Products formed during decomposition of dead roots and other plant residues also sometimes affect root development, and the microorganisms in the rhizosphere likewide influence root growth and function.

SUPPLEMENTARY READING

Böhm, W. (1979). "Methods of Studying Root Systems." Springer-Verlag, Berlin and New York.
Harley, J. L., and Russell, R. S., eds. (1979). "The Soil-Root Interface." Academic Press. New York.
Hook, D. D., and Crawford, R. M. M., eds. (1978). "Plant Life in Anaerobic Environments." Ann Arbor Sci. Press, Ann Arbor, Michigan.
Marshall, J. K., ed. (1977). "The Belowground Ecosystem," Range Sci. Dep. Sci. Ser. No. 26. Colorado State University, Fort Collins.
Rice, E. L. (1974). "Allelopathy." Academic Press, New York.
Russell, E. W. (1973). "Soil Conditions and Plant Growth," 10th ed. Longmans, Green, New York.
Russell, R. S. (1977). "Plant Root Systems." McGraw-Hill, New York.
Also see references at end of Chapter 5.

Water Movement in the Soil–Plant–Atmosphere Continuum

INTRODUCTION

Preceding chapters discussed the properties of water and solutions, cell water relations, soil water, and roots and root systems. Succeeding chapters will discuss water absorption, the ascent of sap, transpiration, and water stress and its effects on plant processes and growth. This chapter deals with water movement in general.

The success of plants depends on the existence of a vascular system capable of moving from roots to shoots the large quantities of water lost by transpiration and the presence of a control or feedback mechanism that keeps absorption and transpiration approximately in balance. Before discussing these problems in detail, we will present a brief review of the general principles governing the movement of water and the relationship to other types of transport.

General Transport Laws

Water movement can be described by special forms of the general transport law governing the movement of heat, liquids, and electricity, and movement by diffusion. In general terms,

$$\text{Flux} = -K \times \text{driving force}$$

where K is a proportionality factor referring to the properties of the system being studied. It has a minus sign because net transport is "downhill." Heat flow is described by Fourier's law, which states that the rate of heat conduction is proportional to the temperature gradient and the conductivity of the system. Thus, ΔT is the driving force and K varies with the material, the thermal conductance of metals being much higher than that of wood, for example. The movement of electricity is described by Ohm's law, which states that current flow is proportional to the gradient in electrical potential or voltage and inversely proportional to the resistance [see Eq. (7.3)]. Flow of water is governed by Poiseuille's or Darcy's law. According to Darcy's law, flow in porous media such as soil is proportional to the pressure gradient and the conductance of the medium or inversely proportional to the resistance of the medium. Poiseuille's law applies more particularly to flow through capillaries such as the xylem elements of plants. Flow in tubes is proportional to the pressure drop and the fourth power of the radius of the tube and inversely proportional to the viscosity of the liquid.

Applications of Poiseuille's and Darcy's laws are discussed in Hillel (1980a, Chapter 8), and the former appears in discussions of flow through the xylem. Fick's law, describing diffusion, is another example of a general transport law. It was discussed in Chapter 1.

Terminology. Fiscus (1983) pointed out that terms such as resistance, resistivity, conductance, and conductivity are often used loosely and even indiscriminately with respect to water movement in soil and plants. An example is the term Lp in Eq. (7.11). It usually is termed the hydraulic conductivity, but it is really the conductance because the dimensions of the system are not specified. The conductance of heat or electricity varies among metals, but conductivity refers to conductance through a conductor of specified dimensions under a given driving force. Electrical resistance is a general term, but resistivity refers to measurement of resistance made in a system of known dimensions, such as a wire of specified length and cross section. Conductivity of soil extracts, cell sap, and other liquids is measured in reciprocal ohms (mhos) in cells with electrodes of known area and spacing. The terms resistance and conductance apply to the capacity of stomata to conduct gases, because no distance term is used, but merely a velocity.

THE SOIL–PLANT–ATMOSPHERE CONTINUUM CONCEPT

Water movement through the soil–plant–atmosphere system is best treated as a series of interrelated, interdependent processes. For example, the rate of water absorption is affected both by the rate of water loss and by the rate at which water can move from the soil to the root surface. The rate of transpiration depends not only on stomatal aperture and the atmospheric factors affecting evaporation, but also on the rate of water absorption, as will be shown later.

Water Transport Equations

Water transport can be described by simple equations. In general terms:

$$\text{Flux} = \frac{\text{driving force}}{\text{resistance}} \tag{7.1}$$

More specifically:

$$\text{Water flux} = \frac{\text{difference in water potential } (\Delta\Psi)}{\text{resistance } (r)} \tag{7.2}$$

This is similar to the equation for Ohm's law, describing the flow of electricity:

$$\text{Current} = \frac{\text{voltage}}{\text{resistance}} \tag{7.3}$$

Because of the similarity of these equations, this treatment of water movement is often termed the Ohm's law analogy. It is a convenient concept to use in describing water movement through the soil–plant–atmosphere continuum (SPAC) because factors affecting water movement can be discussed in terms of their effects on either the driving forces ($\Delta\Psi$) or the resistances (r). However, other treatments are possible, and perhaps even preferable (Fiscus, 1983). According to Richter (1973), this concept was proposed by Huber in 1924, and it was developed further by Gradmann (1928) and by van den Honert in 1948, but it did not come into general use until the 1960s, when it was discussed and applied by Slatyer and Taylor (1960), Rawlins (1963), Cowan (1965), and others. Van den Honert (1948) wrote an expression for steady-state flow as follows:

$$\text{Flow} = \frac{\Psi_{soil} - \Psi_{root\ surface}}{r_{soil}} = \frac{\Psi_{root\ surface} - \Psi_{xylem}}{r_{root}}$$

$$= \frac{\Psi_{xylem} - \Psi_{leaf}}{r_{xylem} + r_{leaf}} = \frac{\Psi_{leaf} - \Psi_{air}}{r_{leaf} + r_{air}} \tag{7.4}$$

This treatment provides a useful model for teaching and research in which the various plant and environmental factors that affect water movement at each stage in its progress through the plant can be treated as affecting either the driving forces or the resistances, or sometimes both, as in drying soil (Fig. 3.12). An example of such a flow diagram or model is shown in Fig. 7.1.

Although useful for teaching and modeling water flow, Eq. (7.4), like most generalizations, is an oversimplification. First, it assumes a steady state, something that seldom exists in plants, although water movement is sometimes treated as a sequence of quasi-steady states (Cowan and Milthrope, 1968; Kaufmann and Hall, 1974). It is therefore not strictly applicable to situations in which measurements are made over time intervals shorter than those required to establish a steady state of water flow in the plant. This problem is discussed in the sections on capacitance and root–shoot relations. Second, it assumes constant resistances in parts of the pathway such as the roots, where the resistance sometimes appears to vary with the rate of flow. Third, water moves as a liquid through the soil and the plant but changes to vapor at evaporating surfaces in the leaves. The problem of apparent variations in root resistance will be discussed later. Although the driving force for movement is always the difference in chemical potential, liquid flow is directly proportional to $\Delta\Psi$, whereas vapor flow is proportional to the differences in vapor pressure or vapor concentration. Water potential is related to vapor pressure by an equation including a logarithmic term, as shown in the following equation:

$$\Psi = RT \ln e/e^o \tag{7.5}$$

Conversion of humidity gradients from vapor pressure to equivalent water potentials greatly exaggerates the drop in potential in the vapor phase. For example, a leaf water potential of -3 MPa is equivalent to a relative humidity in

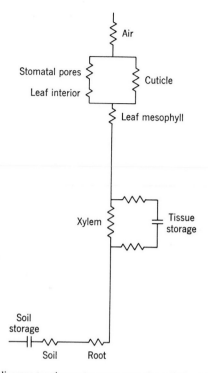

Fig. 7.1. Simplified diagram treating water movement through the soil-plant-atmosphere continuum as analogous to movement of an electrical current through a conducting system containing resistances and capacitances. (Derived from Cowan, 1965.) (From Kramer and Kozlowski, 1979, by permission of Academic Press.)

the leaf of about 98% and a vapor pressure of about 4.15 kPa at 30°C. Reducing the relative humidity of the air from 98 to 50% decreases the vapor pressure nearly 50%, but reduces the equivalent water potential gradient nearly 25-fold. Therefore, according to van den Honert's application of Ohm's law, the resistance in the vapor pathway should be 25 times that in the liquid pathway. Actually, the total leaf and air resistance depends on stomatal opening, vapor pressure deficit, and air movement (Slatyer and Bierhuizen, 1964). Thus, it is preferable to use vapor pressure or vapor concentration rather than water potential in dealing with movement of water in the vapor stage.

Control of Water Movement

There is no doubt that movement of water through and out of plants is ordinarily controlled primarily by cuticular and stomatal resistances at the leaf–air interface. On the other hand, it is also clear that drying soil, cold soil, and inadequate aeration of the root medium all drastically reduce transpiration because they decrease absorption of water and produce a leaf water deficit that causes stomatal closure (Chapter 9). Slatyer (1967, pp. 224–226) explained this as shown in Fig. 7.2. In this diagram, it is assumed that at zero time a plant in darkness with roots in water culture ($\Psi_{root} = 0$, $\Psi_{leaf} = 0$) is illuminated. Transpiration rises to a steady state in about 30 min, and after 1 hr transpiration and absorption are occurring at the same rates and the $\Delta\Psi$ between leaf and root medium is stabilized as at (A). The temperature of the root medium is then reduced enough to increase r_{root} and decrease water absorption. As absorption lags behind transpiration, leaf Ψ_w decreases to level B and water absorption returns to its original rate because the driving force, $\Delta\Psi$, is increased, as shown in condition (B). It is assumed that the decrease in leaf Ψ_w was not great enough to cause stomatal closure. However, if the root medium is cooled more, further reducing water absorption, before $\Delta\Psi$ has fallen low enough to bring water absorption back to the original level, leaf Ψ_w will have decreased so much that partial closure of stomata occurs. This reduces transpiration until it is in balance at a lower level with the new reduced rate of absorption and a much greater $\Delta\Psi$, as seen in situation (C). In actual experience, stomatal closure caused by restricted water absorption often overshoots, resulting in recovery of turgor and even a reopening of stomata, and sometimes cycling in stomatal opening occurs (Chapter 11). In summary, according to this concept, (1) water flow through plants is controlled primarily in the vapor phase between the evaporating surfaces

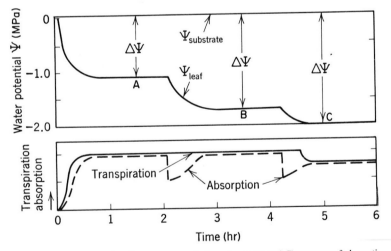

Fig. 7.2. Effects of cooling the root system in two stages (B and C) on rates of absorption and transpiration, on leaf water potential (Ψ_{leaf}), and on the differences between Ψ_{leaf} and $\Psi_{substrate}$ ($\Delta\Psi$). (After Slatyer, 1967.)

and the bulk air; (2) the stomata are the principal regulators of water movement through plants; and (3) increase in resistance to water flow elsewhere, especially in the soil and the roots, reduces transpiration indirectly by reducing leaf turgor enough to cause stomatal closure. Thus, although transpiration is controlled directly by stomatal aperture, plant turgor is often reduced by decreased absorption of water.

Levitt (1966) questioned the correctness of this concept of control over water movement. However, Slatyer and Lake (1966), Cowan and Milthorpe (1967), and others reaffirmed that final control over water movement can occur only in the vapor phase, between the evaporating surfaces in the leaves and the outside air. Trial and error during the evolution of land plants culminated in such a control system in the form of an epidermis covered with a water-resistant cuticle and containing stomata. The fact that the transport equation states that the resistance should be there is merely a rationalization to explain what is obvious from a consideration of leaf structure and stomatal functions.

Other Terms in the Flow Equation

Thus far, we have emphasized driving forces and resistances as shown in Eqs. (7.1), (7.2), and (7.3), and we will continue to do so, both in this chapter and throughout this volume. However, comparison of water flow in the SPAC to the flow of electricity in conducting systems requires consideration of two other terms, conductance and capacitance.

Conductance versus Resistance

There are two ways of expressing the limitation on movement of water in a conducting system with a constant driving force, the resistance in s cm^{-1} or the conductance in cm s^{-1}. A restatement of Eq. (7.1) gives:

$$\text{Resistance} = \frac{\Delta\Psi}{\text{flow}} \tag{7.6}$$

$$\text{Conductance} = \frac{1}{\text{resistance}} \tag{7.7}$$

It is fairly easy to measure electrical resistances in ohms and to calculate the conductance in reciprocal ohms or mhos. Likewise, resistance to water flow can be estimated from measurements of water potential at various points in the system and total flow through the system. Most of the data on water flow are expressed as resistances. However, it can be argued that conductance is preferable to resistance for expressing data on diffusion of CO_2 and water vapor through stomata, because it gives a more accurate representation of the relationship between stomatal aperture and diffusion of gases. Burrows and Milthorpe (1976) plotted the approximate values for conductance and resistance shown in Fig. 7.3. Diffusion porometers are calibrated in resistance units of s cm^{-1}, and most of the older data in the literature are expressed in resistance units. We will generally use resistance units, but we caution the reader to remember that an increase in stomatal resistance from 5 to 15 s cm^{-1} is associated with a much smaller change in stomatal aperture than a change from 0.5 to 1.5 cm s^{-1}, as shown in Fig. 7.3.

Capacitance

Another property of some electrical circuits is the storage of electricity in capacitors or condensers in such a manner that it can be released later into the circuit. Storage of water in parenchyma cells can be regarded as analogous to the

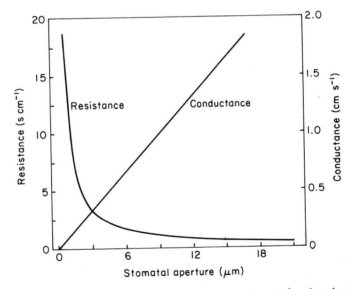

Fig. 7.3.   Relationship between stomatal aperture, resistance in s cm^{-1}, and conductance in cm s^{-1}. Notice that the scales for conductance and resistance are different. (Modified from Burrows and Milthorpe, 1976; the values are only approximations.)

storage of electricity in capacitors, because it is readily available to replace water lost by transpiration. The quantity of water stored in the parenchyma cells of leaves and stems is large because the average herbaceous plant is over 80% water, and even tree trunks contain 50 or 55% of water. Herbaceous plants sometimes lose 25 or 30% of their original water content on hot, sunny days (see Table 12.1), but this is normally replaced at night. Considerable water is removed from the sapwood of transpiring trees, and the water content of tree trunks gradually decreases during the summer. This is discussed in more detail by Waring and Running (1978) and in Chapter 11.

Existence of this capacitance factor prevents a strict application of the Ohm's law analogy because flow through different parts of large plants often occurs at different rates. For example, when transpiration increases rapidly after sunrise, absorption lags behind transpiration and there is a reduction in leaf and stem water content. In trees there is sometimes a progressive reduction in diameter of the bole from the top downward (Dobbs and Scott, 1971), and in the early morning sap flow is more rapid in the branches than in the lower part of the bole. This occurs because the resistance to removal of water from turgid parenchyma cells is usually lower than the resistance to flow of water through the root system. As the water content and water potential of the leaves and upper stem decrease, an increasing proportion of the water replacing that lost by transpiration enters through the roots. When transpiration decreases in the afternoon, the parenchyma cells regain water, causing absorption to continue for a time at a higher rate than transpiration (see Fig. 8.1). The capacitance factor is also involved in the slow equilibration of leaf water potential with changing rate of transpiration, mentioned later in this chapter.

Modeling

The van den Honert equation and the soil–plant–atmosphere continuum concept (SPAC) made important contributions by stimulating interest in modeling various aspects of plant water relations. These range from models of stomata (DeMichele and Sharpe, 1973) and leaves (Gates *et al.*, 1965) to whole plants and stands of plants (Baker *et al.*, 1972; Lemon *et al.*, 1971). An early example is the model of Cowan (1965) for water flow through the SPAC, a simplified version of which is shown in Fig. 7.1. Molz and Ferrier (1982) reviewed several mathematical studies of cell and tissue water relations.

Kaufmann and Hall (1974) and Kaufmann (1976) rearranged components of the van den Honert equation to analyze the impact of soil factors on the water balance of citrus:

$$\Psi_{leaf} = \Psi_{soil} - (flux)\,(r_{soil\ to\ leaf}) \qquad (7.8)$$

Flux is the transpiration rate and $r_{soil\ to\ leaf}$ is the resistance to flow in the liquid phase from soil to leaf. This equation emphasizes the dependence of leaf water potential on rate of transpiration, soil water potential, and resistances to flow in the liquid system. Elving *et al.* (1972) established that the ratio of vapor pressure deficit to leaf diffusion resistance provides a satisfactory estimate of transpiration from leaves on the north side of citrus trees. This is also true for some desert shrubs (Sanchez-Diaz and Mooney, 1979). Deviations from the transpiration rate predicted by this formula in healthy plants indicate that soil factors are limiting water uptake. Sterne *et al.* (1978) used this approach to interpret the wilting of avocado infected with *Phytopthora* as being caused by decreased flow in the soil–plant system, even in moist soil. Duniway (1977) emphasized the importance of increased root resistance as the cause of wilting of plants suffering from root rot diseases. By analyzing a large amount of data on transpiration and leaf water potential, Kaufmann and Hall (1974) determined that when the soil water potential falls below -0.03 MPa, leaf water potential is reduced regardless of the rate of transpiration. They also found that soil temperatures below 15°C reduce leaf water potential, the critical point being about 13.5°C for citrus.

Other applications of SPAC models are discussed by Shawcroft *et al.* (1974), particularly SPAM. It is a model intended to simulate soil–plant–atmosphere interactions and can be used to study crop responses to cultural treatments, environmental factors, and even such details as stomatal number and size and leaf angle. Models to control the timing of irrigation are discussed in Chapter 4. However, it seems clear that more information concerning the interaction of plants with their environment is required to improve the usefulness of these models. Perhaps one of the most important benefits of model building is that it identifies important deficiencies in our knowledge of important physiological processes and of how they are affected by the environment. An example is the paper by Running *et al.* (1975), discussing the information required to model water flow in trees. Models do not create new information, but they often put existing information into a form in which it can be used more effectively. Curry and Eshel in Raper and Kramer (1983) pointed out that more information is needed before simulation models will be useful in managing crop water problems, and Passioura (1973) presented some cautions concerning the limitations of simulation models.

DRIVING FORCES AND RESISTANCES

The driving force causing water movement is the difference in chemical potential. However, this can be expressed in terms of osmotic (Ψ_s), pressure (Ψ_p), matric (Ψ_m), and gravitational (Ψ_g) potentials, as shown in the following equation:

$$\Psi_w = \Psi_s + \Psi_p + \Psi_m + \Psi_g \qquad (7.9)$$

We will now discuss how the various driving forces operate and which are predominant at various stages in the soil, roots, stems, and leaves; the nature of the resistances in various parts of the pathway will also be considered. Many details of water movement, the structure of the conducting system, and the process of transpiration will be discussed in later chapters.

Water Movement in Soil

Driving Forces. All of the four kinds of driving forces just mentioned operate in soils, but different ones dominate at different water contents and water potentials. In soil that is wetter than field capacity, gravity causes saturated flow or drainage downward through noncapillary pore space until the weight or tension of the water column is balanced by the matric forces in the soil. As mentioned in Chapter 3, the matric forces consist principally of capillary forces in the small capillary pores where the surface tension of the water menisci balances the tension exerted by gravity and attractive forces on the surface of soil particles and at broken edges of clay particles.

In soils with water contents ranging from field capacity to the permanent wilting percentage, water movement occurs largely by unsaturated flow as liquid in films and through capillary pores, along gradients of decreasing matric potential. These gradients are produced by evaporation of water from the soil surface and absorption by plant root systems.

Resistances. As the water content approaches the wilting point, water movement as liquid decreases rapidly because the number and size of channels through which water can move decrease and the difference in water potential between soil and roots decreases. Thus, there is an increase in resistance or a decrease in conductance to water flow and a decrease in driving force, as shown in Fig. 3.14. Near the permanent wilting percentage, most water movement occurs in the vapor phase, as pointed out in Chapter 3. It has been proposed from time to time that as drying soil and roots shrink, a gap may develop between soil and roots. Philip (1958) suggested that such a gap might explain some peculiarities of absorption in saline soil, but these also occur in water culture (Slatyer, 1961). However, as pointed out later in this chapter, situations in which uptake of water is limited by resistance in the soil or at the soil–root interface are probably unusual until the soil water content approaches the permanent wilting percentage.

Water Movement from Soil into Roots

Two kinds of driving forces cause movement of water from soil into roots: osmotic movement in slowly transpiring plants and mass flow in rapidly transpiring plants, caused by tension or negative pressure in the xylem sap. Tension develops because the water columns of the xylem are anchored by inbibitional or matric forces in the evaporating surfaces of the cells of the leaves. Hence, the tension may be regarded as matric in origin, but it is treated as tension or negative pressure in the xylem sap.

For our purposes, the principal forces involved in water absorption may be described as follows:

$$\text{Absorption} = \frac{(\Psi_m + \Psi_s)_{\text{soil}} - (\Psi_p + \Psi_s)_{\text{root}}}{r_{\text{soil}} + r_{\text{root}}} \qquad (7.10)$$

The term Ψ_m refers to the matric potential, Ψ_s to the osmotic potential, Ψ_p to the pressure potential, all expressed in MPa, and r to the resistance to water flow in s cm^{-1}. Uptake of water can be described as osmotic or passive (nonosmotic), depending on which term is dominant in the equation.

Osmotic or Active Absorption.   Root systems in moist, warm, well-aerated soil often behave as osmometers because accumulation of solutes in the root xylem lowers the water potential below that in the soil. As a result, water diffuses inward and positive pressure (root pressure) develops in the xylem, resulting in guttation or exudation from stumps of detopped plants. This is discussed in detail in Chapter 8 in the section on osmotic absorption and root pressure.

Passive Absorption.   As the rate of transpiration increases, the increasing mass flow of water through the roots dilutes the root xylem sap (see Fig. 7.4) until the osmotic mechanism becomes ineffective and absorption is controlled by the pressure potential in the xylem sap. In rapidly transpiring plants in drying soil, this often falls as low as -1.5 to -2.0 MPa, producing a much steeper gradient in water potential from soil to roots than can be produced by the osmotic mechanism. This is shown by the fact that intact plants can absorb water from drier soil and more concentrated solutions than detopped root systems (Army and Kozlowski, 1951; Jäntti and Kramer, 1957; McDermott, 1945). Eventually, as the soil dries, its water potential decreases and approaches the lowest water potential that can be produced in a water-stressed plant. Absorption slows and finally ceases because of lack of a sufficient gradient in water potential from soil to roots. Increasing resistance to water flow in the soil and in roots (decreasing conductance), and possibly decreasing contact between roots and soil (Tinker,

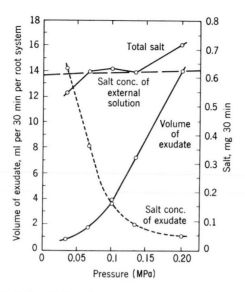

Fig. 7.4. Movement of water through a tomato root system under pressure. The decrease in xylem sap concentration indicates the presence of an ion barrier in roots. Xylem sap concentration decreases in the root xylem as the rate of transpiration increases, decreasing the osmotic component of the absorption mechanism. The changing volume–pressure relationship has been attributed to change in root resistance, but it is probably caused by change in the principal driving force from osmotic to pressure flow (Fiscus, 1975). (From Lopushinsky, 1964a.)

1976), also decrease the rate of water movement to the soil–root interface. The absorption mechanisms are discussed in more detail in Chapter 8.

Transition from Osmotic to Pressure Mass Flow. As indicated in the preceding section, there is a gradual transition from osmotic flow to a largely nonosmotic, passive pressure or mass flow in transpiring plants caused by tension in the xylem. This change is demonstrated quantitatively in experiments such as those of Mees and Weatherley (1957) and Lopushinsky (1964a), in which water was forced through root systems by increasing the pressure on the water solution in which the roots were immersed. An example of the results from such an experiment is shown in Fig. 7.4.

A simplified equation for osmotic and pressure flow is:

$$J_v = Lp \, (\Delta\Psi_p + \sigma\Delta\Psi_s) \tag{7.11}$$

Here J_v is the total volume flow, Lp the hydraulic or pressure flow conductance, $\Delta\Psi_p$ the pressure difference between root xylem and outside solution, σ the reflection coefficient, and $\Delta\Psi_s$ the difference in osmotic potential between the xylem sap and the external solution.

When $\Delta\Psi_p = 0$, water enters roots entirely by osmosis. However, as ΔP

increases, the increasing water flow quickly dilutes the xylem sap and osmotic movement, caused by $\Delta\pi$, is gradually replaced by pressure flow caused by ΔP. The line representing total flow is curved in the region where both osmotic and pressure flow are occurring, but it becomes linear where pressure flow becomes the dominant driving force (Fig. 7.4). Calculations by Fiscus indicate that a plot of osmotic movement over $\Delta\pi$, the difference in $\Psi\pi$ between the root xylem sap and the external solution, is a straight line over a considerable range of $\Delta\pi$ (Fiscus and Kramer, 1975). Although osmotic movement and pressure flow can be described separately by straight lines, when a transition from one to the other is occurring a curved line results, as shown in Fig. 7.4. This is often interpreted erroneously as indicating a change in root resistance. Fiscus (1975) and Dalton *et al.* (1975) made mathematical analyses of this problem that include the effects of Lp, change in external osmotic potential, and change in salt uptake on the degree of nonlinearity between pressure and water flow. Their models explain phenomena such as the change in apparent negative resistances and apparent non-osmotic uptake of water (Fiscus and Kramer, 1975). We will return to the problem of changing resistance later in this chapter.

WATER MOVEMENT THROUGH PLANTS

We are concerned here with driving forces and resistances in various parts of the pathway. The structure and processes involved and the rise of sap in tall trees will be discussed in more detail in later chapters.

Pathways and Driving Forces in the Xylem

After water crosses the cortical cells of the roots, it enters the xylem and moves longitudinally through the various roots to the base of the stem, where it passes through the root–shoot transition zone into the stem xylem. From the stem xylem it is distributed into branches and finally into leaves. The general plan of the conducting system of a young plant is shown in Fig. 7.5. From the xylem in the leaf veinlets, water moves out across the mesophyll cells to the evaporating surfaces. Richter (1973) discussed the complications that arise in attempting to estimate fluxes and resistances in these branched pathways. In transpiring plants, the driving force for water movement is the tension or negative pressure created in water by evaporation from exposed cell walls, chiefly in the leaves. This creates imbibitional or matric forces in the cell walls that produces tension in the xylem sap. This tension is represented by Ψ_p in Eq. (7.10), and it has a negative

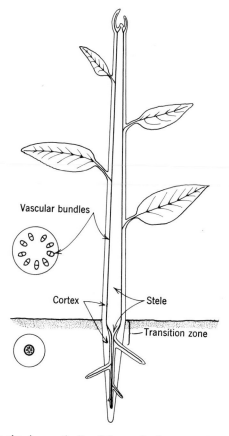

Fig. 7.5. Diagram showing continuity of the conducting system in a young dicotyledonous plant. Cross sections show the arrangement of vascular tissues in roots and shoots. (Adapted from various sources.)

value in transpiring plants. Occasionally, as in slowly transpiring plants that exhibit root pressure, it is positive. Thus, the driving force for water seems to be clear and presents no special problems.

The xylem pathway is also straightforward. The conducting system consists of dead elements from which the protoplasts have disappeared. In angiosperms many end walls also disappear, forming vessels 20–700 or 800 μm in diameter and several centimeters to several meters in length. In gymnosperms water moves through tracheids, which are single cells about 30 μm in diameter and 3–5 mm in length. Pits or thin areas in the walls of adjacent elements facilitate water movement from element to element, while preventing the spread of gas bubbles. Xylem vessels are sometimes blocked by gas bubbles and by tyloses caused by protrusion of the protoplasts of adjacent parenchyma cells into the vessels. Some resistance is also developed by constriction in leaf petioles and by thickenings on the inner surfaces of vessel walls, but on the whole, the xylem provides a pathway for rapid movement of large volumes of water over long distances. Occasional exceptions are reported. One is the high resistance to water flow through flower stalks of onion which results in much more severe water stress in the flowers than in the leaves (Millar *et al.*, 1971). Another is the high resistance to water flow in the upper part of the stem in a wilty corn mutant, caused by poor development of xylem vessels (Dube *et al.*, 1975).

Water Movement Outside of the Xylem

Uncertainties exist in connection with the pathways and the resistances to water movement outside of the xylem, both in roots and in leaves, and we will now discuss some of these. One difficulty is lack of reliable information concerning the relative resistances to water movement through cell walls versus across protoplasts or through the symplast (Newman, 1976).

The general magnitude of the resistance to water movement across a mass of parenchyma cells versus resistance to movement in the xylem is indicated by the following example. If the average length of tracheids is 5 mm, water must traverse 200 tracheid walls per meter of xylem. However, if the average diameter of a parenchyma cell is assumed to be 50 μm, water must cross 20,000 cells per meter. Thus, if the resistance to flow were the same per wall in both kinds of tissue, it would be 100 times greater through the parenchyma cells than through the tracheids. However, the added resistance of protoplasmic membranes in parenchyma cells makes the difference in resistance even greater. The comparison is more striking in angiosperms, where vessel elements are usually at least a few centimeters and sometimes several meters in length. These calculations indicate that resistance to radial water movement across the root cortex or out of the veins in leaves is much greater per unit of distance than the resistance to longitudinal movement in the xylem.

It should be pointed out that inward movement of water across the root cortex occurs much more slowly than movement through the xylem. For example, Russell and Woolley (1961) made the reasonable assumption that a corn plant possesses 2 m^2 of leaf surface (both surfaces), 2 m^2 of roots functioning as absorbing surface, a stem 3 cm in diameter with a xylem cross section of 0.2 cm^2, and a transpiration rate of 200 g/hr. Assuming these dimensions, water would enter the roots at the same velocity at which it evaporates from the leaves, or about 0.01 cm/hr. However, it must move through the constricted xylem passageway at the base of the stem at a velocity of 1000 cm/hr to replace the water lost by transpiration. Thus entrance and exit of water are occurring at rates explainable in terms of diffusion, but movement through the xylem can only be by mass flow. Rates of conduction through the xylem in woody plants often exceed 2500 cm/hr (Table 10.2).

Relative Importance of Various Resistances

The existence of substantial resistances to water flow in the soil–plant system is indicated by the development of leaf water potentials in transpiring plants that are considerably lower than those in the medium in which the plants are growing. For example, leaf water potentials of -1.0 to -2.0 MPa are common on hot, sunny days, even in plants growing in nutrient solutions or moist soil where the water potential is only -0.1 or -0.2. However, there has been considerable discussion concerning the location of the dominant resistance, and especially whether it is in the soil, in the root, or even at the soil–root interface.

Resistances in Plants. The resistance to water flow into roots must be large because when roots are removed the rate of absorption increases in plants ranging from sunflowers (Kramer, 1938) to pine trees (Running, 1980). The lag of absorption in relation to transpiration is also significantly reduced, as shown in Fig. 7.6. One of the few direct attempts to measure the relative resistances in various parts of a plant was made by Jensen *et al.* (1961). They arranged plants

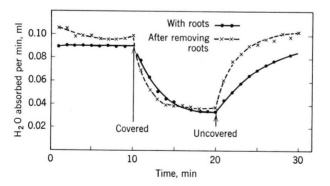

Fig. 7.6. Diagram showing that removal of the root system decreases the lag of water absorption behind transpiration in sunflower plants. Absorption by intact plants was measured for 10 min with shoots exposed. The shoots were then covered for 10 min to reduce transpiration and uncovered for another 10 min. The root systems were then removed, and the experiment was repeated. The lag in response of absorption to change in rate of transpiration was reduced by removal of the roots because they offer a significant resistance to water absorption. (From Kramer, 1938.)

in a three-compartment chamber (see Fig. 7.7) so that water could be forced through them under pressure. By cutting off either the roots or the leaves, the resistance in each part could be estimated. Although the method has some defects, the results seem reasonable. The data are presented in Table 7.1, where it is seen that the root resistance of sunflower is 1.6 that of the leaves and nearly 4.0 times that of the stem. The total resistance of tomato is considerably lower than that of sunflower, but the root resistance is 1.4 times that of the leaves and 4.0 times that of the stems. Incidentally, water was found to move equally well in either direction in these plants.

Neumann *et al.* (1974) estimated resistances to water movement from measurements of leaf water potential at various levels on the stem and at various rates of transpiration. They estimated that stem resistance was 15–25% of the total plant resistance in corn but only 8% in sunflower, leaf resistances were 15–30 and 20%, respectively, and root resistance was 50–70% of the total in corn and 70% in sunflower. The size of the root resistance varied so much from plant to plant that it was impossible to draw conclusions concerning species differences.

Resistances in Soil and at the Soil–Root Interfaces.   There continues to be considerable difference in opinion regarding the relative importance of resistance to water movement in the soil, at the soil–root interface, and in the plant. At one time, it was assumed that water movement through the soil is so slow that roots of a rapidly transpiring plant might actually dry out the soil in their immediate vicinity, leaving them surrounded by cylinders of soil much drier than the bulk soil. Gardner (1960) developed equations for flow toward roots indicating that only a small gradient in water potential is required to move water at the required rate at a soil Ψ_w of -0.5 MPa, but a much steeper gradient at -1.5 MPa because of reduced hydraulic conductance. Later, Gardner and Ehlig (1962, 1963) found

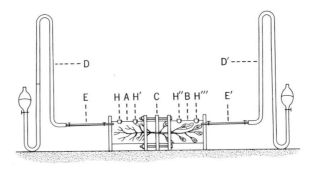

Fig. 7.7.   Apparatus for measuring water flow through plants. A and B are closed chambers, C is a partition separating roots from shoots, E and E′ are calibrated capillary tubes attached to mercury manometers D and D′, and H, H′, H″, and H‴ are stoppered openings. (From Jensen *et al.*, 1961.)

TABLE 7.1 Estimated Relative Resistances to Water Movement
through Roots, Stems, and Leaves of Sunflower and Tomato[a]

	Resistance	
	Sunflower	Tomato
Whole plant	1.60	1.40
Leaves	0.66	0.60
Stem	0.26	0.24
Roots	1.00	1.00

[a] From data of Jensen *et al.* (1961).

experimentally that the total resistance in the soil–plant system began to increase at a soil water potential below −0.06 MPa, and they concluded that at a soil water potential lower than 0.1 or 0.2 MPa, soil resistance became limiting. Cowan (1965) also developed equations for water flow which indicated that soil resistance would exceed plant resistance at a soil water potential only a few bars below zero. Newman (1969) reviewed the literature on root density and found that it is usually much greater than the density used by Gardner and by Cowan. He concluded that with the root densities usually existing, soil resistance in the rhizosphere or immediate vicinity of the root does not become limiting for most plants until the soil water content is near the permanent wilting percentage.

Herkelrath *et al.* (1977a,b) found a rapid decrease in absorption at soil water potentials below 0.01 MPa and attributed this to decreasing contact between roots and soil as the soil dried, a view also held by Weatherley and his associates. For example, Faiz and Weatherley (1978) located the larger part of the soil–plant resistance at the soil–root interface. This is based on the assumption that as the soil dries it shrinks, decreasing contact with the soil. It is also well established that roots can shrink 25% or more in diameter as they become dehydrated, further reducing contact with the soil (Huck *et al.*, 1970; Cole and Alston, 1974). However, Tinker (1976) reviewed the shrinkage problem and pointed out several uncertainties. For example, roots maintain partial contact with the soil and soil water films even when both soil and roots shrink. Also, roots hairs presumably maintain some degree of contact, and the mucilaginous coating (mucigel) found on many roots probably aids in maintaining contact for water movement between roots and soil. If they do not collapse, root hairs might be much more useful when soil and roots are shrinking than when there is complete contact between roots and soil. It has also been suggested that water might move across gaps from soil to roots by diffusion in the form of vapor (Bonner, 1959; Cowan and Milthorpe, 1968; Philips, 1958). However, calculations by Bernstein *et al.* (1959) indicate that diffusion as vapor is so much slower than movement in liquid films that it could not supply enough water. Tinker concluded that under field conditions resistance at the soil–root interface is seldom a serious problem. However, root and soil shrinkage deserve more study.

Several investigators, including Reicosky and Ritchie (1976) using sweet corn and sorghum and Taylor and Klepper (1975) using cotton, reported that root resistance exceeds soil resistance. Blizzard and Boyer (1980) found the plant resistance in soybean to be greater than the soil resistance over a soil water potential range from -0.02 to -1.1 MPa, at which point transpiration had decreased to 10% of the original rate. They suggest that some investigators have neglected the apparent increase in plant resistance that occurs in drying soil. Possible causes for this include decrease in root growth and increase in suberization (Kramer, 1950), cavitation in the xylem (Milburn and Johnson, 1966), and decreased root–soil contact (Herkelrath *et al.*, 1977b).

Resistance to Longitudinal Flow

Resistance to longitudinal flow in roots is usually regarded as negligible in comparison to radial movement through the soil. However, Wind (1955) claimed that water can rise more rapidly through soil than through roots of some grasses, and Passioura (1972) observed high resistance to longitudinal movement in the seminal roots of wheat. Herkelrath *et al.* (1977b), however, found no evidence of any effective resistance to longitudinal movement in wheat roots, Stone and Stone (1975a) reported very low resistances to water movement through roots of red pine, and Sands *et al.* (1982) found negligible resistances in bean and pine roots. Meyer and Ritchie (1980) found similar resistances in shallow- and deep-rooted cotton and concluded that the longer roots had a lower resistance per unit of length, probably because the xylem was better developed in the long roots. In general, resistance to longitudinal movement through healthy roots is seldom a dominant factor in water supply to the shoot.

The Control System

We have discussed the nature of the driving forces and resistances involved in water movement through the soil and the plant. The next questions are how water absorption and transpiration are kept in balance and how water flow through the plant is regulated. Quantitatively, water flow is regulated chiefly by the loss of water by transpiration because this accounts for over 95% of the water absorbed by most plants. However, a small quantity is also used in cell expansion and biochemical processes in which water serves as a reactant. It was pointed out earlier that evaporation of water from leaf cells creates matric or imbibitional forces (Ψ_m) in the cell walls that cause water to move into them from the xylem. This produces tension or negative pressure in the xylem sap (Ψ_p) that is transmitted to the roots and causes inflow of water, as discussed under passive absorption in Chapter 8. Thus, increasing transpiration causes increased absorption and decreasing transpiration causes decreased absorption. Conversely, if absorption is materially reduced by factors such as cold or drying soil or inadequate aeration, the water potential in the roots is reduced and this reduction is transmitted to the leaves in the xylem sap, where decreasing water potential causes stomatal closure and reduction in transpiration. Thus, an effective feedback mechanism operates between roots and shoots through the continuous, cohesive water system of the xylem and the cells in contact with it. The operation of this system is also discussed in Chapter 10.

The necessity for such a control system and the role of the xylem sap in providing it were recognized by Renner (1912), who stated that the cohesive water columns in the xylem constitute the only mechanism ensuring that water loss from leaves will be balanced by an equal rate of absorption through the roots. Unfortunately, critics of the cohesion theory of the ascent of sap overlooked this essential role of the sap stream for several decades.

Growth also requires water. The amount used is small quantitatively but very important qualitatively. Cell expansion can create sufficient tension in the xylem sap to lift the small quantity of water required for growth. The absorption of water by growing plant tissue is indicated by the fact that leaf tissue rarely shows zero water potential at zero transpiration, as seen in Fig. 12.20. Thus, although water flow in the plant is dominated by transpiration, in the absence of transpiration cell expansion will create some water movement.

VARYING RESISTANCES IN ROOTS

Study of water absorption is complicated by variations in root resistance. These include differences in permeability of roots of different ages and at different distances from the apex on a single root, changes with dehydration and temperature, changes in apparent resistance with rate of water flow, and diurnal changes in root resistance.

Age and Development of Roots

Differences in permeability of roots with increasing age and maturation are discussed in Chapters 5 and 8. An example of the changing permeability along a root is shown in Fig. 5.8, and the effects of root development on permeability of an entire root system are shown in Fig. 6.2. Measurements of root permeability made at different regions on roots vary widely and do not necessarily indicate the average condition of the root system. These differences can be explained by changes in structure that occur during maturation of roots, discussed in Chapter 5.

Change in Resistance with Change in Rate of Water Flow

Brower (1953) reported that the absorbing region shifted away from the root tip as the rate of absorption increased (Fig. 7.8). It also has been observed by several investigators that the rate of water flow through roots under pressure increases more rapidly than the pressure. This is shown in Fig. 7.4 and in experiments such as those of Mees and Weatherley (1957). This indicates that the root resistance apparently decreases with increasing rate of water flow. It has

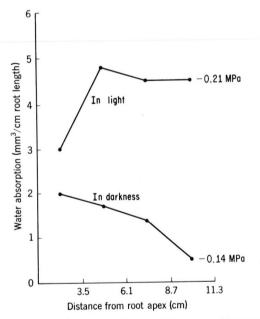

Fig. 7.8. Shift in the region of the root through which most rapid water absorption occurred when the rate was increased by increasing the rate of transpiration. In darkness the water potential in the root was estimated to be 0.14 MPa, and in the light 0.21 MPa. At the low rate of transpiration absorption was probably largely osmotic; at the high rate it was largely by mass flow. (After Brouwer, 1953.)

also been reported from experiments in which the water flow into the roots was varied by varying the rate of transpiration and the leaf water potential was measured and plotted over the rate of transpiration, as shown in Fig. 7.9.

The variation in results in Fig. 7.9 is difficult to explain. In some experiments leaf water potential decreased as the rate of transpiration increased, but not in others. Failure of the leaf water potential to decrease with increasing water flux suggests that the root resistance was decreasing. The curves obtained in different experiments for the same kind of plant had quite different slopes, as seen for maize (D and L), sunflower (C, E, H, and N), and cotton (O and R). Some of the differences in slope among plants of the same species might be caused by different rates of osmotic or active water absorption occurring at low rates of transpiration in plants with different past treatments. Kaufmann (1976, p. 316) mentions large differences in exudation rates of sunflowers. Part of the failure of leaf water potential to approach zero at zero water loss may result from continued

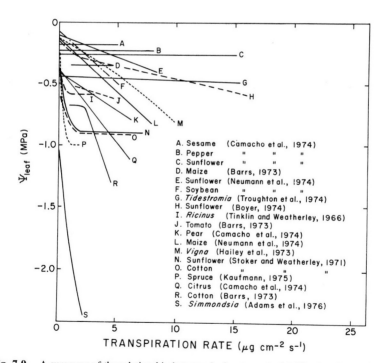

A. Sesame   (Camacho et al., 1974)
B. Pepper   "   "   "
C. Sunflower   "   "   "
D. Maize   (Barrs, 1973)
E. Sunflower (Neumann et al., 1974)
F. Soybean   "   "   "
G. *Tidestromia* (Troughton et al., 1974)
H. Sunflower (Boyer, 1974)
I. *Ricinus*   (Tinklin and Weatherley, 1966)
J. Tomato (Barrs, 1973)
K. Pear (Camacho et al., 1974)
L. Maize (Neumann et al., 1974)
M. *Vigna* (Hailey et al., 1973)
N. Sunflower (Stoker and Weatherley, 1971)
O. Cotton   "   "   "
P. Spruce (Kaufmann, 1975)
Q. Citrus (Camacho et al., 1974)
R. Cotton (Barrs, 1973)
S. *Simmondsia* (Adams et al., 1976)

Fig. 7.9. A summary of the relationship between leaf water potential (Ψ_{leaf}) and rate of transpiration of well-watered plants at normal growing temperatures. Curves D, I, J, M, N, O, and R were obtained with plants growing in solution culture; the remainder are for plants growing in well-watered soil, sand, or perlite. Curves A, B, C, D, and G show no decrease in leaf Ψ_w with increase in transpiration, suggesting an increase in root permeability. See the section on varying resistances in roots for discussion of these data. (From Kaufmann, 1976, by permission of Academic Press.)

absorption of water by growing tissue (Molz and Boyer, 1978), although this can scarcely explain the value of 1 MPa for *Simmondsia*.

There are various possible explanations for the differences in relationship between leaf water potential and rate of water flow (transpiration). As mentioned earlier in this chapter, Fiscus (1975) and Dalton *et al.* (1975) suggested that some discrepancies are caused by transition from active or osmotic absorption to passive absorption. Boyer (1974) suggested that the apparent variation in resistance really occurs in the leaves. He interpreted his data to indicate that there are two pathways for water movement in leaves, a high-resistance pathway for water moving into protoplasts and a low-resistance pathway around the protoplasts, used by the transpiration stream. At low rates of transpiration most of the water is entering the growing cells by a high-resistance pathway, but with increasing transpiration most of the water bypasses the protoplasts and follows a low-resistance pathway. Thus, there might be a considerable change in water flow without a corresponding change in water potential because most of the additional water movement occurs in a low-resistance pathway. In this model, leaf water potential is controlled more by the rate of water movement into the protoplasts than by the rate of transpiration. The high protoplast resistance means that leaf cells equilibrate slowly after changes in rate of transpiration, as suggested by Bunce (1978a) (see the section on root–shoot relations). However, a high resistance to water movement into protoplasts seems inconsistent with the rapid recovery of turgor by wilted leaves when the stem of a badly wilted plant is cut under water. Apparently, more research is needed on the rate of equilibration of leaf parenchyma cells. Powell (1978) suggested that changes in turgor of the endodermal cells might cause changes in their hydraulic conductivity that could explain both the nonlinear relationship between leaf water potential and water absorption and diurnal cycles in water uptake. However, there is no solid evidence to support this idea.

Dehydration and Root Resistance

In most of the experiments cited here, soil water was not limiting. However, if the soil is allowed to dry toward the wilting point, root permeability may be reduced, and actual wilting can cause serious reduction in root permeability. Tomato, tobacco, and sunflower plants kept wilted overnight, rewatered, detopped, and attached to a vacuum line, showed a reduction in water uptake of over 35% as compared with unwilted controls. Rewatering should have eliminated any effects of soil or root shrinkage, and the reduced water uptake was attributed to decreased root elongation and increased suberization of roots (Kramer, 1950). Tomatoes kept wilted for 4 days recovered 4 days after they were rewatered, but sunflowers did not (Boyer, 1971). Ramos and Kaufmann

(1979) found that root resistance of rough lemon was increased by water stress. The increase was apparently caused by decrease in root permeability rather than by decrease in root growth.

Diurnal Changes in Resistance

Several investigators have reported diurnal variations in root permeability and resistance, the highest permeability occurring near midday, the lowest near midnight (Barrs and Klepper, 1968; Skidmore and Stone, 1964). Diurnal variation in the rate of root pressure exudation has also been reported, with maximum rates near noon and minima near midnight (Grossenbacher, 1939; Vaadia, 1960). Hagan (1949) even found a diurnal cycle in the rate at which water supplied to stumps moved back through root systems into dry soil (negative exudation). Parsons and Kramer (1974) measured water movement through cotton roots under pressure or vacuum for periods of several days and found that the rate under constant pressure was two to three times greater near noon than near midnight, indicating that the resistance was two or three times greater at midnight than at noon. This cycle gradually disappeared 2 or 3 days after plants were detopped, and was prevented by exposure to 8 or more days of continuous light before the shoots were removed. It could not be explained by variation in volume of root pressure exudation because there was no significant cycling in total salt per unit of time. The cycling appeared to be related in some way to signals from the shoot because the cycle could be reset by reversing the light–dark cycle several days before starting. This diurnal cycling cannot be fully explained. Perhaps it is relevant to report that diurnal cycling in respiration has also been observed in root systems of intact corn and soybean plants, but not in detached roots (Huck *et al.*, 1962). Neales and Davies (1966) observed diurnal cycling of respiration in wheat roots that was related to photoperiod, and suggested that it was a response to a varying supply of photosynthate from the shoot. Thus far, there is no satisfactory explanation for diurnal variations in apparent root resistance.

Root–Shoot Relations

As mentioned earlier, much of the evidence for short-term variations in root resistance depends on the response of leaf water potential to change in transpiration. Research by Bunce (1978a) on soybean suggests that in some of these experiments the measurement time was too short for leaf water potential to attain a steady state. As shown in Fig. 7.10, leaf water potential measured 1 hr after

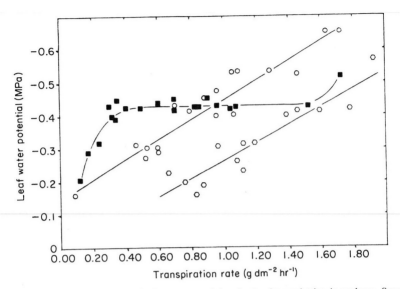

Fig. 7.10. Relation between leaf water potential and rate of transpiration in soybean. Squares represent leaf water potential measured 1 hr after the transpiration rate became constant at 500 μE m^{-2} s^{-1}; open circles 3 hr or more after the transpiration rate became constant at 500 μE m^{-2} s^{-1}; hexagons 3 hr or more after the transpiration became constant at 1500 μE m^{-2} s^{-1}. Transpiration is measured per unit of root surface area. Average ratio of leaf area (one side) to root surface was 0.8 to 1.0. (From Bunce, 1978a.)

transpiration became constant was similar over a wide range of transpiration rates, but if it was measured 3 hr or more after transpiration became constant, it decreased with increasing rate of transpiration, as might be expected. It was also found that low light and decreased leaf area resulted in lower leaf water potentials at a given rate of transpiration than high light and normal leaf area, suggesting increased root resistance with low light. Bunce's data suggest that the shoots are affecting the apparent root resistance and complicating interpretation of the relationship between leaf water potential and transpiration in terms of root resistances. In contrast, M. R. Kaufmann (private communication) observed no decrease in leaf water potential in sesame, pepper, or sunflower over a period of several hours. The problem of apparent changes in root resistance clearly deserves more study.

SUMMARY

Water movement through the soil–plant–atmosphere continuum is a series of interrelated, interdependent processes which can be treated as analogous to the flow of electricity through a conducting system. Thus, water movement can be described by an analogy to Ohm's law, which states that flow is proportional to the driving force, the water potential, and inversely proportional to the resistances in the flow path. Factors affecting water movement can therefore be discussed in terms of effects on water potential and on resistances.

The xylem provides an efficient pathway for water movement by mass flow from roots to leaves. However, there is uncertainty concerning the exact pathway of water movement outside the xylem, where it must move from the epidermis to the root xylem and from the xylem in leaf veins to the cells bordering the intercellular spaces and substomatal cavities and to the epidermis. Movement might occur across the vacuoles, through the symplast, or in the walls, or probably by all three pathways, and present information does not clearly indicate which pathway is most important.

The driving forces can be described as osmotic, matric, pressure, and gravitational. All four operate in soil, their relative importance depending on the water content. In slowly transpiring plants absorption is chiefly by osmosis, but in rapidly transpiring plants tension or negative pressure in the xylem sap is the chief driving force, and this brings about both water absorption and the ascent of sap. Loss of water in the form of vapor (transpiration) is controlled by the gradient in water vapor pressure between evaporating surfaces and the outside air and by stomatal and cuticular resistances. In moist, warm soil water movement through plants is controlled by the rate of transpiration, but resistances in the soil and roots often become important.

The relative importance of soil and plant resistances is of particular importance. It has been claimed that soil resistance becomes limiting in relatively moist soil, but most experiments indicate that plant resistances exceed soil resistances until soil water content has decreased to the vicinity of the permanent wilting percentage. Sometimes shrinking soil and roots may produce an increase in resistance at the soil–root interface. Root resistance is also increased by dehydration and cooling. Results of some experiments have indicated apparent decreases in root resistance with increasing rate of water flow, but these probably result from failure to take into account all the forces involved in water absorption or from failure to equilibrate leaf water potential.

More information is needed concerning the pathway of water and solute movement across masses of cells in roots and leaves. We also need more information concerning effects of shoots on root functioning with respect to such phenomena as diurnal variations in root resistance and root pressure exudation.

SUPPLEMENTARY READING

Blizzard, W. E., and Boyer, J. S. (1980). Comparative resistance of the soil and the plant to water transport. *Plant Physiol.* **66,** 809–814.

Herkelrath, W. N., Miller, E. E., and Gardner, W. R. (1977). Water uptake by plants I. II. *Soil Sci. Soc. Am. J.* **41,** 1033–1043.

Newman, E. I. (1976). Water movement through root systems. *Philos. Trans. R. Soc. London, Ser. B* **273,** 463–478.

Nobel, P. S. (1974). "Introduction to Biophysical Plant Physiology." Freeman, San Francisco, California.

Richter, H. (1973). Frictional potential losses and total water potential in plants: A reevaluation. *J. Exp. Bot.* **24,** 983–984.

Slatyer, R. O. (1967). "Plant Water Relationships." Academic Press, New York.

Tinker, P. B. (1976). Transport of water to plant roots in soil. *Philos. Trans. R. Soc. London, Ser. B* **273,** 445–461.

van den Honert, T. H. (1948). Water transport as a catenary process. *Discuss. Faraday Soc.* **3,** 146–153.

Zimmerman, M. H., and Brown, C. L. (1971). "Trees: Structure and Function." Springer-Verlag, Berlin and New York.

8

The Absorption of Water and Root and Stem Pressures

INTRODUCTION

The continuous absorption of water is essential to the growth and even the survival of most plants. Only a few xeromorphic types have rates of water loss so low that they can survive for more than a day or two without absorbing measurable quantities of water. The daily loss of water by transpiration often exceeds the water content of a plant. For example, a corn plant may lose 2–4 liters of water on a hot summer day, or up to twice the weight of water in the plant. Unless most of the water lost is replaced immediately, rapidly transpiring plants would die of desiccation in a single day.

Absorbing Organs

Most of the water used in plants is absorbed through roots, but under some conditions water and solutes are absorbed through leaves and stems.

Absorption through Leaves.   When wetted, the cuticle of leaves is moderately permeable, permitting the entrance of water and solutes, including minerals and herbicides. Absorption of water and solutes also occurs through lenticels and leaf scars on stems. Conversely, rain and sprinkler irrigation can leach minerals out of leaves (Madgwick and Ovington, 1959; Tukey *et al.*, 1965). Some early work on absorption of liquid water through leaves was reviewed by Miller (1938, pp. 188–190) and by Williams (1933).

Dew and Fog.   There has been considerable difference of opinion concerning the absorption of dew and fog by plants and its importance (Stone, 1957). Duvdevani (1953, 1957), Waisel, (1958), and Gindel (1973) claim that dew is important for the survival and even for the growth of plants in Israel, and Stone and Fowells (1955) showed that in greenhouse experiments dew prolonged the survival of ponderosa pine seedlings. On the other hand, Monteith (1963) claimed that the amount of dew deposited on leaves is ordinarily too small to be of much practical significance. Breazeale and Crider (1934) claimed that water can be absorbed from the air, move through the plant, and escape into the soil, and Gindel claimed that Aleppo pine also absorbs water from the air and transports it to the soil. However, it seems possible that the increase in soil water content under trees observed by Gindel occurs because the cooler soil in the shade of the trees acts as a "sink" for soil water moving as vapor. It is claimed that *Prosopis tamarugo,* growing in the desert of northern Chile, where rainfall is rare, depends chiefly on fog absorbed through its leaves for water (Went, 1975).

Slatyer (1967, pp. 231–236) discussed this problem in detail. He pointed out that there is such a small vapor pressure deficit, even in wilted plants, that the inward movement of water would be slow even if the stomata were open, which is unlikely. However, it might aid somewhat in restoring leaf turgor and might reduce transpiration during the early morning. The effects of fog and fog drip in reducing water loss in the coastal fog belt of the United States Pacific states are more obvious and are shown by a change in species as one moves out of the fog belt. Fritschen and Doraiswamy (1973) reported that dew sometimes amounts to 15–20% of the water loss from a Douglas-fir tree growing in a lysimeter near Seattle, Washington, but this probably is unusually high. Chaney's (1981) brief review of the literature indicates that fog and fog drip are sometimes significant sources of water, and dew may be, although its importance is more controversial. The importance of dew and fog probably depends on local conditions.

ABSORPTION MECHANISMS

All absorption of water occurs along gradients of decreasing water potential from the medium in which the roots are growing to the root xylem. However, the gradient is produced differently in slowly and in rapidly transpiring plants. This results in two absorption mechanisms: active absorption, or osmotic absorption in slowly transpiring plants where the roots behave as osmometers, and passive absorption in rapidly transpiring plants where water is pulled in through the roots, which function merely as passive absorbing surfaces.

Differences between Slowly and Rapidly Transpiring Plants

Conditions in the water-conducting systems of slowly transpiring plants are quite different from those existing in rapidly transpiring plants. When the soil is moist and warm and little transpiration is occurring, water in the xylem is often under positive pressure, as indicated by the occurrence of guttation and the exudation of sap from cuts made in the xylem. When transpiration is rapid, the water in the xylem is usually under tension, and no guttation occurs. The difference can be demonstrated by immersing a portion of the stem of an herbaceous plant in a dye such as acid fuchsin and cutting into the stem beneath the surface of the dye. If the plant is transpiring even moderately rapidly, dye will rush into the cut xylem elements and almost instantly stain the stem above and below the cut to a distance of many centimeters. If the plant has been in moist soil and a humid atmosphere before the cut is made, the dye will not enter the opened xylem elements, but in many species sap will begin to exude from them. Such experiments indicate that the water in the xylem of rapidly transpiring plants is at less than atmospheric pressure, but in very slowly transpiring plants it is under positive pressure, the root pressure.

Long ago, Renner (1912) noted that the water absorption responsible for root pressure occurs only in healthy, well-aerated root systems and depends on the presence of living cells in the roots; he therefore termed it active absorption. In contrast, uptake of water by transpiring shoots can occur through anesthetized or dead roots, or even in the absence of roots (Kramer, 1933). In transpiring plants the roots seem to function as passive absorbing surfaces through which water moves by mass flow, so Renner called the process passive absorption. Renner's terminology is fairly well established, but readers are warned that the term active absorption, as used by Renner, does not mean that there is active transport or nonosmotic uptake of water. For this reason, it seems preferable to use the term osmotic absorption in place of active absorption. In osmotic absorption the roots function as osmometers because the water potential of the xylem sap is lowered by accumulation of solutes, and the pressure on the root xylem sap is positive, as seen in Table 8.1. During passive absorption by rapidly transpiring plants the solute concentration of the root xylem sap is low and the Ψ_w of the xylem sap is lowered chiefly by decreased pressure or tension caused by transpiration. The water potential of the xylem sap can be represented by the following equation:

$$\Psi_{\text{xylem sap}} = \Psi_s + \Psi_p \tag{8.1}$$

In slowly transpiring plants, Ψ_s is often -0.1 or -0.2 MPa and dominates water absorption and Ψ_p is positive and 0.1–0.15 MPa. In rapidly transpiring plants (passive absorption), the xylem sap is dilute and Ψ_s is negligible, whereas Ψ_p is negative and dominates water absorption. Examples of the potentials in slowly and rapidly transpiring plants are shown in Table 8.1.

Passive Absorption by Transpiring Plants

The forces bringing about the absorption of water by transpiring plants originate at the evaporating surfaces in the shoots and are transmitted to the roots through the sap stream in the xylem (Chapter 10). In rapidly transpiring plants, the roots act merely as absorbing surfaces through which water is pulled in by the tension developed in the sap stream. As water evaporates from the leaves, the reduction of water potential in the leaf cells causes water to move into them from the xylem of the leaf veins. Removal of water from the xylem reduces pressure on the xylem sap and decreases its water potential. This reduction is transmitted as reduced pressure or tension through the continuous cohering water columns of the xylem elements and their water-saturated walls down to the root system.

TABLE 8.1   **Relative Water Potentials in Soil, Root Cortex, and Xylem Sap of Slowly and Rapidly Transpiring Plants in Soil at Approximately Field Capacity**[a]

	Slowly transpiring plant, osmotic absorption		Rapidly transpiring plant, passive absorption	
Soil	Cortex	Xylem sap	Cortex	Xylem sap
$\Psi_s - 0.01$	-0.5	-0.2	-0.05	-0.05
$\Psi_m - 0.02$	—	—	—	—
$\Psi_p -$	-0.4	-0.05	0.10	-0.5
$\Psi_w - 0.03$	-0.1	-0.15	-0.4	-0.55

[a] Values are estimates in megapascals. The positive pressure in the xylem of the slowly transpiring plant often results in guttation.

Reduction of water potential in the root xylem produces a gradient along which water moves from the root surface across the intervening tissues and into the xylem. Under these conditions, water can be regarded as moving through the plant by mass flow as a continuous cohesive column, pulled by the matric or imbibitional forces developed in the evaporating surfaces of the leaf cells (see also Chapter 10).

As shown in Fig. 8.1, even when the roots are in moist soil, absorption tends to lag somewhat behind transpiration, indicating the existence of resistance to water flow. As mentioned in Chapter 7, there are resistances to water flow at several points in the soil–plant–atmosphere system, but the largest resistance seems to be in the living cells of the roots. This statement is based on the observation that killing roots greatly reduces the resistance to water flow through root systems under a pressure gradient (Renner, 1929; Kramer, 1932; Brouwer, 1954). Also, removal of the root system is accompanied by a temporary increase in absorption and by reduction in the lag in absorption when the rate of transpiration is changed, as shown in Fig. 7.6. The relative importance of the resistances in various parts of the system is discussed in Chapter 7.

Path of Radial Movement of Water in Roots.   As pointed out in Chapters 5 and 7, there are three possible parallel pathways for water movement from the epidermis to the xylem in roots: (1) across the vacuoles of the cortical cells, (2) through the symplast, and (3) through the cell walls. In the 1960s investigators favored movement through the cell walls, at least as far as the endodermis, as the principal pathway but later the preference seemed to shift back to a symplastic pathway because the resistance to movement through the walls seemed too high (Newman, 1976; Tyree and Yianoulis, 1980). Some water probably moves along all three pathways, but data on their relative permeabilities are too uncertain to

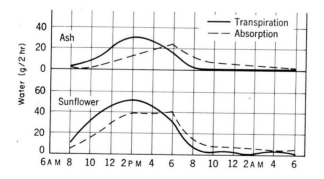

Fig. 8.1.   Rate of transpiration and absorption of a woody and an herbaceous species on a bright, hot summer day. The plants were rooted in autoirrigated pots, similar to that shown in Fig. 4.10. (From Kramer, 1937.)

indicate which is the major one (Dainty, 1976; Newman, 1976; Tyree *et al.*, 1981).

Whatever the pathway, movement of water is reduced by treatments that decrease the permeability of protoplastic membranes, such as chilling, inadequate aeration, and respiration inhibitors. Water movement is also increased if the protoplasmic membranes are killed. These facts indicate that at some point water passes through living cells. This is usually assumed to be at the endodermis in young roots, although Ginsburg and Ginzburg (1970) demonstrated that a cylinder of cortical tissue removed from the stele will function as an osmotic membrane. This and other observations suggest that in at least some kinds of roots the epidermis and cortical parenchyma constitute a protoplasmic barrier to water and solute movement.

Osmotic Absorption and Root Pressure

The exudation of sap from cut or broken stems has been observed from very early times. In Europe birch trees were tapped for sap several centuries ago (Evelyn, 1670), and in the Far East sap has been obtained from palms since before the beginning of written history. Hales (1727) made measurements of root pressure in the early eighteenth century, and there was interest in the problem during the nineteenth century which has continued to the present. In fact, few plant processes have attracted more attention or provoked more different explanations than root and stem pressure in plants. Some confusion resulted from the tendency to lump together such diverse phenomena as exudation of organic solutes from nectaries of flowers, bleeding from wounded stems, and guttation. The various explanations of root pressure can be classified in three groups. One group consists of theories that assume secretory activity by root cells, a second includes various electroosmotic theories, and a third assumes that roots behave as osmometers. The older literature on root and stem pressures was reviewed by Kramer (1949, Chapter 7).

Secretion Theories. Over the past century, various writers have suggested that root pressure is produced by the secretion of water into the xylem, resulting from a higher permeability to water on the inner than on the outer side of the cells. Ursprung (1929), for example, claimed that the water potential is lower on the inner than on the outer side of the endodermal and stelar parnchyma cells. However, it seems very unlikely that differences in water potential could be maintained on opposite sides of a cell because of mixing by diffusion and cyclosis. Most solutes can diffuse across ordinary cells in less than 1 s. Because of the difficulty in developing a plausible theory of secretion, interest in this theory waned for a time.

However, in the 1940s and 1950s interest in nonosmotic movement of water was revived, with respect to both cells (Chapter 2) and roots. Some of the literature was reviewed by Kramer (1956). The evidence for nonosmotic uptake of water by roots is based chiefly on observations that the osmotic potential of the exudate from detopped root systems is higher (less negative) than the osmotic potential of the solution in which the root systems must be immersed to stop exudation (van Overbeek, 1942; Broyer, 1951; House and Findlay, 1966). Ginsburg and Ginzburg (1970) reported that the osmotic potential of the external solution must be 0.05–0.2 MPa lower than that of the exudate to stop exudation from corn root segments. They interpreted the difference as indicating the occurrence of nonosmotic water movement, and introduced a term for nonosmotic movement of water (J_v*) into the equation for water movement by osmosis, so that it read as follows:

$$J_v = Lp\ \sigma\Delta\pi + J_v* \qquad (8.2)$$

Bennet-Clark *et al.* (1936) also introduced a nonosmotic term into their equation. Later, Ginsburg (1971) concluded that differences in reflection coefficients (σ) between the inner and outer membranes might explain the discrepancy without invoking a nonosmotic component. Russian investigators (Zholkevich *et al.*, 1980; Mozhaeva and Pil'shchikova, 1980) revived the claim that nonosmotic movement of water is important in the development of root pressure, based again on observations that the somotic potential of the exudate is much higher than the osmotic potential of the solution required to stop exudation. The reason for the discrepancy observed by them is not clear. They propose that water is absorbed osmotically by the root cells, which then inject it into the xylem by contraction. This is reminiscent of the spongiole theory of A. P. de Candolle (1832), who mistook the root caps for contractile absorbing organs, and of the claim by Bose (1927) that "peristaltic waves of pulsation" occur in roots. It is also related to the work of Sabinin, as described by Maximov (1929).

Other investigators supported an osmotic theory and explained the reported discrepancies in osmotic potential between the exudate and the solution required to stop exudation without invoking nonosmotic forces. Roots are not completely impermeable to solutes, i.e., they have a reflection coefficient of less than 1, so the concentration of the external solution would not necessarily be equal to that of the internal solution at equilibrium, at least in short-term experiments. Also, considerable salt is removed from the xylem sap as it moves up into the stem, so the concentration in the absorbing region may be appreciably higher than at the stump where the exudate is collected (Klepper and Kaufmann, 1966; Oertli, 1966). An example of this is shown in Fig. 8.2.

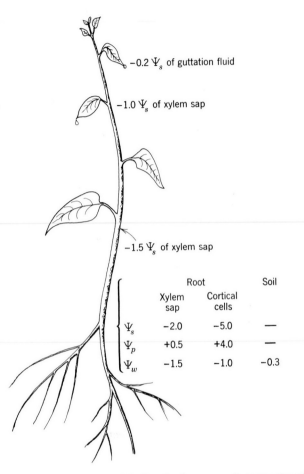

Fig. 8.2. Decrease in osmotic potential, Ψ_s, of xylem sap as it moves upward, caused by transfer of salt to living cells adjacent to the xylem. The amount of decrease is based on data from Klepper and Kaufmann (1966) and Oertli (1966). The tabular data illustrate how, in slowly transpiring plants, a gradient in water potential can occur from moist soil to root xylem across the root cortex, which has a much lower osmotic potential than either soil or root xylem but an intermediate water potential. (From Kramer, 1969.)

Electroosmotic Theories. Several investigators have attributed root pressure to the electroosmotic transport of water into the xylem. Water can be moved across a membrane under the influence of an applied electric current, the direction of flow being toward the pole with the same polarity as the membranes. The interior of roots is negative to the exterior and the cellulose membranes are negatively charged, so water movement should be inward. However, attempts to cause water flow into roots by applying electric potentials have been unsuccessful. Interest in electroosmosis was revived by Fensom (1958), who observed correlations between cycles in root pressure exudation and in bioelectric currents, and formulated an electrokinetic theory of transport to account for the correlations. However, it is doubtful if electoosmosis can cause significant net movement of water in plant tissue, because the high permeability of plant cells allows water to leak back almost as rapidly as it is moved inward (Dainty, 1963; Slatyer, 1967, pp. 174–175).

Osmotic Theories. The most satisfactory explanation of root pressure assumes that it is an osmotic process in which root systems function as osmometers because of accumulation of solutes in the xylem sap. As shown in Table 8.1 and Fig. 8.2, water can move by osmosis from a dilute soil solution with a Ψ_w of perhaps -0.03 MPa across the turgid cortical cells into the xylem sap with a water potential of -0.1 or -0.2 MPa. Such a situation would result in enough positive pressure in the xylem sap to cause guttation in intact plants or exudation from stumps of detopped plants.

It is possible to demonstrate rapid reversal of osmotic water flow into and out of detopped root systems by dipping them in water until sap exudes on the stump and then transferring them to a dilute salt or sugar solution, whereupon the sap on the stump rapidly retreats into the stem. This process can be repeated indefinitely.

The principal problems in connection with an osmotic explanation of root pressure are (1) the location of the differentially permeable membrane permitting solute accumulation in the stele and (2) the source of the solutes in the xylem sap. Most botanists assume that in roots which have not undergone secondary growth, the endodermis functions as the differentially permeable membrane. This may be true in general, but root pressure has been demonstrated in maize roots from which the cortex was removed, destroying the endodermis (G. H. Yu, personal communication). Also, Ginsburg and Ginzburg (1970) removed cylinders of cortical tissue from maize roots and used them as osmometers to produce root presure. Thus, it appears that either the cortex or the pericycle and stelar parenchyma can function as a multicellular osmotic membrane, although probably not as effectively as intact roots. No attention has been given to the location of the differentially permeable membrane in roots that have lost their cortical parenchyma and endodermis by secondary growth. It must be present, however, because grape and conifer root systems that are completely suberized develop root pressure, and we have found that water flow through suberized woody roots is increased more than salt movement by application of pressure to the root systems (Chung and Kramer, 1975). We suggest that the cambial region probably functions as the salt barrier because its cell walls are too thin to permit significant outward leakage. However, this problem deserves further investigation.

It is generally assumed that the solutes in the xylem sap of herbaceous plants are predominantly inorganic salts moved in by an active transport mechanism. However, Triplett *et al.* (1980) reported that although $NO_3{}^-$ was the principal anion for K^+ in exudate from wheat seedlings in KNO_3 solution, malate was the principal anion in exudate from root systems in K_2SO_4. It is believed that much of the inhibition of root pressure caused by respiration inhibitors is produced by reduction in salt transport into the xylem, plus possibly some damage to protoplasmic membranes. It has also been suggested that part of the solutes are released by disintegrating protoplasts during the differentiation of xylem vessels (Priestley, 1922; Hylmö, 1953; Scott, 1965; Anderson and House, 1967). However, the quantity of solutes in the sap from rapidly "bleeding" roots usually is too great to have been supplied by disintegrating xylem initial cells. Also, it was demonstrated by Wieler (1893) nearly a century ago that root pressure developed in roots from which the apical 5 cm had been removed. Organic substances including sugar, nitrogenous compounds, and growth regulators also occur in the root xylem sap, but no one has attempted to explain how these compounds are moved from living cells where they presumably were synthesized into the nonliving xylem elements.

It is well established that there is a correlation between salt accumulation in roots and the occurrence of root pressure. It develops only if root systems are healthy, well aerated, provided with a dilute supply of minerals and nutrients, and kept at a moderate temperature. Root pressure stops when roots are subjected to low temperature, inadequate aeration, dry soil, or prolonged immersion in distilled water, because of reduction in the supply of salt required to keep the osmotic system in operation.

Relative Importance of Osmotic and Passive Absorption

There are two views concerning the importance of root pressure. In the past it was regarded as important, and Fensom (1957), Minshall (1964), Rufelt (1956), and others claimed that its importance is underestimated. Rufelt (1956) and Brouwer (1965, for example, claimed that active absorption operates in series with passive absorption, and others have suggested that it plays a role in refilling xylem vessels that have filled with gas during periods of rapid transpiration or after being frozen during the winter (Zimmermann and Brown, 1971, p. 210). Palzkill and Tibbits (1977) reported that root pressure is necessary to provide sufficient calcium to nontranspiring tissue such as that in heads of cabbage and lettuce.

The writer regards root pressure as the fortuitous result of the accumulation of

TABLE 8.2 A Comparison of Exudation with the Rate of Transpiration prior to Removal of the Tops[a,b]

Species	Number of plants	Transpiration, milliliters of water per plant per hour		Exudation, milliliters of water per plant per hour		Exudation as percent of transpiration[d]
		First hour	Second hour	First hour	Second hour	
Coleus	6	8.6	8.7	0.30	0.28	3.2
Hibiscus	5	5.8	6.7	−0.01	0.05	0.7
Impatiens	6	2.1	1.9 Tops	−0.22	−0.06	
Helianthus	8	4.3	5.0 removed	0.02	0.02	0.4
Tomato (1)	6	10.0	11.0	−0.62	0.07	0.6
Tomato (2)	6	7.5	8.7	0.14	0.27	3.1

[a] Rapidly transpiring plants usually show absorption of water through the stumps during at least the first half hour after the tops are removed, exudation beginning only after the water deficit in the root system is eliminated.

[b] From Kramer (1939).

[c] A minus sign indicates absorption of water by the stump instead of exudation.

[d] Percentage relations are based on transpiration and exudation rates for the second hour.

solutes in the stele of roots, and doubts if it has any essential role. In general, the volume of exudate is only a small percentage of the volume of water lost by transpiration, as shown in Table 8.2. Also, intact transpiring plants can absorb water from drier soil and more concentrated solutions than can detopped root systems (Jäntti and Kramer, 1957; McDermott, 1945). Finally, no root pressure can be demonstrated in the xylem of transpiring plants, but instead water is absorbed through the stumps if the tops are removed, as shown in Fig. 8.3. Occasionally, the volume of exudate is much greater than that shown in Table 8.2. Minshall (1968) reported that the volume of exudate from tomato root systems supplied with an abundance of nitrogen was 25–50% of the loss by transpiration, but this is exceptionally high. Triplett *et al.* (1980) reported that the volume of exudate was twice as great from wheat root systems in KNO_3 as from those in K_2SO_4.

In conclusion, it can be stated that roots generally function as passive absorbing surfaces through which water is pulled in by forces developed in the evaporating surfaces of the shoots. However, when the rate of transpiration is low and the soil is moist, warm, and well aerated, roots operate as osmometers, producing root pressure which sometimes causes guttation and bleeding from wounds and leaf scars.

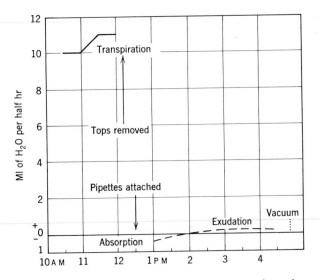

Fig. 8.3. Behavior of root systems of rapidly transpiring tomato plants after removal of the shoots. Because of a tissue water deficit, the root systems absorbed water for over an hour after removal of the shoots. The rate of exudation was very low compared with the rate of transpiration prior to removal of the shoots. The rate of exudation under a vacuum of 64 cm Hg was 3.5 times the previous rate. These plants were rooted in soil at approximate field capacity. (From Kramer, 1939.)

ROOT AND STEM PRESSURES

A great deal of information has accumulated during the past century on the occurrence of root and stem pressures. Unfortunately, early writers indiscriminately lumped together all examples of "bleeding" or "weeping." For example, Wieler (1893) listed nearly 200 species in nearly 100 genera distributed among ferns, flowering plants, and conifers that showed bleeding, but his list included true root pressure, guttation, sap flow from wounds, and even secretion from glandular hairs. It is necessary to distinguish between sap flow caused by root pressure, as in tomato, grape, and birch, and that caused by stem pressure, as in maple; or by wounding, as in agave and palm.

We will present some additional information about root and stem pressures because even though they are not important in the water economy of plants, they are interesting processes and of some economic importance. First, it should be emphasized that although most exudation is caused by root pressure, localized stem pressures that are independent of root pressure are developed in some plants, including maples, palms, and agaves.

Root Pressure Phenomena

There are large differences among species with respect to magnitude of pressure, volume of exudate, and periodicity of exudation.

Species Differences. It seems that any kind of plant that can accumulate salt in the root stele should develop root pressure if other conditions are favorable. However, some species, such as corn, sugarcane, tomato, and sunflower, produce larger volumes of exudate from stumps of detopped root systems than others, such as some legumes. For example, Yu (1966) found that corn roots produced more exudate containing a higher concentration of salt than broad bean roots. She attributed the difference to the larger diameter of roots, more and larger vessels, and less ion leakage from corn roots. Relatively few woody plants show root pressure, and then usually only in the spring before leaves develop. However, Parker (1964) reported copious exudation from black birches in New England in October and November, after leaf fall. There is said to be extensive use of birch sap to produce wine and health drinks in the Ukraine (Sendak, 1978), and its use in Canada and the United States seems to be increasing. Reports of root pressure in conifers have been rare until recently, although a couple of investigators produced it by immersing root systems in a concentrated nutrient solution and then transferring them to water (Kramer, 1949, Chapter 7). White *et al.* (1958) reported exudation from the basal ends of severed roots of white spruce and white pine, and O'Leary and Kramer (1964) reported exudation from apical segments of loblolly pine and white spruce roots. Recently, Lopushinsky (1980) consistently obtained exudation from stumps of detopped seedlings of *Abies grandis, A. procera, A. amabilis, Pinus contorta, P. ponderosa, Picea engelmanni, Pseudotsuga menziesii,* and *Larix occidentalis.* Some seedlings had new root growth, but the root systems of others were completely suberized. Exudation persisted for as long as 79 days after detopping. The volume of exudate varied widely (from 0.1 to 8.0 ml per root system) and was not related to the size of the root system. The osmotic potential of the exudate ranged from 0.024 to 0.078 MPa, which is lower than the values reported for exudates of various herbaceous species.

It is believed that consistent root pressure exudation occurred in Lopushinsky's experiments because the seedlings were stored for 2 months in a cold room at $1°-2°C$. Probably the cold storage favored conversion of starch to sugar, producing conditions in the roots favorable for exudation. Clarkson (1976) reported that exposure of roots of rye and barley to $8°C$ for a few days resulted in an increase in sugar content, ion uptake, and root pressure exudation. It is now certain that root pressure sometimes occurs in conifers. This supports the earlier statement that it ought to occur in the roots of all plants that accumulate salt in their root xylem sap.

Magnitude.   Hales (1727), who made the first recorded measurement of root pressure, observed a pressure of about 0.1 MPa in grape, and pressures of 0.2–0.3 MPa were reported for birch in New England by Merwin and Lyon (1909). Stem and root pressures for red maple and river birch are shown in Fig. 8.4. There are various reports of pressures ranging from 0.05 to 0.19 MPa in herbaceous species, but the highest pressure ever recorded, over 0.6 MPa, was measured on excised tomato roots growing in culture solution (White, 1938). Higher pressures have been reported in stems of woody plants, but they are

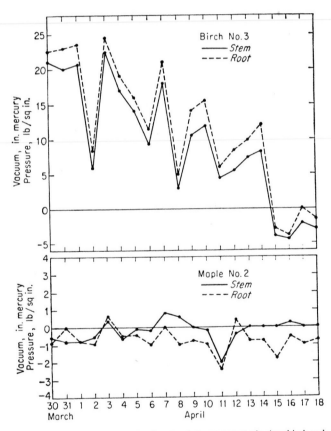

Fig. 8.4.   Simultaneous measurements of root and stem pressures in river birch and red maple. In birch, root pressure exceeds stem pressure, but the two are closely correlated. In maple, root pressure was usually absent, even when positive stem pressure existed. (From Kramer and Kozlowski, 1979, by permission of Academic Press.)

believed to have been caused by wounding rather than by root pressure (see Kramer, 1949, pp. 163–164).

Composition of Exudate. As mentioned earlier, the solutes in the xylem exudate from herbaceous root systems usually consist chiefly of minerals, but sap from woody root systems often contains appreciable amounts of organic matter, chiefly sugars and nitrogen-containing compounds. Xylem sap of both herbaceous and woody plants also contains growth regulators, especially cytokinins and gibberellins. The composition of the xylem sap varies with the plant species, the season, and the amount of fertilization. For example, van Overbeek (1942) reported an osmotic potential of −0.13 MPa in exudate from tomato root systems growing in Hoagland solution, but only −0.04 MPa in exudate from root systems in distilled water. Eaton (1943) reported osmotic potentials of −0.15 to −0.24 MPa in cotton exudate, and Stocking (1945) reported an osmotic potential of −0.19 MPa in squash exudate.

Bollard (1960) reviewed the extensive literature on the composition of xylem sap and discussed its role in the translocation of minerals and nitrogen compounds. Several investigators attempted to use the composition of root pressure exudate as a guide to the mineral nutrient needs of plants (Pierre and Pohlman, 1934; Lowry *et al.*, 1936), but direct measurements on soil solutions are better. Nangju (1980) claimed that the organic nitrogen content of soybean xylem exudate can be used as a measure of the nitrogen-fixing capacity of root systems, but this was questioned by Rufty *et al.* (1982).

Much effort has been spent explaining how salt is transported into the xylem, but the transport of organic molecules has never been explained adequately. How organic nitrogen compounds, growth regulators, and sugars get from the surrounding living cells into the nonliving xylem vessels constitutes a puzzling problem that deserves study.

Volume of Exudate. The volume of exudate is generally greater from larger root systems, but it also varies with the species and with environmental factors such as soil moisture and temperature. The volume of exudate obtained from root systems of similar age and treatment often varies widely for no apparent reason. Exudation usually ceases in cold, dry, or flooded soil, although Thut (1932) reported root pressure in the submerged root systems of some aquatic plants. Birch trees are reported to yield 20–100 liters of sap in a spring and one large birch produced 675 liters, but yield is not proportional to tree size (Kramer and Kozlowski, 1979, p. 453). Sugarcane stools have exuded 1 liter of sap in a week and corn plants over 100 ml/day for 15 days. Crafts (1936) observed volumes of exudate from squash root systems in 24-hr periods greater than the volume of the root systems. G. H. Yu (unpublished data) estimated that exudation from apical segments of corn roots represented a turnover of xylem vessel contents as often as three times per hour. Minshall (1968) found that supplying urea or KNO_3 to the soil in which tomato plants were growing greatly increased exudation from detopped root systems. He observed rates of up to 80 ml/day per plant for tomatoes in the 16- to 18-leaf stage, a rate much greater than usually reported.

Periodicity. Over a century ago, Hofmeister observed a diurnal periodicity in root pressure, and in recent decades several investigators have observed a maximum in pressure and volume during the day and a minimum at night. Hagan (1949) even observed a diurnal periodicity in negative exudation when water was supplied to the stumps of detopped, wilted sunflowers in dry soil. Greatest flow of water through the roots out into the soil occurred near midnight and least near midday, just the reverse of the cycle for positive exudation. The periodicity in positive exudation has been attributed to greater translocation of salt into the xylem during the day than at night (Hanson and Biddulph, 1953; Vaadia, 1960; Wallace *et al.*, 1966), but Parsons and Kramer (1974) reported little diurnal variation in the amount of salt in cotton root exudate.

Diurnal fluctuations in apparent root resistance or permeability also occur, with the lowest resistance at midday and the highest at night (Skidmore and Stone, 1964; Barrs and Klepper, 1968; Parsons and Kramer, 1974). The cycles in exudation and in root resistance can be reset by reversing the light–dark cycle under which the plants are grown, suggesting that the cycles are controlled by signals from the shoots (Parsons and Kramer, 1974). This is supported by observations of Bunce (1978a) who reported that an increase in apparent root resistance was correlated with decreased rates of root elongation. According to Ivanov (1980), Russian investigators proposed that the daily periodicity in root pressure exudation is related to the daily periodicity in cell division and growth of roots. More research is required before a satisfactory explanation can be offered for diurnal variations in root resistance and exudation.

Guttation. The most obvious evidence of root pressure is the occurrence of guttation, the exudation of liquid water from leaves and occasionally from leaf scars. Examples are the droplets of water on blades of grass and along the margins of some leaves in the morning. Burgerstein (1920) reported guttation in plants of 333 genera, and others have since been added to the list. Guttation usually occurs from hydathodes, which are stomate-like pores located over intercellular spaces in the leaf tissue (Fig. 8.5). The xylem of a small vein usually terminates among the thin-walled parenchyma cells below each hydathode, and when root pressure develops, water is forced into the intercellular spaces and flows out of the hydathodes. Guttation sometimes occurs through stomata, as in some grasses, and from twigs and branches of trees. Friesner (1940) reported exudation from stump sprouts of red maple in early spring in Indiana, apparently

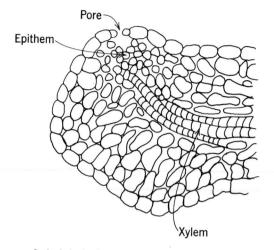

Fig. 8.5. Diagram of a hydathode showing a pore, the underlying epithem, and termination of xylem. The epithem is merely a mass of thin-walled parenchyma with large intercellular spaces through which water can move readily. Hydathodes often resemble incompletely differentiated stomata with nonfunctional guard cells. They usually occur at the tips and along the margins of leaves. (Adapted from several sources.)

from lenticels, and Raber (1937) reported sap flow from leaf scars of deciduous trees in Louisiana in the autumn. Exudation of liquid from the surfaces of uninjured roots has also been reported (Head, 1964), and this might be called root guttation.

The quantity of liquid exuded varies from a few drops to many milliliters in the case of some tropical plants, such as *Colocasia*. The composition varies from almost pure water to a dilute solution of organic and inorganic solutes. Leaves of certain saxifrages become encrusted with calcium salts left by evaporation of guttation water, and deposits are sometimes reported on leaves of grasses, but they usually disappear after the next rain (Duell and Markus, 1977). Curtis (1944) reported an instance in which grass leaves became encrusted with glutamine following heavy fertilization of a lawn. In general, the concentration of solutes in guttation fluid is lower than in the exudate from stumps of similar plants. Guttation is most common at night and usually ceases during the day because transpiration reduces the pressure in the xylem sap. It decreases or ceases in dry or poorly aerated soil and is increased by moderately heavy fertilization.

Although guttation is an interesting phenomenon, it is of little importance to plants. Occasionally, injury is caused by accumulation of salts on leaf margins, by evaporation of guttated liquid, or by reabsorption into the intercellular spaces.

Pathologists suggest that guttation water creates conditions favorable for invasion of leaves by fungi and bacteria. More information about guttation can be found in Kramer (1969, pp. 165–167).

Stem Pressures

Although exudation from herbaceous plants and some woody plants such as birch and grape is caused by root pressure, exudation from sugar maple, palm, and a few other plants is caused by stem pressures that develop independently of root pressure, often in the phloem.

Maple Sap Flow. The best-known example of exudation from stems in North America is the flow of sap from maple trees, chiefly from *Acer saccharum* Marsh and *A. nigrum* Michx. W. S. Clark (1874, 1875), Jones *et al.* (1903), Stevens and Eggert (1945), Johnson (1945), Marvin (1958), and many others have written about maple sap flow. Sap flow can occur from late autumn to early spring, any time that freezing nights are followed by warm days with temperatures above freezing, but the best flows are obtained in the spring. Over 60% of the flow occurs before noon, and it often ceases in the afternoon because transpiration from the branches decreases the pressure in the xylem sap. The yield varies widely, usually falling in the range of 35–70 liters per tree in a season, although occasionally twice that amount is produced. The sugar content, all sucrose, is usually 2–3% but varies from 1 to 7%. The distinctive flavor of maple syrup is produced by heating, which changes certain nitrogen compounds present in it (Pollard and Sproston, 1954). Reheating usually restores the flavor to syrup which has lost it during storage.

The flow of maple sap is caused by stem pressure, not by root pressure. Sap flow will occur from isolated tree trunks placed in tubs of water and subjected to alternating temperatures (Stevens and Eggert, 1945). Simultaneous measurements of root and stem pressure indicate that no root pressure can be detected at the time maple sap is flowing (Fig. 8.4). In contrast, root pressure is always observed at times when sap flow occurs in birch and grape, and the flow appears to be caused directly by root pressure. Marvin (1958) reviewed the literature on maple sap flow and added further observations.

Sap is obtained by drilling a hole through the bark into the sapwood and installing a spout. This allows sap to flow out by gravity into a container. In recent years, the yield has been increased greatly by applying a vacuum to the spouts. According to Sauter (1971), carbon dioxide produced in respiration collects in the intercellular spaces in the stem during the day, and the resulting pressure forces sap out. At night the carbon dioxide dissolves in the xylem sap reducing the pressure and causing upward movement of water from the roots, refilling the xylem vessels. However, the process is not fully explained.

Other Examples of Stem Pressure. In India and tropical Asia, large amounts of sap are obtained from palms, chiefly coconut, date, and Palmyra palms, and used as a source of sugar and palm wine. Sap is usually obtained by cutting out the inflorescence, and the flow can be maintained at rates of 6 or 8 to as high as 20 liters/day for several months by recutting or otherwise rewounding. Sap is obtained from the stems of some palms by making incisions into them, somewhat as in maple. However, the sap comes from the phloem, instead of the xylem as in maple, and the sugar was probably originally mobilized for use in stem tips or developing inflorescences.

In Mexico, large quantities of sap, containing sucrose, are obtained from agaves by cutting out the young infloresence and scooping out a cavity in the top of the stem. Sap collects in the cavity at the rate of 1 liter or more per day for 10 or 15 days and is removed daily, mostly to be fermented into pulque. Sap flow in agave and palms was originally attributed to root pressure, but it is actually caused by local pressure in the phloem. Exudation from agave and palms is discussed in detail by Van Die and Tammes, in Zimmermann and Milburn (1975), and by Milburn and Zimmermann (1977). The latter concluded that beating or massaging the tissue is not essential for induction of flow, but may increase it. They measured phloem sap pressures up to 0.76 MPa in *Cocos nucifera* L. The high yield of sap indicates that palm phloem has a very high transport capacity.

Wounding of stems often results in development of local high pressures. MacDougal (1926) reported exudation pressures in holes bored in stems of large cacti, *Pinus radiata, Juglans regia,* and various oaks. Occasionally, sap flows from cracks and other wounds in trees and is fermented by bacteria and yeasts, causing slime flux. According to Carter (1945), slime flux in elm is caused by bacterial activity in the heartwood. Stem pressures high enough to blow the cores out of increment borers have been observed in trees with decaying heartwood (Abell and Hursh, 1931). In some instances the gas escaping from the holes will burn when ignited, probably because it contains methane produced by the organisms causing decay. The production of methane in trees was discussed by Zeikus and Ward (1974).

SUMMARY

Absorption of water occurs along gradients of decreasing water potential from the soil of other root medium to the root xylem. This gradient is produced differently in slowly and rapidly transpiring plants. This results in two absorption mechanisms: active absorption, or osmotic absorption in slowly transpiring plants where roots act as osmometers, and passive absorption in rapidly transpiring plants where water is pulled in by the decreased pressure or tension produced in the xylem sap. Osmotic absorption is responsible for root pressure, guttation, and most of the exudation of sap that occurs from wounds in the stems of grape, birch, and some other woody species in the spring. In a few plants, including maple, palms, and agave, sap flow is caused by local stem pressures rather than by root pressure. The cause of this pressure is not fully understood.

Root pressure has been attributed to secretion of water into the root xylem, electroosmosis, and osmosis. It is believed to be a simple osmotic process, caused by accumulation of sufficient solutes in the xylem to lower the water potential of the xylem sap below that of the substrate. The reduction in root pressure caused by deficient aeration, low temperature, and respiration inhibitors is attributed to reduction in salt accumulation in the root xylem and to changes in root permeability, rather than to inhibition of any nonosmotic water transport mechanism. There is often a well-defined diurnal periodicity in root pressure, with the highest values at midday and the lowest at midnight. The reasons for this periodicity are not fully understood.

Osmotic or active absorption is an interesting physiological process but plays no essential role in the water economy of plants. The volume of water absorbed is much less than the volume required by transpiring plants, and the process becomes inactive in rapidly transpiring plants. It is simply the fortuitous result of the accumulation of solutes in the root xylem.

SUPPLEMENTARY READING

Gindel, J. (1973). "A New Ecophysiological Approach to Forest Water Relationship's in Arid Climates." Junk, The Hague.
Kramer, P. J. (1956). Physical and physiological aspects of water absorption. *Encycl. Plant Physiol.* **3,** 124–129.
Renner, O. (1912). Versuche zur Mechanik der Wasserversorgung der Pflanzen. *Ber. Dtsch. Bot. Ges.* **30,** 576–580, 642–648.
Stone, E. C. (1957). Dew as an ecological factor. *Ecology* **38,** 407–413, 414–422.

Factors Affecting the Absorption of Water

INTRODUCTION

Unimpeded absorption of water is essential for successful growth of plants, because if water absorption does not balance water loss, reduction in turgor occurs, causing cessation of growth and eventual death by dehydration. The factors affecting the rate of absorption are therefore of major importance in any discussion of the water economy of plants. Using the Ohm's law analogy, they can be divided into two groups: (1) those affecting the driving force, which is the gradient in water potential from soil into roots, and (2) those affecting the resistance to water movement through the soil and the roots. A few factors, such as soil water content, affect both driving force and resistance to movement. Water absorption can be described by the following equation:

$$\text{Absorption} = \frac{\Psi_{soil} - \Psi_{root\ surface}}{r_{soil}} \tag{9.1}$$

$$= \frac{\Psi_{root\ surface} - \Psi_{root\ xylem}}{r_{root}}$$

Water absorption by plants growing in moist, warm, well-aerated soil is controlled largely by the rate of transpiration, as described in Chapters 7 and 8. Unfortunately, in the field water absorption is often limited by decreasing soil water content, and occasionally by excess salinity, cold soil, or inadequate aeration. It is also limited by the extent and efficiency of root systems. Transpiration will be discussed in Chapter 11 and the effects of inadequate absorption in Chapter 12. The effects of inadequate root systems and various environmental factors will be discussed in this chapter.

EFFICIENCY OF ROOT SYSTEMS AS ABSORBING SURFACES

The efficiency of root systems depends both on their extent and total surface and on the permeability of their surface. The latter varies widely with age and stage of development.

Extent of Root Systems

The variation in size of root systems was discussed in Chapter 6. The larger the volume of soil occupied by a root system, the larger the volume of water available to it and the longer the plant supplied by it can survive without replenishment of soil water by rain or irrigation. Thus, in general, deep-rooted plants survive droughts better than those with shallow roots. For example, Hurd (1974) pointed out that extensive root development is an important factor in drought tolerance of wheat and alfalfa, and this is probably true of many other crops. Root branching, often described in terms of root length density in centimeters of root per cubic centimeter of soil, is also important. For example, an increase in root length density from 1 to 2 cm/cm^3 greatly improves the capacity of sorghum to extract water from the soil (Jordan and Miller, 1980), although such a high density seems unnecessary for soybeans and cotton in more humid areas (Taylor, 1980).

The usefulness of deep, widespread root systems depends on the resistance to longitudinal movement in the roots being lower than that in the soil. Although

Gardner and Ehlig (1962), Passioura (1972), and Wind (1955) cite instances of considerable resistance to longitudinal water movement in roots, these seem to be unusual situations. In general, resistance to longitudinal flow seems to decrease with increasing length. Passioura (1972) reported that the diameter of the central metaxylem vessel in wheat roots increased from 50 μm at 30 cm to 75 μm at 50 cm and 1000 μm at 90 cm from base of the plant. Meyer and Ritchie

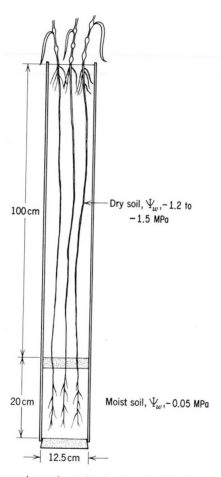

100 cm

Dry soil, Ψ_w, −1.2 to −1.5 MPa

20 cm

Moist soil, Ψ_w, −0.05 MPa

12.5 cm

Fig. 9.1. Equipment used to study uptake of water and salt by deep roots of *Phalaris tuberosa.* The lower compartment of the tube was separated from the upper compartment by a rubber stopper containing holes through which roots grew into the lower compartment. Watering the upper compartment was then discontinued, and after the soil in the upper compartment had dried to −1.2 or −1.5 MPa, tritiated water or ^{32}P was supplied to the roots in the lower compartment. Appearance of tracer in the shoots indicated uptake by roots from the lower compartment. (After McWilliam and Kramer, 1968; from Kramer, 1969.)

(1980) reported that the resistance per unit of length decreased toward root tips in sorghum roots, compensating for the increase in distance and a similar situation was found in roots of red pine by Stone and Stone (1975a). Research on fruit and forest trees indicates that water is absorbed from distances of several meters as readily as from nearby areas (Veihmeyer and Hendrickson, 1938; Hough *et al.*, 1965). Apple trees on a deep, well-aerated loess soil in Nebraska absorbed water from a depth of over 10 m (Wiggans, 1936). It has also been shown that on deep, well-aerated soil, corn can thrive on water absorbed from a depth of 1 or 2 m after the available water in the upper meter is exhausted (Reimann *et al.*, 1946). McWilliam and Kramer (1968) demonstrated that plants of *Phalaris tuberosa* could survive when the water potential in the upper meter of soil was −1.5 MPa because some of their roots had penetrated to a deeper horizon containing readily available water (Fig. 9.1). There is considerable additional evidence that not all of a root system needs to be in soil above the permanent wilting percentage for survival, or even for limited growth. Furr and Taylor (1939) found that good growth of lemon trees could occur when only one-half of the root system was in soil above the wilting percentage.

The pattern of water absorption from soil is quite different in annuals and perennials. Annual plants start with a tiny root system that first absorbs water near the base of the stem, and if no water is added, the volume of drying soil expands rapidly as the roots extend outward and downward. As shown in Fig. 9.2, absorption occurs at progressively greater depths as the roots penetrate the soil. Taylor and Klepper (1973, 1975) describe interesting studies of root development of crop plants, and an example of an expanding root system is shown in Fig. 6.16. Perennial plants start the season with an extensive root system and immediately begin to absorb water from considerable distances and depths, as mentioned earlier. Removal of water from soil is more closely related to root density than to distance from the plant, and water often seems to be absorbed more rapidly near plants and in the surface soil simply because the root density is greatest there. The presence of roots in soil horizons having different water-holding capacities and different water potentials at permanent wilting often makes it difficult to evaluate the amount of absorption at various depths.

Root Surfaces

The importance of the root surface depends on soil moisture and climatic conditions and on whether it is evaluated in terms of survival or good growth. In terms of survival, many vigorously growing plants have an excess of roots, as indicated by the fact that both trees and herbaceous plants often survive removal of one-half or more of their root system. Andrews and Newman (1968) found

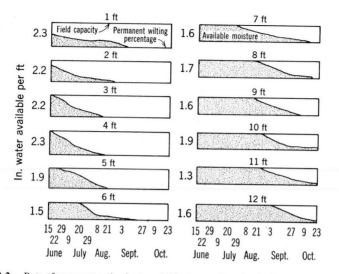

Fig. 9.2. Rate of water extraction by tomato plants at various depths from an unirrigated soil as the roots penetrated progressively to deeper horizons. Growth slowed by early August, when most of the readily available water had been removed from the upper 6 ft of soil. A total of 21 in. of water was removed by October 23. This plot yielded only half as much as an irrigated plot on similar soil. (After Doneen and McGillvray, 1946; from Kramer, 1969.)

that removal of 60% of the roots from wheat plants did not reduce growth measurably, either in drying soil or in soil kept at field capacity. On the other hand, Nutman (1934) regarded the root surface of coffee trees as a limiting factor for water absorption, but perhaps coffee trees have less root surface than some other kinds of plants.

It has been suggested that because of its presumed high resistance to water flow, the limiting surface for water absorption might really be the endodermis (Scott and Priestley, 1928; Nutman, 1934). The latter calculated that because of the small area of xylem in roots of coffee trees, water must enter the xylem 170 times as rapidly as it crosses the epidermis; however, this seems excessively high. A suberized root of *Liriodendron tulipifera* with a diameter of 2.8 mm had a stele with a diameter of 0.8 mm; hence, the velocity of water movement into the stele of this root would be three and one-half times that at the root surface. In secondary roots of *Vicia faba,* the circumference of the epidermis is about four times that of the endodermis, and water must cross the endodermis four times as rapidly as it crosses the epidermis. Near the tips of these roots, the tangential surface of the xylem points is only about 6% of the root circumference; hence, at this point water must enter the xylem with a velocity 16 times that at which it crosses the epidermis.

Resistance to water flow is said to be lower in roots of rice that contain aerenchyma than in roots that contain normal cortical tissue (Tomar and Ghildyal, 1975). Also, Drew *et al.* (1980) reported that development of aerenchyma in roots of maize did not seriously reduce the translocation of ions across the cortex. Apparently, the strands of cells left intact in such roots (see Fig. 6.13) provided an adequate pathway for water and irons.

Readers are reminded that if the root surface is as large as the leaf surface, the average rate of entrance of water, per unit of root surface, may be no greater than the rate of diffusion out of the leaves. Actually, the root surface is often considerably larger than the leaf surface, partially compensating for the fact that the older root surfaces are less permeable to water. Dittmer (1937) reported that the root surface of a winter rye plant 4 months old was about 50 times the leaf area. Newman (1969) summarized the data on root length per unit of soil surface and found values ranging from 35 to over 100 cm cm^{-2} for woody species and 100–4000 for herbaceous species. In most of these species, root surface greatly exceeds leaf surface, which is six or eight times the soil surface. The significance of this fact for water absorption has not been fully evaluated.

Root Permeability

The permeability of roots varies widely, depending on their age and stage of development and on environmental conditions.

Effects of Age and Maturation. Typical root systems consist of roots in various stages of differentiation, ranging from newly formed tips to fully matured secondary roots that have lost their epidermis and cortex and are enclosed in a layer of suberized tissue. Obviously, roots varying so much in structure must vary widely in permeability to water and solutes. Some data on permeability of grape roots of various ages are shown in Table 9.1. In general, it can be stated that the permeability of root hairs and young unsuberized roots is much greater than that of suberized roots. However, as pointed out in Chapter 6, considerable water and salt absorption must occur through the suberized roots of perennial plants because the unsuberized roots constitute a very small percentage of the total root system. For example, Kramer and Bullock (1966) found that even during the growing season, less than 1% of the root surface in the top 12.5 cm of soil in loblolly pine and yellow-poplar forests was unsuberized. The permeability of suberized roots is extremely variable because of differences in thickness and structure of bark, number of lenticels, and breaks caused by death of small branch roots. Kramer and Bullock reported water uptake rates under a constant pressure of 0.04 MPa varying from zero to 30,000 mm^3 cm^{-2} hr^{-1} for yellow-poplar. The permeability of loblolly pine roots varied from 6.6 mm^3 cm^{-2} hr^{-1} for roots 1.33 mm in diameter to 36.6 for roots 3 mm in diameter and 178 for

TABLE 9.1 Relative Permeabilities of Grape Roots of Various Ages to Water and ^{32}P[a,b]

Zone and condition of roots	Relative permeabilities	
	Water	^{32}P
Roots of current season (growing)		
A. Terminal 8 cm, elongating, unbranched, unsuberized	1	1
B. Unsuberized, bearing elongating branches (dormant)	155	75
C. Main axis and branches dormant and partially suberized before elongation completed	545	320
D. Main axis and branches dormant and partially suberized	65	35
Roots of preceding seasons (segments bearing branches)		
E. Heavily suberized main axis with many short suberized branches	0.2	0.4
F. Heavily suberized, thick bark, and relatively small xylem cylinder		
Intact	0.2	0.02
Decorticated	290.0	140.00

[a] From Queen (1967).
[b] Measurements taken under a pressure gradient of about 66 kPa.

unsuberized roots. Other data are given in Table 9.1. Some interesting changes in permeability during development of an entire root system are shown in Fig. 6.2 (Chapter 6).

Environmental Factors. Root systems subjected to severe water stress often show a decrease in permeability that may persist for several days after rewatering. As a result, leaf water potential and processes such as photosynthesis may not return to their prestress rate for a few days. This is discussed in Chapter 13. Exposure of root systems to high concentrations of salt also reduces their permeability. In both instances, the reduction in permeability is probably caused by dehydration of root cells and increased suberization of the existing roots. Inhibition of root growth is probably also important over long periods of time. The effects of salinity and aeration will be discussed in more detail later in this chapter.

Metabolic Activity. There has been some discussion concerning the relationship between metabolic activity and the absorption of water. For example, Henderson (1934) claimed that there is a close relationship between root respiration and water absorption by roots of corn seedlings. However, Loweneck (1930) found no direct correlation between respiration and water uptake, and Wilson and Kramer (1949) found no correlation in tomato roots. It is unlikely that there would be any direct correlation between passive uptake of water by transpiring plants and respiration, but active or osmotic absorption depends on accumulation of salt in the root xylem, a process dependent on the expenditure of energy released by respiration.

It has been shown by several investigators that treatment of roots with respiration inhibitors such as azide, cyanide, and dinitrophenol drastically reduces both active and passive absorption of water (Brouwer, 1954; Lopushinsky, 1964a). This is shown in Fig. 9.3. Passive absorption is decreased by increased resistance to water movement into roots, active absorption by decreased accumulation of solutes in the root xylem, and increased resistance to water movement. Inadequate aeration also reduces water absorption by increasing root resistance. This will be discussed later in this chapter. Currently, there is interest in a possible role of abscisic acid (ABA) in reducing root resistance in plants subjected to water or cold stress (Markhart *et al.*, 1979; Davies *et al.*, 1979), but the data are contradictory. However, ABA seems to increase root resistance at normal temperatures (Fiscus, 1981b).

Species Differences.   There are important differences in the reaction of roots of various species to aeration and temperature, indicating the existence of dif-

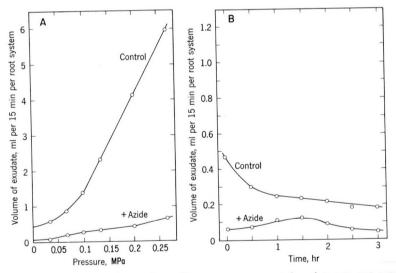

Fig. 9.3.   Effect of 10^{-3} M sodium azide on water movement through tomato root systems treated for 1 hr before measurements were started. (A) Rate of movement through root systems subjected to pressure. (B) Rate of root pressure exudation. (From Lopushinsky, 1964a.)

ferences in their protoplasm. For example, roots of cypress, tupelo, and rice can grow in saturated soil, but dogwood, yellow-poplar, and tobacco are killed by relatively short periods of flooding. The complex nature of these differences in tolerance of flooding is discussed in Chapter 6 and later in this chapter. The differences in effects of low temperature on water absorption through roots of various species, shown in Figs. 9.4 and later in Fig. 9.11, as well as in Table 9.2, also indicate differences in protoplasmic response. It is obvious that collards are affected much less than cotton and watermelon, and white pine less than loblolly pine. The reaction can also be modified by time of exposure, indicating that acclimation can occur. For example, when bean roots were cooled slowly over several days, water absorption was reduced less than when they were cooled rapidly (Böhning and Lusanandana, 1952). Kuiper (1964) reported that if bean root systems were kept at 17°C for 36 hr or more, water absorption was reduced less at low temperatures than if they had been kept at 24°C.

It has been suggested that differences among plants in their reaction to low temperature can be attributed to differences in the lipid composition of their cell membranes, especially the proportions of saturated and unsaturated fatty acids (Kuiper, 1975; Lyons, 1973; Lyons *et al.*, 1979). Markhart *et al.* (1980) investigated this in collards and soybeans and found that the amount of unsaturated fatty acids in roots of both species increased in the new roots produced after the plants were moved from day–night temperature regimes of 29°/23°C to 17°/11°C. Osmond *et al.* (1982) also reported an increase in unsaturated fatty acids in new roots of soybeans produced during chilling.

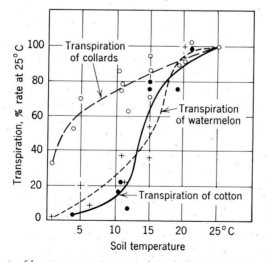

Fig. 9.4. Effects of low temperature on water absorption by plants of 3 species, measured by rates of transpiration. (From Kramer, 1942.)

TABLE 9.2   **Effects of Soil Temperature on Water Absorption by Plants of Various Species**[a,b]

Experiment	Species	Number of plants per experiment	Final soil temperature	Transpiration of cooled plants as percent of control at 25°C
1	Collards(*Brassica oleracea acephala* DC)	6	12.0	63.0
	Cotton (*Gossypium hirsutum* L.)	6	12.0	7.4
2	Collards	6	4.3	53.0
	Cotton	6	4.3	4.3
3	Collards	6	1.0	33.0
	Watermelon (*Citrullus vulgaris* Schrad.)	6	1.0	1.4
4	Loblolly pine (*Pinus taeda* L.)	4	0.5	13.7
	Slash pine (*P. elliotti* Engelm.)	4	0.5	13.9
	White pine (*P. strobus* L.)	4	0.5	37.7
	Red pine (*P. resinosa* Ait.)	4	0.5	25.0
5	Elm (*Ulmus americana* L.)	14	0.5	25.0
6	Privet (*Ligustrum japonicum* Thunb.)	12	2.5	47.0
7	Sunflower (*Helianthus annus* L.)	12	1.0	27.0

[a] Rates of transpiration were measured, and it was assumed that absorption was approximately equal to transpiration over 24-hr periods. The plants were divided into two groups, one of which was cooled by about 5°C per night while the other was kept at 25°C.

[b] Modified from Kramer (1942).

ENVIRONMENTAL FACTORS AFFECTING WATER ABSORPTION

The principal environmental factors affecting the absorption of water are the availability of soil water, the concentration of the soil solution, soil temperature, and soil aeration. Although soil moisture is most often the limiting factor, each of the others is important under certain conditions.

Availability of Soil Water

Transient midday water deficits are common on hot, sunny days, even when plants are growing in soil near field capacity, but long-term water deficits of increasing severity are caused by decreasing availability of soil water. Availability depends chiefly on the soil water potential and the hydraulic conductance, both of which decrease with decreasing soil water content, as shown in Fig. 3.14. The water readily available to plants is usually designated somewhat arbitrarily as that between field capacity and the permanent wilting percentage. As shown in Figs. 3.9 and 9.5 and in Table 9.3, the amount of water available to plants varies widely in different soils. Plants growing in soils that have a low storage capacity, such as Oakley fine sand or Aiken clay loam, will exhaust the readily available water and suffer from drought much sooner than plants growing in soil with a high storage capacity, such as the Salinas clay or the Wooster silt loam. The limitation in water storage capacity is particularly important for shallow-rooted plants and for plants growing on shallow soils.

Figure 9.5 shows how the soil water potential decreases as the water content decreases. This makes the soil water progressively less available, because move-

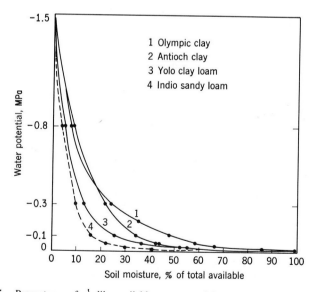

Fig. 9.5. Percentages of readily available water remaining in four soils at various soil water potentials. Curves were constructed from data for soil water potential over soil water content by assuming that available water occurs in the range from −0.015 to −1.5 MPa. (After Richards and Wadleigh, 1952.)

TABLE 9.3 Storage Capacity of Various Soils for Readily Available Water[a]

Soil type	Moisture equivalent	Permanent wilting percentage	Mm of available water per 30-cm depth
Oakley fine sand	3.29	1.33	8.6
Yolo fine sandy loam	16.80	8.93	32.0
Aiken clay loam	31.12	25.70	17.7
Salinas silt clay loam	28.33	12.49	63.2
Salinas clay	34.50	16.80	70.7
Catherine loam	37.90	19.03	77.0
Wooster silt loam	23.36	6.12	72.2
Brockton clay loam	24.51	11.55	49.5
Plainfield fine sand	2.40	1.36	4.3

[a] From Kramer (1969).

ment from soil to roots depends on a gradient in water potential, and the lower limit of water potential in growing crop plants is only -1.0 to -2.0 MPa. During the 1940s and 1950s, there was considerable unprofitable discussion of the relative availability of water between field capacity and the permanent wilting percentage. Veihmeyer and his colleagues claimed that water either is or is not available, and plant growth and plant processes are not affected until the water content approaches the permanent wilting percentage (Veihmeyer and Hendrickson, 1950). However, there is abundant evidence that growth and other processes are progressively reduced as the soil water content and soil water potential decrease below field capacity, and there is no single soil water potential that is consistently limiting for all important processes (Hagan, 1956). Furthermore, most important physiological processes are controlled directly by plant water stress and only indirectly by soil and atmospheric stress. Thus, plants in moist soil are often subjected to periods of water stress because of high midday transpiration, and plants in drying soil may be subjected to relatively low water stress if atmospheric conditions cause a low rate of transpiration.

Movement from Soil to Roots

The rate of movement of water from soil to roots depends on the steepness of the water potential gradient and on the hydraulic conductivity of the soil. Both decrease rapidly as soil water content decreases (Fig. 3.14). On the other hand, the root water potential also decreases as the rate of transpiration increases, although the decrease is not always proportional because of capacitance effects and possibly because of changes in root resistance (see Chapter 7).

Root Water Potential. The roots of slowly transpiring plants seem to behave as osmometers, and their water potential depends on the osmotic potential of the xylem sap, which is seldom lower than -0.2 MPa. As a result, detopped, nontranspiring plants seldom absorb water from soil or solutions with a potential lower than 0.15–0.20 MPa. In transpiring plants the water potential often falls as low as -1.0 or -2.0 MPa, or even -5.0 to -10.0 MPa in some xerophytes subjected to severe water stress. Leaf and root water potentials of pine seedlings at various soil water potentials are shown in Fig. 9.6. There are few measurements of water potential of roots in the soil because of the difficulty of making measurements (Fiscus, 1972). Slavikova (1964, 1967) reported that the water potential in the roots of trees in moist soil usually increased from the base toward the root apex. However, the gradient was reversed in a root in dry soil attached to a plant with most of its roots in moist soil, suggesting that water is translocated from roots in moist soil to roots in dry soil. There are many reports in the literature of translocation of water through roots from moist to dry soil, some of which are cited by McWilliam and Kramer (1968). Such movement can occur whenever the water potential in the dry region falls much below that in the moist region.

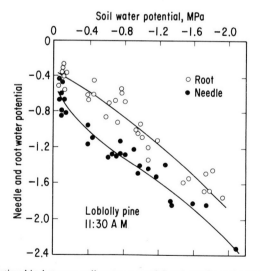

Fig. 9.6. Relationship between soil water potential and needle and root water potentials of loblolly pine seedlings sampled about 11:30 AM. Shoot water potentials were always lower than root water potentials, but as the soil became drier, root and soil water potentials became more similar, indicating conditions unfavorable for water absorption. (From Kaufmann, 1968.)

Transpiring plants wilt temporarily when movement of water into roots lags behind water loss, and they become permanently wilted when the plant water potential equals the soil water potential. Thus, absorption is limited by the lowest value of Ψ_w that can be developed in a plant, and this is approximately equal to the osmotic potential. In drying soil, root water potential must be considerably lower than soil water potential to maintain water flow to roots, because the hydraulic conductivity of soil decreases with decreasing water content.

Soil Conductivity.   Because different soils have different hydraulic conductivities, the same bulk soil water potential is often associated with different rates of movement to roots. Peters (1957) attempted to separate effects of water potential from water content by preparing mixtures consisting of various proportions of sand and silty clay loam that contained different amounts of water at the same potential. He found, as expected, that root elongation of corn seedlings was reduced by decreasing water potential, but it was also reduced as the soil water content decreased at a given water potential. Miller and Mazurak (1958) grew sunflowers in a series of soil fractions composed of particles ranging from 4760 to 2.3 μm in diameter and having pore diameters ranging from 529 to 2.3 μm. The various particle sizes were all maintained at the same water potential, equal to 20 cm of water. As shown in Fig. 9.7, best growth occurred in the intermediate particle size. In the coarse particles there was inadequate contact between particles and roots to provide sufficient water, and in the fine particles aeration was limiting because nearly all of the pore space was filled with water.

Various attempts have been made to describe the movement of water from soil to roots, that of Gardner (1960) being best known. He assumed that the rate of uptake of water from a given mass of soil is proportional to the length of roots in

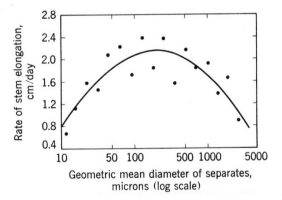

Fig. 9.7.   Rate of elongation of stems of sunflower seedlings growing in a medium composed of different average particle sizes, ranging from 9.25 to 3360 μm in average diameter. All systems were maintained at a water tension of 20 cm. (From Miller and Mazurak, 1958.)

the soil mass, the conductivity of the soil, and the difference between soil water potential and the potential at the root surface. He then developed equations to predict water movement from soil to roots under certain conditions. Figure 9.8 shows the gradient in water potential from soil to roots at two different soil water potentials and indicates that at an assumed uptake rate of 0.1 cm^3/cm of root per day, water can be expected to move to roots from a distance of 4 cm, which is more than adequate because most plants have a root density in excess of 2 cm/ cm^3 and are less than 1 cm apart. Figure 9.8 also indicates that at a soil water potential of -0.5 MPa, a gradient of 0.2 MPa provides sufficient driving force to move 0.1 cm^3/cm of root per day, but at a soil water potential of -1.5 MPa, the driving force required is 1.2 or 1.3 MPa. This increase is necessitated by the decrease in soil conductivity as the soil dries. In general, the driving force ($\Delta\Psi$) required to maintain a given rate of water uptake is proportional to the rate of water uptake and inversely proportional to the soil conductivity. Figure 9.9 shows that water flow to roots decreases rapidly as the bulk soil water potential decreases toward the root water potential, which was set at -2.0 MPa for this graph.

Gardner and Ehlig (1962) estimated that at a soil water potential lower than -0.1 or -0.2 MPa, resistance to water movement through the soil exceeds resistance to movement through plants. Cowan (1965) came to the same conclusion, but Newman (1969) pointed out that both of them used a root density much lower than that characteristic of crop plants. Using more realistic root densities, Newman estimated that root resistance exceeds soil resistance until the soil water potential approaches the permanent wilting percentage. The problem of soil versus root resistance is also discussed in Chapters 7 and 8.

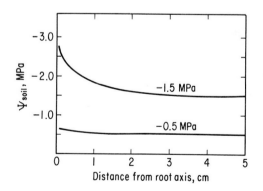

Fig. 9.8. Relationship of soil water potential to distance from the root axis in a sandy soil with an uptake rate of 0.1 cm^3 per centimeter of root length per day. The curves show the difference in water potential between root and soil at various distances from the root, at two levels of soil water potential. (From Gardner, 1960.)

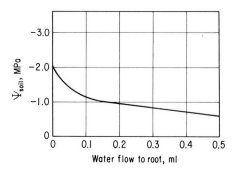

Fig. 9.9. Rate of flow of water to roots in milliliters per centimeter of root length per day at various soil water potentials when the root water potential is set at −2.0 MPa. (After Gardner, 1960.)

Concentration and Composition of the Soil Solution

Large areas of land in the arid regions of the world cannot support crop plants because of the high concentration of salt, and salt accumulation is an increasing problem in irrigated areas. It is rarely a problem in humid areas, where rainfall is adequate to leach out excess salt from the surface soil. Osmotic potentials of −0.35 to −0.4 MPa at permanent wilting seriously reduce the growth and yield of many crop plants (Magistad and Reitemeier, 1943), and at −4.0 MPa only a few halophytes survive. There are important differences in salt tolerance among crop plants, cotton and sugar beets being more tolerant than kidney beans, tomatoes, or deciduous fruits, but the growth of all plants, except a few halophytes, is reduced as the osmotic potential of the soil solution decreases. High salt is said to be more detrimental to seed germination than to established plants (Dunkle and Merkle, 1943; Uhvits, 1946).

Addition of fertilizer occasionally results in osmotic potentials too low for good plant growth. For example, application of 1300 kg/ha of 3–9–3 fertilizer to Norfolk sandy loam temporarily decreased the osmotic potential of the soil to −1.4 MPa, but a similar application to Cecil clay loam decreased it to only −0.3 MPa (White and Ross, 1939). Excessively high concentrations of salt sometimes develop in greenhouse soils and reduce growth (Merkle and Dunkle, 1944; Davidson, 1945). The effect of heavy fertilization on the osmotic potential of a greenhouse soil is shown in Fig. 4.7.

Causes of Reduced Growth.   Early investigations of the effects of high salt concentration on plant growth centered on the influence of various ions and proportions of ions in terms of such indefinite conditions as antagonism, toxicity, physiological balance, and cell permeability, but the results were confusing and unsatisfactory. During the late 1930s and 1940s workers at the United States

Department of Agriculture Salinity Laboratory in Riverside, California, shifted their attention to the total osmotic effects of all the ions and away from specific effects of particular ions. Eaton (1942) pointed out that there was no evidence of a critical concentration for a particular ion, but above an initial low concentration each increase in salt concentration caused by addition of any ion produced a decrease in growth. This was supported by the research of several other investigators, chiefly at the Salinity Laboratory. Furthermore, it appears that similar reductions in growth occur in plants subjected to stress by drying the soil (decreasing Ψ_m), by adding salt to the soil or nutrient solution (decreasing Ψ_s), or by a combination of the two treatments. An example of this effect on beans is shown in Fig. 9.10, and a similar effect on vegetative growth of guayule was reported by Wadleigh *et al.* (1946).

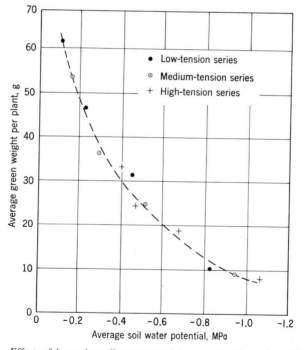

Fig. 9.10. Effects of decreasing soil water potential on growth of bean plants. Plants in the low-tension series were watered when 40–50% of the available water was removed. Medium-tension plants were watered when 60–65% of the available water was removed, and high-tension plants were not watered until 90–100% was removed. Each moisture series was subdivided into four groups which received no salt, 0.1%, 0.2%, or 0.3% salt. Reduction in yield was proportional to decrease in average water potential, whether the decrease was caused by low soil water content, high salt, or a combination of the two. For example, the highest yield was in low-tension soil containing no salt, and the lowest yield was in high-tension soil containing 0.3% salt. (From Wadleigh and Ayers, 1945.)

The reduction in growth observed in saline substrates is usually attributed largely to reduced water absorption caused by reduced water potential in the root environment. However, this is an oversimplification because when plants are transferred from dilute to more concentrated solutions, although they usually wilt at first, they recover after a day or two because absorption of salt lowers the osmotic potential of the roots, producing osmotic adjustment. Slatyer (1961) observed recovery of turgor by tomato plants after 28 hr in nutrient solution to which sufficient KNO_3, NaCl, or sucrose had been added to lower the osmotic potential to -1.0 MPa. In another experiment, Eaton (1942) grew six species of plants in sand culture with the osmotic potential adjusted over a range from -0.07 to -0.6 MPa. He found that the difference in osmotic potential between the substrate and the plants was similar in all concentrations, as shown in Table 9.4. Osmotic adjustment by salt uptake is characteristic of many plants when grown in more concentrated substrates. Reduction in growth therefore cannot be attributed primarily to reduced water uptake caused by a reduced driving force from substrate to roots. Various aspects of osmotic adjustment are discussed in Rains *et al.* (1980) and in Chapters 2 and 13.

There is also considerable reduction in root permeability in concentrated solutions. In short-term experiments, water uptake by corn roots from a solution with an osmotic potential of -0.48 MPa was only 12% of that from a solution with an osmotic potential of -0.08 MPa (Hayward and Spurr, 1943). Solutions of sucrose, mannitol, Na_2SO_4, and NaCl of the same osmotic potential reduced water absorption to the same extent in short-term experiments. Also, root systems preconditioned for several days in concentrated solutions absorbed more water than those transferred directly from dilute to concentrated solutions (Hayward and Spurr, 1943, 1944). Long-term experiments by Eaton (1941), Long

TABLE 9.4 **Effect of Osmotic Pressure of Culture Solution on Osmotic Potential of Plant Sap**[a]

	Osmotic potential of culture solution (MPa)		
Species	-0.072	-0.252	-0.60
Milo	1.03	1.08	1.11
Alfalfa	1.30	1.26	1.04
Cotton	1.31	1.18	0.97
Tomato	0.88	0.83	0.82
Barley	0.92	1.24	1.47
Sugar beet	1.28	1.38	1.50
Average difference	1.12	1.16	1.15

[a] From Eaton (1942).

(1943), and O'Leary (1969) indicate large decreases in permeability of root systems kept in concentrated solutions. This results from a decrease in root permeability, caused by dehydration, and increased suberization and decreased root growth.

Internal Salt Concentration. Apparently, the reduction in growth of crop plants caused by saline habitats is related as much to the accumulation of salt in the plant as to reduced availability of water in the substrate. Thus, if there is "physiological drought" in the sense of Schimper (1903), it is in the plant cells rather than in the environment. The effects of high salt content on metabolic processes need more study, but the high concentration of salt keeps the cell water potential low and probably reduces protein hydration and enzyme activity. The relative importance of excess salt versus water deficits in nonhalophytes is discussed by Greenway and Munns (1980).

In general, salt tolerance seems to depend on how high a salt concentration can be tolerated by the protoplasm without injury (Repp *et al.*, 1959), and plants with a low tolerance suffer in saline soil because of injury to protoplasm from salt accumulation, rather than from desiccation. Tolerance in some instances may be related to the fact that salt is commonly accumulated in the vacuoles to a much higher concentration than occurs in the cytoplasm. This is true in halophytes (Flowers *et al.*, 1977) as well as in nonhalophytes.

Although the discussion has thus far emphasized osmotic effects in general, there are specific ion effects that need to be taken into account (Strogonov, 1964). For example, growth of plants of some species is reduced more by sulfates than by chlorides (Hayward *et al.*, 1946), and it is claimed that chlorides increase the succulence of plant tissue, whereas sulfates decrease it. Van Eijk (1939) attributed the succulence of halophytes to an excess of chlorides, and Boyce (1954) reported that salt spray from the ocean increases the succulence of leaves of coastal vegetation.

In spite of the fact that growth is often reduced in proportion to the reduction in osmotic potential of the substrate, reduced absorption of water is not necessarily the principal cause of reduced growth in saline substrates. Plants subjected to a gradually increasing concentration of salt usually maintain normal turgor because of osmotic adjustment, i.e., increase in osmotic potential. Plants grown in saline substrates are often more succulent than controls (Boyer, 1965; Kreeb, 1965; Meyer, 1931), and succulence (high water content) is common among halophytes. This suggests that they are not suffering from dehydration in the same manner as plants growing in dry soil. Furthermore, plants subjected to high salt concentrations do not recover immediately when restored to normal conditions, as do plants in dry soil (Greenway, 1962). Thus, it appears that the similarity in the effect on growth of similar levels of matric and osmotic potential shown in Fig. 9.10 is somewhat misleading, and the two operate differently.

Nitrogen Deficiency

Little research has been done on effects of mineral deficiencies on water absorption. However, Radin and Boyer (1982) reported that nitrogen deficiency reduces permeability of sunflower roots by nearly 50%, resulting in daytime loss of leaf turgor and severe inhibition of daytime leaf expansion. Additional data of Radin (1982) indicate that this is also true for several other species. In cereals, daytime growth was not inhibited any more than nighttime growth because leaf growth occurs in the basal region and is less affected by daytime water stress. Thus, water stress caused by nitrogen deficiency reduces leaf growth more in dicots than in monocots.

Soil Temperature

It has been known at least since the time of Hales in the early eighteenth century that cold soil reduces water absorption, and in about 1860 Sachs observed that warm-season plants wilt more severely than cool-season plants when the soil is cold. Cold soil is an important ecological factor at high altitudes and high latitudes, and may be a limiting factor for vegetation near the timberline on mountains (Whitfield, 1932; Clements and Martin, 1934; Michaelis, 1934) and in the Arctic (Billings and Mooney, 1968). Local differences in soil temperature also affect both native and cultivated vegetation, and heavy, poorly drained soils are slower to warm than sandy, well-drained soils. Cold soil hinders germination of seed and establishment of seedlings of cotton, cucurbits, and other warm-season crops. Schroeder (1939) found that cold soil in ground beds, aggravated by watering with cold water, caused serious injury to greenhouse cucumbers in the winter, and soil temperatures below 12° or 15°C reduced water absorption by citrus in California (Cameron, 1941; Ramos and Kaufmann, 1979). It has also been reported that cold irrigation water reduces the yield of rice in the Sacramento Valley of California, northern Japan, and northern Italy. Procedures such as warming basins and removal of the warmer water from the surfaces of reservoirs are being used to bring cold water up to physiologically satisfactory temperatures. On the other hand, in some regions, lowering the soil temperature by irrigation with cold water is said to increase the yield of potatoes. The role of irrigation water temperatures was reviewed by Raney and Mihara (1967).

Species Differences in Tolerance of Low Temperatures.   Sachs observed that tobacco and cucurbits wilted more severely than cabbage and turnip plants in soil cooled to 3°–5°C. According to Brown (1939), Bermuda grass (*Cynodon* *dactylon* Pers.), a native of warm climates, wilts when the soil is cooled to 10°C, but bluegrass (*Poa pratensis* L.) which thrives in cool climates, is unaffected. The results of some studies of effects of low temperature are summarized in Table 9.2, and the striking differences between collards, which are grown as a winter crop in the southern United States, and the warm-season crops, cotton and watermelon, are shown in Fig. 9.4. Kozlowski (1943) found that water absorption by loblolly pine (*Pinus taeda* L.), a southern species, was reduced much more by cold soil than water absorption by white pine (*P. strobus* L.), a northern species. Kaufmann (1975) observed that root resistance became limiting for water absorption by citrus at 13.5°C, but not until the temperature fell below 7.5°C for Engelmann spruce, a subalpine species.

Chilling Injury and Water Absorption. It is well known that many plants from warm climates are injured by temperatures below about 15°C. This often is attributed to direct effects of low temperature on the shoots, but there is evidence that some of the injury is related to water deficits caused by decreased absorption of water (Fig. 9.4; see also Fig. 9.13 and Table 9.2) and sometimes by failure of stomata to close (Wilson, 1982). The literature on this subject was reviewed by McWilliam (1983) and by Wilson (1983). It has also been observed that when Arrhenius plots are made of rates of processes over temperature (*l*n rate over reciprocal of absolute temperature), there is often a sharp break in the curves (Fig. 9.11). Several investigators attributed this break to a phase transition from gel to gel in the lipid constituents of cell membranes. However, other investigators claim that phase transitions could not occur in the temperature range of these experiments (Lyons, 1973; Lyons *et al.*, 1979, pp. 1–24, 543–548). When Markhart *et al.* (1979) made an Arrhenius plot of water flow under pressure through broccoli and soybean root systems, he found a discontinuity in the rate for soybean, but not for broccoli. He also found that root systems of both species grown with day and night temperature regimes of 17°/11°C showed less resistance to water flow at low temperatures than root systems grown at 28°/23°C. The data are shown in Fig. 9.11. The propriety of using Arrhenius plots was questioned in Lyons *et al.* (1979), but this does not alter the observation that broccoli is less affected by low temperature than soybean. Later, Markhart *et al.* (1980, 1981) found that although soybean roots contain more unsaturated fatty acids than broccoli roots, there is a much greater increase in unsaturated fatty acids in broccoli than in soybean roots after they are moved to a low temperature. The increase occurs in the new roots grown in the low temperature rather than in preexisting roots. Osmond *et al.* (1982) also found an increase in percentage of unsaturated fatty acids in new roots of soybean. The role of membranes in chilling injury is discussed at length in the book edited by Lyons *et al.* (1979), and Wilson (1983) and McWilliam (1983) discuss the interaction of water and temperature stress. Wilson places particular emphasis on the failure of stomata to

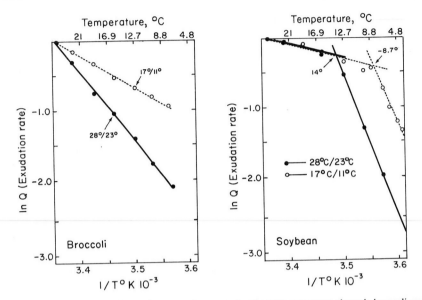

Fig. 9.11. Effects of growth temperature on rate of water movement through broccoli and soybean root systems under a pressure of about 0.5 MPa. Plants were grown at day and night temperatures of 28°/23°C or 17°/11°C. The rates for individual root systems were normalized to the rate at 25°C. The rates are presented as an Arrhenius plot with the natural log of flow (ln Q) on the ordinate and the reciprocal of the Kelvin temperature (1/T) on the abscissa. Celsius temperatures are shown at the top for convenience. Plants grown at lower temperatures are less affected by cooling than those grown at higher temperatures. Also, there are sharp changes in slope in the rates for soybean, indicating a large decrease in permeability. (From Markhart *et al.*, 1979.)

close promptly when plants are chilled, behavior also reported by McWilliam *et al.* (1982). Wilson also reported that drought-hardening plants or gradually exposing them to low temperatures increases their tolerance of chilling.

There is obviously need for much more research on the water relations of chilled plants. There is some evidence of varietal differences in tolerance of chilling in several kinds of crop plants, encouraging the possibility of selection for chilling tolerance. However, it will be helpful to the selection process if we obtain a better understanding of how much injury is caused directly by low temperature and how much is caused by water stress resulting from reduced absorption or reduced stomatal control of water loss.

Causes of Reduced Absorption at Low Temperatures. The marked decrease in absorption of water at low temperatures seen in Fig. 9.4 and Table 9.2 is caused chiefly by the increase in resistance to water flow through roots. The principal factors involved are

1. Increased viscosity of water, which is twice as great near 0°C as at 25°C, as shown in Fig. 9.12.

2. Decreased permeability of roots, shown in Figs. 9.4, 9.11, and 9.12.

3. Decreased metabolic activity, resulting in decreased salt accumulation. This reduces the driving force for active or osmotic uptake of water and is involved in decreased permeability and increased resistance to passive uptake of water.

4. Decreased root growth. This may be important in soils so dry that root extension is a limiting factor in water absorption.

5. Decreased availability of soil water. Richards and Weaver (1944) found that significantly more water was retained at both 0.05 and 1.5 MPa in cold soil than in warm soil. Also, Kramer (1934) found that the water-supplying power of soil, measured with Livingston soil point cones, was nearly twice as great at 25°C as at 0°C.

The combined effects of the decreased permeability of root cells and the increased viscosity of water cause a drastic reduction in water flow through root systems (Figs. 9.4, 9.11, and 9.12). Movement through dead roots is reduced by only about 50% from 25° to 0°C, presumably because the resistance of living membranes is eliminated. The decreased movement through dead roots with decreasing temperature resembles that for movement through collodion and porcelain membranes and is controlled largely by the viscosity of water. Ordin and

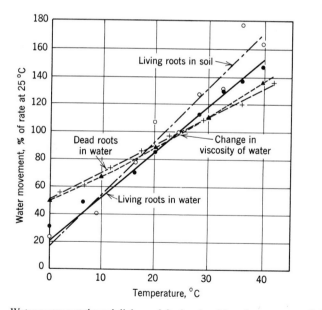

Fig. 9.12. Water movement through living and dead roots subjected to a vacuum of about 0.08 MPa for 1 hr. A different set of six plants was used for each temperature, and rates are plotted as percentages of rates at 25°C. The curve for viscosity is the reciprocal for viscosity plotted as percentage of the value at 25°C. Temperature affects water movement through dead roots much less than through living roots. (From Kramer, 1940a.)

Kramer (1956) and Woolley (1965) also found that killing roots increased the rate of diffusion of labeled water into them about threefold.

According to Kuiper (1964), during the first 30 min of exposure to low temperature, water uptake is controlled by its viscosity, but after that period the added effect of decreasing membrane permeability appears. He restudied data of other investigators and calculated the temperatures at which the permeability of root cell membranes began to be important. Some of his data are summarized in Fig. 9.13, where it can be seen that he estimates the effects on permeability to begin above 20°C for watermelon and cotton, 20°C for loblolly pine, 10°C for citrus, and only 5°C for collards and white pine. Kuiper also reported that the critical temperature could be shifted downward in beans by growing the root systems at 17°C instead of 24°C or by increasing aeration during growth. Markhart *et al.* (1979) were also able to shift the transition point downward by growing soybeans at lower temperatures, as shown in Fig. 9.11.

Brouwer and Hoogland (1964) conducted long-term experiments on the effect of low root temperature on growth of beans and concluded that the reduction in growth was caused by reduced uptake of water. However, other factors, such as

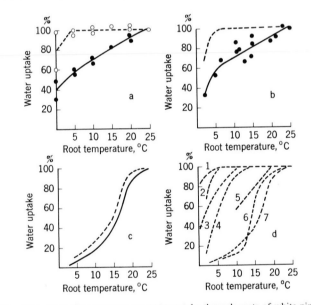

Fig. 9.13. The effect of temperature on water uptake through roots of white pine (a), cabbage (b), and watermelon (c). The solid curves are after Kramer (1942), and the dashed lines are corrected for differences in the viscosity of water at various temperatures. Graph (d) combines corrected curves from various sources: white pine (1), cabbage (2), citrus (3), sunflower (4), loblolly pine (5), cotton (6), and watermelon (7). The rate at 25°C was set as 100%. The critical temperature for reduction in root permeability is much lower in white pine (1) and cabbage (2) than in cotton (6) and watermelon (7). (From Kuiper, 1961.)

mineral uptake, nitrate reduction, and synthesis of growth regulators, may also be involved.

High Temperature and Water Absorption. Few data are available concerning the effects of high temperature on water absorption. Bialoglowski (1936) reported that temperatures above 30°C reduced water absorption by orange trees, and Haas (1936) reported a similar situation in lemon, grapefruit, and Valencia orange trees. Exudation from detopped tomato root systems attained a maximum rate at 24°C and decreased at higher temperatures (Kramer, 1940b). High soil temperatures are injurious to arctic and alpine plants, but most crop plants and native vegetation are relatively tolerant. Only roots in surface soil and in containers exposed to full sun are likely to be subjected to temperatures high enough to interfere with water absorption. The general physiological effects of high temperature were discussed by Gur *et al.* (1972).

Soil Aeration and Water Absorption

It has been known at least since the early nineteenth century that deficient aeration reduces water absorption. Clements (1921) summarized the early literature, beginning with experiments by de Saussure, and some of it is cited by Kramer (1949, Chapter 9).

Species Differences. Differences among species with respect to water absorption through flooded root systems are often conspicuous. Absorption of ions and water by corn and wheat roots is much reduced by inadequate aeration, but rice is little affected (Chang and Loomis, 1945). In one series of experiments, tobacco was most seriously injured by flooding, sunflower least, and tomato was intermediate (Kramer, 1951). In the field, tobacco sometimes wilts so rapidly if the sun comes out after the soil is saturated by a heavy rain that farmers speak of "flopping." In laboratory experiments the absorption of water by cypress was reduced very little by flooding, but that of several species of oak, *P. taeda,* and *Juniperus virginiana* was reduced to less than 50% of the controls (Parker, 1950).

Causes of Flooding Injury. Injury from deficient aeration and flooding is usually attributed to desiccation caused by reduced absorption of water. It has been shown by measurement of water flow under pressure through root systems that both increased carbon dioxide and decreased oxygen concentration cause marked increases in resistance to water flow through roots.

Generally, the first effect of flooding or deficient aeration is a decrease in root

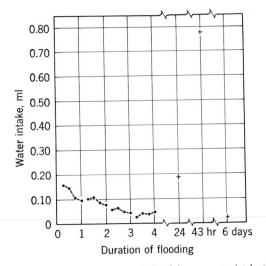

Fig. 9.14.   Effects of flooding for various periods of time on water intake through tobacco root systems attached to a vacuum line and subjected to a pressure of 0.02 MPa. Each point is the average of six root systems. Permeability increased after 24–48 hr because of injury to the roots, but after 6 days the root systems were almost destroyed. (From Kramer and Jackson, 1954.)

permeability, as shown in Fig. 9.14, resulting in decreased water absorption and water deficits and wilting of the shoots. For example, Yelenosky (1964) found that flooding the soil reduced the transpiration of yellow poplar seedlings to 68% of the control rate and produced a leaf water deficit of 47% after 3 days, compared to 11% in the controls. However, some of the effects of flooding cannot be explained by development of water deficits. Epinasty has been observed in several species, and adventitious roots and hypertrophies are developed at the water line on the stems of some plants. These are more characteristic of turgid than of flaccid plants. Thus, other factors, such as formation of ethylene, interference with synthesis of cytokinins and giberellins in the roots, and inhibition of downward translocation of growth regulators and metabolites, may cause some of the symptoms associated with flooding. The effects of flooding on root development were discussed in Chapter 6 and in a review by Kozlowski (1982b).

SUMMARY

Absorption of sufficient water to replace that lost by transpiration is essential to the successful growth of plants. Water absorption by plants in moist, warm, well-aerated soil is controlled largely by the rate of transpiration. In the field, the rate of water absorption is often limited by the extent and efficiency of root systems, by decreasing soil water content, and by soil salinity, low temperature, and deficient aeration.

The efficiency of roots as absorbing surfaces depends on the extent and amount of branching, which determines the amount of surface in contact with the soil, and on root permeability. Permeability is affected by age and stage of maturation, metabolic activity, and various environmental factors.

The various soil factors affecting water absorption operate either by reducing the driving force, i.e., the water potential gradient from soil to roots, or by increasing the resistance to water movement. Some do both. Decreasing soil water content decreases the driving force and increases both soil and root resistance. Increase in salt concentration at first reduces the driving force, but many plants undergo osmotic adjustment that restores the water potential gradient from substrate to roots. High salt also reduces root growth and root permeability, and its accumulation in plant tissue inhibits metabolic processes. Low temperature and deficient aeration reduce absorption by increasing the root resistance to water flow, and deficient aeration has additional effects on plant processes, including metabolism.

There are large differences among species in their tolerance of salinity, low temperature, and deficient aeration. For example, tolerance of high salt concentration by plants, depends chiefly on the ability of their protoplasm to tolerate high internal concentrations of salt. Cool-season plants tolerate low soil temperatures better than warm-season plants, but both undergo some acclimation when subjected to low temperatures.

A better understanding of how soil stresses affect root growth and water and mineral absorption would be very useful to plant breeders who are attempting to increase tolerance of environmental stresses.

SUPPLEMENTARY READING

Hillel, D. (1980). "Applications of Soil Physics." Academic Press, New York.

Hook, D. D., and Crawford, R. M. M., eds. (1978). "Plant Life in Anaerobic Environments." Ann Arbor Sci. Press, Ann Arbor, Michigan.

Kozlowski, T. T. (1982). Water supply and tree growth. II. Flooding. *For. Abstr.* **43**, 145–161.

Kramer, P. J. (1969). "Plant and Soil Water Relationships: A Modern Synthesis," Chapter 6. McGraw-Hill, New York.

Levitt, J. (1980). "Responses of Plants to Environmental Stresses," 2nd ed., 2 vols. Academic Press, New York.

Lyons, J. M., Graham, D., and Raison, J. K., eds. (1979). "Low Temperature Stress in Crop Plants: The Role of the Membrane." Academic Press, New York.

Poljakoff-Mayber, A., and Gale, J., eds. (1975). "Plants in Saline Environments." Springer-Verlag, Berlin and New York.

10

The Conducting System
and the Ascent of Sap

INTRODUCTION

Survival of most land plants depends on enough water moving daily from the roots to the shoots to replace that lost by transpiration. The losses often are large, over 4 liters per day for a corn plant and 200 liters or more for a large tree, and unless this water is replaced immediately, the plants will die of dehydration. The general principles governing water movement were discussed in Chapter 7. This chapter deals with the movement of water in the xylem and the forces that move it upward to the transpiring leaves, also its movement from veins to evaporating surfaces.

The existence of tall land plants became possible only after a vascular system evolved that permitted rapid conduction of water from roots to shoots. In fact, it is very difficult for land plants lacking a vascular system to attain a height of more than 20 or 30 cm because movement of water by diffusion from cell to cell is too slow to keep the tops supplied. In addition to facilitating water movement to the tops of tall plants, the continuity of water in the conducting system provides an essential communication system between roots and shoots that keeps the rates of absorption and transpiration in balance. Thus, when transpiration increases, the demand for an increased water supply to the leaves is transmitted to the roots by a decrease in water potential in the xylem sap, causing an increase in absorption. Conversely, when water absorption is reduced, the information quickly reaches the leaves as a decrease in water potential in the xylem sap, causing loss of guard cell turgor and closure of stomata. This results in a compensating decrease in water loss by transpiration. Actually, there is some lag in response, which was discussed in Chapter 7. On the whole, however, the continuous water system in plants functions fairly effectively as a feedback system that coordinates absorption and transpiration.

THE CONDUCTING SYSTEM

The xylem has been recognized as the principal pathway for upward movement of water at least since the time of Hales in the early eighteenth century. Superficially, the xylem might be compared to a collection of conduits in which the upper and lower ends are separated into a multitude of smaller conduits in the branches and leaves and in the roots. However, the xylem is not a continuous conductor, as is a pipe, but a collection of overlapping vessels and/or tracheids in which water must often pass through hundreds or thousands of crosswalls on its way to the leaves. During maturation, the protoplasts degenerate and disappear from the functional xylem elements. Therefore, in spite of the numerous crosswalls, xylem offers so much lower resistance to water movement than the other tissues of roots and stems that practically all longitudinal movement of water occurs in it. The presence of crosswalls is essential to survival because if the system were continuous it would often be completely blocked by gas bubbles and rendered useless for conduction. Because of the crosswalls, gas bubbles are confined within individual elements instead of spreading under reduced pressure and blocking the entire conducting system.

Readers who wish to know more about the structure of the xylem than is given here should consult textbooks on plant anatomy. Carlquist (1975) has an interesting book on the evolution of xylem, and Zimmermann and Brown (1971) present much information on the structure and functioning of the xylem in woody plants.

Structure of the Conducting System

Root–Shoot Transition.   There is a marked change in arrangement of the vascular tissues in the transition region from root to shoot, especially in herbaceous plants (see Fig. 7.5). The xylem is located in the center of roots, but it splits into a number of vascular bundles that occur in a ring outside the pith in herbaceous dicot stems. Complicated structural patterns often develop on the vascular system at the root–shoot transition region, especially in monocots. In some instances, there appear to be physiological differences above and below the transition zone. For example, in some species the salt content is much higher in the roots than in the shoots. Rush and Epstein (1981) found that although a salt-intolerant type of wild tomato excluded sodium from its shoot, a salt-tolerant type did not.

Woody Stems.   Kozlowski (1961) reviewed the extensive literature on water movement in woody stems. In conifers water moves mostly through tracheids, which are spindle-shaped cells seldom more than 5 mm long and 30 μm in diameter. Their protoplasts die and disintegrate as they mature; the walls are lignified and contain pits. Most of the water movement in angiosperms occurs through vessels formed by the destruction of end walls and disappearance of protoplasts from long rows of cells. The resulting vessels are tubelike structures with diameters ranging from 20 to 800 μm and lengths varying from a few centimeters to many meters. Vessels are relatively short in diffuse-porous species and quite long in ring-porous species, especially in lianas (Greenidge, 1952; Kramer and Kozlowski, 1979). It appears that single, continuous vessels are often differentiated from top to bottom in ring-porous trees (Priestley, 1935), but in many species their effective length is soon reduced by tyloses (Liming, 1934), air bubbles, masses of gum, and other blockages. Some data on vessel length are given in Table 10.1.

Usually, the xylem of ring-porous species ceases to function effectively in conduction after a year or two, whereas that of diffuse-porous species often continues to function for several years. As a result, most of the ascent of sap occurs in the outermost annual ring of ring-porous species, but several to many annual rings are functional in diffuse-porous species and in conifers (Kozlowski and Winget, 1963; Kramer and Kozlowski, 1979, pp. 465–474). Even in ring-porous species, the outermost annual ring might be expected to function as the chief path for sap flow because it is most directly connected to the new leaves and there has been less time for it to become blocked. However, this is not always true. In dormant conifer seedlings up to 4 years of age, water movement occurred in all annual rings. However, experiments with the dye basic fuchsin indicated that less movement occurred through the outermost annual ring than

TABLE 10.1   Apparent Length of Vessels in Trunks of Various Hardwood Trees[a]

Species	Apparent vessel length (cm)		
	Minimum	Maximum	Average
Diffuse-porous species			
Acer saccharum	81	94	88
Betula lutea	86	142	119
Fagus grandifolia	488	556	516
Populus tremuloides	101	132	122
Alnus rugosa	86	122	105
Ring-porous species			
Quercus rubra	853	1524	1188
Fraxinus americana	772	1829	1300
Ulmus americana	518	853	685

[a] From Greenidge (1952).

through the second ring. More water movement also occurred through the large-diameter tracheids of the earlywood than through the smaller tracheids of the latewood (Kozlowski *et al.,* 1966). Swanson (1966) reported that in lodgepole pine and Engelmann spruce, most rapid conduction occurs 10–15 mm in from the surface, and there is relatively little upward movement of water in the outer 5 mm of wood. According to Chaney and Kozlowski (1977), the patterns of water movement in ash (ring-porous) and maple (diffuse-porous) seedlings are similar to the patterns in mature trees of the same species.

The localized conducting system and large diameter of the vessels of ring-porous trees makes them more susceptible than diffuse-porous trees to blockage by gas bubbles and by mechanical injury such as girdling. Huber (1935) questioned if a ring-porous evergreen tree could survive because of extensive blockage of the large xylem vessels by the end of the growing season. Blockage is less important in deciduous species because they have very low rates of transpiration during the autumn and winter, and produce new xylem each spring before their leaves are fully expanded.

The conducting system of herbaceous dicots is often complex, as suggested by the diagram of a node, shown in Fig. 10.1. There is extensive branching and rebranching, with vascular bundles (leaf traces) leading off to each petiole. Because of the interconnections, a local injury does not necessarily present a serious obstacle to sap movement. Interpretation of transport and leaf water potential data is complicated by the fact that adjacent leaves are usually supplied by different xylem strands. This is discussed in the section on nodes and branches and illustrated later in Fig. 10.3.

Fig. 10.1. Vascular system of potato (*Solanum tuberosum*) stem, showing the complex branching at the nodes. Readers are referred to Dimond (1966) for a study of flow in tomato stems, which have a very similar structure. (From ''An Introduction to Plant Anatomy,'' 1947, 2nd ed., by A. J. Eames and L. H. MacDaniels. By permission of McGraw-Hill Book Company.)

Monocots. Most monocots have extremely complicated conducting systems with numerous vascular bundles that are interconnected in a complex pattern. Leaf traces branch, one branch extending into the leaf, one extending upward, and others connecting with other bundles in the stem. Furthermore, the bundles in the center of the stem follow a spiral path. As a result, injury to part of the vascular bundles does not prevent transport of water to leaves above the injury. The complex conducting system of woody monocots is discussed by Zimmermann and Brown (1971), Zimmermann (1973), and Zimmermann and Tomlinson (1974).

Spiral Growth of Xylem. An interesting feature of the xylem of trees is the common tendency to grow in a spiral, resulting in water moving upward in a spiral path rather than in a straight vertical path. Kozlowski and Winget (1963) reviewed the early literature on this topic, and Kozlowski *et al.* (1967) made a further study of spiral structure. An example of spiral ascent of sap is shown in Fig. 10.2. The pitch or degree of spiraling often changes from year to year and sometimes even reverses itself. According to Thomas (1967), in dogwood dye moves spirally around as much as 90% of the circumference per meter of ascent. Rudinsky and Vité (1959) claim that a spiral pattern provides more effective distribution of water to the crown than straight vertical ascent, because over a few years any particular root is connected to several branches and a branch may be connected to several roots. There may also be occasional disadvantages because a toxic substance introduced on one side of a tree with straight grain moves straight upward, but it may become widely distributed in the crowns of trees with spiral grain. For example, Kozlowski and Winget (1963) state that species such as pin oak, in which the transpiration stream spreads out in the crown, suffer more injury from the oak wilt fungus than white oak, in which water usually moves straight upward and the fungus remains localized in a limited area of the crown.

Nodes and Branches. There are often complicated patterns in the vascular system at nodes where connections are made to leaves and branches, as shown in Fig. 10.1. Because of constrictions in the xylem elements at nodes, there is a higher resistance to water flow from main stem to branches than in the stem (Zimmermann, 1978a). Larson and Isebrands (1978) found that in larch there is a reduction in number and size of vessels in the leaf traces, greatly increasing the resistance to flow into the leaves. Zimmermann found a high resistance in petioles of maple, and Dimond (1966) observed a relatively high resistance to water flow in tomato petioles.

Leaf phyllotaxy needs to be taken into account with respect to the water supply of successive leaves on a stem. It is often assumed that adjacent leaves on an herbaceous plant have access to the same water supply, but this is not true of most dicots. For example, tobacco has either $\frac{2}{5}$ or $\frac{3}{8}$ phyllotaxy, meaning that a vascular bundle goes around the stem twice between leaves 1 and 5 or three times between leaves 1 and 8. Thus, if the plant has $\frac{2}{5}$ phyllotaxy, leaves 4 and 6, shown in Fig. 10.3, are on different vascular strands from leaf 5. Fiscus *et al.* (1973) found the resistance to water movement between leaf 5 and either leaf 4 or leaf 6 of tobacco to be much greater than the resistance to water flow from the base of the stem (position zero) to leaf 5. Failure to take this into account in

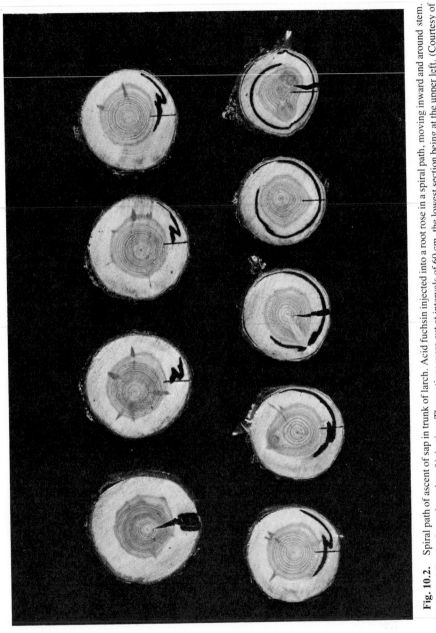

Fig. 10.2. Spiral path of ascent of sap in trunk of larch. Acid fuchsin injected into a root rose in a spiral path, moving inward and around stem. The vertical line is above the point of injection. The sections were cut at intervals of 60 cm, the lowest section being at the upper left. (Courtesy of T. T. Kozlowski.)

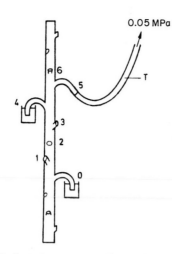

Fig. 10.3. Diagram of leaf attachments on a tobacco plant with 2/5 phyllotaxy. Leaves at positions 0 and 5 are on the same vascular strand, leaves at positions 4 and 6 are on different strands. By supplying different dyes to petioles at positions 0 and 4 and removing sap by suction through petiole 5, it was demonstrated that the resistance to lateral movement from 4 to 5 is much greater than the resistance to vertical movement from 0 to 5. (From Fiscus *et al.*, 1973.)

comparing the water status of adjacent leaves can lead to misinterpretations of data. Likewise, in trees with straight grain, tracers rise straight up from the point of injection, but in trees with spiral grain, tracers injected at the base may move into branches on the side opposite the point of injection or even pass completely around the stem (Fig. 10.2). On the other hand, in monocots there is so much branching and interconnecting of vascular bundles that dye spreads out in all directions. In grasses such as corn and sorghum, the vascular bundles leading to the leaves are so thoroughly interconnected at the lower nodes that the leaves can be regarded as having a common water supply.

Injury to the Conducting System

The conducting system of most plants provides a much larger capacity than is required for survival, and a considerable fraction of the xylem can usually be removed without killing plants. In fact, Jemison (1944) observed that trees on which over one-half of the circumference near the base had been killed by fire made as much growth during the following 10 years as nearby uninjured trees. However, Rundel (1973) reported that there is a strong correlation between basal fire injury and dead tops in redwood trees. New xylem laid down after an injury is often oriented so that it provides a very effective pathway for water flow around wounds (Fig. 10.4).

It has been demonstrated that two horizontal cuts made halfway through the stem, one above the other from opposite sides of the stem, do not prevent the ascent of sap (Elazari-Volcani, 1936; Greenidge, 1955; Preston, 1952). In fact, Postlethwait and Rogers (1958) demonstrated movement of radioactive phosophorus past as many as four cuts made only 15 cm apart in trunks of trees of several species (Fig. 10.5). Scholander *et al.* (1957) reported that air entering xylem elements through cuts is confined to the vessels which have been opened. Although the total resistance to flow increases, water moves around the blockage caused by cuts in the walls and through the xylem elements which are not plugged by air. Thus, the introduction of air into part of the xylem by wounds or other means does not prevent movement of sap; it merely increases the resistance to flow. The vascular bundles in the stems of some herbaceous species are so interconnected at the nodes that many of them can be cut without seriously reducing the flow of water to the leaves. This is true of the large bundles of tomato, but not of the smaller ones (Dimond, 1966).

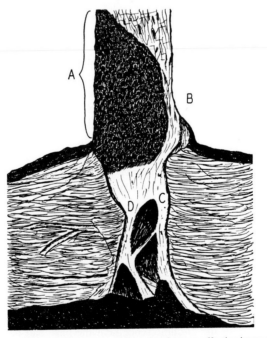

Fig. 10.4. New xylem and phloem are reoriented to form an effective bypass around burned or otherwise injured areas. A is burned area; C and D are roots connected by new conducting tissue to surviving tissue, B. (After Jemison, 1944; from Kramer and Kozlowski, 1979, by permission of Academic Press.)

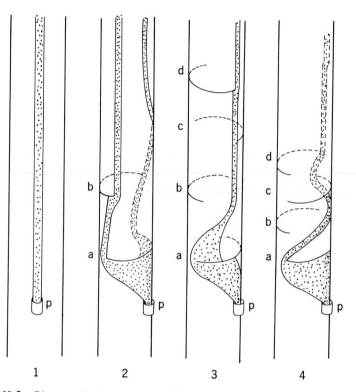

Fig. 10.5. Diagrams showing movement of ^{32}P around cuts in trunks of pine trees. The cuts are designated as a, b, c, and d. ^{32}P was supplied at point p, and the stippled areas indicate the path followed by the isotope. (After Postlethwait and Rogers, 1958; from Kramer and Kozlowski, 1979, by permission of Academic Press.)

Disease. In spite of the large safety factor in the conducting system of most plants, activities of bacteria, fungi, and insects often result in injury and blockage of the xylem; the result is reduction or prevention of the upward flow of water. Bark beetles often introduce a fungus that blocks water transport in the xylem, in addition to causing blue stain in the wood. There has been much discussion concerning the extent to which injury from so-called wilt diseases is caused by blocking of water flow in the xylem and the extent to which it is caused by toxins produced by the fungus (Dimond, 1955; Beckman, 1964). It seems unlikely that the mycelium actually blocks the xylem. However, there is adequate evidence that in at least some diseases, oak wilt for example, the vascular system of infected trees is plugged by tyloses and gums that block the movement of water and cause injury by dehydration (Kozlowski *et al.*, 1962; Kuntz and Riker, 1955). Plugging of the vascular system has often been observed in Dutch elm disease, verticillium wilt of elm and maple, and mimosa wilt. Zimmermann and McDonough (1978) discussed the effects of disease and wounding by insects on xylem transport and attributed most interruptions to embolisms caused by perforation of the walls of the xylem elements. It seems possible that toxins produced by the fungus injure the living cells adjoining the xylem elements and stimulate the formation of gum and tyloses. On the other hand, it was claimed by Gäumann that fusarium wilt of tomato is caused by a toxic substance or substances released by the fungus (see Dimond, 1955, for references). Dimond (1967) regarded toxins as unimportant and attributed the injury largely to blockage of water transport through the xylem. As pointed out by Talboys (see Kozlowski, 1968; Talboys, 1978), injuries from vascular or wilt diseases are complex and cannot always be assigned to a single cause.

Freezing. Winter freezing must often block the ascent of sap. In cold regions tree trunks freeze, and during the freezing process the dissolved gas in the xylem sap is forced out, forming bubbles that may interfere with water movement when the xylem thaws. Zimmermann (1964b) reported that a temperature of $-1°$ or $-2°C$ caused freezing of the xylem sap in deciduous trees, resulting in wilting of leaves above the point at which freezing occurred. Hammel (1967) found that freezing and thawing stem segments of hemlock did not increase resistance to water flow, although it did increase resistance in angiosperms. He suggested that in the tracheids of gymnosperms the tiny gas bubbles are easily reabsorbed during thawing. This may also occur in the smaller vessels of ring-porous angiosperms, but it is unlikely that the large vessels of ring-porous species are refilled. In a few species, such as birch and grape, root pressure may aid in refilling gas-filled xylem vessels with water in the spring. Recovery of the capacity to move sap upward in most ring-porous trees probably depends on production of new xylem each spring before the trees leaf out (Huber, 1935; Zimmermann, 1964a).

Cavitation and Gas Embolisms. When liquids are subjected to pressures lower than their vapor pressure, cavitaton occurs and gas bubbles develop. The bubbles formed by cavitation in xylem sap contain not only water vapor but also gases normally dissolved in the sap. Cavitation in the xylem of transpiring plants, where the sap is under tension, results in filling of some of the xylem elements with gas. Milburn and Johnson (1966) observed progressive cavitation in the petioles of *Ricinus* leaves under increasing water stress by listening with a modified stethoscope to the ''clicks'' produced as the water columns broke. Milburn (1973) discussed some of the problems associated with this technique. In trees there is progressive replacement of sap by gas in the older wood during the life of a tree. Also, every summer some of the sapwood is occupied by gas. Data of Clark and Gibbs (1957) indicate that up to 50% of the water in tree trunks of some species can be replaced by gas (see Fig. 10.12). Fortunately, the gas bubbles are isolated in individual tracheids and vessels because the pores in the pit membranes between xylem elements are so small that gas cannot be forced through them at the tensions ordinarily found in the xylem sap. Thus, even though some passageways are blocked by gas bubbles, enough are left open for the ascent of sap and the cohesive sap columns are not all broken.

Velocity of Sap Flow in Stems

Some measurements of the velocity of sap flow have been made by injecting dyes (Greenidge, 1958) or radioactive tracers (Fraser and Mawson, 1953; Moreland, 1950; Kuntz and Riker, 1955), but measurements by this method suffer from the probability that the rate is modified by cutting into the xylem. Huber (1956) discussed the methods of measuring sap flow. Although somewhat qualitative, the most reliable method seems to be the thermoelectric or heat pulse method devised by Huber and his students (Huber, 1932; Huber and Schmidt, 1937; Skau and Swanson, 1963; Kurtzman, 1966). Heat is supplied to the sap stream by a small heating element attached to or embedded in the stem, and the rate of sap flow is measured by the time required for the warm water to reach a thermocouple or thermistor placed in or on the wood above the heater. Sometimes temperature sensors are installed below as well as above the heater (Skau and Swanson, 1963). Cermak and Kucera (1981) used four sensors. Marshall (1958) discussed the theory and difficulties inherent in this method and concluded that although the sap speed exceeds the speed of the heat pulse, it is a useful method for measuring relative rates of sap flow. Pickard (1973) also discussed the theory. The data of Table 10.2 show that the rate is much greater in ring-porous species, where flow is restricted to a single ring, than in conifers and diffuse-porous species, where many annual rings are involved. Differences in rates in various parts of a single tree are shown in Fig. 10.6 and diurnal variations in an herbaceous plant in Fig. 10.7.

Much interesting information was obtained by use of the heat pulse method by Huber and Schmidt (1936, 1937) and later investigators. In oak and ash the greatest velocities occur at the base of trees (see Fig. 10.6), but in birch the rate increases upward because it has less conducting capacity per unit of leaves in its slender branches. Also, early in the morning, sap movement starts first in the upper part of the tree, near the transpiring leaves, and there often is a lag of 2 or 3 hr before sap movement is measurable in the lower part of the stem. This lag occurs because some water is removed from the stem before enough tension develops in the xylem sap to cause absorption through the roots. In the afternoon, the sap stream slows down in the top before it slows in the base of the stem, because absorption continues until the water removed from the stem is replaced.

TABLE 10.2   Rates of Water Movement in Xylem Measured by Various Methods[a]

Investigator	Method	Material	Velocity (m/hr)
Bloodworth *et al.*, 1956	Heat pulse	Cotton	0.8–1.1
Greenidge, 1958	Acid fuchsin	*Acer saccharum*	1.5–4.5
	Acid fuchsin	*Ulmus americana*	4.3–15.5
Huber and Schmidt, 1937	Heat pulse	Conifers	0.5
	Heat pulse	*Liriodendron tulipifera*	2.6
	Heat pulse	*Quercus pedunculata*	43.6
	Heat pulse	*Fraxinum excelsior*	25.7
Klemm and Klemm, 1964	^{32}P	*Betula verrucosa*	±3.0
Kuntz and Riker, 1955	^{86}Rb	*Quercus macrocarpa*	27.5–60.0
Moreland, 1950	^{32}P	*Pinus taeda*	1.2
Owston *et al.*, 1972	^{32}P	*Pinus contorta*	0.1–0.8
Decker and Skau, 1964	Heat pulse	*Juniperus osteosperma*	0.25

[a] After Kramer and Kozlowski (1979).

This phenomenon was discussed in the section on capacitance in Chapter 7, and the associated changes in stem diameter are discussed in Chapter 12.

Bloodworth *et al.* (1956) and Ladefoged (1960) used the heat pulse method to study the affects of various treatments on rate of sap flow as an indicator of variations in rates of absorption and transpiration. Daum (1967) applied the apparatus to the main stem and the two major branches of a forked ash tree and measured water flow in all three locations. Sap flow started earlier in the morning on the exposed east side of the tree and was more rapid in the morning than in the afternoon in the east side branch. Miller *et al.* (1980), using the heat pulse method, found that sap rises up to four times more rapidly on the sunny side of a tree than on the shady side.

Attempts have been made to apply a method used in measuring blood flow to measurement of sap flow in plants. A uniform magnetic field is applied at right angles to the direction of flow of the liquid. As the liquid passes through the magnetic field, it generates a voltage which is proportional to the rate of flow. Application of this procedure, sometimes termed the magnetohydrodynamic method, was described by Sheriff (1972, 1974).

Direction of Sap Flow

In this discussion, it has been assumed that water movement is upward and outward. This is usually the case because water movement is from regions of

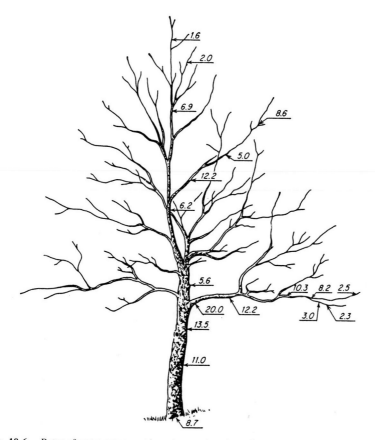

Fig. 10.6. Rates of water movement in meters per hour in various parts of an oak tree at midday, measured by the thermoelectric method. The rate of flow decreases toward the top because the relative conducting surface (ratio of xylem cross section to leaf area) increases toward the top. In birch, relative conductance decreases toward the top, as shown in Fig. 10.10, and rate of flow increases. (After Huber and Schmidt, 1936; from Kramer and Kozlowski, 1979, by permission of Academic Press.)

higher to regions of lower potential, and loss of water by transpiration lowers the water potential in the leaves. However, water can move in the reverse direction. This was demonstrated in experiments by Hales and other early investigators. In fact, as early as 1669, John Ray described experiments showing that water could move either way in branches, thereby dispelling the idea that valves existed in the water-conducting system. Williams (1933) reviewed the earlier work and demonstrated enough water movement from leaves immersed in water to keep other leaves on the same plant in the air turgid for several days. Slatyer (1956) and Jensen *et al.* (1961) also demonstrated that water will flow equally well through plants in both directions. Daum (1967), using the heat pulse method,

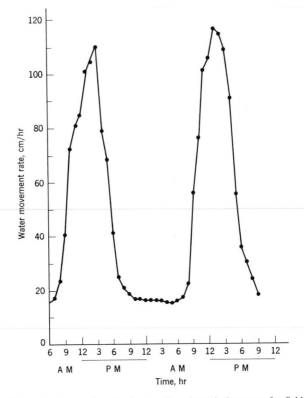

Fig. 10.7. Diurnal variations in rate of water flow through the stem of a field-grown cotton plant, measured by the thermoelectric method. (From Bloodworth *et al.*, 1956.)

demonstrated reversal of sap flow in an ash tree having two major branches. Sometimes there was downward flow in one branch and upward flow in the other, more rapidly transpiring branch. An afternoon rain, following a period of rapid transpiration, resulted in reversal of water flow in the tree trunk, downward movement occurring from the wet leaves toward the roots, which were in dry soil. These and other experiments suggest that there is no more resistance to water movement in the reverse than in the normal direction, and that water is free to move in any direction indicated by the prevailing gradient in water potential.

Efficiency of the Conducting System

Just as there must be some correlation between size of root systems and size of shoots, so there must also be some correlation between the size of the water-conducting system and the leaves supplied by it. One of the earliest attempts to describe some order in the development of the water conducting system of a plant seems to have been made by Leonardo da Vinci about 1500. According to Zimmermann (1978a), da Vinci wrote, "All the branches of a tree at every stage of its height when put together are equal in thickness to the trunk (below them)." This is approximately correct. In this century, research has dealt with the conductance of stems and branches and with the ratio of leaf surface to water-conducting area. Farmer (1918) reported that water conductivity is higher in angiosperms than in gymnosperms, and Huber (1928, and later) made extensive studies. Zimmermann and his colleagues at the Harvard forest have also studied water conduction in woody plants.

Huber (1956) summarized the literature on xylem conductivity and used two methods of expressing the efficiency of the water-conducting system. One measure is specific conductivity, which is the volume of water moved per unit of time under a given pressure through a stem segment of specified length and cross section. The other measure is relative conducting surface, which is the ratio of conducting surface (cross section of xylem) to leaf surface (often leaf fresh weight). Huber expressed the specific conductivity in terms of volume of water moved per hour under a given pressure in a segment of given length and cross section of conducting tissue, for example, ml hr^{-1} cm^{-2} MPa^{-1} over a distance of 1 m. He gave relative values of 20 for conifers, 65–128 for deciduous broadleaf trees, 236–1273 for vines, and even higher values for roots. He reported that the values for branches and twigs are usually lower than those for the trunk. The specific conductivity of tree trunks may decrease or increase from base to top; hence, the resistance to water movement is not proportional to the length of the stem. Conductivity in roots is discussed in Chapter 7.

According to Huber, the relative conducting surface or ratio of conducting surface in square millimeters to transpiring surface in grams of fresh weight is about 0.5 for trees, 0.2 for herbaceous shade plants, and 0.10 for desert succulents, but 3.4 for other desert plants and only 0.02 for certain aquatic plants. Data on the relative conducting surfaces of the stem and branches of a white fir tree (*Abies concolor*), given in hundredths of a square millimeter of xylem cross section per gram of leaf fresh weight, are shown in Fig. 10.8. In this species relative conductivity, or the ratio of conducting surface to transpiring surface, increases from bottom to top of the stem and is higher for the branches than for the main stem. It is possible that this increase in relative conductivity is related to the strong apical dominance of fir, as Huber found it to be reversed in birch, which lacks strong apical dominance.

One difficulty with Huber's method is that he based it on the entire area of the xylem, but in older branches and main stems only the sapwood is involved in water conduction. Thus, the conclusion that specific conductivity decreases from base to top might be changed if the calculations were based on area of sapwood. More recently, there have been several investigations of the relationship between

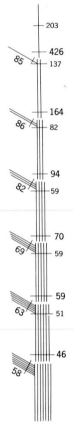

Fig. 10.8. Differences in relative amounts of conducting surface along the main stem and in the side branches of a 6-year-old *Abies concolor* seedling, expressed as hundredths of a square millimeter of xylem cross section per gram of needle fresh weight. The relative conducting surface is lower at the point of attachment of the whorls of branches (numbers in lightface type) than between nodes (numbers in boldface type), but there is a consistent increase in relative conducting surface from base to apex. (After Huber, 1928; from Kramer and Kozlowski, 1979, by permission of Academic Press.)

leaf weight or area and the area of sapwood that supplies it with water. In general, there seems to be a close relationship within most species studied, but large differences among species. For example, Kaufmann and Troendle (1981) found similar ratios between leaf area and sapwood area in the lower and upper parts of the crown in three of the four species they studied. However, the ratio varied from 1.88 m² of leaf area per square centimeter of sapwood in subalpine fir to 0.19 in aspen (see Fig. 10.9). They cite other investigators who found consistent relationships between leaf area and the conducting system that sup-

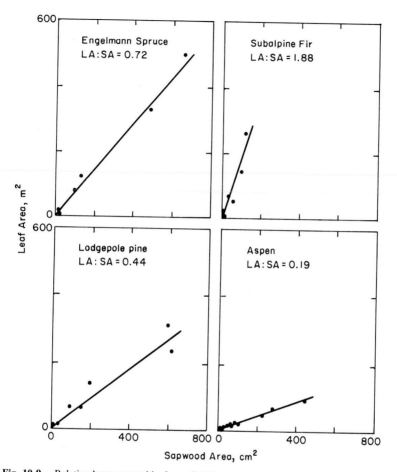

Fig. 10.9. Relation between total leaf area (LA) in square meters and the cross-sectional area of sapwood (SA) in square centimeters at 1.37 m above ground for four subalpine forest tree species. (From Kaufmann and Troendle, 1981.)

plies the leaves with water. A relatively consistent ratio is to be expected if twig extension and leaf development are dependent on an adequate water supply.

The large differences in ratio of leaf area to sapwood among the species studied by Kaufmann and Troendle probably result from differences in efficiency of their conducting systems, i.e., the specific conductivity. Attempts to measure specific conductivity by flowing water through segments of stem have often been thwarted by decreasing rates of flow and highly variable results. Zimmermann (1978a) reported that use of a 5 or 19 m*M* solution of KCl instead of distilled water eliminates this difficulty. He presented his results as leaf specific conductivity, expressed in microliters of water passing through a stem segment under gravity flow per hour per gram of fresh weight of leaves supplied by that stem. He made measurements on maple, white birch, and large-toothed aspen, and found that in general the leaf specific conductivity was greater in the main stem than in the branches. He also found increased resistance to flow at the junction of branches to main stems, just as there is in petioles (see the section on nodes and branches earlier in this chapter). Zimmermann attributes differences in conductivity to the fact that xylem vessel diameter increases from top to bottom in main stems and is smaller in branches than in the main stem. There is also a decrease in vessel diameter at the base of branches which is responsible for the low conductivity at those points, as shown in Fig. 10.10. Huber and Schmidt (1936) also reported that conductivity decreases toward the top in birch, but it increases toward the top in oak (Fig. 10.6) and in *A. concolor* (Fig. 10.3). They also reported low conductivity at the point of attachment of whorls in *Abies,* and the

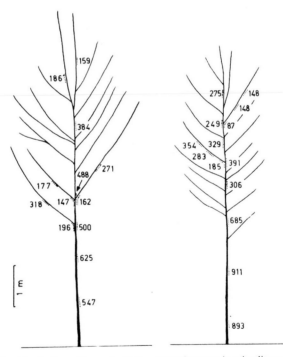

Fig. 10.10. The leaf specific conductivity, or conductance in microliters per hour, per gram fresh weight of leaves supplied, for two paper birch (*Betula papyrifera*) trees. Conductivities are lower in the branches than in the trunk and are considerably lower at the point where branches are attached to the trunk. This is also indicated by the high velocity of flow at the insertion of the lowermost branch in Fig. 10.6. (From Zimmermann, 1978a).

high velocity of sap flow at the point of attachment of the first branch (Fig. 10.6) suggests a high resistance in oak.

Xylem as a Conducting System. As Zimmermann and Brown (1971, Chapter 4) point out, the xylem is not always an ideal conducting system. They estimated the conductivity of the xylem of a number of trees and shrubs to be 15–80% of the estimated conductivity of ideal capillaries having the same diameter, but some lianas had 100% of that for ideal capillaries. Petty (1978) reported the conductivity of birch wood to be about 34% of that for unobstructed capillaries of the same diameters. About one-half of the resistance was in the perforated plates at the ends of the vessels and one-half was in the lumina. Recent research by Jeje and Zimmermann (1979) indicates that resistance to water flow through xylem vessels is strongly affected by the number and size of thickenings of their inner walls. In general, xylem seems to be an evolutionary compromise between efficient conduction and safety against extensive blockage by gas bubbles. Long vessels with large diameters would be most efficient in conduction, but they are also most likely to be blocked by gas bubbles. Tracheids offer more resistance to water flow, but are less subject to blockage by gas bubbles and form the water-conducting system of the tallest trees. The largest and longest vessels are found in vines and in ring-porous deciduous trees that produce a new conducting system every spring.

THE ASCENT OF SAP

The mechanism by which sap reaches the top of trees has been the subject of speculation for several centuries. Root pressure has sometimes been assigned a role in the ascent of sap, but Hales in 1727 recognized that root pressure does not occur in rapidly transpiring plants. Nevertheless, as mentioned in Chapter 8, some physiologists still think it plays a role. During the nineteenth century, several German physiologists thought that the living cells of the stems played some essential role in the ascent of sap, and the Indian physiologist Bose (1923) claimed that rhythmic pulsatory activity in the roots and stems caused the ascent of sap. However, several investigators, from Boucherie (1840) and Strasburger (1891) to Kurtzman (1966), have shown that sap will rise through stems or stem segments killed by heat or poisons. Even though living cells play no direct role in the ascent of sap, their presence may be necessary to prevent drying of the stem and infiltration of air. Also, when xylem cells are killed, there is a tendency for gum formation to block the water-conducting elements, especially at the boundary between living and dead tissue (Kramer, 1933).

The Cohesion Theory

The cohesion theory of the ascent of sap was foreshadowed by Hales, and both Sachs and Strasburger concluded that transpiration produces the pull causing the ascent of sap. Boehm (1893) demonstrated that transpiring branches could raise mercury above barometric pressure, but the demonstration by Askenasy (1895) and Dixon and Joly (1895) that water has considerable tensile strength was necessary to make the cohesion theory acceptable.

The following are the essential features of the cohesion theory of the ascent of sap:

1. Water has high cohesive forces, and when confined in small tubes with wettable walls such as the xylem elements, it can be subjected to a tension of many MPas (3.0 to possibly 30.0 or more) before the cohering columns rupture.

2. The water in a plant forms a continuous system through the water-saturated cell walls and xylem elements from the evaporating surfaces of the leaves to the absorbing surfaces of the roots.

3. Evaporation of water from plant cells, chiefly those of the leaves, lowers the water potential in the cell walls, causing water to move from the xylem to the evaporating surfaces. This reduces the pressure in the xylem sap and produces tension in the cohesive, continuous hydraulic system of the plant.

4. The reduction in water potential at the evaporating surfaces is transmitted to the roots, where it causes inflow of water from the soil. Thus, in transpiring plants, water absorption is controlled by the rate of transpiration, subject to the lag resulting from tissue capacitance.

Under these conditions, in transpiring plants there is continuous mass flow of water from the soil through the roots, up the stems, and out into the leaves to the evaporating surfaces. Dixon (1914) regarded the water as if it were hanging, suspended from the evaporating surfaces and anchored by the imbibitional forces in the cell walls of the evaporating surfaces. This is essentially correct.

Problems with the Cohesion Theory

There was considerable reluctance among plant physiologists to accept the cohesion theory, lasting over half a century (Greenidge, 1957; Lundegardh, 1954; Preston, 1961). It really should not require any defense after more than three-quarters of a century, because, as Renner (1912) observed, a cohesive column of water seems to be the only mechanism by which transpiration and absorption can be coordinated. Unfortunately, the basic necessity of such a mechanism was neglected by the critics.

Most of the objections are concerned with the adequacy of the cohesive forces of water, the instability of water columns under tension, and plugging of xylem elements with gas.

Cohesive Forces of Water. It seems impossible to many people that a column of water, even when confined in a narrow tube, can sustain a pull or tension. It is possible because the theoretical internal cohesive forces in water are very high, amounting to thousands of MPas. Experimentally, Ursprung (1915) measured a tension of 31.5 MPa (315 bars) in annulus cells of fern sporangia, and Briggs (1949) demonstrated a tension of 22.3 MPa in water subjected to centrifugal force. Zimmermann and Brown (1971, p. 200) estimated that a maximum tension of 0.015–0.02 MPa/m is required to lift water to the tops of trees during peak transpiration. This consists of 0.01 MPa/m to overcome gravity plus the tension required to overcome resistance to flow. A tension of about 2.0 MPa should therefore suffice to overcome both gravity and resistance to flow in a tree 100 m in height. Slatyer (1957) measured water potentials of −20.0 MPa in severely stressed privet and −7.7 MPa in cotton. Arcichovskij and Ossipov (1931) found water potentials as low as 14.3 MPa in desert shrubs, and White-man and Koller (1964) reported even lower values for a desert halophyte.

Some measurements of xylem water potential made at two heights in Douglas fir trees (Fig. 10.11) showed water potential gradients in excess of those necessary to lift water to the heights at which the measurements were made. Hellkvist

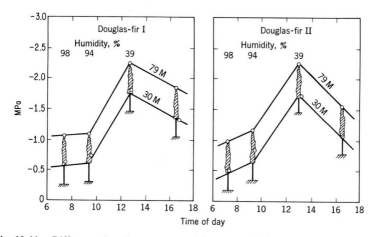

Fig. 10.11. Differences in xylem water potentials, measured with a pressure chamber, of twigs from upper and lower parts of crowns of Douglas-fir trees. (From Scholander *et al.*, 1965.)

et al. (1974) found this to be true in young Sitka spruce, but Connor *et al.* (1977) cited instances in which the stem water potential appeared to be less negative than the required minimum. However, their extensive measurements on eucalyptus 45 and 75 m in height gave gradients of water potential considerably greater than the theoretical minimum of 0.01 MPa/m, except when the foliage was wet. The leaves of a 45-m tree are subjected to a permanent stress of at least 0.45 MPa and those of a 75-m tree to a stress of 0.75 MPa. The vertical profile of decreasing water potential was not accompanied by a decrease in stomatal resistance in these trees, but there was a decrease in leaf size and an increase in thickness of cuticle with height. Long ago, Huber (1923) observed that the leaves at the tops of tall trees are more xeromorphic in structure than those near the base of the crown.

In conclusion, there seems to be no doubt that the cohesive forces of water are adequate to support the pull required to move sap to the top of the tallest trees. Furthermore, high tensions actually exist in trees, and sometimes even in small herbaceous plants subjected to severe water stress.

Cavitation and Gas in the Xylem.   It is true that gas bubbles often develop in xylem elements under tension by cavitation, as mentioned earlier in this chapter. In fact, nearly 50% of the water in the trunks of some trees may be replaced by gas during the summer (Fig. 10.12). However, gas bubbles usually form first in the largest elements and, as mentioned earlier in this chapter, they ordinarily cannot spread beyond the xylem elements (vessels or tracheids) in which they develop. Thus, the entire conducting system is not suddenly blocked by expanding bubbles. There is also a large safety factor in the xylem, and although partial blockage increases the total resistance to flow and the velocity in the remaining pathway, it does not necessarily reduce the volume of flow (Scholander, 1958). The fact that freezing blocks the ascent of sap was mentioned earlier. Bubbles formed during freezing are a more serious threat, but apparently in tracheids and small vessels they are redissolved during thawing. Gas bubbles were considered in more detail earlier in this chapter, and they are discussed at length by Zimmermann and Brown (1971, pp. 206–213).

CONDUCTION IN LEAVES

The final step in water conduction through plants is its movement into leaves and distribution to their various tissues and to the evaporating surfaces. At each node where a leaf is attached, a segment of the vascular system, called a leaf trace, separates from the vascular system of the stem, extends out through the petiole into the leaf blade, and provides a pathway for translocation of water and

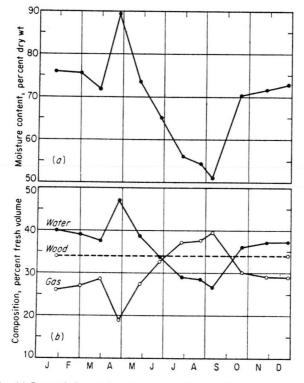

Fig. 10.12. (a) Seasonal changes in water content of yellow birch tree trunks calculated from disks cut from the base, middle, and top of the trunks. (b) Seasonal changes in gas and water content of yellow birch tree trunks calculated as percentages of total volume. (From Clark and Gibbs, 1957.)

solutes. An example of the vascular structure at the node of an herbaceous dicot is shown in Fig. 10.1.

Leaf Venation

The arrangement of the vascular system differs in various kinds of leaves. In most conifers, a single vein extends through the center of the leaf. In grasses, numerous veins extend the length of each leaf, parallel to the midvein, and anastomose at the margins near the leaf tips. These veins are connected by small veinlets extending across the intervening mesophyll tissue. Some dicots have palmate venation, in which a few large veins extend outward from the base of the leaf blade and are connected by a complex system of smaller veins. Other dicot leaves are pinnately veined, i.e., they have numerous branches extending outward from each side of the midvein. These secondary veins are enclosed in bundle sheaths and contain xylem, cambium, and phloem. The vascular system of a mature tobacco leaf is shown in Fig. 10.13. The actual distribution of water to the mesophyll occurs chiefly from the smaller veins that branch off from the secondary veins. These branch and rebranch and lose their cambium, then their phloem, and finally their bundle sheaths, ending as single xylem elements buried in the mesophyll. In some species, xylem elements terminate in epithem tissue near hydathodes. In many species, the small veins anastomose and form complex networks. They are so numerous that most cells of a leaf are only a few cells away from a vein or vein ending. Wylie (1938) found no cell more than 50 μm from a xylem element in a group of mesomorphic species which he studied. There appears to be so much conductive capacity in most leaves that they often survive the cutting of major veins, as shown in Fig. 10.14.

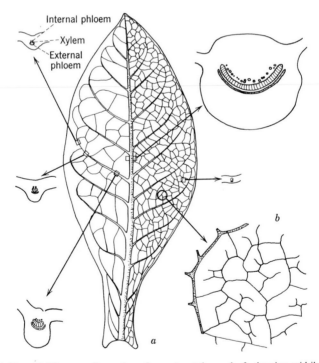

Fig. 10.13. (a) Diagram of venation of a mature tobacco leaf, showing midrib and principal lateral veins, as well as cross sections of midveins and lateral veins of various sizes. Internal phloem is found only in the midrib and principal lateral veins. (b) Enlargement of a small section of leaf blade to show the ultimate network of veins. There were 543 mm of veins per square centimeter of leaf blade on this leaf. (From Avery, 1933.)

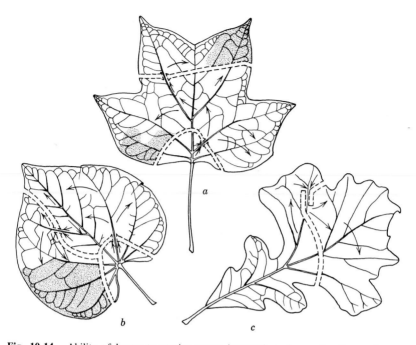

Fig. 10.14. Ability of leaves to survive severe interruption of normal pathways for water conductance by cutting of principal veins. Dotted lines indicate where veins were cut, and arrows show the general direction of water conduction. Shaded areas indicate tissue killed by dehydration. These experiments indicate a large excess of conductive capacity in the small veins and free movement in the reverse direction. Experiments were performed on attached leaves in the open and lasted 20 days. Leaf (a) is *Liriodendron tuplipifera;* leaf (b), *Cercis canadensis;* leaf (c), *Quercus velutina.* (From Plymale and Wylie, 1944.)

Conduction outside the Veins

There is some difference of opinion concerning the pathway by which water is supplied to the epidermis. Long ago, LaRue (1930) reported that the epidermis is normally detached from the mesophyll in *Mitchella repens* and can be detached experimentally over considerable areas in plants of other species without injury. LaRue and, later, Williams (1950) concluded that the epidermis is supplied with water directly from main veins rather than from the underlying mesophyll.

Many vascular bundles have bundle sheath extensions, which are vertical masses of cells extending outward from the bundles to the upper and lower epidermis, forming partitions which divide the air space of the mesophyll into numerous small chambers (Fig. 10.15). Wylie (1952) reported that bundle sheath extensions are most frequent in deciduous leaves and least common in broad-

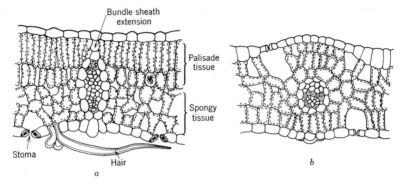

Fig. 10.15. Cross sections of leaves with (a) and without (b) bundle sheath extensions. Section (a) is an oak leaf and shows well-defined palisade and spongy mesophyll tissue, stomata on the lower surface, and an epidermal hair. Section (b) is an oat leaf with stomata on both surfaces. (After Eames and MacDaniels, 1947, by permission of McGraw-Hill Book Company.)

leaved evergreens. He suggested that they form a pathway for water and solute movement from the xylem of the veins to the epidermal cells, at least in mesomorphic species (Wylie, 1943). He also suggested that it might be easier for water to move out to epidermal cells through the bundle sheath extensions and then back to the mesophyll, rather than to move laterally through the mesophyll, because there is limited lateral contact among mesophyll cells.

It has been suggested that a significant amount of water may move to the epidermis as vapor through the intercellular spaces of leaves (Russell and Woolley, 1961). According to Williams (1950), this was disproved long ago because the epidermis remained alive even when the inner surface was covered with a waterproof coating. Furthermore, movement of water as vapor is not likely to be an important source of supply to the epidermis. It is generally supposed that water movement from the veins to the evaporating surface occurs chiefly in the cell walls, but this is sometimes questioned (Tyree and Yianoulis, 1980). This problem was discussed with reference to radial movement of water in roots in Chapter 8. There also is a lively discussion concerning the location of the principal evaporating surfaces that will be considered in Chapter 11.

SUMMARY

Development of a water-conducting system that permits rapid mass flow of water from roots to the evaporating surfaces of shoots was essential to the evolution of large land plants because the resistance to water movement through parenchyma tissue is too high to permit movement of sufficient water through the pith or cortex. The xylem provides an adequate pathway with a conductivity ranging from 15 to 80% of the estimated conductivity of capillaries having the same diameter, and some lianas approach 100% of the theoretical value. In general, xylem represents an evolutionary compromise between the efficient conduction provided by large, continuous vessels and the protection from blockage by air bubbles provided by small vessels and tracheids.

The velocity of sap flow has been measured by introducing dyes and radioactive tracers into the sap stream, but the best method seems to be the thermoelectric or heat pulse method. By a combination of methods, it has been established that in ring-porous species water usually moves only in the current annual ring, and the velocity is high. In diffuse-porous trees, sap ascends in a number of annual rings, but with a lower velocity. In the early morning the velocity of sap flow is greatest in the upper part of the stem, but in the afternoon it is greatest near the base. In oak the greatest velocity of sap flow is in the lower part of the stem, but in birch it is in the upper part because birch has a lower ratio of xylem to leaf area in the upper part of the tree. Most conducting systems have a large safety factor, and plants can survive loss of a considerable part of the xylem.

Water is pulled to the tops of transpiring plants by the matric or imbibitional force developed in the leaf cell walls by evaporation. Removal of water from the leaf xylem creates reduced pressure or tension in the xylem sap that is transmitted through the continuous, cohesive hydraulic system extending from leaves to roots. A tension of 0.015–0.020 MPa/m or less is required to pull water to the tops of transpiring trees, or 2.0 MPa for a 100-m tree, and water will easily sustain such tensions. Parts of the conducting system are often inactivated by accumulation of gas bubbles, but the structure of the xylem prevents bubbles from expanding and enough remains functional to supply the water lost by transpiration. The continuous hydraulic system in the xylem provides a feedback control system that keeps absorption and transpiration approximately in balance. Loss of water from leaves causes a decrease in water potential that is transmitted through the xylem sap to the roots and causes increase in absorption. Decrease in water absorption, caused by cold or drying soil or inadequate aeration, results in a decrease in water potential that is transmitted to the shoots and brings about closure to stomata and reduction in transpiration. Thus, water absorption and water loss are maintained approximately in balance.

SUPPLEMENTARY READING

Carlquist, S. (1975). "Ecological Strategies of Xylem Evolution." Univ. of California Press, Berkeley.

Dixon, H. H. (1914). "Transpiration and the Ascent of Sap in Plants." Macmillan, New York.

Esau, K. (1965). "Plant Anatomy," 2nd ed. Wiley & Sons, New York.

Kaufmann, M. R. (1976). Water transport through plants—current perspectives. *In* "Transport and Transfer Processes in Plants" (I. Wardlaw and J. Passioura, eds.), pp. 313–327. Academic Press, New York.

Kramer, P. J., and Kozlowski, T. T. (1979). "Physiology of Woody Plants." Academic Press, New York.

Nobel, P. S. (1974). "Introduction to Biophysical Plant Physiology." Freeman, San Francisco, California.

Zimmermann, M. H. (1978). Structural requirements for optimal conduction in tree stems. *In* "Tropical Trees as Living Systems" (P. B. Tomlinson and M. H. Zimmermann, eds.), pp. 517–532. Cambridge Univ. Press, London and New York.

Zimmermann, M. H. (1978). Hydraulic architecture of some diffuse-porous trees. *Can. J. Bot.* **56,** 2286–2295.

Zimmermann, M. H., and Brown, C. L. (1971). "Trees: Structure and Function," Chapter 4. Springer-Verlag, Berlin and New York.

11

Transpiration

INTRODUCTION

Transpiration can be defined as the loss of water from plants in the form of vapor. It is therefore basically an evaporation process, dependent on the supply of energy and the vapor pressure gradient between the evaporating surfaces and the ambient air. Viewed merely as an evaporation process, transpiration is quite simple, but when viewed as a plant process, it is extremely complex. Unlike evaporation, transpiration is modified by plant factors such as leaf structure and stomatal behavior, operating in addition to the physical factors that control evaporation. This chapter deals with the nature and importance of transpiration from individual plants and plant communities and with factors affecting it.

Importance of Transpiration

Transpiration can be regarded as the dominant process in plant water relations. Evaporation of water produces the energy gradient that is the principal cause of water movement into and through plants, and it therefore controls the rate of absorption and the ascent of sap. In warm, sunny weather, transpiration causes transient leaf water deficits almost daily, and when drying soil causes water absorption to lag behind loss by transpiration, permanent water deficits develop that result in permanent wilting and finally in death by dehydration. In fact, more plants are injured or killed as a result of water deficits produced by transpiration than by any other cause. If transpiration could be eliminated without stopping photosynthesis, injury from drought would not occur and crop plants could thrive in large areas which are now semidesert.

The quantitative importance of transpiration is indicated by the fact that a Kansas corn plant loses over 200 liters of water during its life, or 100 times its own fresh weight (Miller, 1938), and a field of corn in Illinois transpired over 20 cm, or about 80% of the precipitation during the growing season (Peters and Russell, 1959). Thus, the combined loss from transpiration and evaporation from the soil during a growing season sometimes exceeds the precipitation during the same period. A deciduous forest in the humid southern Appalachians loses 40–55 cm of water per year (Hoover, 1944), or 25–35% of the annual precipitation. Perhaps more important is the fact that several hundred kilograms of water are used per kilogram of dry matter produced by crop plants (see Table 13.3). Of all the water absorbed by plants, about 95% is lost by transpiration and 5% or less is used in metabolism and growth. If it were not for transpiration, a single rain or irrigation might provide enough water to grow a crop, providing evaporation from the soil surface was prevented by a suitable mulch. The relative losses by evaporation and transpiration from a forested and a nonforested watershed are shown in Table 11.1 and from an Illinois cornfield in Table 11.2.

It is sometimes argued that transpiration is beneficial because it cools leaves, causes the ascent of sap, and increases the absorption of minerals (Clements, 1934; D. M. Gates, 1968). Gates, for example, attaches great importance to the cooling effects of transpiration. Although cooling of leaves is beneficial, leaves in full sun are rarely injured by the rise in temperature that occurs when transpiration is reduced by transient wilting and midday closure of stomata. Water moves to the tops of trees as they grow upward, and transpiration merely increases the speed and quantity of water moved. Absorption and translocation of salt are

TABLE 11.1 Amounts of Water Lost in Various Ways by a North Carolina Watershed Covered with a Deciduous Forest (1940–1941) and the Increase in Runoff Which Followed Cutting of All Woody Vegetation and Elimination of Transplantation (1941–1942)[a]

Process	1940–1941	1941–1942
Precipitation	158.0	158.4
Interception	16.6	9.5
Runoff	53.4	93.0
Soil storage	−0.4	9.7
Evaporation	39.7	46.0
Transpiration	48.7	00.0

[a] Data in centimeters. From Hoover (1944).

probably increased by rapid transpiration, but many plants thrive in shaded, humid habitats where the rate of transpiration is very low. Although Winneberger (1958) reported that extremely high humidity reduced plant growth, Hoffman *et al.* (1971) and O'Leary and Knecht (1971) found that growth was generally better in high humidity than in intermediate or low humidities. On sunny, warm days, plants often lose so much water by transpiration that their leaves wilt, growth is reduced or stopped, and stomatal closure reduces photosynthesis. The numerous harmful effects of water deficits are discussed in the next chapter.

Transpiration can best be regarded as an unavoidable evil, unavoidable because of leaf structure and evil because it often produces water deficits and injury by dehydration. Apparently, the evolution of leaf structures favorable for high rates of photosynthesis had more survival value than that of structures favorable to low rates of transpiration, except in very dry habitats. Unfortunately, a leaf structure favorable for entrance of carbon dioxide is also favorable for loss of

TABLE 11.2 Relative Water Losses by Transpiration and Evaporation from an Illinois Cornfield during the Period from Mid-June to Early September[a]

Year	Total evapotranspiration from uncovered plot (cm)	Transpiration from covered plot (cm)	Transpiration as percentage of evapotranspiration	Total precipitation (cm)	Excess of evapotranspiration over rainfall (cm)
1954	32.25	16.5	51%	18.5	13.75
1955	34.50	17.5	51%	23.0	11.50
1957	33.75	15.1	45%	24.0	9.75

[a] From Peters and Russell (1959).

water. Thus, high rates of transpiration are the unavoidable result of the evolution of a leaf structure favorable for uptake of carbon dioxide.

THE PROCESS OF TRANSPIRATION

Transpiration involves two stages: the evaporation of water from cell walls and its diffusion out of the leaves, chiefly through the stomata. These processes can be discussed in terms of resistances or conductances, and driving forces. First, however, we will discuss the location of the surfaces from which most of the evaporation occurs.

Evaporating Surfaces

Some water vapor escapes through lenticels in the bark of twigs (Geurten, 1950; Huber, 1956; Schönherr and Ziegler, 1980), through stomata in the stems of herbaceous plants (Gračanin, 1963), and through the cuticle of leaves. However, most of the water escapes through the stomata of leaves. It has usually been assumed that most of this water evaporates from the surfaces of mesophyll cells bordering on intercellular spaces. This view was developed by Slatyer (1967, pp. 215–221). However, during the past decade, it has been suggested that most of the water evaporates from the inner surfaces of the epidermal cells, largely in the immediate vicinity of the stomata (Meidner, 1975; Byott and Sheriff, 1976). It has even been suggested that much of the water evaporates near or even from the inner surfaces of the guard cells. This is termed peristomatal transpiration by some writers (Maercker, 1965). Rand (1977) constructed a mathematical model of gaseous diffusion of carbon dioxide and water vapor which predicts that practically all water vapor evaporates from cell walls near the substomatal cavity, although carbon dioxide is absorbed throughout the exposed mesophyll cell surfaces of the leaf. Tyree and Yianoulis (1980) made a computer study of the diffusion of water in models of substomatal cavities that led to a similar conclusion. They estimated that about 75% of all the evaporation occurs in the region of the guard cells. This conclusion was based on the assumption that all cell wall surfaces are equally moist, which is at least uncertain and probably not true.

Beginning with von Mohl in 1845, various writers have reported cutinization or suberization of cell walls bordering intercellular spaces. Among the papers indicating the presence of a lipid layer, probably cutin, on mesophyll cell walls are those by Lewis (1945), Scott et al. (1948), Scott (1964), and Sheriff (1977a).

It is not clear from these papers whether cutinization is uniform in thickness over all cell walls, but it seems reasonable to expect a thicker deposit near the guard cells. According to Norris and Bukovac (1968), the cuticle covering the outer surface of the lower epidermis of the hypostomatous leaves of pear extends in through the pores, covering the inner surfaces of the guard cells. They also reported that the inner surface of the lower epidermis is cutinized. They did not examine the inner surface of the upper epidermis. However, it is largely covered by the ends of palisade cells in most leaves (later, see Figs. 11.11 and 11.12) and must have a limited internal evaporating surface even in amphistomatous leaves. Apparently, the location of the principal evaporating surface in leaves cannot be decided until we know more about the degree of cutinization of the mesophyll cell surfaces. There can certainly be little water loss from the internal surface of guard cells in pear or in other plants where the external cuticle extends into the substomatal cavity. Sheriff (1977a) suggested that the inner surfaces of subsidiary and epidermal cells are cutinized in at least some kinds of leaves, because in certain experiments droplets of water formed on them. However, Edwards and Meidner (1978) state that the inner lateral guard cell walls of *Picea* are practically devoid of cuticle. It seems that the importance of peristomatal evaporation should be viewed with caution until more information is available concerning the amount of cuticle on the inner surfaces of the guard cells and mesophyll tissue.

The Water Vapor Pathway

The pathway by which water vapor escapes after leaving the evaporating surfaces is relatively simple. It diffuses through the intercellular spaces and out through the stomatal pores. When the stomata are closed, the only pathway available is through the epidermal cells and the cuticle, a path with a relatively high resistance. The resistances are shown diagrammatically in Fig. 11.1 and numerically in Table 11.3.

Driving Forces and Resistances

The rate of transpiration depends on a supply of energy to vaporize water, the gradient in water vapor pressure or concentration that constitutes the driving force, and the magnitude of the resistances in the pathway. The driving force for liquid water movement through plant tissue is the difference in water potential (Ψ_w), but the driving force for movement of water vapor is the gradient in concentration or vapor pressure.

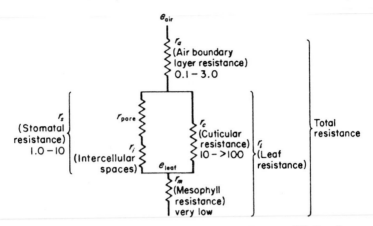

Fig. 11.1. Diagram showing resistances in seconds per centimeter to diffusion of water vapor from a leaf. Stomatal and cuticular resistances vary widely among species and with leaf hydration and atmospheric humidity. The rate of transpiration or transpiration flux density is proportional to Δe, the water vapor pressure gradient, e_{leaf} to e_{air}, and inversely proportional to the resistances in the pathway. (From Kramer and Kozlowski, 1979, by permission of Academic Press.)

Evaporation (E) from a water surface can be described by the following equation:

$$E = \frac{C_{water} - C_{air}}{r_{air}} \quad \text{or} \quad \frac{e_{water} - e_{air}}{r_{air}} \qquad (11.1)$$

E is the evaporation in g cm^{-2} s^{-1} and C_{water} and C_{air} are the water vapor concentrations at the evaporating surface and in the bulk air in g m^{-3}; e_{water} and

TABLE 11.3 Resistances to Movement of Water Vapor through the Boundary Layer (r_a), Cuticle (r_c), and Stomata (r_s) in Leaves of Several Species[a]

Species	Resistances to water vapor (s cm^{-2})		
	r_a	r_s	r_c
Betula verrucosa	0.80	0.92	83
Quercus robur	0.69	6.7	380
Acer platanoides	0.69	4.7	85
Circaea lutetiana	0.61	16.1	90
Lamium galeobdolon	0.73	10.6	37
Helianthus annuus	0.55	0.38	—

[a] From Holmgren et al. (1965).

e_{air} are the water vapor pressures in millibars or kPa; r_{air} is the boundary layer resistance in s cm^{-1}. One millibar = 0.1 kPa or 100 Pa.

Transpiration differs from evaporation because the escape of water vapor from plants is controlled to a considerable degree by leaf resistances that are not involved in evaporation from a free water surface. Therefore, the equation for transpiration (T) requires an additional term to include these resistances in the leaf.

$$T = \frac{C_{leaf} - C_{air}}{r_{leaf} + r_{air}} \quad \text{or} \quad \frac{e_{leaf} - e_{air}}{r_{leaf} + r_{air}} \tag{11.2}$$

where C_{leaf} and e_{leaf} are the water vapor concentration and vapor pressure, respectively, at the evaporating surfaces within the leaf and r_{leaf} is the additional diffusive resistances of the leaf. Some writers use much more complex equations (Cowan, 1977; Cowan and Milthorpe, 1968; Jarvis, 1980).

Equation (11.2) states that the rate of transpiration, T, in g cm^{-2} s^{-1}, the transpiration flux density, is proportional to Δc, the difference in concentration of water vapor, or Δe, the difference in vapor pressure, between the leaf and the bulk air outside, divided by the sum of the resistances to diffusion ($r_1 + r_a$) in the leaf and in the boundary layer adjacent to it. The situation with respect to resistances is shown diagramatically in Fig. 11.1. If most of the stomata are on one surface of a leaf, r_1 will differ significantly for the upper (adaxial) and lower (abaxial) surfaces (Holmgren et al., 1965). Also, the stomata on the upper and lower surfaces often differ in their response to water stress (Davies, 1977) and light (Travis and Mansfield, 1981). The various components of this system will now be discussed.

Energy Utilization. The energy input to leaves and plant stands comes from direct solar radiation, from radiation reflected or reradiated from the soil and surrounding vegetation, and from advective flow of sensible heat from the surroundings. The energy load is dissipated chiefly by three mechanisms: reradiation, convection of sensible heat, and dissipation of latent heat by evaporation of water (transpiration). A small amount of energy is used in photosynthesis, and some is stored in leaves, but these quantities are so small (usually only 2 or 3% of the total) that they are important only for very precise measurements over short periods of time.

It may be helpful to give brief definitions of some of the terms used in discussing the energy balance. Sensible heat transfer refers to conduction of heat between leaves and air by convection or advection. Convection refers to transfer of heat by mass movement of air, usually vertically, such as the flow away from a hot radiator or from a leaf warmer than the air surrounding it. Advection refers to horizontal transfer, as in the oasis effect, in which advective energy transfer of warm air from the hot desert to a small, isolated mass of vegetation results in a rate of transpiration exceeding that accounted for by the incident radiation (see Fig. 11.2). Sensible heat transfer can be positive or negative, depending on whether the leaves are cooler or warmer than the air. Latent heat refers to gain or loss of heat caused by change in state, such as the loss when water evaporates or the gain when dew condenses. Sensible heat transfer refers to changes in the heat content of the air.

The energy load on a leaf can be partitioned as follows:

$$\underbrace{(S+G) - (rS+R)}_{\text{Net radiation}} + H \pm E + A = 0 \qquad (11.3)$$

Net radiation is the radiation actually available to leaves. It consists of the total solar radiation (S), plus long-wave radiation (G) from the environment, minus radiation reflected from leaves (rS) and that reradiated (R). H is the sensible heat exchange from the environment by convection and advection, E the latent heat lost by evaporation or gained by condensation, and A the energy used in photosynthesis. The last term is small, only 2 or 3% of the total, and is usually neglected. As all of the terms in Eq. (11.3) are variable, the energy relations of leaves are rather complex. When leaf and air temperatures are the same, the entire heat load is dissipated by reradiation and transpiration. If transpiration is

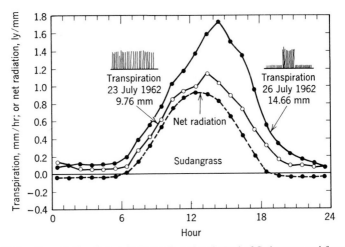

Fig. 11.2. Transpiration from a lysimeter in a closed stand of Sudangrass and from the same lysimeter 3 days later after the surrounding crop had been removed, exposing the plants in the lysimeter. Radiation was essentially the same on the 2 days, but additional energy supplied by advection caused much higher transpiration after the plot was explored. (From van Bavel *et al.*, 1963.)

curtailed by stomatal closure, leaves become warmer than the air and the heat load is dissipated by reradiation and sensible heat transfer, but usually all three mechanisms operate. At night leaves are often cooled below air temperature by radiation to the cold sky, resulting in flow of sensible and latent heat to them and the deposition of dew. The complex energy balance of leaves is discussed in more detail by D. M. Gates (1968, 1976, 1980), Knoerr (1967), and Nobel (1974). Gates (1976) suggested that water loss from plants could be explained in terms of the energy balance if all the relevant plant and environmental factors could be measured. Lack of space prevents further discussion of leaf energy balance, but readers are reminded that leaf temperature is interrelated with the energy dissipation mechanisms, reradiation and convection, and, through its effect on vapor pressure, the rate of transpiration. It also affects photosynthesis and respiration.

Vapor Pressure Gradients from Leaf to Air. The difference in water vapor pressure between leaves and air constitutes the driving force causing the movement of water vapor out of plants. This difference depends on two variables: the vapor pressure of the bulk air surrounding the leaves and the vapor pressure at the evaporating surfaces in the leaves. The vapor pressure of the bulk air depends on its absolute humidity and temperature, whereas that at the evaporating surfaces depends on the temperature and the water potential. If it is assumed that the water potential at the evaporating surface is zero, the vapor pressure is the saturation vapor pressure of water at the leaf temperature. However, this assumption is strictly correct only for turgid cells with a water potential of zero. Cells in transpiring leaves often have water potentials of -1.0 to -5.0 MPa, but this is of minor importance because a large reduction in cell water potential results in only a small reduction in water vapor pressure. For example, as shown in Table 11.4, at 30°C and a cell Ψ_w of -1.5, -3.0, and -6.0 MPa,

TABLE 11.4 The Vapor Pressure of Water (e), the Vapor Pressure at Cell Surfaces at Three Cell Water Potentials, and the Difference in Vapor Pressure (Δe) between Cells and Air at Relative Humidities of 80 and 50%[a]

Ψ_w of water and of mesophyll cells (MPa)	e at cell surfaces (kPa)	Δe at 80% relative humidity (kPa)	Δe at 50% relative humidity (kPa)
0.0	4.243	0.849	2.121
-1.5	4.200	0.806	2.079
-3.0	4.158	0.764	2.037
-6.0	4.073	0.679	1.952

[a] The relative effect of a low Ψ_w diminishes as the relative humidity decreases. All vapor pressures are for 30°C.

the vapor pressure at the cell surfaces will be approximately 99, 98, and 96%, respectively, of the saturation values.

From these data, it can be seen that the decrease in leaf water potential or osmotic potential caused by leaf water deficit will have little effect on the vapor pressure of cell surfaces and therefore will not materially reduce the rate of transpiration. Nevertheless, there is evidence that the vapor pressure at the evaporating surfaces is sometimes lower than would be expected from measurement of bulk leaf water potentials. For example, Shimshi (1963) and Whiteman and Koller (1964) reported water potentials in the intercellular spaces of rapidly transpiring plants of -9.0 to -32.0 MPa. Jarvis and Slatyer (1970) also reported evidence that the water vapor pressure at the evaporating surfaces may be lower than the saturation vapor pressure in rapidly transpiring cotton. A possible explanation for this is offered in the section on resistances in leaves.

An increase in temperature without an increase in water content of the air or the absolute humidity increases the rates of evaporation and transpiration because it increases the vapor pressure gradient from evaporating surfaces to the surrounding air (Δe), as shown in Table 11.5 and Fig. 11.3. For example, an increase in temperature from 10° to 20°C nearly doubles Δe, and increasing it from 10° to 30°C more than triples it. Other things being equal, the rate of evaporation or transpiration will be proportional to the vapor pressure gradient (Bange, 1953; Cole and Decker, 1973). An increase in leaf and air temperature with no change in the absolute humidity of the air is accompanied by a decrease in relative humidity, as shown in Table 11.6. However, the changes in relative humidity are merely coincidental and the evaporation gradient is really controlled by the difference in vapor pressure between the evaporating surface and the ambient air. An increase in air temperature without any change in absolute humidity causes a slight increase in atmospheric vapor pressure, which, according to the gas laws, is proportional to the change in absolute temperature. Adjustment of atmospheric moisture in growth chambers and closed containers

TABLE 11.5 Effect of Temperature on the Saturation Vapor Pressure of Water, on the Vapor Pressure of Air at 70% Relative Humidity, and on the Vapor Pressure Gradient (Δe) from Water Surface to Air

Temperature (°C)	Vapor pressure at saturation (kPa)	Vapor pressure of air at 70% relative humidity (kPa)	Δe (kPa)
0	0.610	0.427	0.183
10	1.227	0.858	0.369
20	2.337	1.635	0.701
30	4.243	2.970	1.273
40	7.377	5.163	2.214

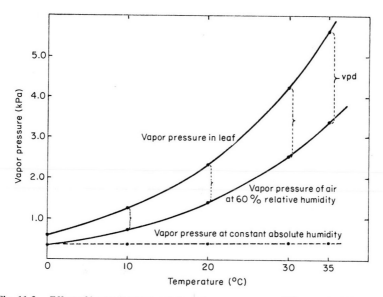

Fig. 11.3. Effect of increasing temperature on the vapor pressure difference (vpd) between leaf and air if the air in the leaf is assumed to be saturated, leaf and air temperatures are similar, and the external air is at 60% relative humidity at each temperature. The dashed line shows water vapor pressure in the atmosphere if the absolute humidity is kept the same at all temperatures.

to maintain similar evaporating conditions at various temperatures requires that all vapor pressure differences be the same. For example, if the relative humidity is set at 40% at 20°C, it must be set at approximately 70% at 30°C to maintain the same vapor pressure deficit in the air. If the humidity were set at 70% at both 20° and 30°C, the vapor pressure deficit and the rate of evaporation would be nearly twice as great at 30°C as at 20°C, as shown in Table 11.5. Some investigators prefer to express these gradients in terms of absolute humidity, expressed as weight of water per unit volume of air (g/m³ or µg/cm³). The two terms are interconvertible because the partial pressure of water vapor is proportional to its concentration in air, 1 mg of water vapor per liter having a vapor pressure of 1 mmHg.

TABLE 11.6 Effect of Increasing Temperature of Leaf and Air on Vapor Pressure Gradient from Leaf to Air, with No Change in Absolute Humidity from That at 10°C

	10	20	30
Leaf and air temperature (°C)	10	20	30
Relative humidity of air, assuming no change in absolute humidity from that at 10°C (%)	80	43	25
Vapor pressure at evaporating surface of leaf (kPa)	1.227	2.337	4.243
Vapor pressure in air at indicated temperatures	0.981	1.015	1.050
Vapor pressure gradient from leaf to air	0.246	1.322	3.193

Resistances to Diffusion in Leaves. The various resistances to diffusion are shown in Fig. 11.1. The total leaf resistance, r_1, is composed of the cuticular resistance, r_c; stomatal resistance, r_s; mesophyll cell wall resistance, r_m; resistance in the intercellular spaces, r_i; and resistance of the stomatal pores, r_p. Because stomatal and cuticular resistances are in parallel, an equation for them involves their reciprocals:

$$\frac{1}{r_1} = \frac{1}{r_c} + \frac{1}{r_s} \qquad (11.4)$$

Likewise for amphistomatous leaves:

$$\frac{1}{r_1} = \frac{1}{r_{upper}} + \frac{1}{r_{lower}} \qquad (11.5)$$

The components of stomatal resistance are in series and can be described by the following equation:

$$r_s = r_m + r_i + r_p \qquad (11.6)$$

The cuticular resistance varies widely, as shown in Table 11.3, but is usually between 10 and 40 s cm^{-1} for crop plants. When stomata are open, r_s is so much lower than $r_{cuticle}$ that nearly all of the water vapor escapes through the stomata. Thus, transpiration is usually controlled largely by stomatal aperture, as shown later in Fig. 11.18. Stomatal behavior is discussed later.

Resistance in the intercellular spaces, as measured with pressure flow porometers, appears to be negligible in wheat (Milthorpe and Spencer, 1957), but it is measurable in some other species (Bange, 1953; Heath, 1941). It increases as leaves are dehydrated because of leaf shrinkage and changes in internal geometry. Measurements with diffusion porometers suggest that r_i can be a significant component of r_s when stomata are open (Jarvis and Slatyer, 1966). It can be expected to be higher in thick than in thin leaves and higher in leaves with small intercellular spaces than in leaves containing large intercellular spaces.

There has been much discussion concerning the importance of the mesophyll resistance, r_m, with respect to transpiration (Kramer, 1969, pp. 306–309; Slatyer, 1967, pp. 256–260). It was reported early in this century that the rate of transpiration often decreased with no change in stomatal aperture. Livingston and Brown (1912) suggested that this occurred because during periods of rapid transpiration the water menisci retreat into the pores in the cell walls, reducing the vapor pressure at the evaporating surfaces. However, several investigators were unable to find any decrease in rate of transpiration with decreasing leaf water content until stomatal closure begins. Slatyer (1966) pointed out that extremely low water potentials would be required to cause retreat of water from the cell wall surface. However, as mentioned earlier in this chapter, several investigators have reported that mesophyll cell surfaces are covered with a hydrophobic lipid layer, usually regarded as cutin. This suggests that water must evaporate through pores in that layer, lengthening the pathway and lowering the conductance, so that the vapor pressure might be lower than would be expected from measurements of bulk leaf water potential. Some evidence for this was reported by Shimshi (1963) and by Jarvis and Slatyer (1970), but neither Weatherspoon (1968) nor Jones and Higgs (1980) could find significant mesophyll resistance in several mesophytic leaf types.

The mesophyll resistance for carbon dioxide is much higher than for water, and much higher than r_s when stomata are open (Gaastra, 1959), and this difference is the basis for the assumption that partial closure of stomata should reduce transpiration more than it reduces photosynthesis. This situation will be discussed in Chapter 13 in the section on water use efficiency.

External Resistance.   The external resistance depends chiefly on wind speed and leaf size and shape. Leaf size and shape will be discussed later in connection with plant factors.

The effects of wind on rate of transpiration are rather complex. Increased velocity of air movement increases transpiration by removing the boundary layer of water vapor that surrounds leaves in quiet air, reducing r_{air} in Eq. (11.2). Most of the effect occurs at low velocities, as shown in Fig. 11.4, and it has been reported that at higher velocities stomatal closure occurs, either because of mechanical effects or because of dehydration. Knoerr (1967) pointed out that

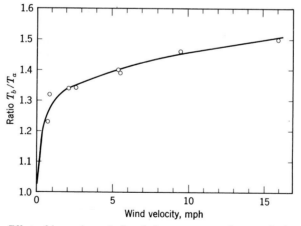

Fig. 11.4.   Effect of increasing velocity of air movement on the transpiration rate of potted sunflowers growing in a sunny greenhouse. Ordinate is the ratio of rate of plants exposed to wind (T_b) to the rate of plants in quiet air (T_a). Most of the effect occurs at velocities of less than 2 mph. A velocity of 1 mph equals 44.69 cm/s. (After Martin and Clements, 1935; from Kramer and Kozlowski, 1979, by permission of Academic Press.)

although a breeze should increase transpiration at low or moderate radiation, at high levels of radiation, where leaves tend to be warmer than the air, a breeze may decrease transpiration by cooling leaves, as shown in Fig. 11.5.

The actual behavior of different kinds of plants in wind seems to be quite variable, as shown in Fig. 11.6. The transpiration rates of alder and larch were higher in the wind than in quiet air up to a velocity of 20 m s^{-1}, but those of the other species decreased with increasing air movement. The differences observed by Tranquillini (Fig. 11.6) can probably be explained by differences in stomatal behavior and cuticular transpiration. Caldwell (1970) found that exposure to wind caused immediate closure of stomata of rhododendron but had little effect on stone pine. Davies *et al.* (1974) also reported differences among seedlings tested in an illuminated wind tunnel at air speeds of 5.8 to 26.0 m s^{-1}. Wind increased transpiration of white ash at all speeds, decreased transpiration of sugar maple, and had no significant effect on red pine.

Other supposed effects of wind, such as ventilation of the intercellular spaces by flexing the leaves and increasing air flow through amphistomatous leaves, are debatable. Woolley (1961) thought that they were of minor importance, and Rushin and Anderson (1981) doubted if wind can increase bulk air flow in the hypostomatous leaves of quaking aspen. However, Shive and Brown (1978) claimed that wind significantly increased gas exchange in the amphistomatous leaves of *Populus deltoides* and Rushin and Anderson (1981) cite considerable literature on this topic.

Incidentally, the interest in quaking aspen illustrates the current tendency to search for a useful function for characters such as leaf quaking. Many botanists seem to have the misconception that all characters in a plant must be beneficial or

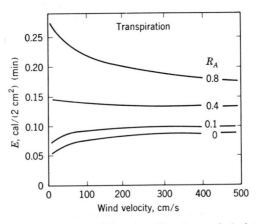

Fig. 11.5. Curves showing theoretical latent heat (E) exchange of a leaf at various wind speeds and net radiations (R_A). At high levels of radiation, wind may decrease transpiration by cooling leaves; at low levels, it may increase transpiration by supplying energy. (After Knoerr, 1967; from Kramer and Kozlowski, 1979, by permission of Academic Press.)

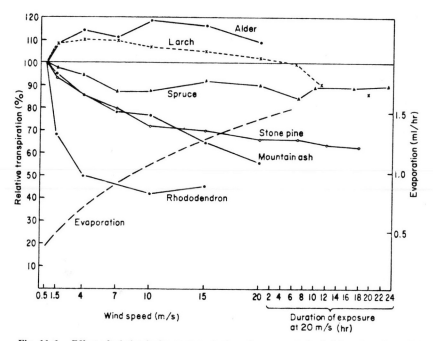

Fig. 11.6. Effect of wind velocity on transpiration of young potted subalpine plants in a wind tunnel at air temperature of 20°C, soil temperature of 15°C, and light intensity of 30,000 lux. Evaporation was measured with a green Piche atmometer. Although evaporation increased steadily with increasing wind speed, transpiration of most species decreased. Alder is *Alnus viridis;* larch is *Larix decidua;* spruce, *Picea abies;* stone pine, *Pinus cembra;* mountain ash, *Sorbus aucuparia;* rhododendron, *Rhododendron ferrugineum.* (After Tranquillini, 1969; from Kramer and Kozlowski, 1979, by permission of Academic Press.)

they would have disappeared. This is not true! Williams (1966), in his book "Adaptation and Natural Selection," warned that the presence of a character is not proof that it is presently essential or even beneficial. Leaf quaking probably never conferred any significant benefit, but the flattened petioles that cause it survived because they had no negative effect on natural selection.

PLANT FACTORS AFFECTING TRANSPIRATION

Leaves

The size, shape, surface characteristics, and position of leaves materially affect absorption and reflection of incident energy and leaf temperature, which affect e_{leaf}. Leaf size and shape also affect the external resistance, r_a. The internal structure affects r_1, the total leaf resistance to water vapor movement. Some of these factors are discussed by Kriedemann and Barrs (1982).

There has been considerable study of the relationship between leaf area per unit of land area (termed the leaf area index), leaf orientation, and the rate of photosynthesis and crop yield. Presumably, leaf exposure favorable for photosynthesis is also favorable for transpiration. However, selection for what is believed to be optimum leaf orientation has not resulted in important increases in yield. Loomis *et al.* (1971) discussed some of the research in this area.

Leaf Area. Plants and stands of plants with large leaf areas usually transpire more than those with smaller leaf areas. However, removal of part of the leaf area does not necessarily result in a proportional reduction in transpiration rate. Miller (1938) cited several experiments in which removal of part of the leaves resulted in decreased transpiration per plant but increased transpiration per unit of leaf area. The increase in rate per unit of leaf area is caused by better exposure of the remaining leaves and increased air movement causing a decrease in r_a, and perhaps sometimes the increase in root–shoot ratio provides a better water supply for individual leaves (Bialoglowski, 1936; Parker, 1949). Some woody perennials, such as creosote bush and buckeye, shed most or all of their leaves when water-stressed, greatly reducing the transpiring surface. The curling or rolling of wilting leaves seen in maize, sorghum, and other grasses greatly reduces the exposed surface and increases resistance to diffusion of water vapor. Stålfelt (1956) cited work indicating that rolling of leaves reduces transpiration by about 35% in mesophytes and by 75% in some xerophytes. Developmental and seasonal changes in the transpiring surfaces of a number of species were discussed by Killian and Lemée (1956), and Kozlowski (1974) discussed shedding of leaves.

As shown in Table 11.7, there are wide variations among species in transpiration per unit of leaf surface. However, differences in total leaf area sometimes compensate for differences in rate per unit of leaf area, as with loblolly pine in Table 11.7.

TABLE 11.7 Transpiration Rates per Unit of Leaf Surface and per Seedling of Loblolly Pine and Hardwood Seedlings for the Period August 22–September 2[a]

	Loblolly pine	Yellow-poplar	Northern red oak
Transpiration (g/day/dm²)	5.08	9.76	12.45
Transpiration (g/day per tree)	106.70	59.10	77.00
Average leaf area per tree (dm²)	21.00	6.06	6.18
Average height of trees (cm)	34.00	34.00	20.00

[a] Average of six seedlings of each species.

Root–Leaf Ratio. The ratio of roots to leaf surface is more important than the leaf surface alone, because if absorption lags behind transpiration, plant water deficits develop. Parker (1949) found that the rate of transpiration per unit of leaf area of northern red oak and loblolly pine seedlings growing in moist soil increased as the ratio of roots to shoots increased, but red pine did not respond. The relationship between transpiration and root–shoot surface of loblolly pine is shown in Fig. 11.7. Pereira and Kozlowski (1977) reported that sugar maple seedlings with a large leaf area developed more severe stress than partly defoliated seedlings. Tree seedlings are more likely to suffer from an unfavorable root–shoot ratio than grasses and some other herbaceous plants with their extensive root development. In general, plants with deep, extensively branched root systems are less likely to suffer from drought than plants with shallow, sparsely branched systems.

Loss of roots during transplanting results in an unfavorable root–shoot ratio in both herbaceous and woody seedlings. The tops of transplanted trees and shrubs

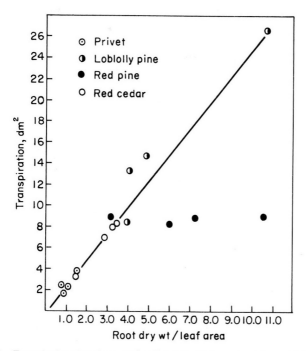

Fig. 11.7. Transpiration plotted over ratio of roots to leaf area (root dry weight in grams over leaf area in square decimeters) for four species of tree seedlings. An increase in ratio of roots to leaf area was accompanied by an increase in rate of transpiration per unit of leaf area for three species. The exception, red pine, had an extensively branched root system and densely clustered needles. (After Parker, 1949; from Kramer and Kozlowski, 1979, by permission of Academic Press.)

are often pruned or partially defoliated to compensate for the loss of roots, and Allen (1955) found that clipping the needles of longleaf pine planting stock back to 12.5 cm in length reduced mortality after transplanting. Lopushinsky and Beebe (1976) found that survival of outplanted Douglas-fir and ponderosa pine seedlings in a region with dry summers was improved by large root systems and high root–shoot ratios. Sometimes transpiration is reduced by dipping or spraying the shoots with an antitranspirant. Use of antitranspirants is discussed in Chapter 13.

Leaf Orientation.   The rate of transpiration is affected by leaf orientation because leaves exposed at right angles to the sun are warmer than those more or less parallel to incident radiation. The leaves of most plants tend to be oriented more or less perpendicular to the average incident radiation. This is favorable not only for photosynthesis but also for transpiration. However, the leaves of a few kinds of plants, including species of *Lactuca* and *Silphium,* turkey oak (*Quercus laevis*), and jojoba (*Simmondsia chinensis*) are oriented nearly parallel to the average incident radiation. It has been claimed that such an orientation increases survival of seedlings in hot, dry habitats, but this is doubtful because other plants lacking this characteristic survive in the same environment. Pine needles that occur in bundles shade each other, probably reducing transpiration and certainly reducing photosynthesis per unit of leaf surface (Kramer and Clark, 1947). As mentioned earlier, wilting, rolling, and changes in leaf orientation also reduce the amount of solar radiation received by leaves. The leaflets of *Stylosanthes humilis* are normally perpendicular to incident radiation, but when water stress causes loss of turgor in the pulvini they become parallel to radiation.

Leaf Size and Shape.   The size and shape of leaves affect the rate of transpiration per unit of surface because they affect the air resistance, as shown in Fig. 11.8. It is about three times greater for a cotton leaf 10 cm wide in quiet air than for a grass leaf 1 cm wide. Small leaves, deeply dissected leaves, and small leaflets of compound leaves all tend to be cooler than large leaves because their thinner boundary layers permit more rapid sensible and latent heat transfer. This should decrease transpiration by decreasing e_{leaf}, but the lower boundary layer resistance, r_a, of small leaves is favorable to water vapor loss, so that the two effects tend to compensate for each other (Raschke, 1976). According to Smith (1978), the small leaves characteristic of most Sonoran desert perennials have temperatures close to those of the air and moderate rates of transpiration. However, when supplied with water by rains, the larger leaves of *Hyptis* and *Encelia* transpire so rapidly that their temperatures are as much as 8–18°C below air temperature. Smith suggested that the cooling effect of transpiration results in a higher rate of photosynthesis than would be possible in leaves at or above the high air temperatures characteristic of the desert.

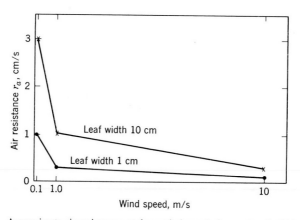

Fig. 11.8. Approximate air resistances at three wind speeds for a cotton leaf 10 cm wide and a grass leaf 1 cm wide. (From data of Slatyer, 1967.)

Parkhurst and Loucks (1972) made an elaborate analysis of leaf size in relation to environment, based on the assumption that natural selection should have resulted in leaves with the maximum water use efficiency in their environment. However, it is doubtful if water use efficiency has been a major selective force in humid and subhumid environments (see Cowan, 1977), and it is not surprising that they found many exceptions to their model. Success in evolution usually depends on possession of a combination of favorable adaptations. Furthermore, the importance of differences in leaf size and shape tends to decrease in closed stands.

Leaf Surface Characteristics. The outer surfaces of leaves, stems, fruits, and even flower petals are often covered with a relatively waterproof layer, the cuticle. It is composed of wax and cutin and is anchored to the epidermis by a layer of pectin (Fig. 11.9). Permeability to water is apparently controlled by the wax content rather than by the thickness of the cuticle because transpiration often increases severalfold after removal of the wax (Radler, 1965; Jeffree *et al.*, 1971). The wax is probably synthesized in the epidermal cells as droplets and passes out through the cell walls in liquid form. Some is deposited in the cutin, and some passes through and accumulates as rods and plates on the surface, as shown diagrammatically in Fig. 11.9. Wax deposits form the "bloom" characteristic of some leaf and fruit surfaces. According to Chambers *et al.* (1976), in eucalyptus wax molecules probably move to the surface in oil solution through lamellar channels in the cutin, and are deposited in various configurations as the solvent evaporates. Examples of wax deposits on leaves are shown in Fig. 11.10. The amount of wax varies widely, depending on the species and environment, but in general more cutin and wax are found on leaves developed in dry air

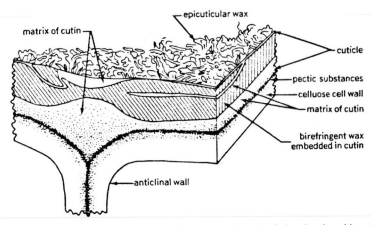

Fig. 11.9. Diagram of the upper surface of epidermis of pear leaf, showing deposition of cutin and wax. The cutin is anchored to the cellulose cell wall by a layer of pectin. Various kinds of wax deposits are shown in Fig. 11.10. (After Norris and Bukovac, 1968; from Kramer and Kozlowski, 1979, by permission of Academic Press.)

(Pallardy and Kozlowski, 1980). According to Chatterton *et al.* (1975), the rates of transpiration and photosynthesis are both lower for sorghum plants with bloom than for those without bloom, but transpiration is reduced more than photosynthesis.

Schönherr (1976) confirmed earlier observations that the cuticle becomes relatively permeable to water when wet. It is fairly permeable to a wide range of substances, including mineral nutrients, radioactive elements in fallout, herbicides, and growth regulators. However, when dry, the cuticle is relatively impermeable to water vapor, as shown by the low rates of cuticular transpiration of many plants (see Table 11.8). Cuticle is generally thin on leaves of shade plants and mesophytes and much thicker on leaves developed in the sun and on xerophytes. There is some evidence that the permeability of cuticle increases with increasing temperature (Holmgren *et al.*, 1965; Schönherr *et al.*, 1979). The early work on cuticle was reviewed by Stålfelt (1956), and Kolattukudy (1981) recently reviewed the synthesis of cutin and suberin.

Leaves of a few kinds of plants are covered by thick layers of epidermal hairs. The role of leaf hairs was reviewed by Ehleringer (1980) with special reference to *Encelia*. The white, densely pubescent leaves of the desert shrub, *E. farinosa*, reflect much more radiation than the green leaves of *E. californica*, a native of the moist coastal region. The midday temperature is about 5°C lower in *E. farinosa*, and transpiration is reduced. Unfortunately, so much light is reflected by the coat of hairs that net photosynthesis of *E. farinosa* is also reduced. Overall, Ehleringer concluded that at high air temperatures the gain in net photosynthesis from reduced leaf temperature is more valuable than the reduction in

Fig. 11.10.   Variations in patterns of wax deposit on leaves of four species of trees. (A) American elm; (B) white ash; (C) sugar maple; (D) redbud. (Photographs by W. J. Davies.)

transpiration. He estimated that the boundary layer resistance is increased by about 50%, from 0.3 to 0.45 s cm^{-1}. This probably has only a small effect on diffusion of water vapor, and, because of the high mesophyll resistance for carbon dioxide, a negligible effect on entrance of carbon dioxide. Although hairs may have some adaptive importance for *Encelia,* other desert plants thrive without them, and in general they are obviously not essential to survival. Also, some mesophytes, such as mullein (*Verbascum thapsus*), are pubescent. Thus, pubescence seems to have evolved in some plants growing in conditions in which no

**TABLE 11.8   Cuticular Transpiration of Plants of
Several Species under Similar Evaporating
Conditions Expressed in Milligrams per Hour per
Gram of Fresh Weight**[a]

Species	Transpiration
Impatiens noli-tangere	130.0
Caltha palustris	47.0
Fagus sylvatica	25.0
Quercus robur	24.0
Sedum maximum	5.0
Pinus sylvestris	1.53
Opuntia camanchica	0.12

[a] From Pisek and Berger (1938).

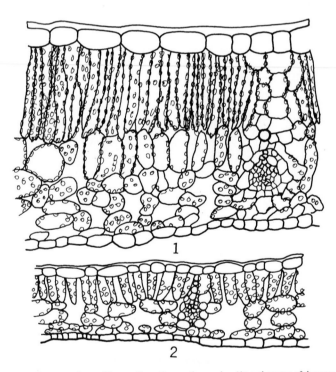

Fig. 11.11.   Cross sections of leaves from the southern edge (1) and center of the crown (2) of an isolated sugar maple (*Acer saccharum*). (After Weaver and Clements, 1938; from Kramer and Kozlowski, 1979, by permission of Academic Press.)

benefit can be ascribed to it, unless it discourages predation. The ecological role of hairs was discussed by Johnson (1975), who concluded that they can no longer be regarded simply as an adaptation to arid environments, but must have developed under a variety of selection pressures. Perhaps hairs, like some other alleged adaptations, developed as a result of random mutation with little or no selection pressure. Mullein might be a good example.

Leaf Anatomy. Leaf structure can be greatly modified by the environment (Stålfelt, 1956). An example in agriculture is the shading of tobacco to produce the large, thin leaves desired as cigar wrappers. Sun leaves are usually smaller, have smaller cells and interveinal areas and thicker cutin, and are thicker than shade leaves of plants of the same kind (Fig. 11.11). Sun leaves also often possess an extra layer of palisade cells. The upper leaves of trees are said to be more xeromorphic than the lower leaves because they are exposed to more sun and wind. Leaves from dry habitats (xeromorphic type) resemble sun leaves, and leaves from moist habitats (mesomorphic type) resemble shade leaves. Patterson *et al.* (1977) found significantly more mesophyll tissue and a higher rate of photosynthesis per unit of leaf area in field cotton than in the thinner leaves of cotton grown in controlled environments. Van Volkenburgh and Davies (1977) reported that field-grown leaves of cotton and soybean were thicker than leaves grown in greenhouses or growth chambers, and cool nights increased the thick-

TABLE 11.9 Variations in Mean Thickness and Weight of Cotton and Soybean Leaf Tissue Produced in Various Environments[a]

Crop	Type of environment	Palisade thickness (μm)	Mesophyll thickness (μm)	Combined epidermal thickness (μm)	Total leaf thickness (μm)	Specific leaf wt. (g dm²)
Soybean	Field	64 X[b]	41 X	15 X	119 X	0.646 X
	Greenhouse	50 Y	24 Y	10 Z	84 Y	0.343 Y
	Chamber 30°/26°C	28 Z	20 Y	11 Z	58 Z	0.232 Y
	Chamber 28°/17°C	67 X	35 X	13 Y	115 X	0.610 X
Cotton	Field	71 A	46 A	21 C	139 A	0.573 A
	Greenhouse	33 B	29 B	18 C	81 B	0.466 AB
	Chamber 30°/26°C	50 C	31 B	19 C	101 C	0.398 B
	Chamber 28°/17°C	62 D	43 A	28 A	133 A	0.470 AB

[a] From Van Volkenburgh and Davies (1977).

[b] Means followed by the same letter within a column are not significantly different at the 1% level.

ness of leaves grown in chambers (see Table 11.9). Nobel (1980) summarized considerable data on variations in ratio of mesophyll cell surface to leaf surface, and this will be discussed in Chapter 13 in connection with water use efficiency. Examples of mesomorphic and xeromorphic types of leaves are shown in Fig. 11.12. The differences in structure are usually attributed to differences in light intensity during leaf development (Nobel, 1980), but light intensity may operate in part through differences in temperature and water stress. Increased concentration of carbon dioxide sometimes causes development of thicker leaves and increase in leaf area.

It is sometimes assumed that water loss from mesomorphic leaves such as beech, maple, or red oak, with their thin cutin and large internal air spaces, must be greater than from the thicker, more heavily cutinized xeromorphic types such as magnolia or holly. Long ago, Maximov (1929) pointed out that this is not necessarily true for well-watered plants, although it may be true when plants are sufficiently stressed to close the stomata. Table 11.10 shows that the transpiration rates of well-watered *Ilex glabra* and *Gordonia lasianthus* with thick, heav-

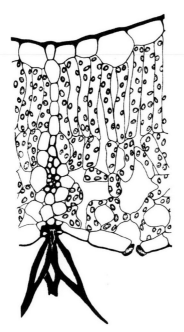
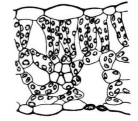

Fig. 11.12. Cross sections through leaves of post oak (left) and American beech (right). Post oak (*Quercus stellata*) leaves are representative of the xeromorphic type, with thick cutin, a double layer of palisade cells, bundle sheath extension, and a high ratio of internal to external surface. Beech (*Fagus grandifolia*) leaves are typical of the mesomorphic type, being thinner, with a thinner cutin, a single layer of palisade cells, and a lower ratio of internal to external surfaces. (Drawings courtesy of J. Philpott.)

TABLE 11.10   Midsummer Transpiration Rates of Various Species of Trees Expressed as Grams of Water Lost per Square Decimeter of Leaf Surface per Day, Compiled from Various Sources[a]

Species	Location	Season	Duration (days)	No. of plants	Av. transpiration $(g/dm^2/day)$
Liriodendron tulipifera	Columbus, Ohio	August	1	7	10.11
Liriodendron tulipifera	Durham, N.C.	August	3	4	11.78
Quecus alba	Durham, N.C.	August	3	4	14.21
Quercus rubra	Durham, N.C.	August	3	4	12.02
Quercus rubra	Fayette, Mo.	July	14	6	8.1
Acer saccharum	Fayette, Mo.	July	14	6	12.2
Acer negundo	Fayette, Mo.	July	14	6	6.4
Platanus occidentalis	Fayette, Mo.	July	14	6	8.8
Pinus taeda	Durham, N.C.	August	3	4	4.65
Clethra alnifolia	Durham, N.C.	August	3	4	9.73
Ilex glabra	Durham, N.C.	August	3	4	16.10
Myrica cerifera	Durham, N.C.	August	3	4	10.80
Gordonia lasianthus	Durham, N.C.	July	23	4	17.77
Liriodendron tulipifera	Durham, N.C.	Aug. 26– Sept. 2	12	6	9.76
Quercus rubra	Durham, N.C.	Aug. 26– Sept. 2	12	6	12.45
Pinus taeda	Durham, N.C.	Aug. 26– Sept. 2	12	6	5.08

[a] All seedlings were growing in pots in soil near field capacity in full sun. From Kramer (1969).

ily cutinized leaves are higher per unit of leaf surface than the rates of oak, maple, and yellow poplar. Swanson (1943) reported that on sunny days American holly transpired more than lilac, coleus, or tobacco, but on cloudy days when the stomata were closed, it transpired less.

The high rates of transpiration of xeromorphic leaves are usually attributed to the higher ratio of internal to external surfaces in sun and xermorphic leaves caused by the larger amounts of palisade tissue (Turrell, 1936, 1944). Swanson found ratios of 12.9, 7.1, and 4.6 in American holly, tobacco, and coleus. A high ratio of internal to external surface presumably provides more evaporating surface per unit of leaf surface. However, when stomata are closed by water stress, transpiration from xeromorphic leaves is generally lower than from mesomorphic leaves because the former are more heavily cutinized and have lower rates of cuticular transpiration. This explanation is based on the assumption that evaporation occurs from the entire mesophyll surface. However, as pointed out earlier in this chapter, Meidner (1975, 1976), Tyree and Yianoulis (1980), and others claim that most of the water is evaporated from the immediate vicinity of the stomata, including the inner surfaces of the epidermis and guard cells. The relatively high rates of transpiration from leaves with a high ratio of internal to external surface suggest that evaporation occurs from the entire mesophyll. Apparently, the relative importance of various evaporating surfaces in leaves requires further study.

The cause of the xeromorphic characteristics of leaves of shrubs growing in bogs is uncertain. The leaves of North Carolina bog shrubs are thicker, contain more palisade tissue and more small veins, and are more heavily cutinized than leaves of related shrubs growing in the mountains (Philpott, 1956). At one time, the xeromorphic structure of leaves of bog plants was attributed to the "physiological dryness" of bogs, referring to the cold, poorly aerated root environment, but this is doubtful (Caughey, 1945). Mothes (1932) and Müller-Stoll (1947) attributed the xermorphic structure of leaves of bog plants to nitrogen deficiency, Albrecht (1940) suggested calcium deficiency as a cause, and Loveless (1961) proposed phosphorus deficiency. This problem was discussed by Christensen *et al.* (1981). None of these ideas seems to have been investigated experimentally. It was claimed by van Eijk (1939) that the succulence characteristic of halophytes is caused by an excess of chloride ions, and Boyce (1954) reported that the chloride in salt spray increases succulence in coastal vegetation. In general, it appears that leaf succulence is increased by chlorides and decreased by sulfates.

The degree to which variations in leaf structure affect transpiration has not been fully assessed. Stålfelt (1956) reported that the internal air space varies from 70% of the total leaf volume in shade leaves to 20% in sun leaves. A large internal volume of air should reduce r_i, which according to calculations by Jarvis *et al.* (1967) is about equal to the stomatal pore resistance of open stomata in cotton. Of course, when stomatal closure begins, pore resistance begins to increase and r_s soon becomes the dominant factor in r_{leaf} (see Fig. 11.1).

Disease.   The effects of leaf diseases on plant water relations seldom receive much attention. However, Durbin (1967) stated that when sporulation occurred, transpiration from bean leaves infected with rust increased by as much as 50%. It also increased when barley leaves were infected with powdery mildew. Bushnell and Rowell (1968) stated that shoots of wheat plants badly infected with rust died from water deficit caused by loss of absorbing capacity by the roots, probably because they suffered from lack of carbohydrate. Arntzen *et al.* (1973) reported that toxin produced by *Helminthosporium maydis,* the cause of corn leaf blight, caused immediate closure of stomata of corn plants possessing the Texas male-sterile cytoplasm but no closure on plants lacking that character. The effects of diseases that attack roots or the vascular system are mentioned elsewhere. The interaction of water status and plant diseases is discussed in several chapters in "Water Relations of Diseased Plants," Volume V (1978) of the series edited by Kozlowski (1968–1981).

Stomata

Stomata are important because most of the water lost by transpiration escapes through them and most of the carbon dioxide used in photosynthesis enters through them. Thus, the widest possible opening (lowest r_s) is desirable for photosynthesis, but it is undesirable in terms of conserving water. This problem will be discussed later in the section on optimization. It is impossible to cite all the literature on stomata, and readers are referred to the reviews by Zelitch (1969) and Raschke (1975), a symposium edited by Jarvis and Mansfield (1981), and a discussion of stomatal response to environmental factors by Hall *et al.* (1976). The discussions of stomatal behavior by Cowan (1977), Farquhar and Sharkey (1982), and Raschke (1979) are particularly interesting.

Origin and Structure. Development of the two guard cells that surround a stomatal pore is described by Esau (1965, pp. 163–166) and by Palevitz, in Jarvis and Mansfield (1981). They are formed by transverse division of stomatal mother cells that are said to differ morphologically and biochemically from their neighbors (Stebbins and Shah, 1960). The middle lamella disintegrates and the two walls separate, leaving an opening between them, the stomatal pore. Guard cells usually contain chloroplasts, but their function is uncertain. Usually, the walls bordering the pore become considerably thickened, and sometimes they develop curious ledges and projections that extend into the pores. These are generally cutinized. It was long supposed that the thickenings on the inner wall played an essential role in stomatal opening, but this has been questioned. Aylor *et al.* (1973) concluded that the radial micellar structure of the guard cell walls is more important than the thickening of the inner wall in causing bulging and separation of turgid guard cells. Shoemaker and Srivastava (1973) proposed a similar explanation for stomatal opening in grasses, and DeMichele and Sharpe (1973) developed a model for stomatal behavior consistent with the new explanation. Cowan (1977) also discussed stomatal structure and behavior.

The guard cells of most dicots are somewhat semicircular when closed and crescent- or kidney-shaped when open, but those of monocots are more elongated, and many modifications in shape occur (see Esau, 1965, and Fig. 11.13). Stomata are often approximately flush with the leaf surface, but they are some-

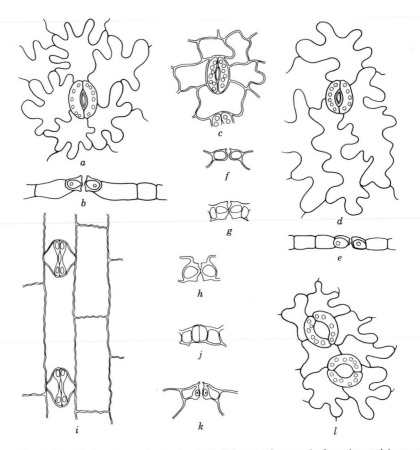

Fig. 11.13. Various types of stomata, (a,b) *Solanum tuberosum* in face view and in cross section; (c) apple; (d,e) *Lactuca sativa;* (f) *Medeola virginica;* (g) *Aplectrum hyemale;* (h) *Poly-gonatum biflorum;* (i,j,k) *Zea mays.* Part (i) is a face view; (j) a cross section near the ends of guard cells; (k) cross section through the center of a stoma; (l) a face view from *Cucumis sativus.* (From Eames and MacDaniels, 1947; by permission of McGraw-Hill Book Company.)

times in pits or furrows that increase the length of the diffusion path and increase the apparent r_{leaf}. This is true of conifer needles, pineapple, and some grasses. Two types of guard cells are shown in Fig. 11.14 and others in Fig. 11.10.

Number and Size.   The number of stomata per unit of leaf surface varies widely with species and environmental conditions, ranging from 6000 to 8000/ cm² in corn, 8000 to 15,000 in sorghum, about 15,000 in alfalfa and clover, and 30,000 in apple to 100,000 in scarlet oak. Other data on stomatal frequency can be found in Meyer *et al.* (1973) and Miller (1938). Stomata occur on both

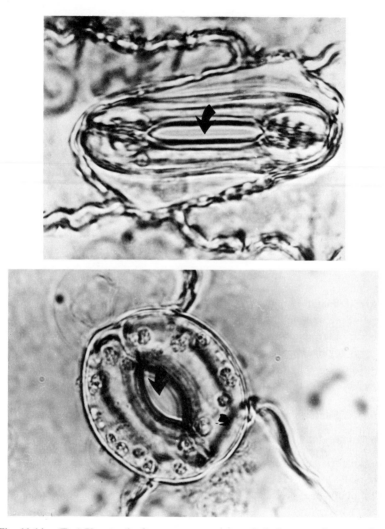

Fig. 11.14. (Top) Photograph of an open corn stoma, typical of stomata of grasses. (Bottom) Photograph of an open stoma of bean, typical of stomata of most other plants. (Courtesy of J. E. Pallas, U.S. Department of Agriculture.)

surfaces (amphistomatous) of monocot and herbaceous dicot leaves, but are usually considerably more abundant on the lower surface. They occur only on the lower surfaces (hypostomatous) of the leaves of most woody plants, but excep-- tions are willow and poplar. When wide open, stomatal pores are usually only 3–12 μm wide and 10–30 or occasionally 40 μm long. The total area of the pores when fully open is no more than 3% of the total leaf surface, and often much less.

The Stomatal Mechanism.   Few things in plant physiology have undergone more change during the past 2 decades than views concerning the stomatal mechanism. Many details remain unsettled, and what is written here is likely to be revised by further investigation. However, some generalizations seem possible. Stomata usually open in the light and close in darkness, although the reverse occurs in crassulacean acid metabolism plants such as pineapple and many succulents. They also tend to open in low concentrations of carbon dioxide and to close in high concentrations. In addition, other factors, such as humidity, temperature, growth regulators produced in both roots and shoots, and rate of photosynthesis, affect stomatal opening, resulting in a diversity of behavior.

The degree of opening of the stomatal pores depends on the turgor of the guard cells. When they are turgid, the pore between them is large and r_s is low; when they lose turgor, the pore decreases in size and r_s increases. For many years, changes in guard cell turgor were attributed to changes in the proportions of starch and sugar in the guard cells (Sayre, 1926). However, it now seems to be established that the changes in turgor are caused chiefly by uptake or loss of K^+. There also are complex metabolic changes involving organic acids, presumably derived from the starch that disappears during stomatal opening, plus the effects of growth regulators. Thus, stomatal opening and closing involve a complex series of processes that are not fully understood. Current views are discussed by several authorities in the book edited by Jarvis and Mansfield (1981).

Internal Factors.   Explanation of stomatal behavior is complicated by the observation that it seems to be affected by growth regulators such as abscisic acid (ABA) and cytokinins. For example, the failure of stomata to close in the wilty tomatoes studied by Tal is attributed to their inability to synthesize ABA (Livne and Vaadia, 1972). The latter suggested that stomatal closure in water-stressed plants may be caused by decrease in cytokinins supplied from the roots and by increase in ABA.

Several investigators have shown that there is an increase in ABA in the leaves of water-stressed plants, and it has also been shown that application of exogenous ABA to leaves causes stomatal closure in the absence of water stress (Davies *et al.*, 1979). However, the stomatal closure of some kinds of plants is not well correlated with the ABA content of the leaves (Ackerson, 1980). It has also been suggested that other substances are involved, such as farnesol (Mansfield *et al.*, 1978) and phaseic acid (Sharkey and Raschke, 1980). The fact that stomata sometimes remain open in leaves high in ABA, close before the concentration increases, or stay closed or partly closed after the ABA concentration has been reduced cause difficulties in interpreting the role of ABA, as Davies *et al.* (1979) point out. It is also difficult to reconcile the noon closure and after-noon reopening often observed during hot, sunny weather with changes in ABA content. Possibly the role of ABA in stomatal closure has been overemphasized, as claimed by Trewavas (1981).

Another peculiarity of stomatal behavior is the cycling or oscillation between the open and closed conditions (see Fig. 11.15). Cycling occurs chiefly in water-stressed plants and has a periodicity ranging from minutes to hours, but most of it occurs in the range from 15 to 120 min. It has been observed in various herb-aceous species (Barrs, 1971), and Levy and Kaufmann (1976) observed it in citrus trees growing in the field. Raschke (1975) suggested that it might develop when a negative feedback signal is delayed and reaches the guard cells at a time such that it reinforces the initial response instead of weakening it. Cycling can be caused by shocks such as a sudden change in temperature, humidity, or light intensity, usually in combination with a high root resistance. Cowan (1972) suggested that stomatal cycling optimizes the conflicting requirements for carbon dioxide uptake and control of water loss. There are also endogenous rhythms that cause stomatal opening and closing to continue for several days when plants are moved to continuous darkness from normal day–night conditions.

Environmental Control of Stomatal Opening. The principal environmen-tal factors affecting stomatal behavior are light, carbon dioxide, humidity, and temperature. In spite of the fact that stomatal opening and closing of both C_3 and C_4 plants usually are strongly correlated with light intensity, there has been uncertainty concerning the direct action of light (Raschke, 1975, pp. 322–324). It was not clear whether it operates directly or principally through the changes in internal CO_2 concentration caused by photosynthesis. The latter view was sup-

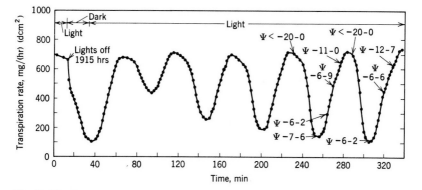

Fig. 11.15. Cyclic variations in transpiration and water potential of a cotton leaf, resulting from stomatal cycling initiated by a 20-min dark period in the afternoon. This cycling was attributed to oscillations in water deficit caused by high root resistance in rapidly transpiring plants. (From Barrs and Klepper, 1968.)

ported by the fact that stomata can be caused to open in darkness by CO_2-free air and close in the light by high CO_2. However, Sharkey and Raschke (1981) reported that in bean, cotton, *Perilla,* and *Xanthium,* stomatal response is caused chiefly by direct response to light and influenced to only a small extent by the internal CO_2 concentration. The stomata of many kinds of plants, when well watered, either never completely close in darkness or reopen after a few hours (Miller, 1938, pp. 435–436). Louwerse (1980) suggested that the stomata of some plants stay open independently of irradiance and external CO_2 concentration, others keep the internal CO_2 concentration constant, regardless of irradiance and external concentration, and still others maintain a relatively constant ratio between internal and external CO_2 concentration. The long-term increase in atmospheric CO_2 concentration now underway is likely to decrease significantly stomatal aperture (increase r_s) and probably will decrease the rate of transpiration more than it decreases photosynthesis.

In hot, sunny weather, leaf water deficits often develop that override the effects of light and associated changes in internal CO_2 concentration. Midday closure of stomata, accompanying loss of leaf turgor, is common in both herbaceous and woody plants (e.g., Turner and Begg, 1973). It was long assumed that this hydropassive stomatal closure was caused by reduced water potential and turgor of the entire leaf. However, it has now been demonstrated that exposure of the epidermis to dry air causes closure of stomata in at least some species of plants (Lange *et al.,* 1971; Schulze *et al.,* 1972; Sheriff, 1977b). The data of Hall and Kaufmann (1975) for sesame and those of Kaufmann (1976b) for spruce indicate that low humidity caused increased closure of stomata in these species, and Lawlor and Milford (1975) reported that stomata of sugar beets can be kept open in humid air even when a water deficit exists in the leaf as a whole. Some examples of the effects of humidity are shown in Fig. 11.16. Aston (1976) demonstrated an increase in r_1 with increasing vapor pressure deficit without any change in leaf water content. Thus, stomatal closure can be caused either by dehydration of the entire leaf or by dehydration of the guard cells and adjacent epidermal cells by dry air, as in the experiments of Lange *et al.* (1971). Other investigators have found less response to humidity in a variety of plants, including camellia and privet (Wilson, 1948), and *Atriplex halimus* and *Kochia* (Whiteman and Koller, 1964). It is uncertain how much of the difference in response is real and how much results from differences in experimental methods and previous treatment of the plants.

The responsiveness of stomata to light, CO_2, and water stress depends on leaf age, temperature and other environmental conditions, and past treatment. As leaves become older, the stomata often become less responsive and may open only partly, even at midday (Slatyer and Bierhuizen, 1964; Brown and Pratt, 1965). According to Ackerson and Krieg (1977), stomata of corn and sorghum close when water-stressed during the vegetative stage but not during the re-

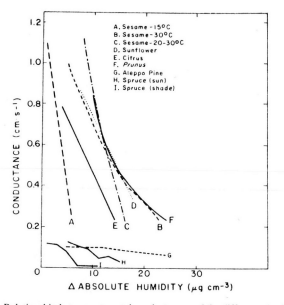

Fig. 11.16. Relationship between stomatal conductance and the difference in absolute humidity between leaves and air. Conductances for conifers are much lower than those for other species studied. (After Kaufmann, 1976.)

productive stage of growth. The stomata on older leaves of mulberry also are said to remain open (Tazaki *et al.*, 1980). It has also been reported that stomata of plants grown in greenhouses and growth chambers close at a higher leaf water potential than those of plants grown in the field (Jordan and Ritchie, 1971; Davies, 1977). However, exceptions occur. Kaufmann (1976a) reported that stomata of previously stressed citrus seedlings were more sensitive to atmospheric vapor pressure deficit than those of unstressed seedlings. Levy (1980) found that although the stomatal resistance of well-watered citrus trees generally increased with increasing vapor pressure deficit, there was a tendency for the stomata to open at very low humidities in spite of the large vapor pressure difference.

In general, there are large differences among species with respect to the degree of water stress at which stomatal closure occurs and in their sensitivity to humidity (see Fig. 11.16). The leaf water potential for closure seems to range from −0.7 to −0.9 MPa in tomato to −1.2 or −1.6 MPa in grape (Hsiao, 1973, p. 525). Federer (1977) reported closure at from −1.5 MPa in birch to −2.3 MPa in cherry; Lopushinsky (1969) found that closure occurred at −1.5 or −1.6 MPa in ponderosa and lodgepole pine, but at −2.5 MPa in grand fir. Jordan and Ritchie (1971) found stomata of field-grown cotton open at −2.7 MPa, but Davies (1977) reported that the abaxial leaf resistance of both greenhouse and field grown cotton had doubled at a leaf water potential of -2.0 MPA. The stomata of brigalow, a xerophytic acacia, are said to remain partly open to leaf water potentials as low as -5.0 MPa (van den Driesche *et al.*, 1971) and those of jojoba (*Simmondsia chinensis*) to about -4.0 MPa (Adams *et al.*, 1978). In plants of some species there appears to be a fairly definite threshold at which stomatal closure begins to occur. For example, the stomata of onion open and close within a narrow range of -0.3 to -0.7 MPa (Miller *et al.*, 1971). However, in other plants it occurs over a wide range of decreasing leaf water potential and saturation deficits, as shown in Fig. 11.17. It is reported that the water potential at which stomata close in wheat decreases as the plants grow older, at least partly because of osmotic adjustment (Teare *et al.*, 1982). Kozlowski (1982a) doubts if so-called threshold values of leaf water potential for stomatal closure should be taken seriously because they are modified so much by past and

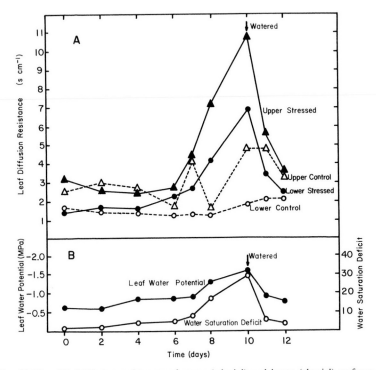

Fig. 11.17. (A) Diffusion resistances of upper (adaxial) and lower (abaxial) surfaces of the seventh and eighth leaves of stressed and unstressed corn, measured in the morning with a diffusion porometer. The plants were growing in a controlled environment chamber. (B) Leaf water potential and saturation deficit of stressed corn plants. (From Sanchez-Diaz and Kramer, 1971.)

present environmental conditions. Figure 11.17 also shows the well-known fact that the stomata on the upper (adaxial) surfaces of leaves usually have a higher resistance than those on the lower (abaxial) surfaces (Davies, 1977).

Mineral deficiencies also appear to affect stomatal behavior, although the evidence is contradictory. Desai (1937) reported that deficiencies of nitrogen, phosphorous, and potassium reduced the responsiveness of stomata of several kinds of plants, and Pleasants (1930) reported that stomata of nitrogen-deficient bean seedlings were less responsive to water stress than those of plants receiving adequate nitrogen. In contrast Radin and Parker (1979) found that nitrogen deficiency made cotton stomata close sooner in response to water stress. Further research by Radin and Boyer (1982) indicated that nitrogen deficiency increases root resistance and causes leaf water stress, and this probably is the cause of early stomatal closure.

The effects of temperature on stomatal aperture generally are modest in the growing range and often can be disregarded, as in Kaufmann's work with forest trees (Kaufmann, 1982b). On the other hand, Wuenscher and Kozlowski (1971) reported that in 5 species of tree seedlings studied by them, leaf resistance increased as the temperature increased from 20° to 40°C, the increase being greatest in species normally growing on dry sites. This tends to reduce transpiration when the vapor pressure gradient is steepest. At low temperature, responses to light and humidity are slower (Wilson, 1948) and stomatal resistance usually increases. However, there are species differences, there being no increase in collards at 5°, but a large increase in cotton and bean (McWilliam *et al.*, 1982). It has been reported that chilling "locks open" the stomata of some tropical plants (Wilson, 1982) and slow closure was observed by McWilliam *et al.* (1982) for cotton and bean, but not in soybean at 10° by Musser *et al.* (1982).

It is difficult to generalize about stomatal behavior because so many contradictory reports occur in the literature. This is not surprising, because stomatal activity is affected by numerous internal and external factors, including leaf water status, leaf temperature, internal and external CO_2 concentration, growth regulators, irradiance, and ambient humidity. These interact, often in complex ways that sometimes are overlooked by investigators. The past history of plants also can have important effects on stomatal reactions. For example, stomata of plants subjected to one or more periods of water stress might respond quite differently from those not previously stressed. Perhaps if more attention were given to the control of environmental conditions preceding and during measurements some of the apparent contradictions would disappear. After Kaufmann (1982a, 1982b) studied stomatal behavior of four tree species in a subalpine forest for an entire season, he concluded that stomatal opening is controlled primarily by irradiance, measured as photosynthetic photon flux density and the difference in absolute humidity between the leaves and the air. He concluded that factors such as water stress, temperature, and CO_2 operate only intermittently on plants growing in their normal habitat. The general applicability of this simplifying concept deserves more study.

Stomatal Control of Transpiration.   Although stomata constitute the principal transpiration control system, and 80–90% of the water usually escapes through them, under some circumstances they do not exercise very good control. According to Ackerson and Kreig (1977), stomatal control of transpiration of maize and sorghum is very good during the vegetative stage but becomes ineffective during the reproductive stage because the stomata do not close even when the leaves are severely stressed. On the other hand, Kaufmann (1982b) followed stomatal resistance of four species of subalpine forest trees throughout the growing season and found no important changes as the leaves aged.

Occasionally mutant plants are reported that wilt severely because the stomata do not close when water stress develops. Examples are the "wilty" tomato studied by Tal (1966) and the "droopy" potato discussed by S. A. Quarrie in 1982. Both seem to be deficient in ABA, although application of ABA does not cause complete closure of stomata of droopy potatoes. Quarrie suggested that these mutants might provide good material for study of the role of ABA in stomatal closure.

Earlier in this century there was much unproductive discussion concerning the effect of partial closure of stomata on the rate of transpiration. The pioneer experiments of Brown and Escombe (1900) were conducted in quiet air where the boundary layer resistance (r_a) was as great as stomatal resistance (r_s), and their results indicated that large changes in stomatal aperture should have little effect on transpiration. However, later investigators found that in moving air, where r_a is low relative to r_s, there is a relatively high correlation between r_s and transpiration (Bange, 1953; Stålfelt, 1932, 1956), as shown in Fig. 11.18. Slatyer (1967, pp. 260–269) discussed in detail why the effect of partial closure of stomata on transpiration depends on the relative values of r_s and r_a.

Optimization.   In recent years there have been attempts to develop theories explaining stomatal behavior in terms of optimization of cost to benefit, costs being loss of water and benefits carbon assimilation. Ideally, optimal efficiency will result when stomatal aperture during the day varies in a manner that results in minimum transpiration for maximum photosynthesis. This seems to require a constant ratio of transpiration (E) to photosynthesis (A), i.e., dE/dA = a constant. According to Wong *et al.* (1979), in at least some plants, stomata adjust in a manner that maintains a fairly constant ratio of internal to external concentration of CO_2. They claim that there is a feedback mechanism relating stomatal aperture to the photosynthetic capacity of the mesophyll cells. It is argued that

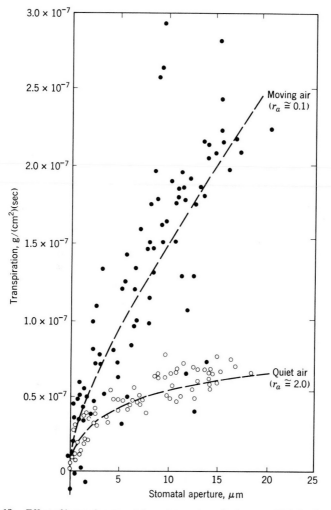

Fig. 11.18. Effect of increasing stomatal aperture on transpiration rate of Zebrina leaves in quiet and moving air. The effect of increasing stomatal aperture is very large in moving air but small in quiet air, where air resistance (r_a) or boundary layer resistance is often as large as stomatal resistance (r_s). (From Slatyer, 1967; after Bange, 1953.)

the stomatal control system produces an optimum degree of opening for the entrance of CO_2, which in turn results in efficient control of water loss. However, it appears that the amount of water saved by optimal stomatal behavior is small, and perhaps the concept is of more philosophical than practical importance. The optimization concept cannot be developed adequately here, but interested readers are referred to papers by Cowan (1977, 1982), Cowan and Farquhar (1977), and Farquhar and Sharkey (1982) for more details.

Measurement of Stomatal Aperture or Resistance

Because of its importance in controlling water loss and uptake of CO_2 there has been much interest in measuring stomatal aperture. Direct observation of stomatal aperture under a microscope is difficult, so Lloyd (1908) stripped off bits of epidermis and fixed them in absolute alcohol before observing them under a microscope. This method was used by Loftfield (1921) for his classical studies but it is difficult to strip epidermis from some leaves and stripping sometimes causes changes in stomatal aperture. A better method is to make impressions of the epidermis in collodion (Clements, 1934), silicone rubber (Zelitch, 1961), or even Duco cement or nail polish, and examine them under the microscope. However, Pallardy and Kozlowski (1980) warned that epidermal impressions are unreliable where cuticular ledges develop over the guard cells and mask changes in stomatal aperture. Shiraishi *et al.* (1978) studied cyclic variations in stomatal aperture with a scanning electron microscope and reported that the aperture of the outer ledge often changes independently of the pore aperture. This increases the danger of misinterpreting observations on stomatal aperture.

Infiltration. Another method, apparently first mentioned by Haberlandt in 1905, but popularized by Molisch (1912), depends on measuring the time required for liquids of various viscosities to infiltrate leaves. Alvim and Havis (1954) used mixtures of paraffin oil and *n*-dodecane, and investigators in Israel used mixtures of paraffin oil and turpentine or benzol, or kerosene alone. Fry and Walker (1967) described a pressure infiltration method for use on pine needles. It indicates whether stomata are open or closed, but does not permit calculation of leaf resistances (Lassoie *et al.*, 1977).

Porometers. These methods are now largely supplanted by porometers that measure the approximate rate of diffusion of water vapor from leaves and are calibrated in such a manner that readings can be converted into diffusion resistance of the leaf in s cm^{-1} or its reciprocal, conductance in cm s^{-1}. This method measures the total leaf resistance, but if cuticular transpiration is low it is a reasonable approximation of the stomatal resistance.

Francis Darwin, the plant physiologist son of Charles Darwin, is credited with developing the porometer and Darwin and Pertz (1911) introduced simple porometers that measured gas flow through leaves under a slight pressure. Numerous modifications have appeared, including recording porometers (Gregory and

Pearse, 1934; Wilson, 1947) and portable models such as that developed by Alvim (1965). Gas or viscous flow porometers have disadvantages, including the fact that if a pressure greater than 10 cm of water is used stomatal aperture may be affected (Raschke, 1975). Also they cannot be used on leaves with stomata on one surface and vascular bundle extensions (heterobaric structure) that prevent lateral diffusion of gases in the intercellular spaces. E. L. Fiscus (private communication) designed a viscous flow recording porometer that can be used in the field to measure changes in stomatal resistance for several successive days. Diffusion porometers have been developed that measure the time required for a predetermined change in humidity to occur in a small chamber attached to a leaf. They have gone through extensive developments by Heath (1959), Slatyer and his colleagues (Slatyer and Jarvis, 1966; Jarvis *et al.*, 1967), Wallihan (1964), van Bavel *et al.* (1965), Kaufmann and Eckard (1977), and others. Beardsell *et al.* (1972) described a null point porometer which measures the steady-state rate of transpiration of a leaf enclosed in a ventilated cuvette by blowing in sufficient dry air to maintain a constant humidity in the cuvette. This eliminates some calibration problems and the lag caused by water absorption on the cuvette walls. Some of the problems encountered in using cuvettes are discussed by Berkowitz and Hopper (1980), Hack (1980), Morrow and Slatyer (1971), and Slavik (1974).

Most measurements are now made with various modifications of the instru-

Fig. 11.19. A porometer for measuring the diffusion resistance of leaves using a cuvette (upper left) which is momentarily clamped on a leaf. (From Kramer and Kozlowski, 1979, by permission of Academic Press. Photograph courtesy of Lambda Instruments Corporation.)

ment described by Kanemasu *et al.* (1969), using a humidity sensor connected to an electronic circuit and meter (see Fig. 11.19). Some have electronic timers to measure the time required for a given change in humidity and other modifications that increase the speed and accuracy of measurements. Kaufmann (1981) described a system using large ventilated cuvettes (approximately 15 liters in volume) in which tree branches can be enclosed and monitored over an entire growing season. A computer controls acquisition of relative humidity, temperature, and solar radiation measurements at hourly intervals and computes values for leaf conductance and transpiration.

INTERACTION OF FACTORS AFFECTING TRANSPIRATION

The important environmental factors affecting transpiration are usually described as irradiance, vapor pressure deficit of the air, temperature, wind, and water supply. Plant factors include leaf area, leaf structure and exposure, stomatal behavior, and the effectiveness of the root system as an absorbing surface. The complex interactions among these factors can be described in terms of their effects on the resistance and energy terms in Eq. (11.2).

Changes in irradiation affect r_{leaf} by affecting stomatal aperture and e_{leaf} because change in temperature affects the vapor pressure gradient, Δe, from leaf to air, as shown in Fig. 11.3 and Tables 11.5 and 11.6. Martin (1943) found a close relationship between transpiration of plants kept at constant temperature in darkness and the vapor pressure of the atmosphere, and Cole and Decker (1973) observed that within limits transpiration is a linear function of Δe. However, when the humidity and Δe fall too low, stomata close, reducing transpiration. Wind increases transpiration directly by reducing r_{air} and removing water vapor from the boundary layer, and reduces it by cooling leaves and reducing e_{leaf}. Other effects of wind are uncertain, although Shive and Brown (1978) reported that flexing of cottonwood leaves by wind caused bulk flow of gas through them and decreased the total resistance by about 25%. Leaf orientation affects temperature and e_{leaf}; leaf structure and stomatal behavior affect various components of r_{leaf}. Decreasing water supply to roots caused by drying or cold soil results in stomatal closure, increasing r_s.

The important fact is that a change in one of the factors affecting transpiration does not necessarily produce a proportional change in rate of transpiration because the rate is controlled by more than one factor. A good example is the effect of a breeze, shown in Fig. 11.5. If initially the leaves are cooler than the air, a breeze may increase transpiration; if they are warmer, it may decrease transpiration by cooling them.

MEASUREMENT OF TRANSPIRATION

Transpiration has been measured on surfaces varying in area from part of a leaf to entire fields and forests, and the methods used have varied equally widely. Originally, most measurements were made on individual plants, but in agriculture and forestry interest has turned toward study of the water balance of large stands of plants. We will discuss measurement of transpiration from leaves and plants in this section and deal with plant stands later.

Gravimetric Methods

From the time of Hales (1727) to the present, investigators have grown plants in containers and measured transpiration by weighing the containers at appropriate intervals. It is necessary to grow the plants in waterproof containers and to cover the soil surface to prevent loss by evaporation. Soil moisture must be replenished frequently so that water supply does not becoming limiting, a common defect of many early experiments (Raber, 1937). The size of the plants studied is limited by the size of the containers and the capacity of the scales used to weigh them. Veihmeyer (1927) measured the transpiration of fruit trees growing in tanks of soil weighing 450 kg, and Nutman (1941) measured the transpiration of coffee trees with a balance having a capacity of 225 kg and sensitivity of 25 g. The ultimate in size with this method seems to be the lysimeter constructed by Fritschen *et al.* (1973) to weigh a block of soil 3.7 × 3.7 × 1.2 m containing a Douglas-fir tree 28 m in height. The container, soil, and tree weighed 28,900 kg, and the weighing apparatus had a sensitivity of 630 g. Lysimeters are discussed in Chapter 4. Occasionally, change in soil water content is monitored by absorption of radiation, as described in Chapter 4.

The containers in which plants are growing should be protected from direct sun to prevent overheating, and the ideal arrangement is to have them set with the tops flush with the surrounding soil in the habitat where they would normally grow. This was done by Biswell (1935) and Holch (1931) in their studies of the transpiration of tree seedlings in sun and shade in Nebraska, and their procedure is generally followed.

The Cut-Shoot Method. Measurements have also been made by scores of investigators on detached leaves or twigs weighed at intervals of a minute or two on a sensitive balance. Such measurements can proceed for only a few minutes after cutting because the transpiration rate tends to decline with decreasing leaf water content. Sometimes there is a transient increase in rate shortly after detaching a leaf or branch, the Ivanov effect, probably resulting from release of tension in the xylem. The method has several inherent errors caused by detaching the plant organ and by measuring transpiration in an environment different from that of its location on the plant. Also, there are often large differences in rates among individual leaves from the same plants and among leaves from different plants (Hölzl, 1955). In spite of its inherent errors, the cut-shoot method has been used to measure differences in transpiration rate among species and differences among leaves from the bottom and top and north and south sides of trees. It was also used by Hygen (1953), Kaul and Kramer (1965), and others to determine when stomatal transpiration ceases and water loss occurs only through the cuticle. An example is shown in Fig. 11.20. Franco and Magalhaes (1965) and Slavik (1974) discuss this method in detail.

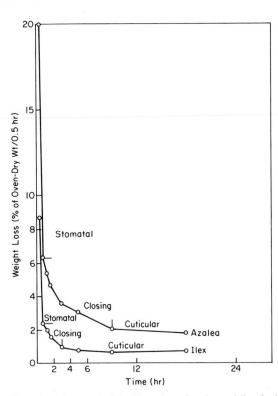

Fig. 11.20. Water loss from excised leaf branches of azalea and ilex for half-hour intervals, expressed as percentage of oven-dry weight. The stomatal, closing, and cuticular phases of transpiration, as described by Hygen (1953), are indicated for both species. The stomata of ilex close more rapidly, and this species has a lower rate of cuticular transpiration than azalea, which suffers more from drought injury. (From Kaul and Kramer, 1965.)

Volumetric Methods

Occasionally, transpiration is determined by measuring the volume of water absorbed by a detached leaf or branch. However, water uptake by detached plant organs is often reduced by plugging of the conducting system with air or debris. Also, the detached branch is no longer in competition for water with the remainder of the shoot and may be subjected to less stress. More reliable results are obtained when entire root systems can be placed in a porometer and the volume of water absorbed can be measured. However, absorption through root systems grown in soil is likely to be seriously reduced if the roots are suddenly immersed in water. Ladefoged (1963) and Roberts (1977) made useful measurements of transpiration of trees up to 16 m in height by cutting them and placing their bases in containers of water with their tops supported in their normal position in the canopy. Knight *et al.* (1981) made numerous measurements of transpiration of lodgepole pine trees up to 30 cm in diameter by this method. They found that absorption slowed after 2 or 3 days but could be restored by recutting the base of the trunk.

Measurement of Water Vapor Loss

Measurements of transpiration can be made by monitoring the change in humidity of an air stream passed through a container enclosing the plant material. The containers are usually made of plastic and vary from tiny cuvettes holding one leaf or part of a leaf (Slavik, 1974, Chapter 5) to those holding a branch (Kaufmann, 1981), and plastic tents enclosing an entire tree (Decker *et al.,* 1962; see Fig. 11.21). This method eliminates errors caused by detaching leaves or branches, but it imposes a somewhat artificial environment on the leaf or plant enclosed in the container. The difference in leaf and air temperature, humidity, and wind speed inside and outside the chamber can cause important differences between transpiration rates measured by these methods and rates of fully exposed plants. One remedy is to provide continuous compensation for changes in humidity and CO_2 concentration, as was done by Koller and Samish (1964) and Moss (1963). Apparatus for control of the environment and continuous, simultaneous measurements of stomatal resistance and water vapor and CO_2 exchange of leaves was described by Jarvis and Slatyer (1966).

Some of the errors caused by prolonged enclosure of leaves can be minimized or eliminated by short-term measurements of water loss. Grieve and Went (1965) described use of cuvettes containing a humidity sensor to enclose a single leaf for short-term measurements, and this method has been developed into equipment

Fig. 11.21. An Aleppo pine 7.2 m in height enclosed in a plastic tent supported by air pressure. The enclosure was ventilated by a blower. Air entered near the bottom and left through apertures near the top. Transpiration rates were calculated from the difference in humidity of air entering and leaving the enclosure (Decker and Skau, 1964). (Photograph courtesy of J. P. Decker; from Kramer and Kozlowski, 1979, by permission of Academic Press.)

that can make a measurement of transpiration and stomatal resistance in less than 1 min. Examples are instruments described by Kanemasu *et al.* (1969), Kaufmann and Eckard (1977), and Beardsell *et al.* (1972). Several porometers are described by Meidner in Jarvis and Mansfield (1981), and an automated system is described by Kaufmann (1981).

Velocity of Sap Flow

Attempts have been made to estimate the rate of transpiration from measure-ments of the velocity of sap flow by the heat pulse method described in Chapter 10. Ladefoged (1963) used a diathermy machine as a source of heat and cali-brated the method by measuring absorption of water by tree trunks cut off and placed in a container of water. Decker and Skau (1964) found a good correlation between simultaneous measurements of sap velocity and transpiration of trees in enclosed plastic tents. Swanson and Lee (1966) found the heat pulse method useful to indicate when transpiration stops and starts and how it is affected by environmental conditions. They found a lag in sap flow behind transpiration in the morning, probably because water stored in the upper part of the trunk is used before sap flow in the lower part is affected by increasing transpiration. It is generally difficult to obtain quantitative data by this method.

Bases for Calculating Transpiration Rates

After deciding on an acceptable method of measuring transpiration, investigators must select a method for expressing the rate. Although it seems reasonable to express transpiration in terms of leaf surface, should this be based on both surfaces or only one? We have generally used one surface, but the stomate-bearing surfaces are sometimes used. Many European measurements have been expressed as units of water lost per unit of leaf fresh weight. However, the fresh weight varies from morning until afternoon and from day to day. Furthermore, the ratio of leaf area to weight varies among species and between sun and shade leaves of the same species. Examples of the differences in relative rates of transpiration expressed on a leaf area basis and per unit of fresh weight are shown in Fig. 11.22.

It seems that leaf area is the best basis for calculating rates of transpiration and photosynthesis because the energy received is more closely related to the area than to the weight of the leaves. However, Nobel (1980) pointed out that leaf thickness and amount of mesophyll tissue affect the water relations of leaves and

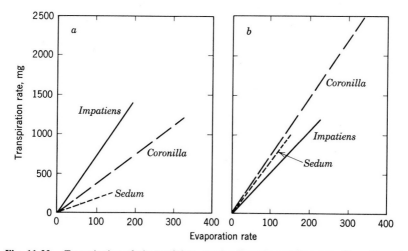

Fig. 11.22. Transpiration of plants of three species (*Impatiens noli-tangere, Coronilla varia,* and *Sedum maximum*). Rates are expressed in milligrams per gram of leaf fresh weight (a) and milligrams per square decimeter of leaf surface (b). Transpiration is plotted over evaporation from a filter paper atmometer. The transpiration rate of *Sedum* was affected most by the method used to express it because it has a low ratio of surface to mass. (Adapted from Pisek and Cartellieri, 1932.)

their capacity to carry on photosynthesis. Use of leaf area introduces the necessity of measuring or estimating it. Where leaves can be harvested the area can be measured quickly on a leaf area machine, and if they cannot be removed the area can usually be estimated from a formula relating area to leaf length or breadth (Wiersma and Bailey, 1975).

Transpiration from stems is generally disregarded, but Gračanin (1963) reported that the rate of water loss per unit of surface from stems of several herbaceous species is high and that stem transpiration sometimes constitutes a significant fraction of the total water loss. Huber (1956) cited some data on leaf and stem transpiration, indicating that transpiration from the bark of trees is negligible compared with foliar transpiration. Perhaps water loss from large stems of herbaceous plants such as tobacco deserves further investigation.

Rates of Transpiration

After reviewing the errors inherent in various methods of measuring transpiration, it may seem doubtful if such measurements have much relationship to rates

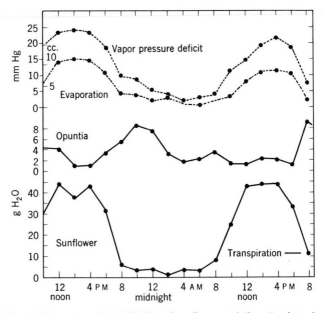

Fig. 11.23. Daily course of transpiration of sunflower and *Opuntia* plants in soil at field capacity on a hot summer day. Note the midday decrease in transpiration of sunflower the first day, probably caused by loss of turgor and partial closure of stomata. Also, note that the maximum rate of transpiration of *Opuntia* came at night, a characteristic of plants with Crassulacean acid metabolism. (After Kramer, 1937.)

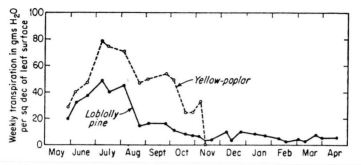

Fig. 11.24. Seasonal course of transpiration of potted seedlings of an evergreen and a deciduous species growing outdoors in Durham, North Carolina. (From Kramer and Kozlowski, 1979.)

in the field. However, differences among species under similar conditions and seasonal variations are of interest, and a few data are presented in Table 11.10 and in Figs. 11.23 and 11.24. Perhaps the data are more reliable than might be expected. For example, transpiration of potted seedlings of *Liriodendron tulipifera* in full sun in August averaged 10.1 g dm^{-2} day^{-1} at Columbus, Ohio, and 11.7 g dm^{-1} day^{-1} at Durham, North Carolina. Comparison of transpiration of six species of strees by four investigators using two different methods indicated that birch had the highest and spruce the lowest rate (Kramer and Kozlowski, 1979, p. 437). Measuring transpiration of orange seedlings by the cut-shoot method and weighing potting plants gave similar results on mild days, but on hot, dry days the transpiration rate of potted plants was lower (Halevy, 1956). The principal difficulty is in extrapolating such measurements to field conditions.

Curves for the daily course of transpiration of sunflower and of *Opuntia*, which has crassulacean acid metabolism (CAM), are shown in Fig. 11.23. CAM plants such as *Opuntia* and pineapple (Joshi *et al.*, 1965) usually have a lower rate of transpiration during the day than at night, when their stomata are open (see Fig. 13.1). The seasonal course of transpiration of an evergreen and a deciduous species is shown in Fig. 11.24. Readers are reminded that because of the large leaf area of conifers, the transpiration per tree may be as great as for deciduous species, as shown in Table 11.7.

EVAPORATION FROM STANDS OF PLANTS

Thus far, the discussion of transpiration has dealt chiefly with leaves or individual plants, but in nature plants usually occur in stands or communities, and one must consider the extent to which what has been said about individual plants applies to plants growing in stands. The general principles of Eq. (11.2) apply to water loss from all kinds of evaporating surfaces. However, in a stand of plants the principal evaporating surfaces are leaves of various ages and physiological conditions, located at various levels in the canopy and exposed to different microenvironments. Thus, the water loss from a stand is an average of the loss from many leaves with different exposures and stomatal behaviors. Sometimes the evaporating surface is treated as the surface of the stand, but this is an oversimplification, because both environmental and leaf conditions vary from top to bottom of the canopy. Examples of vertical differences in leaf characteristics are given by Turner (1974) and others. The amount of water lost by evaporation from the soil varies widely, depending on the amount and kind of plant cover.

Measurement of total water loss by evapotranspiration per unit of land surface is more important in agriculture and forestry than measurement of transpiration alone. Such measurements are especially important for the timing of irrigation and the estimation of water loss from watersheds. There are four general methods of measuring water loss from land areas: (1) determination of the evaporation term (E) in the water balance equation (4.1); (2) the energy balance method; (3) determination of net upward flow of water vapor above the surface; and (4) estimates from pan evaporation or from meteorological data. Lack of space prevents discussion of their advantages and disadvantages, and readers are referred to articles by Tanner, in Hagan *et al.* (1967) and in Kozlowski (1968) for details.

The amount of water lost from a unit area of land depends first of all on the energy available to evaporate water. On a bright summer day, 320–380 calories of incident energy per square centimeter are available, but 570 calories are required to evaporate 1 cm^3 of water at 20°C, so the maximum possible evaporation is approximately 6 mm day^{-1}. However, the maximum actual evapotranspiration seldom exceeds 80% of the theoretical maximum. If evaporation exceeds this, it is because incident energy is supplemented by advection or horizontal flow of energy from the surroundings, as in the case of exposed trees or fields surrounded by desert, the oasis effect shown in Fig. 11.2. Although the leaf area of plant stands (leaf area index) is often four to six times that of the soil on which they are growing, water loss cannot exceed that from an equal area of moist soil or a water surface receiving the same amount of energy. As the soil begins to dry, causing stomatal closure, transpiration will be reduced below the potential rate. Evaporation from drying soil is also reduced because water movement is slow through the dry surface layer. Thus, evaporation is often far below the potential rate calculated from pan evaporation.

It has been estimated that evaporation from soil under forests, meadows, and cultivated crops is approximately 10, 25, and 45% of the total water loss, respectively. Data for losses by evaporation and transpiration from a forest and a field of corn are shown in Tables 11.1 and 11.2, and the effect of removing forest

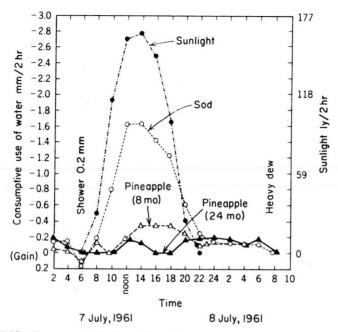

Fig. 11.25. Evapotranspiration from a Bermudagrass sod and from stands of pineapple 8 and 24 months after planting. All plants were growing in lysimeters in Hawaii. (From Ekern, 1956.)

cover is shown in Table 11.1. In theory, water losses ought to be similar from various types of plant cover receiving similar amounts of radiation, but variations in albedo, roughness of the canopy, depth of rooting, and other factors result in significant differences. Some data on evapotranspiration from various kinds of plant cover are given in Kramer (1969, pp. 337–342) and in Kramer and Kozlowski (1979, pp. 439–441). An extreme example of species differences in water loss by evapotranspiration is shown in Fig. 11.25. Pineapple, a CAM plant, has such a low rate of transpiration that when a mature stand fully shades the ground, the rate of evapotranspiration is reduced below that of a younger stand, which leaves the soil partly exposed.

SUMMARY

Transpiration is the loss of water from plants in the form of vapor. Although basically an evaporation process, it is affected by leaf structure and stomatal opening, as well as by the environmental factors that affect evaporation and stomatal aperture. Transpiration is the dominant process in plant water relations because it produces the energy gradient that causes water movement into and through plants. Excessive transpiration produces midday water deficits and temporary wilting, and when drying soil causes absorption to lag behind transpiration, permanent wilting and death by dehydration occur. Probably more plants are injured or killed by excessive transpiration than by any other cause.

The rate of transpiration depends on the supply of energy available to evaporate water, the difference in water vapor concentration or vapor pressure between leaves and air that constitutes the driving force, and the resistances in the water vapor pathway. The chief resistances are the stomata, the cuticle, and the boundary layer surrounding leaves.

The rate of transpiration is affected by leaf size, shape, and orientation, but it is controlled primarily by stomatal resistance, which depends on the degree of stomatal opening. Stomatal opening is affected by internal and external CO_2 concentration, light intensity, atmospheric humidity, and temperature. Increasing light intensity usually causes stomatal opening; decreasing light causes closure. The interaction between light intensity and CO_2 concentration with respect to control of stomatal opening and closing is complex and varies among species. Stomatal closure can also be caused by dehydration of leaves or, in some plants, by dehydration of epidermal and guard cells by low humidity. Abscisic acid and cytokinins are thought to be involved in stomatal activity. The responsiveness of stomata to light, CO_2, and water stress varies with leaf age and past and present environmental conditions.

Several problems deserve more research. Among them are the location of the principal evaporating surfaces within leaves, the pathway of water movement from veins to evaporating surfaces, the number of species in which low atmospheric humidity can cause stomatal closure independently of the bulk leaf water status, and the role of ABA in stomatal closure and reopening in water-stressed plants.

In agriculture and forestry, measurement of total water loss per unit of land area by evapotranspiration is important. Such measurements are especially important for the timing of irrigation and the estimation of water losses from watersheds and various vegetation types.

SUPPLEMENTARY READING

Burrows, F. J., and Milthorpe, F. L. (1976). Stomatal conductance in the control of gas exchange. *In* "Water Deficits and Plant Growth" (T. T. Kozlowski, ed.), Vol. 4, pp. 103–152. Academic Press, New York.

Cowan, I. R. (1977). Stomatal behavior and environment. *Adv. Bot. Res.* **4**, 117–228.

Cowan, I. R. (1983). Regulation of water use in relation to carbon gain in higher plants. *Encycl. Plant Physiol., New Ser.* **12B,** 535–562.

Gates, D. M. (1980). "Biophysical Ecology." Springer-Verlag, Berlin and New York.

Hall, A. E., Schulze, E.-D., and Lange, O. L. (1976). Current perspectives of steady-state stomatal responses to environment. *In* "Water and Plant Life" (O. L. Lange, L. Kappen, and E.-D. Schulze, eds.), pp. 169–188. Springer-Verlag, Berlin and New York.

Jarvis, P., and Mansfield, T. A., eds. (1981). "Stomatal Physiology." Cambridge Univ. Press, London and New York.

Raschke, K. (1975). Stomatal action. *Annu. Rev. Plant Physiol.* **26,** 309–340.

Raschke, K. (1979). Movements of stomata. *Encycl. Plant Physiol., New Ser.* **7,** 383–441.

Slavik, B. (1974). "Methods of Studying Plant Water Relations." Springer-Verlag, Berlin and New York.

12

Water Deficits and Plant Growth

INTRODUCTION

Plants are subjected to a variety of environmental stresses, including abnormal temperatures, unfavorable chemical and physical soil conditions, and various diseases and insect pests. However, in the long run, water deficit reduces plant growth and crop yield more than all the other stresses combined, because it is ubiquitous. Water deficit or water stress refers to situations in which plant water potential and turgor are reduced enough to interfere with normal functioning. The exact cell water potential at which this occurs depends on the kind of plant, the stage of development, and the process under consideration. For example, cell enlargement usually begins to decrease at a water potential of only -0.2 to -0.4 MPa, but stomatal closure does not begin until the water potential falls to -0.8 to -1.0 MPa in some plants (see later Fig. 12.13B) and much lower in others (Hsiao, 1973). A summary of responses of various processes to water stress is shown in Fig. 12.1.

Water deficits vary in intensity from small decreases in water potential, detectable only by instrumental measurements, through transient midday wilting, to permanent wilting and death by dehydration. They are characterized by decreases in water content, turgor, and total plant water potential, resulting in wilting, partial or complete closure of stomata, and decrease in cell enlargement and plant growth. If severe, water deficits cause cessation of growth, decrease or cessation of photosynthesis, disturbance of many metabolic processes, and finally death.

The preceding chapter discussed the loss of water from plants. This chapter

process affected	sensitivity to stress very sensitive ——————→ insensitive reduction in tissue Ψ_w required to affect the process 0 · · · · · · 1 · · · · · · 2MPa	references
cell growth $(-)$		Acevedo *et al.*,1971; Boyer,1968
wall synthesis[†] $(-)$		Cleland,1967
protein synthesis[†] $(-)$		Hsiao,1970
protochlorophyll formation[‡] $(-)$		Virgin,1965
nitrate reductase level $(-)$		Huffaker *et al.*,1970
ABA synthesis $(+)$:		Zabadal,1974; Beardsell &
stomatal opening $(-)$:		Cohen,1974
(a) mesophytes		reviewed by Hsiao,1973
(b) some xerophytes		Van den Driesche *et al.*,1971
CO₂ assimilation $(-)$:		
(a) mesophytes		reviewed by Hsiao,1973
(b) some xerophytes		Van den Driesche *et al.*,1971
respiration $(-)$		
xylem conductance[§] $(-)$		Boyer,1971; Milburn,1966
proline accumulation $(+)$		
sugar level $(+)$		

[†] fast growing tissue
[‡] etiolated leaves
[§] should depend on xylem dimension

Fig. 12.1. Summary of sensitivity to stress of a number of plant processes. Length of the horizontal bars represents the range of stress within which a process is first affected, and the broken portion is that part of the range within which the response is not well established. In the left column, $(+)$ indicates an increase and $(-)$ a decrease. (From Hsiao *et al.*, 1976.)

will discuss how water deficits develop, how they affect physiological processes and modify the quantity and quality of growth, and how they can be measured.

CAUSE AND DEVELOPMENT OF WATER DEFICITS

In simple terms, water deficits develop during periods when water loss in transpiration exceeds absorption. This occurs to some extent every sunny morning in plants that are transpiring at normal rates, but it becomes more severe when transpiration becomes rapid on hot, sunny days, when water absorption is limited by drying, cold, or poorly aerated soil, or when a combination of the two situations occurs. Turner and Begg (1981, p. 100) point out that absorption sometimes exceeds transpiration in the afternoon and at night, as shown in Fig. 12.2, although a water deficit still exists. This does not change the fact that the deficit originated during a period when absorption lagged behind transpiration. The cause of water deficits is clear, but the manner in which they develop is complex and will now be discussed.

Development of Midday Deficits

Water loss usually exceeds absorption as transpiration increases in the morning because (1) there is considerable resistance to water flow from the soil into the root xylem and (2) there is an appreciable volume of readily available water in the turgid parenchyma cells of the leaves and stems. This stored water is usually treated as capacitance in models based on an analogy to an electrical circuit, but it really seems more analogous to electricity stored in a battery. Usually, there is considerably less resistance to water flow out of vacuoles of turgid parenchyma cells to the evaporating surfaces than there is to inward flow through the roots.

Therefore, when the morning increase in transpiration lowers the water potential of cells from which water is evaporating, water flows to them first from the nonevaporating parenchyma cells because the latter have a high water potential and a low resistance to water loss. Removal of water decreases their turgor and water potential, and by noon on a hot sunny day the leaves are losing their turgor and the bulk leaf water potential is so low that most of the water used in transpiration is entering through the roots. In the afternoon, as the stomata begin to close, transpiration decreases, but absorption continues at a relatively rapid rate until the parenchyma cells are refilled and their water potential rises too high

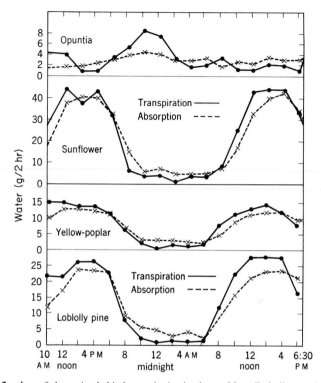

Fig. 12.2. Lag of absorption behind transpiration in plants of four dissimilar species growing in soil supplied with water by an autoirrigation system (see Fig. 4.10, Chapter 4) that permitted measurement of water uptake. Note the midday decrease in transpiration of sunflower and the evening maximum in transpiration of *Opuntia*. (From Kramer, 1937.)

to produce the driving force necessary for water movement. This change in relative rates of absorption and transpiration is shown in Fig. 12.2.

Development of Long-Term Deficits

The development of long-term water deficits in plants that are progressively reducing the supply of available soil water was carefully analyzed by Slatyer (1967). It begins with a daily cycle similar to that just described, but this is altered by the eventual inability to recover at night as plants reduce the water content and potential of the soil mass in which they are growing. Figure 12.3 shows diagrammatically how the daily cycles in leaf and root water potential

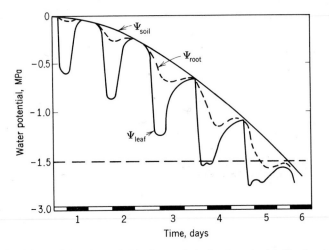

Fig. 12.3. Diagrams showing probable changes in leaf water potential (Ψ_{leaf}) and root water potential (Ψ_{root}) of a transpiring plant rooted in soil allowed to dry from a Ψ_w near zero to a Ψ_w at which wilting occurs. The dark bars indicate darkness. (From Slatyer, 1967.)

decrease as the soil water potential decreases. At first, the leaf water potential returns at night to a value equal to the soil water potential, but as the soil dries, less and less recovery is possible. Daytime wilting occurs when movement of water toward the roots in the drying soil becomes too slow to replace daytime water loss. This is the situation on days 3 and 4 in Fig. 12.3. Permanent wilting occurs when the soil water potential decreases to the leaf water potential at which wilting occurs (day 5 in Fig. 12.3), and leaves do not recover turgor at night.

This analysis emphasizes three factors in the development of water deficits, the rate of transpiration, the rate of water movement through the soil to the roots, and the relationship of the soil water potential to the leaf water potential. As pointed out in Chapter 3, the soil water potential at which permanent wilting occurs depends on the water potential at which leaves wilt. Thus, it is not really a characteristic of soil but is determined by the osmotic characteristics of plants. However, it usually occurs at a soil water potential of about -1.5 MPa because plants usually wilt at about that water potential (Slatyer, 1957, 1967).

Transpiration. The degree of water stress developed in plants is strongly dependent on the rate of transpiration, which in turn is strongly dependent on irradiance. Daytime growth of plants is often reduced by water deficits caused by excessive midday transpiration, and plants growing in moist soil or aerated nutrient solutions sometimes wilt on hot, sunny days. In general, midday water stress is more severe in dry climates, but it occurs in the humid tropics and it is sometimes severe enough in humid Japan to reduce photosynthesis, even in irrigated rice (Tazaki *et al.*, 1980). An example of midday shrinkage in various organs of an avocado tree is shown in Fig. 12.4. In Iowa, Thut and Loomis (1944) found that corn often grows more at night than during the day because daytime growth is inhibited by water deficit. However, at Davis, California, the nights are so cool that most of the growth of corn occurs during the day in spite of a large depression in leaf water potential (Hsiao *et al.*, 1976, pp. 294–296). There are important interactions among temperature, water deficit, and growth.

The effect of these transient midday water deficits on crop yield has never been fully evaluated, but it must be substantial. Boyer *et al.* (1980) reported that higher yields are correlated with lower afternoon water deficits in soybean. It has even been suggested that intermittent mist irrigation might be beneficial because it would prevent midday water deficits and lower leaf temperatures (Lawlor and Milford, 1975; also Chapter 4).

Availability of Soil Moisture. The literature on agriculture, forestry, and horticulture is filled with papers describing the effects of decreasing soil water content on plant growth. In general, there is a reduction in growth, as shown in various reviews, from Richards and Wadleigh (1952) to Kozlowski (1982a). For example, Bassett (1964) calculated the available water in the root zone of a

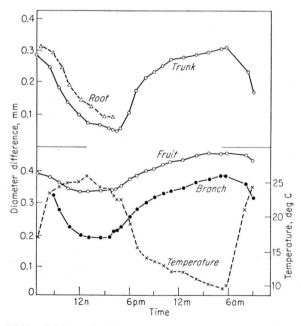

Fig. 12.4. Midday shrinkage of various parts of an avocado tree caused by rapid transpiration. (From Schroeder and Wieland, 1956.)

mixed pine stand in Arkansas from rainfall and evaporation data for every day of the growing season over a period of 20 years. He found that the actual growth in volume was nearly the same as the volume calculated from the available soil water content. Figure 12.5 shows how reduction in soil water potential over periods of a few weeks reduces diameter growth of avocado.

Under some circumstances, there are poor correlations among rainfall, soil moisture, and plant growth over a growing season. This was observed by Coile (1936) for forest trees. More growth occurred in a cool summer with below-normal rainfall than in a hot summer with above-normal rainfall, but also above normal transpiration. Glock and Agerter (1962) reported other examples of poor correlations. However, according to Zahner (1968), 80–90% of the variation in width of annual rings of trees in humid climates and 90% in arid climates can be attributed to water stress. As a result it is possible to establish the rainfall pattern of past centuries from the varying width of annual rings of old trees and from timbers in ancient buildings (Fritts, 1976). This is known as dendrochronology.

The interrelationship between soil water and plant growth is affected by atmospheric factors that influence the rate of transpiration, chiefly because of the high temperature and high vapor pressure deficit accompanying a high level of irradiance. For example, Denmead and Shaw (1962) found that a higher soil water content was required to prevent water stress in corn on sunny days than on cloudy days, because on cloudy days the rate of transpiration was low. This is because as the soil dries the soil water potential and conductivity decrease, slowing water movement to the roots. Loomis (1934) and Thut and Loomis (1944) found that

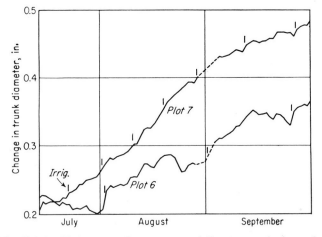

Fig. 12.5.   Relationship between soil water stress and diameter growth of avocado trees. Plot 7 was watered whenever the soil matric potential fell to −0.05 MPa, and plot 6 when the matric potential fell to −1.0 MPa. The short vertical bars indicate irrigations. Growth ceased and there was even shrinkage of tree trunks before irrigation of the dry plot. (From Richards *et al.*, 1958.)

during an Iowa summer, water deficits in the regions of stem and leaf elongation, caused by high transpiration, inhibited the growth of corn and several other plants more often than soil water deficits. Perhaps tree growth suffers less from short periods of water stress than growth of annual plants, in which 7 to 10 days of water stress at a critical period can seriously reduce yield. This will be discussed later.

Plant Water Balance

The terms water balance and water economy arise from an analogy between plants and personal and corporate finance, in which the dollar balance depends on the relationship between income (water absorption) and expenditures (water loss by transpiration). Generally, the water balance or water status of plants growing in moist soil depends chiefly on the atmospheric factors that control the rate of transpiration, but in dry soil it depends chiefly on the availability of soil water.

In this section we will discuss variations in water content, competition among organs for water, and bases for expressing water content.

Variations in Water Content. There are wide variations in water content in various parts of a plant, as well as among plants of different species and in different stages of development. Data on the water content of various plant organs were given in Table 1.1 (Chapter 1), and changes in water content of woody plants are discussed in detail by Kramer and Kozlowski (1979, pp. 474–482). The changes in water content are large enough to cause measurable changes in the diameter of tree trunks (Fig. 12.5). Figure 12.6 shows relatively large daily variations in the water content of leaves, stems, and roots of sunflower. Because of the large volume of pith parenchyma in the stems of many herbaceous plants, they function as significant water reservoirs (see Chapter 13). However, except in a few succulents with large storage capacity and a low rate of transpiration, this reservoir is relatively small in terms of the daily transpiration, which can equal or exceed the weight of the plant. Tree trunks also function as reservoirs, as indicated by large variations in water content. Waring and Running (1978) regard this as relatively important in the northwestern coniferous forest, but the storage capacity of a Douglas-fir forest is estimated to be equal only to one-half of a summer day's transpiration.

In general, the sapwood of trees has a higher water content than the heartwood, and its water content is usually highest in late spring or early summer and lowest in late summer. Seasonal variations in water content of tree stems are not only interesting physiologically but important economically because they affect

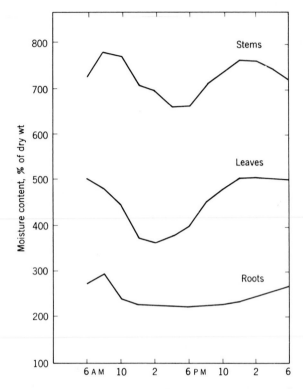

Fig. 12.6. Daily variation in water content of roots, leaves, and stems of sunflower plants growing in moist soil on a hot summer day. (From Wilson *et al.*, 1953.)

the rate of drying, the flotation of logs, and the cost of transport. For example, in eastern Canada, birch and poplar logs contain so much water in the spring that they cannot float, but by September loss of water causes a decrease in density to 0.60–0.75 (Gibbs, 1935). Leaving the leafy tops attached for several days after felling trees reduces the water content of the trunks of most species.

There are large differences in water content of various parts of the same plant. An example is seen in Fig. 12.6, where the water content of the leaves of 7-week-old sunflower plants is twice that of the principal roots and that of the stems is nearly three times that of the roots. More puzzling is the fact that mature tomato seeds in a ripe tomato fruit consisting of nearly 95% water contain only 50% water (McIlrath and Abrol, 1963). Likewise, bean and soybean seeds usually begin to dry out on plants still containing 75 or 80% of water in their vegetative organs. Prokof'ev *et al.* (1981) state that during maturation the Ψ_w of the placenta becomes lower than that of the seed. Presumably, this occurs because of conversion of osmotically active constituents into inactive forms such as starch and lipids. The differences in water content between root or stem tips and older parts of the stem, between heartwood and sapwood, and between old and young leaves are well-known examples of the variation in water content in a plant. However, readers are reminded that these structures must have fairly similar water potentials, at least at night.

Competition for Water among Organs.   Because of differences in stage of growth and exposure, different parts of plants lose water at different rates. Because of internal resistances to water flow and differences in osmotic potential, differences in water potential also develop, leading eventually to redistribution of water within plants. Upper leaves are more exposed than lower leaves and transpire more rapidly, lowering their Ψ_w and causing water to move to them from the shaded lower leaves. Both young leaves and growing fruits compete for water at the expense of older leaves, which often die prematurely from desiccation on water-stressed plants. Stem tips of some kinds of plants even continue to elongate while the older leaves are wilting (Slatyer, 1957; Wilson, 1948). Some data on the water relations of stem apices and leaves of stressed wheat are presented by Barlow *et al.* (1980).

Leaves compete successfully with older fruits, which often show reduced growth or even shrinkage at midday, when transpiration is rapid (Kozlowski, 1968). Although young oranges (Rokach, 1953) and young cotton bolls (Ander-

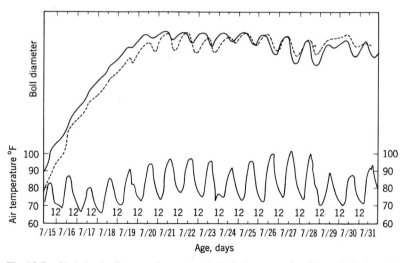

Fig. 12.7.   Variation in diameter of two cotton bolls during maturation. Young bolls do not lose water and shrink during midday water deficits, but older bolls show the midday shrinkage often observed in other kinds of fruits. Rain prevented shrinkage on July 29. (From Anderson and Kerr, 1943.)

TABLE 12.1   Changes in Water Content of Leaves During a Day[a]

Plant	Water content as % of fresh weight		Decrease in H₂O, assuming no change in dry weight		Water content as % of dry weight	
	Max.	Min.	g/100 g fresh weight	% of original H₂O	Max.	Min.
Amaranthus	86	79	33.4	38.8	614	376
Nicotiana	85	80	25.0	29.4	567	400
Physalis	90	87	23.1	25.3	900	669
Euphorbia	85	81	21.0	24.7	567	426
Helianthus	83	78	23.0	27.7	500	365
Zea mays	72	67	15.2	21.0	253	203

[a] Adapted from Curtis and Clark (1950) and other sources.

son and Kerr, 1943) do not lose water during periods of rapid transpiration, older fruits show daily decreases in diameter (Fig. 12.7).

Bases for Expressing Water Content.   The rapid variations in water content and even in dry weight that occur in leaves and other plant organs create sampling problems in connection with the determination of water content. The fresh weight basis is unsatisfactory because the water content often varies greatly within a few hours, especially in leaves (Table 12.1). Unfortunately, dry weight is not stable either, because photosynthesis, respiration, and translocation can cause significant short-term changes in dry weight, whereas long-term increase in leaf dry weight can indicate that the water content is decreasing when no change is occurring (Ackley, 1954; see later Fig. 12.19). This problem will be discussed in more detail in the section on measurement of plant water deficits.

EFFECTS OF WATER DEFICITS

Water deficits affect every aspect of plant growth, including the anatomy, morphology, physiology, an biochemistry. Some major effects will be discussed in this section, but there is more information available than can be assimilated and summarized. Also, some topics are discussed in other chapters. Readers are referred to papers by Hsiao (1973), Slatyer (1969), Begg and Turner (1976), Turner and Begg (1978, 1981), Parsons (1982), papers in Part 4 of Lange *et al.* (1976) and in Taylor *et al.* (1982), and the six volumes of "Water Deficits and Plant Growth," edited by Kozlowski (1968–1981), for more information.

General Effects

The most obvious general effects of water stress are reduction in plant size, leaf area, and crop yield.

Slatyer (1969) pointed out that most studies of the effects of water stress concentrated either on developmental processes or on metabolic processes such as photosynthesis, but research would be much more productive if physiological processes were studied at various stages of development. Furthermore, the degree of water stress has not always been monitored quantitatively, although this is improving. It might be added that water stress is sometimes developed too rapidly under experimental conditions, producing results that are not comparable to those obtained in the field, where severe stress usually develops more gradually (Turner and Begg, 1978, pp. 62–63).

Research on water stress would be more productive if the projects were more comprehensive, including measurements of growth, water status, and physiological and biochemical processes, as well as the relevant soil and atmospheric factors. Unfortunately, financial and bureaucratic limitations make it difficult to organize the broad-scale, interdisciplinary research projects that would be most productive.

Water Stress in Relation to Ontogeny

The amount of injury caused by water stress depends to a considerable extent on the stage of plant development at which it occurs. This is particularly true of annual plants in which the effects of a short period of stress at a critical stage of development are more likely to be important than in perennial plants. The life cycle can conveniently be divided into three stages: seed germination and seedling establishment, vegetative growth, and the reproductive stage. C. T. Gates (1968) discussed the effects of water stress on herbaceous plants in detail, and we will give only a few examples and some generalizations.

Seed Germination and Seedling Establishment. Many seeds are tolerant of dehydration and can be kept in a dry condition for years. The embryos of seeds of wheat and some grasses and desert plants can survive repeated wetting and drying without injury, but changes in moisture content are injurious to others. Russian investigators claim that wetting and redrying before planting increase the drought tolerance of several species (Henckel, 1964), but Salim and Todd (1968) found no consistent evidence of this. After the roots appear and cells become vacuolated, the tissue is usually much more susceptible to injury from dehydration. Under field conditions, seed germination and establishment are often inhibited by soil water deficits, resulting in poor stands.

Vegetative Growth.   Vegetative growth, in general, and leaf expansion, in particular, are severely inhibited by relatively moderate water stress (e.g., see Figs. 12.9 and 12.10). This will be discussed in detail in the section on cell division and enlargement.

Reproductive Growth.   The reproductive stage of plant growth is particularly sensitive to water stress. The effects of water deficit on wheat and other grains have been studied all over the world, but it is difficult to compare results because of differences in timing and degree of stress. However, it seems safe to state that in general, water stress during initiation of flower primordia and anthesis is especially injurious to wheat (Fischer, 1973; Sionit *et al.*, 1980b) and some other plants. Water stress is said to increase the percentage of cleistogamous flowers on *Stipa leucotricha* (Brown, 1952). According to Moss and Downey (1971), water stress reduces female expression in corn by causing abortion of embryo sacs, delay in silking, and development of male flowers on ears, but Damptey and Aspinall (1976) reported that water stress increased the number of ears. Loss of synchrony in development of silks and tassels, caused by delay in silking is particularly damaging in corn (Castleberry, 1983). Subjecting soybeans to water stress during flower induction shortens the flowering period and causes flower abortion, whereas stress during pod filling reduces seed number and weight (Sionit and Kramer, 1977). Water stress during the growing season reduces the germinability of seeds of some varieties of peanuts (Pallas *et al.*, 1977). Cotton sheds bolls when water-stressed, and some kinds of trees shed fruits, the ''June drop'' of apples being an example. Water stress at the time of bud formation inhibits bud formation of apricots (Uriu, 1964) and probably reduces it in other kinds of trees. In contrast, Alvim (1960) reported that coffee trees must be subjected to water stress before flowering can be induced by rain or irrigation. Kaufmann (1972) reviewed the literature on effects of water deficits on reproductive growth.

Effects of Water Deficits on Plant Growth

The quantity and quality of plant growth depend on cell division, enlargement, and differentiation, and all are affected by water deficits, but not necessarily to the same extent (Barlow *et al.*, 1980). The most common effect of water deficit is reduction in plant size and yield, and we will discuss some of the causes of this reduction.

Cell Division and Enlargement. Both cell division and enlargement are reduced by water deficit, but there is some uncertainty concerning the relative sensitivity of the two processes. Gardner and Nieman (1964) concluded from measurement of DNA content of radish cotyledons that some cell division continued at water potentials of −0.4 to −1.6 MPa (Fig. 12.8). However, Kirkham *et al.* (1972) concluded that cell division in radish cotyledons was inhibited at a higher turgor than cell enlargement, and McCree and Davis (1974) found that cell division and cell enlargement were equally important in determining the size of leaves on sorghum stressed to various degrees (Fig. 12.9). Hsiao (1973, pp. 540–541) concluded that in general, cell division is inhibited less than cell enlargement. In some instances, cell division may be checked by failure of cells to enlarge, because at least some kinds of cells apparently must attain a certain size before they can divide (Doley and Leyton, 1968). It seems that the effect of water deficit on cell division deserves more investigation.

Although cell and leaf expansion clearly depend on the existence of some minimum turgor pressure, the relationship is rather complex, depending in part on the age of the tissue and its past history. For example, Bunce (1977) found that over a period of 72 hr there was a linear relationship between elongation rate and turgor in leaves of field and growth chamber plants of soybean. However, leaves of plants in the drier field environment required less turgor for elongation than those grown in the more humid growth chamber. Bunce also found that mild water stress increased leaf area by increasing the number of epidermal cells. Severe stress also increased the number of epidermal cells but reduced cell enlargement so much that leaf area was not increased. Wenkert *et al.* (1978) concluded that in the field, within the usual daily range of water deficits, mean

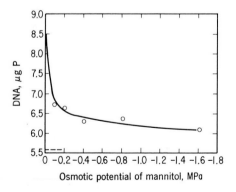

Fig. 12.8. DNA content of detached cotyledonary radish leaves grown for 24 hr on media of various osmotic potentials. The dashed line at the lower left indicates DNA content at the beginning of the experiment. Although DNA synthesis was materially reduced at −0.1 MPa, measurable synthesis occurred at −1.6 MPa, indicating that some cell division was occurring. (From Gardner and Nieman, 1964.)

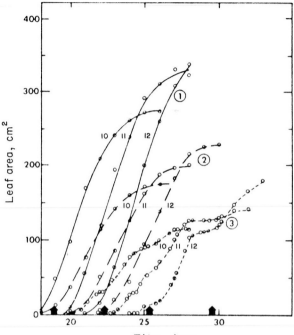

Fig. 12.9. Effect of water stress on increase in area of leaves 10, 11, and 12 of sorghum. Numbers in circles identify the experiments, and the large arrows at the bottom indicate times of watering in experiment 3. Plants in experiment 1 were watered daily and kept in warm, humid air; those in experiment 2 were also watered daily, but the tops were kept at 41°C. Plants in experiment 3 were in hot, dry air, and the soil was allowed to dry to −0.4 MPa five times from the 10th to the 34th day. Afternoon leaf Ψ_w values were −0.8 MPa in experiment 1, −0.9 in experiment 2, and −1.7 in experiment 3. Growth rates during the linear phase were 62, 37, and 26 cm²/day. (From McCree and Davis, 1974.)

leaf growth over long periods is not limited primarily by turgor pressure, but by metabolic processes. In both investigations, it was found that reduction in growth rate caused by water deficits was made up by rapid "compensatory" growth when leaves regained turgor (see Fig. 12.11). Watts (1974) pointed out that in grasses leaf elongation is controlled by the water status in the embryonic region at the base of the leaves, which may be quite different from that in the older, more exposed central and terminal regions. This situation should be kept in mind in research on water relations of corn, sorghum, and other grasses. Barlow *et al.* (1981) also pointed out that in graminaceous plants leaf expansion is not necessarily related to bulk leaf turgor. This problem does not exist in dicot leaves, but there are sometimes differences in the water status of different parts of large dicot leaves, probably chiefly because of differences in exposure of various regions of the blades. Differences in reaction between dicot and monocot leaves to daytime water stress is discussed by Radin (1982).

Hsiao (1973) and Turner and Begg (1978) give many examples of reduction in cell enlargement and vegetative growth caused by water stress. Leaf enlargement is often reduced or stopped before photosynthesis is much reduced (Fig. 12.10). New leaves develop more slowly (increased plastochron), and old leaves senesce more rapidly. Thus, there is likely to be a serious reduction in the photosynthetic area of stressed plants as well as a reduction in rate of photosynthesis per unit of leaf area. Bunce (1978b) reviewed the differing views concerning the importance of reduced leaf area relative to decreased rate of photosynthesis in reducing plant growth. In his experiments with cotton and soybean, low humidity (40% at 23°C) did not decrease CO_2 uptake even though stomatal resistance increased. Plant dry weight was reduced only in experiments in which leaf area was reduced.

In some instances, especially if water stress develops rather slowly, sufficient osmotic adjustment may occur to enable growth to continue at a lower water potential than would be possible otherwise. Barlow *et al.* (1980) suggested that osmotic adjustment might be common in young, relatively unexpanded tissue such as that of root and stem tips, and Sharp and Davies (1979) found evidence of osmotic adjustment in roots of several species. Munns and Weir (1981) reported that osmotic adjustment occurs in both expanding and fully expanded regions of wheat leaves.

In some experiments, when water becomes available after a short period of stress, growth is very rapid for a short time, so that no net reduction is caused by the stress. This has been observed in *Nitella* cells (Green *et al.*, 1971), corn plants (Acevedo *et al.*, 1971), sugar beets (Owen and Watson, 1956), tomato (Gates, 1955), alfalfa, barley, and corn (Kemper *et al.*, 1961), and stems of pine seedlings (Miller, 1965). An example is shown in Fig. 12.11. It is sometimes

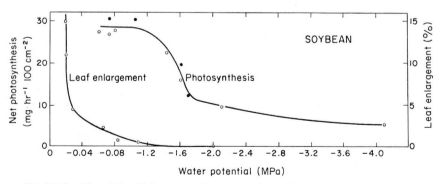

Fig. 12.10. The relationship between leaf water potential, leaf elongation, and photosynthesis in soybean. Leaf elongation practically ceased before there was much reduction in photosynthesis. See also Fig. 1.4. (From Boyer, 1970a.)

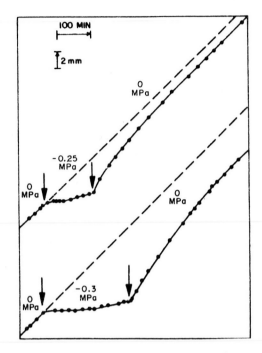

Fig. 12.11. Rapid resumption of elongation by maize leaves after temporary inhibition by transfer of roots from nutrient solution to a polyethylene glycol solution with an osmotic potential of −0.25 or −0.3 MPa. (From Acevedo *et al.*, 1971.)

termed stored or compensatory growth, because it is suggested that metabolites accumulated during the period when cell enlargement is inhibited by lack of turgor are available for cell wall synthesis and other processes associated with growth after turgor is restored. Kleinendorst (1975) attributes it to a greater reduction of cell enlargement than of cell division in moderately stressed plants. If stress is prolonged, cell walls may become too rigid to resume expansion when turgor increases. Both Boyer (1970a) and Acevedo *et al.* (1971) reported that if plants are stressed severely or for more than a few hours, the rate of growth does not return to normal. However, Miller (1965) reported that compensatory growth in loblolly pine continued for several days after rewatering. The literature on the mechanics of cell growth and of readjustment after a period of stress is reviewed by Hsiao (1973, pp. 535–539). Some relevant material is also given in Chapter 2.

Differentiation. There are two aspects of differentiation: that of vegetative structures and that of reproductive structures. Water stress not only reduces leaf area but often increases leaf thickness, thereby increasing the weight per unit area or specific leaf weight. Also, it is often accompanied by increase in the thickness of cutin and sometimes in the amount of wax on the leaf surfaces. In addition, Ehleringer (1980) reported that water stress increases the production of hairs on leaves of *Encelia farinosa,* and this probably is true of some other plants. Light also has important effects on leaf structure (Nobel, 1980).

The effects of water stress on reproductive differentiation are even greater. Spike elongation and spikelet formation of wheat are inhibited by water stress, and if it occurs at early anthesis, flowers are injured and the number and size of seeds are reduced (Fischer, 1973; Sionit *et al.,* 1980b). Stress during flower induction and at flowering of soybeans shortens the flowering period and causes flower abortion (Sionit and Kramer, 1977). Stress did not affect development of flower primordia in sunflowers, but it increased the time from seed germination to flowering (Yegappan *et al.,* 1980).

It has been reported that flower bud differentiation and flowering are sometimes accompanied by perturbations in stomatal conductance, transpiration, and plant water stress. The literature on this topic was reviewed by Longstreth and Kramer (1980). They could find no evidence from their experiments of any change in stomatal conductance, leaf water potential, or plant water stress at flower induction or flowering of either cocklebur or soybean growing in a controlled environment.

Root–Shoot Ratios. Water deficits not only reduce the dry matter production of plants and stands of plants but also change the partitioning of carbohydrates among organs. In general, shoot growth is reduced more than root growth because more severe water deficits develop in the transpiring shoots and probably persist longer, although there seem to be few data on the latter point. Thus, root–shoot ratios are generally increased by water stress, although the absolute weight of roots usually decreases. Also, the root–shoot ratio is generally supposed to be larger for plants of arid regions than for plants of humid regions, but there seems to be some uncertainty about this (Kummerow, 1980, p. 60). Occasionally, there is an absolute increase in the amount of roots formed on mildly stressed plants, possibly because there is more effective osmotic adjustment in roots than in shoots (Sharp and Davies, 1979). An example of this unusual occurrence is shown in Fig. 12.12. It seems likely that it occurs when mild water stress inhibits shoot growth more than it inhibits photosynthesis, making a surplus of carbohydrate available for root growth.

Physiological Effects of Water Stress

Under this heading we will consider effects of water stress on a few important processes, including photosynthesis, dark respiration, translocation, and the par-

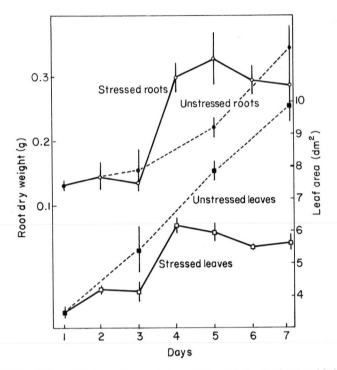

Fig. 12.12.   Effect of drying soil on root dry weight and leaf area of maize. Moderate water stress stopped increase in leaf area but resulted in increased root growth. (From Sharpe and Davies, 1979.)

titioning of metabolites. Stomatal closure was dealt with in Chapter 11, but it will be discussed briefly here in connection with photosynthesis.

Photosynthesis.   Water deficits can reduce photosynthesis by reduction in leaf area, closure of stomata, and decrease in the efficiency of the carbon fixation process. Reduction in leaf area by water stress is an important cause of reduced crop yields because the reduced photosynthetic surface persists after the stress is relieved. Reduction in rate of photosynthesis in water-stressed plants has usually been attributed chiefly to stomatal closure, but this is now being questioned. It is true that rates of water loss and CO_2 uptake often decrease at the same rate in plants subjected to increasing water stress, as shown in Fig. 12.13A (Barrs, 1968a; Brix, 1962). After porometers became available, it was found that stomatal conductance, transpiration, and photosynthesis often decrease to similar degrees, as shown in Fig. 12.13B. However, there is increasing evidence that water stress severe enough to cause stomatal closure simultaneously causes inhibition of CO_2 fixation because of injury to the photosynthetic machinery. It also has

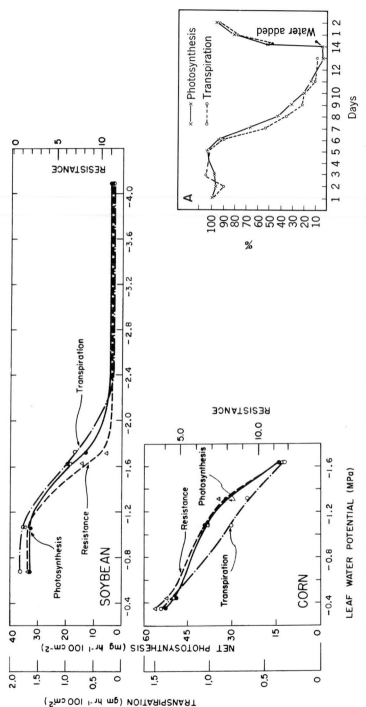

Fig. 12.13. (A) Changes in rates of transpiration and photosynthesis of loblolly pine seedlings in gradually drying soil and after rewatering. The small increase in rates before the decrease began was attributed by Stälfelt to wider opening of stomata accompanying a slight loss in turgor. Rates are given as percentages of rates in soil at field capacity. (B) Changes in net photosynthesis, transpiration, and diffusive resistance in corn and soybean with decreasing leaf water potential. The data here and in (A) suggest that the decrease in photosynthesis can be explained by stomatal closure, but other works cited in this chapter indicate that the photosynthetic machinery is also being inhibited. [(A) From Brix, 1962; (B) from Boyer, 1970b.]

been suggested that accumulation of carbohydrates when growth is reduced by water stress may cause reduction in photosynthesis, but Bunce (1982) found no evidence of this.

Keck and Boyer (1974) reported that both cyclic and noncyclic photophosphorylation and electron transport of isolated sunflower chloroplasts are reduced by a water stress of 1.0 or 1.1 MPa. Boyer (1976) suggested that with increasing water stress, the limitation on carbon fixation might shift from electron transport to photophosphorylation. Mooney et al. (1977) concluded that nonstomatal inhibition of photosynthesis was more important than stomatal inhibition in water-stressed Larrea. Farquhar and Sharkey (1982) concluded that although stomatal closure limits water loss, it limits CO_2 fixation only marginally and indirectly because the latter is being limited by the same factors causing stomatal closure. This rather extreme view must be qualified by the fact that closure of stomata caused by low atmospheric humidity or added ABA causes direct reduction in photosynthesis. The nonstomatal effects of water stress on photosynthesis are discussed by Osmond et al. (1980, pp. 355–367).

Water deficits also affect some of the enzyme-mediated steps of the dark reaction of photosynthesis. Several investigators reported that the activity of such important enzymes as ribulose-1,5-bisphosphate carboxylase, ribulose-5-phosphate kinase, and phosphoenolpyruvate carboxylase are reduced by water stress. However, Björkman et al. (1980, p. 155) found the ribulose bisphosphate carboxylase concentration of Nerium oleander little affected by water stress, although photosynthetic electron transport and photophosphorylation were significantly reduced. Mayoral et al. (1981) reported that the carboxylating enzymes of a wild wheat remained active to a much lower water potential than those of cultivated wheat.

There are numerous other evidences of injury by water stress to the photosynthetic apparatus. One is the decrease in the fluorescence transient observed by Govindjee et al. (1980). Björkman et al. (1981) reported that water stress, increases the sensitivity of the photosynthetic mechanicm in Nerium leaves to photoinhibition. They speculated that this might be the principal means by which water stress damages the photosynthetic apparatus, but other kinds of damage also occur. This also is discussed in Chapter 13 in the section on recycling of CO_2. Water potentials lower than −0.5 MPa retard the development of chlorophyll in jackbean by reducing the rate of formation of the chlorophyll a/b protein and retarding accumulation of chlorophyll b (Alberte et al., 1975). Bhardwaj and Singhal (1981) report a similar effect on greening of barley. Chlorophyll destruction is hastened by water stress. In corn, a C_4 plant, most of the chlorophyll loss occurs from the mesophyll cells and little from the bundle sheath chloroplasts (Alberte et al., 1977). The loss consists of the lamellar, light-harvesting chlorophyll a/b protein complex, a major component of chloroplast membranes, which seems to be a specific target of water stress.

Vieira da Silva *et al.* (1974) found that when cotton leaf disks were floated on polyethylene glycol solution with a water potential of -1.0 MPa there was a sharp increase in acid phosphatase and alkaline lipase in the chloroplasts and serious disorganization of fine structure (see Fig. 12.14). Disorganization of chloroplasts has been reported by other investigators (Giles *et al.*, 1976). The loss of chlorophyll *a/b* protein and disorganization of fine structure seem to provide adequate explanation for the decrease in electron transport and the reduction in quantum yield reported in the literature (Mohanty and Boyer, 1976). Fellows and Boyer (1978) observed breakage of the plasma membrane and tonoplast in stressed cells, but little damage to the thylakoid membranes. They regard the severe damage reported by some investigators as an artifact caused by rehydration during fixation and attribute reduction in chloroplast activity to changes in spatial relationships in membranes.

It is difficult to generalize about the effects of water or other stresses on the rate of photosynthesis because there are differences among species and within a species, depending on past treatment. For example, plants previously stressed usually continue to carry on photosynthesis to a lower leaf water potential than

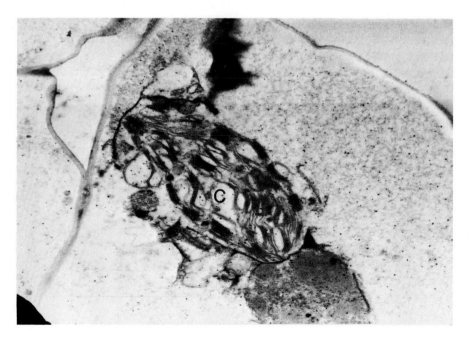

Fig. 12.14. Electron micrograph of leaf cell of *Gossypium hirsutum* after 11 hr of stress in polyethylene glycol solution, showing disorganization of chloroplasts. (Courtesy of J. B. Vieira da Silva, Laboratoire d'Ecologie Générale et Appliquée, Université Paris VII.)

those not previously stressed. Thus there often are significant differences be-tween field- and greenhouse-grown plants. Also, there are important differences among species, especially between mesophytes and xerophytes. Some of these differences may result from differences in leaf thickness, resulting in differences in the amount of photosynthetic apparatus per unit of leaf area (Patterson *et al.*, 1977; Nobel, 1980, for example), but others may result from differences in stress tolerance of the photosynthetic apparatus itself.

Dark Respiration.   The early studies on water stress and respiration were somewhat contradictory, at least partly because the plant water stress was not measured. More recent research indicates that in general the rate of respiration decreases with increasing water stress, as shown in Fig. 12.15. However, it is notable that the rate of soybean respiration showed no further decrease when the leaf water potential decreased from -1.6 to -4.0 MPa. This suggests that the enzyme systems involved are relatively tolerant of dehydration. The greater reduction in photosynthesis than in respiration is likely to decrease carbohydrate reserves. Lawlor (1979) and Osmond (1980) discussed the effects of water stress on carbon metabolism in detail.

Occasionally, a temporary increase in respiration occurs with decreasing water content. Brix (1962) and Parker (1952) observed this in conifers and Mooney (1969) in Mediterranean broadleaf evergreens. An example of this unusual be-havior is shown in Fig. 12.16. One explanation is that water stress causes hydrolysis of starch to sugar, providing more substrate for respiration. The problem deserves more research.

Translocation.   It has been demonstrated that translocation of ^{14}C-labeled photosynthate out of leaves is reduced by leaf water deficit (Hartt, 1967;

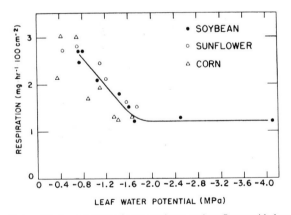

Fig. 12.15.   Rates of dark respiration of corn, soybean, and sunflower with decreasing leaf water potential. (From Boyer, 1970a.)

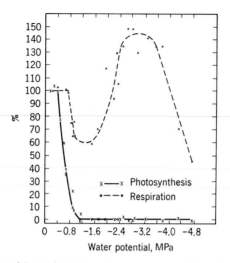

Fig. 12.16. Effect of increasing water stress on rates of photosynthesis and respiration of loblolly pine seedlings in drying soil. Rates are expressed as percentages of rates when seedlings were in soil at field capacity. (From Brix, 1962.)

Roberts, 1964; Wiebe and Wirheim 1962). Water deficits also reduce the translocation of herbicides out of leaves (Basler *et al.*, 1961; Pallas and Williams, 1962). Bunce (1982) reported that daytime translocation of carbohydrate out of leaves was reduced and nighttime translocation increased in water-stressed corn and soybeans. However, the translocation system seems surprisingly resistant to dehydration. Sung and Krieg (1979) found translocation in cotton and sorghum more resistant to leaf water deficit than photosynthesis, and neither process was completely stopped at −2.7 MPa. The velocity of translocation in wheat and *Lolium temulentum* is not reduced by water potentials as low as −2.0 or −3.0 MPa, although the amount of material translocated is reduced. Thus the translocation system seems to be operating at a water potential that severely inhibits photosynthesis in crop plants. Perhaps the amount of material translocated is limited by the reduction in photosynthesis, or by the reduction in sink size caused by decrease in cell enlargement. Wardlaw's (1968) conclusion still seems valid. He stated that reduction of translocation in water-stressed plants is caused more by reduced source or sink activity than by direct effects on the capacity of the conducting system to function. More research is needed to separate the direct effects of water stress on the translocation process itself from indirect effects on sink and source strength and even on the loading and unloading processes.

Partitioning of Photosynthate. The preceding statement concerning the control of translocation serves to introduce the too often neglected problem of partitioning. In the long run the manner in which the products of photosynthesis are partitioned among the various organs of plants determines their survival and their economic value. By selection man has developed plants that have a high harvest index, i.e., plants that convert a larger proportion of their food into economically valuable structures such as seeds, fruits, leaves, roots, or tubers than did their wild ancestors. Although the harvest index is said to be fairly stable (Sinclair *et al.,* 1981), water stress sometimes changes the pattern of partitioning of photosynthate at the expense of the quantity or quality of economic yield. It has already been mentioned that if it occurs during seed filling water stress reduces the number and size of seed of soybean, and it sometimes causes abortion of fruits and seeds. It also decreases leaf size and often results in changes in leaf structure and in root–shoot ratio. Perhaps the most important contribution that could be made toward increasing plant productivity would be sufficient understanding of the control of partitioning so more photosynthate could be channeled into economically important sinks such as seeds and fruits.

Ion Uptake.   The information on ion uptake and translocation in water-stressed plants is rather limited. It was stated in Chapter 2 that ion uptake by some algal cells seems to be controlled by cell turgor, but there is no clear evidence of the operation of this mechanism in cells of seed plants. It is unlikely that moderate water stress has any serious direct effect on movement of ions from cell to cell. However, because long distance movement from roots to shoots occurs in the transpiration stream, reduction in rate of ion transport might occur in plants sufficiently stressed to reduce transpiration. Absorption may be reduced because the movement of minerals is slow in drying soil, root extension is decreased, and suberization decreases root permeability. Slatyer (1969, p. 61) concluded that limitation of mineral nutrient supply may sometimes limit growth of stressed plants. However, after a lengthy review of the literature, Viets (1972) concluded that mineral deficiency is seldom a cause of reduced growth in water-stressed plants. The only exception might be boron deficiency on soils already deficient in that element. He suggests that most plants contain sufficient reserves of minerals to carry them through ordinary droughts, especially since vegetative growth is reduced or even stopped. Pitman (in Paleg and Aspinall, 1981) reviewed this topic and stated that it is difficult to determine how much of the reduction in ion uptake is caused by reduced growth and how much is the direct effect of drying soil and stressed roots.

Biochemical Effects of Water Stress

This section deals chiefly with enzyme-mediated processes, such as carbohydrate and nitrogen metabolism, and growth regulators. Papers on enzyme activity are too numerous to cite. Furthermore, it is often difficult to distinguish between direct effects on enzyme activity and indirect effects resulting from decreased synthesis of the enzyme, increased degradation, or decreased use of endproducts (Hanson and Hitz, 1982). An extreme example of complicated interrelations may be the loss of nitrate reductase activity in water-stressed corn leaves, caused by decreased flux of nitrate to the leaves as the rate of transpiration decreases.

Carbohydrate Metabolism. It has been known since the early part of this century that water deficit often causes a decrease in starch content and sometimes an increase in sugar. It was reported by Spoehr and Milner (1939) and by Eaton and Ergle (1948) that amylase activity is greatly increased in water-stressed leaves. The latter also reported that cotton leaves allowed to wilt daily contained only one-half as much total carbohydrates and one-third as much starch as well-watered controls. The effects of repeated water stress on the carbohydrate content of tomato are shown in Fig. 12.17. It is likely that if plants are subjected frequently to sufficient water stress to cause significant closure of stomata, they may suffer from insufficient carbohydrates. Actual starvation probably is rare although Adjahossou and Vieira da Silva (1978) reported that it occurs in oil palms in which the stomata are closed for months during the dry season. They also reported that water stress causes conversion of starch to sugar in oil palms.

Nitrogen Metabolism. Protein synthesis is closely related to production of new tissue, which is the principal sink for nitrogen compounds, and it is not surprising that when water stress inhibits growth, nitrogen metabolism is dis-

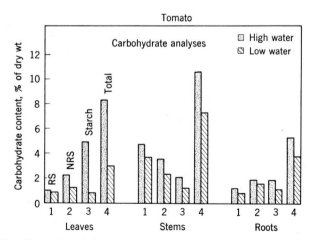

Fig. 12.17. Changes in carbohydrate content of tomato plants growing in soil near field capacity (high water) and plants subjected to four periods of soil water stress during a period of 8 weeks (low water). RS, reducing sugars; NRS, nonreducing sugars. (From Woodhams and Kozlowski, 1954.)

turbed. In general, water stress results in hydrolysis of proteins and accumulation of amino acids, especially proline. The role of proline in drought tolerance is discussed in Chapter 13. It is not always clear how much of the increased amino acid pool found in water-stressed plants is derived from proteolysis and how much represents accumulation resulting from decreased use in formation of new protoplasm. According to Barnett (see Naylor, 1972), some free amino acids are synthesized in stressed tissue, although protein synthesis is reduced. Shah and Loomis (1965) reported that in sugar beet, synthesis of RNA and protein is reduced before wilting occurs, but Gates and Bonner (1959) reported that RNA synthesis continued in water-stressed tomato. However, there was such rapid breakdown that the total RNA decreased. This is consistent with several reports of increased ribonuclease activity in stressed plants, cited by Todd (1972). Changes in enzyme activity also are involved in the accumulation of betaine and proline discussed in Chapter 13.

Growth Regulators. One of the developments in the study of water stress is the increasing attention given to growth regulators. Larson (1964) suggested that the premature shift from earlywood to latewood formation in water-stressed trees occurs because of reduction in auxin formation in tips of twigs. It has also been shown that water deficits caused by various means result in reduced cytokinin production in roots and reduced supply to the shoots. Since the effects of water stress resemble those of senescence and the latter can be postponed by the application of the synthetic cytokinin benzyladenine, some investigators attribute at least part of the effects of water stress to reduction in cytokinin activity in the shoots. The literature dealing with this was reviewed by Livne and Vaadia (1972), but the manner in which growth regulators operate remains speculative (Trewavas, 1981).

Recently, attention has shifted to the role of abscisic acid because its concentration increases rapidly in water-stressed leaves. It was also shown by several investigators that spraying plants with a dilute solution of ABA causes closure of stomata. This led to the conclusion that the increase in ABA in stressed leaves brings about stomatal closure. This was discussed in the section on stomata in Chapter 11. Attempts have been made to extend the role of ABA to root permeability and control of root growth (Davies *et al.*, 1979). It is claimed by some investigators that ABA increases the permeability of roots to water and ions, but the evidence is contradictory. Markhart *et al.* (1979) found that ABA reduced the hydraulic conductance of roots, and Fiscus (1981b) found that it increased ion movement into roots but decreased hydraulic conductance. Mansfield *et al.* (1978) and Davies *et al.* (1979) reviewed the literature on this subject, and Milborrow in Paleg and Aspinall (1981) reviewed the literature on ABA.

Aspinall (1980) reviewed the role of ABA, cytokinin, and ethylene in water-stressed plants. He concluded that although there is strong evidence that ABA acts as a water-deficit sensor and control system during intermittent stress, he doubts if it has any major role in plants subjected to severe or prolonged stress. He also doubts if there is adequate evidence that reduced export of cytokinins from roots of water-stressed plants has a major role in relation to senescence, protein synthesis, or stomatal opening. The increased production of ethylene in stressed leaves might be caused by increase in ABA. The overall role of ABA and of phaseic acid derived from it requires more study, as pointed out by Milborrow in Paleg and Aspinall (1981) and by Trewavas (1981).

Effects of Water Stress on Protoplasm

The physiological and biochemical effects discussed thus far operate in the protoplasm. Thus, there has been a lively interest, especially among European physiologists, in the effects of dehydration on the physical condition of the protoplasm itself, especially permeability and viscosity. Their ideas are set forth in reviews by Iljin (1957), Stocker (1960), and Henckel (1964). Stocker stated that dehydration occurs in two stages: first the reaction phase and then the restitution and hardening phase. The reaction phase is characterized by increased permeability to water, urea, and glycerol, increased proteolysis, and increased respiration. If water stress continues, there is a restitution phase involving increased viscosity, decreased permeability to water and urea, and decreased respiration. This course of events is said to occur during gradual dehydration, and if it does not go too far the processes are reversible. According to Henckel (1964), Russian plant physiologists support this two-stage concept of reaction to water stress. Henckel and his collaborators also seem to place emphasis on elasticity of the protoplasm as important for tolerance of dehydration. Stadelmann (1971) also emphasized the importance of protoplasmic viscosity and membrane permeability, and Leopold *et al.* (1981) used leakage of solutes as an index of injury to plant cell membranes.

Fine Structure. Stocker and others attributed death of plants from dehydration to disorganization of the fine structure of the protoplasm. Today, the nature of the fine structure is better understood and the points at which injury occurs can be better identified. The disorganization of thylakoid membranes in chloroplasts and destruction of the chlorophyll *a/b* protein have already been mentioned, and the decrease in respiration is correlated with reduction in mitochondrial activity (Koeppe *et al.*, 1973). Fellows and Boyer (1978) reported breakage of the plasma membrane and tonoplast and loss of organelles in cells of sunflower leaves, beginning at a water potential of -1.5 MPa and becoming increasingly irreversible below -2.0 MPa. Today there is a tendency to attribute injury from dehydration, chilling, and heat to damage to membrane structure. This includes the internal membrane structure of organelles such as chloroplasts and mitochondria, as well as the membranes of the protoplast, the plasma membrane and the tonoplast. Examples are the research on dehydration and heat tolerance at the Carnegie Laboratory (Mooney et al., 1977; Björkman et al., 1980) and the book edited by Lyons et al. (1979) dealing with effects of chilling on cell membranes.

Poljakoff-Mayber (in Paleg and Aspinall, 1981) reviewed the effects of water stress on fine structure of cells and cell organelles. She concluded that the chief visible effects are rearrangement of chromatin in large masses around the nucleoli and displacement of lipids from membranes, forming droplets in the cytoplasm. She also cited work indicating that structural changes occurring during the dehydration of maturing seeds are similar to those occurring in other plant tissues undergoing dehydration. It is interesting and puzzling to consider that although these changes are reversible in seeds when germination occurs, they generally are irreversible in the vegetative tissues of the plants that produce the seeds. This situation seems to present an opportunity for research on the factors responsible for tolerance of dehydration by protoplasm.

A General Theory of Stress Response. Russian investigators here tended to emphasize the similarity of damage from dehydration, heat, and cold (Genkel, 1980; Henckel, 1964; Maximov, 1929), and Levitt (1980) devoted a chapter to discussion of generalized stress responses. Parker (1972) also discussed the similarities between dehydration and heat injury. According to Castleberry (1983), corn breeders seldom attempt to separate heat and drought tolerance, probably because drought usually is accompanied by high temperature. It was mentioned in the preceding section that there is a tendency to attribute injury from dehydration, chilling, and high temperature to damage to membranes. This may include effects on both protein and lipid constituents. In addition, Henckel (1964) cited Russian work indicating that the high temperature accompanying drought causes release of ammonia from decomposition of proteins that injures plant tissue. Incidentally, it is well established that plants normally release small amounts of amonia (Weiland and Stuttle, 1980).

Some investigators have reported that development of water stress also increases freezing tolerances (Cloutier and Siminovitch, 1982) or at least accompanies it (Parsons, 1978). However, other investigators claim that development of freezing tolerance is independent of dehydration tolerance (Obloj and Kacperska, 1981, for example). Steponkus (1980) discussed the possibility of a unified theory of stress injury in detail. He agreed that drought, freezing, and high salinity all cause injury by dehydration, but doubted if this supports a unified theory of injury because each type of environmental stress probably produces specific cellular stresses in addition to dehydration. Furthermore differences are to be expected among different kinds of protoplasm, resulting in differences among species and cultivars.

Water Stress in Relation to Disease and Insect Attacks

In considering the role of water in relation to plant diseases it is easier to find examples of the effects of pathogens on plant water relations than to find examples of effects of plant water stress on susceptibility to attack by pathogens. Parker (1965) and Durbin (1978) treat some of the effects of water stress as physiological or abiotic disease. Sometimes water stress causes premature shedding, cracking, or pitting of young fruits or internal necrosis such as bitter pit of apple. Blossom end rot of tomato and blackheart of celery and head lettuce seem to involve an interaction of water stress and calcium deficiency (Durbin, 1978, p. 113; Tibbits, 1979).

Ayres (1978) reviewed the water relations of diseased plants in detail and pointed out that injury to root systems, blockage of xylem, disturbance of normal stomatal functioning, and damage to the cuticle occur. Some of these effects have been discussed elsewhere in this book.

It is difficult to separate the direct effects of an excess or deficit of water in the environment on susceptibility to disease from the indirect effects caused by change in water status. For example, root systems in wet soil are often damaged by fungi, and humid air is favorable for infection of leaves by spores of a number of pathogens (Yarwood, 1978). It is generally believed that water-stressed plants are more susceptible than unstressed plants to attacks by pathogens, and Schoeneweiss (1978) gives a number of examples. However, he concluded that until the nature of the host defense mechanism is better understood, it would be impossible to explain why water-stressed plants are more susceptible to attack. Various aspects of water and plant disease are discussed in Volume V of ''Water Deficits and Plant Growth,'' edited by Kozlowski (1978).

Host–Parasite Relationships

The osmotic relations of parasites and their hosts were studied by Harrison and co-workers. The osmotic potential of sap expressed from plants of Jamaican Loranthaceae was usually lower than that of the host plants (Harris and Lawrence, 1916). Later, Harris and Harrison (1930) found that the osmotic potential of sap from the desert mistletoe, *Phoradendron,* was usually lower (-2.7 MPa) than that of its hosts (-2.0 to -2.3 MPa). They also suggested that the parasite seldom occurs on creosote bush because its osmotic potential is too low. Mark and Reid (1971) found that the xylem water potential of dwarf mistletoe was consistently lower than that of the lodgepole pine branches on which it was growing. Host–parasite water relations probably deserve more study.

Insect Attacks.   There is some evidence that attacks on conifers by boring insects that live in the inner bark and outer wood are more severe in dry seasons than in seasons when little water stress develops. Vité (1961) stated that infestation of ponderosa pine by beetles is more severe in trees suffering from water deficits than in well-watered trees, and Ferrell (1978) reported that engraver beetles cannot establish themselves unless the water potential is lower than -1.5 MPa.

The susceptibility of loblolly pine trees to attack by bark beetles is influenced by the duration and severity of water stress (Lorio and Hodges, 1968). Probably the high oleoresin pressures of well-watered trees are unfavorable to the establishment of beetles. The interaction between insect attacks and plant water status deserves further study.

Beneficial Effects of Water Stress

Water stress is not always injurious. Although it reduces vegetative growth, it sometimes improves the quality of plant products. As shown in Fig. 12.18, although mild water stress reduces the fresh weight of guayule, the rubber content is increased significantly. Water stress increases the desirable aromatic properties of Turkish tobacco (Wolf, 1962), but it also increases the nitrogen and nicotine content, which is undesirable (van Bavel, 1953). Water deficit is said to increase the alkaloid content of *Atropa belladonna, Hyoscyamus muticus,* and *Datura* (Evenari, 1960), but it was reported to reduce the alkaloid content of *Cinchona ledgeriana.* Evenari (1960) also stated that water stress increases the oil content of mint and olive, and Miller and Beard (1967) reported that it increased the percentage of oil in soybean but decreased the yield per acre. However, Sionit and Kramer (1977) found no effect of stress on either oil or protein content of soybean seed. Moderate water stress is said to improve the quality of apples, pears, peaches, and prunes (Richards and Wadleigh, 1952), and the protein content of hard wheat is said to be increased by moderate water stress. Kaufmann (1972) discussed some effects of water stress on the quality of vegetables and fruit, and pointed out that an excess of water can cause cracking and splitting of fruits.

A moderate degree of water stress is beneficial to seedlings if applied before lifting for transplanting, and the root systems of forest tree seedlings are often

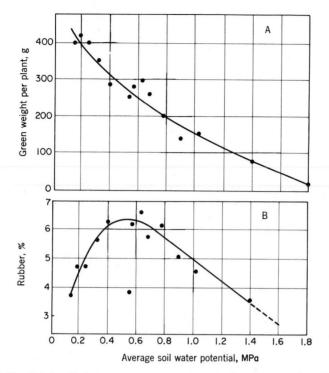

Fig. 12.18. Relationship between average water stress developed between irrigations and (A) the fresh weight of guayule plants and (B) the percentage of rubber in the plants. The yield of rubber per plant is about the same at average soil water potential of −0.2 as at −0.5 MPa, although fresh weight is much reduced. (From Wadleigh *et al.*, 1946.)

mechanically pruned in the seed bed in order to provide compact root systems suitable for transplanting (Rook, 1973). The Russians recommend presowing hardening of wheat by soaking seeds for 2 days and then air-drying them. This is claimed to increase the hydrophilic properties of protoplasmic colloids and to produce seedlings that are more drought-tolerant (Henckel, 1964). However, attempts to repeat these experiments in England (May *et al.*, 1962) and in the United States (Salim and Todd, 1968) showed no definite increase in dehydration tolerance.

There have been several reports indicating that water-stressed plants are injured less than unstressed plants by exposure to air pollutants such as ozone and SO_2. This is probably because closure of stomata in water-stressed plants hinders entrance of pollutants. This topic was reviewed by Olszyk and Tibbits (1981) and is discussed in Mudd and Kozlowski (1975).

Although water deficit almost always decreases the yield of vegetative structure such as hay, it sometimes increases the yield of clover and alfalfa seed. In general, it appears that low humidity and moderate soil water stress are favorable for seed production by alfalfa (Abu-Shakra *et al.*, 1969). According to Hagan *et al.* (1959), limiting the irrigation of Ladino clover increases the yield of harvestable seed by preventing premature seed germination. The beneficial effects of moderate water deficits are sometimes the indirect results of decrease in disease or insect injury or even of an increase in pollinating insects. For example, in the corn belt of the United States a rainy summer is said to decrease the yield of red clover seed because it reduces the population of bumblebees essential for pollination. It is also said that high humidity reduces fruit set of tomato because the pollen becomes too sticky to be windborne. The beneficial effects of moderate water stress in increasing tolerance of more severe stress and of transplanting are discussed in Chapter 13.

MEASUREMENT OF PLANT WATER STRESS

Importance of the water status of plants has been recognized for centuries, but until recently it could be described only in such general terms as wilted or unwilted. Lack of satisfactory methods of measuring plant water stress and failure to use the available methods have led to much inconclusive and even contradictory research.

What Should Be Measured

Attempts to estimate plant water stress from measurements of soil water content are useful for some purposes, but they are inadequate for study of the effects of water deficits on physiological processes. As pointed out earlier, leaf water status and transpiration are often better correlated with atmospheric conditions than with soil moisture. The only reliable indicator of the water status of plants is measurements made on the plants themselves. Unfortunately, there has been considerable uncertainty concerning what to measure and how to measure it. Those problems were discussed in detail by Barrs (1968a), Boyer (1969), Slavik (1974), and Turner (1981).

In the first third of this century, plant water status was measured most frequently in terms of the osmotic pressure or osmotic potential of expressed sap (Fitting, 1911; Harris, 1934; Korstian, 1924; Miller, 1938, pp. 39–45). However, questions concerning the validity of measurements on expressed sap began to discourage use of this method. Also, it became apparent by the 1930s that water movement in plants is controlled by what is now termed the water potential (see Chapter 2), but there was no generally applicable method for measuring water potential. Interest therefore shifted toward various measures of water content, such as relative turgidity, relative water content, and saturation deficit, or even stomatal aperture, as indicators of plant water status. Unfortunately, in many experiments no measurements of plant water status were made. Today the technology is available, and the chief problem is to determine what measurement will be most useful for the objectives of the investigator.

A satisfactory method of monitoring plant water status should have most of the following characteristics:

1. There should be a good correlation between rates of physiological processes and the degree of water stress measured by the method.
2. A given degree of water stress measured by the selected method should have similar physiological significance in a wide range of plant materials.
3. The units employed to express water status should be applicable to plant material, soil, and solutions.
4. The method should be as simple, rapid, and inexpensive as possible.
5. It should require a very small amount of plant material for a measurement.

The three principal possibilities are measurement of water potential, osmotic potential, or water content expressed as a percentage of the water content at saturation, the relative water content or saturation deficit of some writers. Measurements of osmotic potential have been used extensively as indicators of plant water status, but they are less satisfactory than measurements of water potential. Osmotic potential varies widely with changes in solute concentration, and an osmotic potential which is low for mesophytes might be high for halophytes. Thus, a given level of osmotic potential has less physiological significance than a given level of water potential. Also, osmotic potential is usually an inadequate indicator of soil water status. However, measurements of osmotic potential are necessary if turgor is to be calculated from the difference between water potential and osmotic potential. We will return to this problem later.

Measurement of water saturation deficit is useful in relation to processes controlled primarily by cell turgor. However, as with osmotic potential, a given water saturation deficit does not have the same physiological significance in different kinds of plants (see later Fig. 12.26). Furthermore, plant water status measured as relative water content or water saturation deficit cannot be related numerically to any measure of soil water status.

In contrast, water potential is a measure of the free energy status of water in plant tissue, soil, and solutions, and it can be related to atmospheric moisture by Eq. (12.3)(see later). Furthermore, water movement into and through plants occurs along gradients of decreasing water potential. Thus, measurements of water potential seem to have maximum utility. However, knowledge of the osmotic and pressure potentials is needed to understand fully plant water relations and should often be added to measurements of water potential (Chapter 2).

Sampling Problems

A major question in monitoring plant water status is to decide where, when, and what to measure. Leaves are most often sampled, but occasionally the water potential of roots (Fiscus, 1972; Slavikova, 1967; Hellkvist *et al.,* 1974; Adeoye and Rawlins, 1981) and even of stem tissue (Kaufmann and Kramer, 1967; Wiebe *et al.,* 1970) is measured. The time of day, location on stem, exposure, and age of leaves affect both their water content and water potential. This makes it essential to use leaves of similar age and exposure when comparing the effects of various treatments on plant water status. Large errors can be introduced by comparing samples of different ages, different exposures, or collected at different times of day. Different areas of large leaves can differ in water status because of unequal exposure to the sun (Rawlins, 1963; Slavik, 1963). Differences also can exist between the fully expanded and the elongating zones of grass leaves. Errors can also occur as a result of careless handling of samples. They should be placed in watertight containers, kept in the shade, and measured as soon as possible after collection.

Water Content

The oldest method of measuring the water status of plants is in terms of water content, expressed as a percentage of either fresh or dry weight.

Fresh Weight.   Although water content is often expressed as a percentage of fresh weight, this has the disadvantage that water content often varies widely from morning to evening, as shown in Fig. 12.6. Furthermore, fresh weight is insensitive to changes in water content. For example, a decrease from 85 to 80% on a fresh weight basis seems small, but it indicates a loss of nearly 30% of the original water content (Table 12.1). This is equivalent to a decrease from 567 to 400% on a dry weight basis, assuming no change in dry weight. Data on changes in water content of several species are given in Table 12.1. The assumption that there is no change in dry weight is unlikely, as will be seen in the next section.

Dry Weight.   Water content, expressed on a dry weight basis, is usually based on dry weight after oven drying at 60°–85°C. Higher temperatures can cause excessive loss of volatile material. Unfortunately, the dry weight of fully expanded leaves undergoes both short- and long-term changes. For example, Miller (1938, p. 564) reported an increase in dry weight of corn leaves per unit of leaf area of over 15% from 7 AM to 5 PM. Over the same time period, the water content decreased by 6.4%. The daytime increase in dry weight consists mostly of carbohydrates which are translocated out overnight and leaf shrinkage can also cause an increase in dry weight per unit of area. Some investigators have attempted to minimize these errors by extracting the soluble and easily hydrolyzable constitutents, leaving only the cell walls and other inert materials. This is termed the residual dry weight. Figure 12.19, derived from Weatherley (1950), shows that even residual dry weight increases over a period of time, creating an apparent change in water content on a dry weight basis, although the actual water content undergoes little change during leaf maturation. The increase in residual dry weight results chiefly from cell wall thickening.

Relative Water Content

Because of the difficulties experienced in using fresh or dry weight as a base, some investigators turned to expression of leaf water content as a percentage of turgid water content. Stocker (1929) placed intact leaves in water in a moist

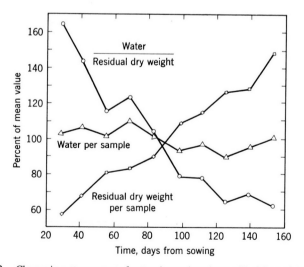

Fig. 12.19. Change in water content of cotton leaves based on residual dry weight. The increase in dry weight per leaf creates the illusion that water content per leaf is decreasing, although it actually remains essentially unchanged. (From Weatherley, 1950.)

chamber until they reached constant weight and calculated what he termed the water deficit as follows:

$$\text{Water deficit} = \frac{\text{turgid weight} - \text{field weight}}{\text{turgid weight} - \text{oven-dry weight}} \times 100 \qquad (12.1)$$

This is sometimes called the water saturation deficit. Although Hewlett and Kramer (1963) obtained good results with entire leaves, there have been complaints that intact leaves take too long to attain equilibrium. Weatherley (1950, 1951) proposed the use of disks of leaf tissue, which can be floated on water, but this introduced errors from infiltration along the cut edges. To eliminate this error, Čatsky (1965) placed the disks on pieces of water-saturated polyurethane foam in a moist chamber. It has been shown that water uptake can be divided into two phases, the first associated chiefly with elimination of the water deficit and the second associated with growth. The first phase requires only a few hours, whereas the second can continue for days, as shown in Fig. 12.20. Barrs and Weatherley (1962) reduced the water uptake time to 4 hr, eliminating the need for duplicate samples to correct for loss of dry weight by respiration. The shorter period also minimizes uptake of water by enlarging cells. Growing tissue does not become fully saturated, and it does not attain zero water potential because continued cell expansion causes water uptake (Boyer, 1969, p. 357). Millar (1966) reported that considerable errors can be caused by measuring water uptake at a temperature very different from the temperature at which the tissue was growing.

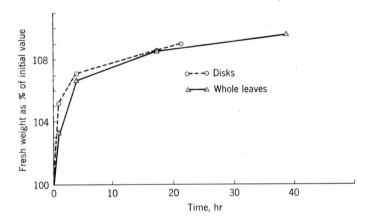

Fig. 12.20. Uptake of water over time by entire leaves and disks cut from leaves. The initial rapid increase is associated with elimination of leaf water deficits; the long, slow uptake is associated with continued growth (cell expansion). (From Barrs and Weatherley, 1962.)

Weatherley calculated what he termed relative turgidity, better termed the relative water content (RWC), by the following equation:

$$RWC = \frac{\text{field weight} - \text{oven-dry weight}}{\text{turgid weight} - \text{oven-dry weight}} \times 100 \qquad (12.2)$$

Some investigators eliminate the time-consuming oven drying and estimate RWC from the ratio of field fresh weight to turgid fresh weight. This is based on the assumption that negligible changes occur in turgid fresh weight and in dry weight between sampling times. This shortcut requires calibration for possible changes in the water content of turgid tissue and leaf dry weight over time (Smart and Bingham, 1974). The occurrence of changes in leaf dry weight was mentioned earlier in this chapter.

Water deficit and RWC are related; RWC = 100 − water deficit, or RWC + water deficit = 100%.

The exact procedure most likely to give a reliable measure of full turgidity varies with the species. Clausen and Kozlowski (1965) and Harms and McGregor (1962) found the use of entire needles satisfactory for several species of conifers. Hewlett and Kramer (1963) found entire leaves more satisfactory than disks for some species. In all procedures, care in sampling and in handling samples is necessary. One source of error is from infiltration of intercellular spaces with water; another is in drying the leaf surfaces before weighing. Measurements should be started as soon after collection as possible.

Measurements of RWC or water deficit form a convenient method for following changes in water content without errors caused by changing dry weight. Unfortunately, a given water deficit or RWC does not represent the same level of water potential in leaves of different species or ages, or from different environments. This is shown later in Figs. 12.25 and 12.26. The data in Fig. 12.25 indicate a decrease in water deficit (increase in RWC) of dogwood leaves at a given leaf water potential as the season progressed, and Zur *et al.* (1981) reported that there was an increase in RWC at a given water potential in maturing soybean leaves. This is probably related to the increasing thickness and decreasing elasticity of cell walls in maturing leaves. Comparisons of values for different kinds of tissue are therefore questionable.

Indirect Measurements of Water Content

Attempts have been made to estimate the water content of leaves and other plant tissues by measurements of electrical conductivity, capacitance, and absorption of radiation (Slavik, 1974, pp. 123–131). The most successful method is use of a so-called beta gauge. A source of beta radiation is placed on one side of the leaf and a radiation detector on the other side. The amount of radiation absorbed by the leaf tissue changes with change in mass per unit of area, and over a short period of time the change in mass (really leaf thickness) depends on change in water content (Mederski, 1961; Mederski and Alles, 1968; Nakayama and Ehrler, 1964). Leaves differ in thickness, so that calibration is required for different kinds of leaves, although Jarvis and Slatyer (1966) described a method that minimizes these difficulties. It and other methods are discussed in Slavik (1974).

Qualitative Estimates.   A number of indirect methods are available to indicate development of water deficits. The simplest is the change in color that occurs because of change in leaf orientation with decreasing turgor. This can be seen in a variety of plants, either by the eye or by photography with infrared film. According to Wenkert (1980), zero turgor in maize leaves is indicated by perceptible loss of sheen and development of dull, pale color long before leaf rolling begins. Leaf rolling and death of leaf tips is said to be a reliable indicator of differences in water stress among rice cultivars (O'Toole and Moya, 1978). Idso *et al.* (1977) cite a number of attempts to use temperature measurements of crop canopies by remote sensing to detect water stress. They developed a formula based on the difference between leaf and air temperatures during head development that correlates well with the yield of durum wheat and can be modified for use by remote sensing.

Premature stomatal closure in the morning is a relatively sensitive indicator of developing water stress because guard cells are very responsive to loss of turgor. Several methods of monitoring stomatal aperture are described by Slatyer and Shmueli (1967), but porometers are used most frequently. They are discussed in Chapter 11. The oleoresin exudation pressure in pine (Lorio and Hodges, 1968) and latex pressure in rubber trees (Buttery and Boatman, 1966) are qualitative indicators of the water status. However, such methods are useful only as indicators of the existence of water deficit and give no quantitative measure of its severity.

Measurement of Water Potential

As suggested earlier in this chapter, the most useful single measurement of the degree of water stress is the water potential. It has been measured by liquid or vapor equilibration, by change in concentration or density of the solution in which the tissue is immersed, by thermocouple psychrometers, and by the Scholander pressure chamber method.

Liquid Equilibration.   The first method used to measure water potential was to immerse tissue in a series of solutions with osmotic potentials extending from well below to above that likely to be encountered in the tissue. Theoretically, the osmotic potential at which the tissue neither gains nor loses water is equal to its water potential. Sometimes this is determined by measuring changes in length of strips of tissue after immersion in the test solutions, but this is feasible only for tissue composed of thin-walled cells with no major veins. In the gravimetric version, pieces of tissue are weighed, immersed in a series of solutions for an hour or two, removed, blotted dry, and reweighed. The water potential is equal to the osmotic potential of the solution in which no change in weight occurs. Actually, weights or lengths are plotted over osmotic potential of the solution, and the water potential is taken as the value of the osmotic potential where the plot of weight or length intersects the zero line. The gravimetric method is suitable only for large masses of tissue from which numerous disks or cubes of tissue can be cut, and errors result from failure to dry the pieces of tissue uniformly and from loss of weight during handling and weighing. The solutions most often used are sucrose or mannitol, or sometimes polyethylene glycol of high molecular weight. Ideally, the solute should be nontoxic, cell membranes should be relatively impermeable to it, and it should not be used in metabolism. None of those mentioned is perfect.

A liquid equilibration method that avoids some of the problems inherent in the gravimetric method was developed in Russia. Samples of plant tissue are placed in a series of sucrose or mannitol solutions with osmotic potentials covering the range of expected tissue water potentials. The tissue loses water to solutions with a lower water potential, diluting them, and gains water from solutions with a higher potential, concentrating them. The tissue water potential is assumed to be equal to the osmotic potential of the solution which undergoes no change in concentration. The changes in concentration of the sucrose solution can be measured with a refractometer or by the Shardakov dye method shown in Fig. 12.21. Although largely supplanted by the methods to be described later, it is very inexpensive and can be used in the field. It is discussed in detail by Brix (1966), Knipling (1967a), Knipling and Kramer (1967), and Slavik (1974).

Vapor Equilibration.   Some of the errors of the liquid equilibration method can be avoided by allowing weighed samples of soil or tissue to equilibrate over solutions of known osmotic potential. The material is supported over the solution in airtight containers immersed in a constant-temperature bath for some hours and is then reweighed. The solution over which no change in weight occurs is

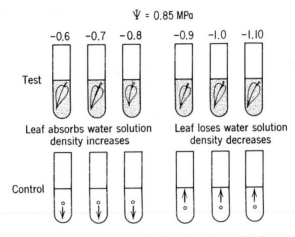

Fig. 12.21. Diagram showing operation of the Shardakov dye method for measurement of leaf water potential. The water potential is assumed to be between that of the solution in which the drop rises and that of the solution in which it sinks, in this instance −0.85 ± 0.05 MPa. (From Knipling, 1967a.)

assumed to have the same potential as the tissue. Slatyer (1958) used this method successfully for leaves and soil. Roberts (1969) warned that the tissue must be shielded from light, which might cause an increase in temperature and raise its vapor pressure.

Thermocouple Psychrometers. The modern thermocouple psychrometer or hygrometer seems to have evolved from the thermoelectric vapor pressure osmometer described about 1930 by A. V. Hill in England. In recent years, psychrometers have come into general use for measurement of water potential. The samples are enclosed in small containers and immersed in a water bath kept at a constant temperature, and the relative humidity is measured with a thermocouple psychrometer. Water potential, Ψ_w, is related to relative humidity by the following equation:

$$\Psi_w = \frac{RT}{\bar{V}} \ln e/e^\circ \tag{12.3}$$

where R is the gas constant, T the absolute temperature, $\bar{V}$ the partial molal volume of water, and e/e° the relative humidity. One type of thermocouple psychrometer, suggested by Spanner (1951) and introduced by Monteith and Owen (1958), uses the Peltier effect to condense water on a thermocouple junction. The rate of cooling depends on the humidity of the air in the chamber, and the current generated during cooling is measured. In the Richards and Ogata

(1958) version, a drop of water is placed on the junction and the current generated by evaporative cooling is measured. Both types must be calibrated over a range of solutions of known osmotic potentials. The Spanner type, using Peltier cooling, is used most frequently because its operation can be automated. Boyer and Knipling (1965) proposed the use of an isopiestic method, using a modified Richards psychrometer with removable thermocouple and obtaining equilibrium readings with drops of solution of three or four different concentrations on the junction. This eliminates the need for calibration and errors caused by diffusive resistances, and can be used for tissues with very low water potentials, but it cannot be automated. The construction and use of isopiestic psychrometers were discussed by Boyer, in Brown and Van Haveren (1972).

There are several sources of error in the psychrometer method, including resistance to diffusion of water vapor into or out of leaves, heat generated by tissue respiration, adsorption of water on the walls of the container, temperature drift in the water bath, thermal effects in electrical connections, and the effects of excision of leaves from plants. The diffusion resistance problem is most important with heavily cutinized leaves and can be eliminated by the isopiestic method described by Boyer and Knipling (1965) and by Boyer, in Brown and Van Haveren (1972). When large masses of plant tissue are placed in psychrometer vessels, measurable heat is sometimes generated (Barrs, 1964). This can be measured by adding a second dry junction, but the small disks used today in most psychrometers eliminate this problem. Absorption of water can be minimized by constructing the vessels of nonadsorptive material or coating the walls with a thin layer of Vaseline. The effects of excision of leaves and release of xylem tension were studied by Baughn and Tanner (1976). They found the water potential of excised leaves of four herbaceous species to be lower than when measured on the plant with an *in situ* psychrometer, but the value for one species was higher.

Various adaptations have been made of thermocouple psychrometers for monitoring soil water potentials, the water potential in tree trunks, attached roots, and attached leaves. Measurements on attached leaves are of particular interest, and readers are referred to papers by Boyer (1968), Campbell and Campbell (1974), and Neumann and Thurtell (1972) for more details. The problems involved in measuring soil moisture are discussed by Rawlins (1976) and Wiebe and Brown (1979). It is difficult to measure water potentials lower than about -7.5 MPa with a thermocouple psychrometer, but Wiebe (1981) described a method for measuring water potentials of dry seeds and other materials down to about -500 MPa.

It is impossible to cite the voluminous literature on thermocouple psychrometers, but much information can be obtained from reviews by Barrs (1968a), Boyer (1969), and Wiebe *et al.* (1970, 1971), and from books edited by Brown and Van Haveren (1972) and Slavik (1974). A diagram of a thermocouple psychrometer is shown in Fig. 12.22.

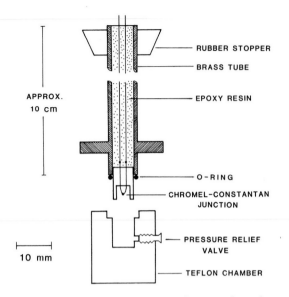

APPROX.
10 cm

RUBBER STOPPER
BRASS TUBE
EPOXY RESIN
O-RING
CHROMEL-CONSTANTAN
JUNCTION
PRESSURE RELIEF
VALVE
TEFLON CHAMBER

10 mm

Fig. 12.22. Diagram of a section through a Peltier-type thermocouple psychometer suitable for measuring the water potential of disks of leaf tissue and the other small samples. The relief valve is left loose until the cap and cup are joined together, to prevent development of pressure in the cup. (Courtesy of Dr. Nasser Sionit.)

Osmotic pressure can also be measured by the psychrometer. After the water potential is measured, the tissue is killed by freezing with dry ice or by immersion in a bath at $-5°C$ or lower, and the equilibrium vapor pressure is measured again. As turgor pressure has been eliminated by death of the tissue, the new measurement gives the osmotic potential. The chief problem with this method is that the vacuolar sap is diluted by cell wall sap. This may constitute a significant error with some material (Tyree, 1976; Markhart *et al.*, 1981), but it is probably not very important in thin-walled tissue. This problem is also discussed in Chapter 2.

Pressure Equilibration.   Scholander and his colleagues (1964, 1965) introduced a method based on work of Dixon (1914, Chapter 10) that is widely used. A leafy shoot or single leaf is sealed in a pressure chamber with the cut surface protruding, as shown in Fig. 12.23. Pressure is applied to the shoot from a tank of compressed gas until xylem sap appears at the cut surface. The amount of pressure that must be applied to force water out of the leaf cells into the xylem until it is refilled is regarded as equal to the tension originally existing in the xylem sap and approximately equal to the water potential of the cells. This value is often termed the xylem pressure potential. Waring and Cleary (1967) described apparatus that can be carried into the field. The equipment is commer-

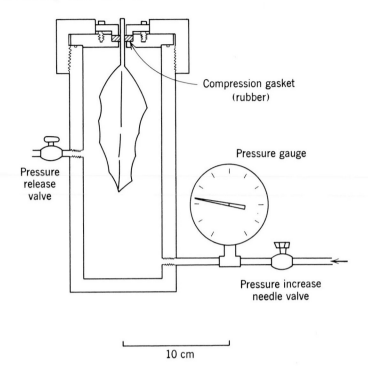

Fig. 12.23. Diagrammatic cross section through a pressure chamber for measurement of leaf water potential by pressure equilibration. (From Kaufmann, 1967.)

cially available, and the method is discussed in detail by Slavik (1974, pp. 70–74) and by Ritchie and Hinckley (1975).

The chief operational problem is to establish an airtight seal without causing collapse of the xylem. The seal has been adapted by various investigators to accommodate leaf blades and pine needles. Firm stems or petioles are much easier to seal than soft tissue. It is also desirable to line the pressure chamber with moist paper or to humidify the air in order to prevent dehydration of the plant material. Excessively rapid increase in pressure causes an undesirable increase in temperature in the chamber. Measurements on conifers with resin ducts and plants containing latex ducts are sometimes difficult or impossible because these substances exude and obscure the appearance of xylem sap. Kaufmann (1968b) discussed this and other problems.

A number of precautions are listed by Ritchie and Hinckley (1975). One is never to recut the stem or petiole because this will change the volume of xylem that must be refilled. Some investigators report that the amount of stem protruding from the chamber affects the reading, and the amount of stem inside the pressure chamber may also affect it.

The pressure chamber provides approximate measurements of water potential, but there is likely to be more variability than with psychrometer measurements. For example, Boyer (1967) found good agreement between potentials of yew and sunflower measured by the two methods, but not for rhododendron. Kaufmann (1968b) also found discrepancies in some tree species, but not in others. Examples are given in Fig. 12.24. The reliability of the pressure chamber method should be tested by comparison with psychrometer measurements when starting research on a new type of tissue.

Ishihara and Hirasawa (1978) reported that leaf water potentials of rice measured with a psychrometer and xylem water potentials measured with apressure chamber were similar for slowly transpiring plants, but the xylem water potential was 0.3–0.5 MPa lower than the leaf water potential in rapidly transpiring plants. Camacho-B *et al.* (1974) also found the xylem water potential about 0.3 MPa lower than the leaf water potential in rapidly transpiring pear trees, but there was no difference in slowly transpiring plants. Ishiwara and Hirasawa suggested that water in the vacuoles of mesophyll cells is isolated from the transpiration stream, water movement occurring chiefly through bundle sheath extensions to the epidermis, but this is uncertain. As pointed out earlier, there is uncertainty concerning the relative amounts of water moving through the apoplast and symplast.

The pressure chamber can also be used to estimate osmotic potential by the pressure–volume method described by Tyree and Hammel (1972) and Cheung *et al.* (1975). A leafy twig is placed in a pressure chamber and subjected to

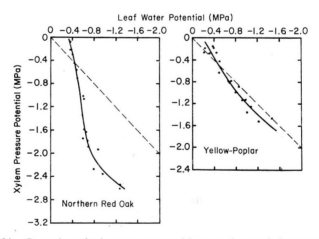

Fig. 12.24. Comparison of xylem pressure potential measured on a twig in a pressure chamber and leaf water potential measured with a thermocouple psychrometer. Measurements by the two methods usually agree fairly well, but exceptions such as that shown for northern red oak are reported occasionally. (From Kaufmann, 1968.)

increasing increments of pressure, and the volume of sap exuded with each increment is measured. Finally, the branch is weighed, dried, and reweighed. From these data, a pressure–volume curve is plotted that indicates the original osmotic potential and the volume of apoplastic water (see Fig. 2.9, Chapter 2). The method is discussed by Boyer (1969), and its application is described by Roberts and Knoerr (1977). It is also discussed briefly in Chapter 2. It avoids the dilution error of psychrometer measurements but is a time-consuming method requiring rather large samples.

A variation of the pressure chamber method has appeared that is more rapid than the Scholander method, is more suitable for soft tissue, and eliminates the need for compressed air. Leaf segments are laid on a sheet of soft rubber, covered with a thick sheet of transparent plastic, and subjected to pressure with a hydraulic jack until water appears at the cut edges. The pressure at this point is taken to be equal to the leaf water potential. Jones and Carabaly (1980) obtained a good correlation ($r = 0.90$) between values obtained with this method and those obtained with a Scholander pressure chamber for four species of tropical grasses. Bristow *et al.* (1981) did not find as good a correlation for wheat, but they regarded the method as usable. Radulovich *et al.* (1982) reported that the hydraulic press method does not measure the water status of cotton accurately at a water potential below −0.9 MPa. Shayo-Ngowi and Campbell (1980) reported that measurements of matric potential made by this method on frozen tissue were similar to those made by the Scholander method, used by Boyer (1967), and required less time. The underlying theory is not as well established for this method as for the Scholander method. However, it deserves further investigation because of its low cost and ease of operation.

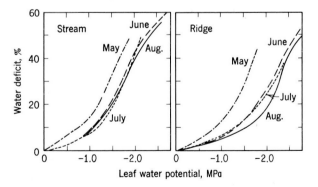

Fig. 12.25. Comparison of the relationship between water deficit and water potential for dogwood leaves of increasing age from a moist and a dry habitat. The change with increasing age was greater for leaves from the dry ridge habitat than for leaves from the moist stream side habitat. (From Knipling, 1967b.)

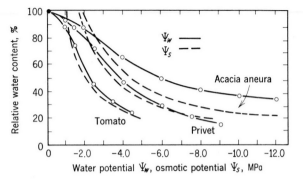

Fig. 12.26. Relationship between leaf water potential and relative water content of three dissimilar species. (From Slatyer, 1967.)

Indirect Estimation of Water Potential

At a time when water potential measurements were more difficult to make than today there was a strong desire to estimate water potential from some simple measurement such as relative water content. Weatherley and Slatyer (1957) suggested that perhaps relative water content–water potential curves might be constructed for various kinds of tissue. Unfortunately, this relationship changes with age of leaves and habitat, as shown in Fig. 12.25. The differing relationships of water potential and osmotic potential to relative water content for three different species are shown in Fig. 12.26. They are too variable to permit estimation of one value from another. These differences also might provoke some thought concerning the complex nature of cell water relations and the difficulty of evaluating plant water status and drought tolerance from any one type of measurement.

SUMMARY

Plant growth is reduced more often by water deficits than by any other factor. Temporary, midday water deficits often occur on hot, sunny days because water loss temporarily exceeds absorption, but long-term deficits are caused by decreasing availability of soil water. Water deficits affect practically every aspect of plant growth, including the anatomy, morphology, physiology, and biochemistry. Plant size is reduced by a decrease in cell enlargement; photosynthesis by a decrease in leaf area, closure of stomata, and damage to the photosynthetic apparatus. Dark respiration apparently is reduced less than photosynthesis, and the capacity for translocation is not greatly reduced, but the amount of material translocated is reduced. Carbohydrate and protein metabolism are disturbed, often leading to accumulation of sugars and amino acids. However, it is uncertain how much of the accumulation results from increased breakdown and how much results from decreased utilization of metabolites in growth. Biochemical effects are attributed largely to damage to membrane structure of cells and changes in enzyme activity.

Water deficit often increases susceptibility to attacks by pathogens and insects, but it decreases susceptibility to organisms that require moist leaf surfaces or wet soil for infection.

Research on effects of plant water stress require quantitative measurements of the water status of plants. Water potential seems to be the best single measure of plant water status, because it is a measure of the chemical potential of water, it controls water movement in the soil–plant–atmosphere system, and it can be measured in plants and soils. The most reliable measurements of water potential are made with thermocouple psychrometers, but the pressure chamber method is rapid and can be used in the field as well as in the laboratory. Both methods are subject to certain errors of which investigators should be aware. A complete understanding of plant water relations requires knowledge of the osmotic and pressure potentials as well as the total water potential.

SUPPLEMENTARY READING

Barrs, H. D. (1968). Determination of water deficits in plant tissues. *In* "Water Deficits and Plant Growth" (T. T. Kozlowski, ed.), Vol. 1, pp. 236–368. Academic Press, New York.

Boyer, J. S. (1969). Measurement of the water status of plants. *Annu. Rev. Plant Physiol.* **20,** 351–364.

Brown, R. W., and Van Haveren, E. P., eds. (1972). "Psychrometry in Water Relations Research." Utah Agric. Exp. Stn., Utah State University, Logan.

Fischer, R. A., and Turner, N. C. (1978). Plant production in the arid and semiarid zones. *Annu. Rev. Plant Physiol.* **29,** 277–317.

Kozlowski, T. T., ed. (1968–1981). "Water Deficits and Plant Growth," Vols. 1–6. Academic Press, New York.

Levitt, J. (1980). "Responses of Plants to Environmental Stresses," 2nd ed., Vol. 2. Academic Press, New York.

Ritchie, G. A., and Hinckley, T. H. (1975). The pressure chamber as an instrument for ecological research. *Adv. Ecol. Res.* **9,** 165–254.

Slavik, B. (1974). "Methods of Studying Plant Water Relations." Springer-Verlag, Berlin and New York.

Turner, N. C. (1981). Techniques and experimental approaches for the measurement of plant water stress. *Plant Soil* **58,** 339–366.

Turner, N. C., and Kramer, P. J., eds. (1980). "Adaptation of Plants to Water and High Temperature Stress." Wiley, New York.

13

Drought Tolerance and Water Use Efficiency

INTRODUCTION

Drought, permanent or temporary, limits the growth and distribution of natural vegetation and the yield of cultivated plants more than any other enviornmental factor. Over one-third of the earth's surface is classified as arid or semiarid because it is subject to permanent drought. Equally important is the fact that most of the humid temperate regions, where much of the world's food supply is produced, are often subjected to periods of severe drought. As a result, investigators all over the world have been and are concerned with the improvement of drought tolerance and increase in the efficiency of water use by plants.

This chapter deals with the nature of drought, how plants react to drought, some adaptations that increase tolerance of drought, and some factors that affect the efficiency of water use by plants.

Terminology

It seems desirable to define some of the terms used in this chapter because there is considerable variation among writers regarding their usage. In general, the term "resistance," used with respect to ability to survive drought, cold, or other stress, has been replaced in this book by "tolerance" because the latter word more nearly describes the manner in which plants react to stresses. The terminology follows that used in Chapter 1 of Turner and Kramer (1980).

Drought. In this book drought is treated as a meteorological and an environmental event (May and Milthorpe, 1962), defined as absence of rainfall for a period of time long enough to result in depletion of soil water and injury to plants. The length of time without precipitation necessary to cause injury depends on the kind of plant, the water-storage capacity of the soil in the root zone, and the atmospheric conditions affecting the rates of evaporation and transpiration. Drought may be essentially permanent, as in desert areas; seasonal in areas with well-defined wet and dry seasons; or random, as in many humid areas. We will return to this in discussing the frequency of droughts.

Broad application of the term drought to include plant and atmospheric conditions causes unnecessary confusion. A recent paper began, "We define drought as the occurrence of a substantial water deficit either in the soil, atmosphere, or plant." The confusion caused by using drought with reference to both the environment and the plant is illustrated by the following quotation from a paper in a scientific journal: "Growth drought tolerance, defined as the plant drought that is just sufficient to halt the increase of seedling leaf area, varied among seedlings from about −20 to −40 bars. Growth drought avoidance, defined as the difference between plant drought and the drought of the shoot environment is related to total leaf diffusion resistance, maximum capacity for water uptake, and seedling leaf area." This is meaningless! A number of writers, following Levitt (1972, 1980, Chapter 4), term plants that tolerate drought without serious dehydration drought avoiding, but this is also misleading because the plants are subjected to meteorological drought. They are really drought tolerant, not drought avoiding.

Tolerance, Stress, and Strain. Levitt (1980, Volume 2, Chapter 1) recommended applying the terminology of the physical sciences to discussion of stress in living organisms. In engineering and the physical sciences, stress is usually defined as the force applied per unit area, and the result is strain. However, in biology stress is usually described as any factor that disturbs normal functioning of an organism, and a drought is an environmental stress that produces plant water deficit or plant water stress sufficient to disturb internal physiological processes. Perhaps the effects of internal stresses could be treated as secondary or tertiary stresses, but it is unlikely that the term plant water stress will be replaced soon by plant water strain.

Adaptations. An important but dangerous term in biology is adaptation. Adaptations are *heritable* modifications in structures and processes that increase the probability of organisms surviving in a given environment. The term is dangerous because it tempts scientists to indulge in unfounded speculation concerning the adaptive value of various characters and sometimes to write as though organisms have acquired certain characters in order to survive. There is a tendency to assume that any special character that commonly occurs in plants exhibiting tolerance of water stress is a beneficial adaptation. For example, the thick cuticle, hairiness, and responsive stomata common in many plants growing in dry habitats are regarded as important adaptations. They often may be, but other plants lacking these characteristics also grow in dry habitats, and hairy plants such as mullein also occur in humid regions.

It was suggested by Stewart and Hanson (1980) that a given character may be beneficial (i.e., adaptive), injurious, or merely incidental. For example, accumulation of betaine and proline occurs in water-stressed plants incidental to disturbance of the normal course of nitrogen metabolism, and it may or may not be beneficial in increasing tolerance of water stress. It often is difficult to determine if a character has adaptive value in a particular environment because survival usually depends on an optimum combination of characters and seldom on a single character (Bradshaw, 1965). As Osmond (1980) pointed out, natural selection operates on the entire physiological system of an organism rather than on a single component.

The difficulties encountered in demonstrating the adaptive value of specific characters are discussed in the section on metabolic adaptations later in this chapter.

Innumerable modifications in structures and processes have occurred during the evolution of plants as a result of random mutations and recombinations. Most were deleterious and disappeared, but a few enabled plants to grow and reproduce more successfully and were preserved by natural selection. As a result, plants subjected to frequent water stress accumulated modifications with adaptive value such as thick cuticle, extensive root systems, low osmotic potential, or tolerance of dehydration that increased the probability of their survival. However, these modifications did not originate *in order to improve survival,* but from random mutations and recombinations that occurred in plants growing in both moist and dry habitats but were usually preserved by natural selection only in dry habitats. There is no assurance that the adaptations necessary for survival in a changing environment will appear, and many organisms have become extinct because the necessary adaptations did not appear. Also, some characters with no apparent function have survived because they have not been subjected to either positive or negative selection pressure.

Acclimation. This term refers to phenotypic modifications produced by variation in environmental factors. For example, the photosynthetic apparatus of *Larrea divaricata* is capable of acclimation to water stress, photosynthesis of plants grown in a dry environment remaining uninhibited at a water deficit that severely limits photosynthesis in plants grown in a humid environment (Mooney *et al.*, 1977). The difference between adaptation and acclimation is not always clear. The inherited ability to adjust to drought or unfavorable temperature is clearly an adaptation, but the actual occurrence of osmotic adjustment or cold hardening represents acclimation. The difference is illustrated by *Eucalyptus* in the Snowy Mountains of Australia. There is adaptation of photosynthesis to the temperatures characteristic of various elevations, on which is superimposed temporary acclimation to seasonal changes in temperature (Slatyer and Morrow, 1977).

Drought hardening of plants by exposure to moderate water stress is an acclimation process. Doubtless some readers have observed the injury that often occurs when plants grown in shade in a humid greenhouse are suddenly transferred into dry air and full sun. Gardeners and nursery operators are aware that exposure of seedlings to water stress before transplanting them to the field increases survival. Seedlings are hardened by exposure to full sun, decreased frequency of watering, and even by undercutting and loosening of their root systems. Some examples are mentioned in Chapter 12, in the section on beneficial effects of water stress.

Another type of acclimation is cold hardening in the autumn by exposure to gradually decreasing temperatures. This enables plants to survive severe winter weather without injury, but much damage occurs if freezing weather occurs before plants have become hardened. There is evidence that cold hardening often is accompanied by and probably increased by dehydration of tissue. This was discussed in Chapter 12.

Throughout this book mention has been made of changes in various plant characteristics caused by exposure to water stress. These examples of acclimation include reduced leaf area, thicker leaves, less responsive stomata, decrease in osmotic potential (osmotic adjustment), and increased ratio of roots to shoots. Cell size often is decreased and wall thickness increased (Cutler *et al.*, 1977) and turgor is maintained to a lower leaf water potential because of osmotic adjustment (Cutler and Rains, 1978; Cutler *et al.*, 1980). Biochemical effects involving changes in enzyme activity, accumulation of ABA, betaine, and proline, and decrease in starch and protein are well known and are discussed later in this chapter and in Chapter 12. Some of these are merely incidental effects of stress while others increase tolerance of subsequent water stress, but much more research is needed before the relative importance of these changes can be established.

Plant Strategy.   One of the often misused words in plant science is strategy, as applied to modifications such as those just discussed. There are workshops, papers, and even books on strategy in plants, but how can a plant have a strategy? The word refers to a plan of action intended to attain some desired end, and scientists can develop research strategies, but plants cannot develop strategies to insure survival unless we are prepared to credit them with intelligence. Perhaps the users do not intend to credit plants with purposeful behavior, but its use reminds one of the emphasis on purpose in nature characteristic of pre-Darwinian biology. Plant physiology is concerned with mechanisms rather than goals and speculation about the usefulness of adaptations often ends in philosophically and scientifically objectionable teleology. Strategy is a term best left to military planners.

DROUGHT

Drought has already been defined as a period without rain of sufficient duration to cause injury to plants. This definition may seem inadequate to laboratory scientists, but it is quite adequate for experienced farmers and gardeners. Decker (1982) discussed various definitions and bases for estimating the probability of drought and their importance for agriculture. He regarded a definition based on soil water shortage as most useful in agriculture. Van Bavel and Verlinden (1956) defined agricultural drought quantitatively by assuming that a drought begins when the readily available water in the root zone is exhausted, meaning that the soil water content has decreased to approximately the permanent wilting percentage.

Frequency of Drought

In broad terms, drought is permanent, seasonal, or random. Over one-third of the earth's surface is classed as arid or semiarid because it receives too little rainfall for reliable cropping, and about 10% receives so little rainfall (less than

250 mm) that it is treated as desert. Another large portion of the earth's surface experiences well-defined wet and dry seasons, such as areas with a Mediterranean climate, characterized by mild, rainy winters and hot, dry summers. The coastal regions of southern California and Chile, South Africa, and southern Australia are examples, in addition to the Mediterranean area itself. Perennial vegetation in such regions must be capable of surviving long summer droughts, and the best growing season is usually the spring.

Most of the humid regions of the world are subject to at least occasional serious perturbations in amount and distribution of rainfall. For example, England and Western Europe are supposed to have adequate rainfall, but in 1976 they experienced severe midsummer droughts. An example of the seasonal variation in rainfall in England is shown in Table 13.1. Even the humid southeastern United States experiences droughts which can best be described as random, and severe droughts are common in the Great Plains region. At present, it is impossible to predict the occurrence of droughts, but some attempts have been made to estimate their probable frequency. If the amount of readily available water stored in the root zone, the average rate of extraction by evaporation and transpiration, and the average replenishment by rainfall are known for each month, the probability of drought can be estimated. This was done by van Bavel and various colleagues for several areas, and a few data are shown in Table 13.2. This shows that droughts are common in the entire eastern United States.

Plants and Drought

If drought is defined as a purely environmental factor and tolerance is substituted for resistance, the behavior of plants with respect to drought can be classified as follows:

Drought avoidance, when plants are not subjected to meteorological drought

TABLE 13.1 Monthly Rainfall during the Summer at Letcombe Laboratory, Oxfordshire, England[a,b]

	May	June	July	August	September
1975	54	9	36	29	104
1976	38	24	16	18	99
1977	51	89	19	150	20
7-Year avg.	54	60	49	64	66

[a] Rainfall measured in millimeters.
[b] From *Annu. Rep. Agri. Res. Counc. Letcombe Lab. 1977*, p. 92, 1978.

TABLE 13.2   Probable Number of Drought Days 5 Years out of
10, Assuming 2 or 3 Inches of Available Water in the Root Zone[a]

	2 in.	3 in.
Central Virginia	40	20
Central North Carolina	40	20
Central South Carolina	60	40–50
Central Georgia	60–70	40–50
Central Alabama	70–80	—
West Central Mississippi, Southern Arkansas, Northern Louisiana	80–90	70–80
Central Minnesota	—	30–40

[a] From Kramer, 1982.

Drought tolerance
By dehydration postponement (equivalent to Levitt's drought avoidance)
By dehydration tolerance

These three categories are not mutually exclusive, and some kinds of plants, sorghum for example, can be placed in more than one category. They are discussed in the following sections.

Drought Avoidance

The best-known examples of drought avoidance are the short-lived desert annuals that germinate, grow, flower, and produce mature seed in a few weeks after rains have wetted the surface soil. Annual plants of Mediterranean climates also often mature early, before soil water is exhausted. According to Cooper (1963), some Mediterranean grasses grow better in cool weather than their relatives in northern Europe; hence, they complete their development before the summer drought becomes severe. Reitz (1974) stated that each day by which certain varieties of wheat in Kansas and Nebraska matures earlier than the Kharkof variety results in an average increase in yield of 60 kg/ha or more. Where early maturity is important, tolerance of low temperature for seed germination and seedling establishment is a valuable characteristic because it permits planting early enough to ensure maturity before water becomes seriously limiting. Early planting also generally decreases the length of time plants are exposed to atmospheric conditions favorable for high rates of transpiration, and early development of a crop canopy decreases the amount of water wasted by evaporation from the soil surface. Early planting sometimes has other advantages, such as decrease in boll weevil infestation of cotton, but it increases the danger of failure of seed to germinate in cold soil and of injury from late spring frosts.

Drought Tolerance

Plants cannot avoid the random droughts characteristic of most of the humid temperate zone, but many possess adaptations that increase their tolerance. These can be classified into two groups: those that postpone dehydration and those that increase tolerance of dehydration. Some of the latter, such as osmotic adjustment, are really examples of acclimation.

Dehydration Postponement

Injurious dehydration is postponed by morphological or physiological characteristics that either reduce water loss by transpiration or increase water absorption. Thick cuticle, responsive stomata, and leaf rolling reduce water loss, and deep root systems increase water absorption.

Root Systems.   One of the most effective safeguards against drought injury of both native vegetation and crop plants is a deep, wide-spreading, much-branched root system. Shallow-rooted crops such as lettuce, onions, and potatoes suffer sooner from drought than deep-rooted species such as alfalfa, maize, and tomato. According to Hurd (1974), differences in drought tolerance of wheat in the Canadian wheat belt are related principally to differences in root development. Boyer (1982) found that higher root density and lower afternoon water deficits were associated with higher yields of soybeans. It is frequently observed that native plants with shallow roots suffer water deficits during summer droughts, whereas their deeper-rooted neighbors show no evidence of water stress. Also, crops and native vegetation growing on shallow soil suffer more from drought than those growing on deep soil. The benefits of deep tillage to permit deeper root penetration are discussed in Chapter 6 and by Cassell (1983). The only disadvantage of extensive root systems is that they might remove so much water early in the season that not enough is left to mature a crop. Passioura (1972) suggested that where growth of wheat is dependent on stored soil moisture, a limited root system and high root resistance might be beneficial because they slow absorption early in the season and leave more water in the soil until grain filling, when water stress is particularly damaging. Passioura discussed this further in Paleg and Aspinall (1981). On the other hand, Teare *et al.* (1973) reported that although sorghum had twice the root density of soybean, it did not remove soil water as rapidly as soybean because it had better stomatal control of transpiration. In general, extensive root systems are effective in postponing dehydration, especially in deep soils. However, Kummerow (1980) reported that root systems are less important than leaf adaptations in drought tolerance of shrubs of the California chaparral, perhaps because the soils are typically shallow. The role of root density in regions with different rainfall regimes was discussed by Jordan and Miller and by Taylor, in Turner and Kramer (1980). They agreed on the importance of deep rooting, but Taylor doubted if breeding for increased root density would postpone dehydration of cotton and soybean, whereas Jordan and Miller, working with a different rainfall pattern, thought it would be effective for sorghum.

Factors affecting the development of root systems are discussed in Chapter 6, and some data on depth and spread of root systems of various plants are summarized by Taylor and Terrell (1982). Readers are reminded that where crops are irrigated, extensive root systems, which are expensive in terms of photosynthate, are unnecessary. It seems possible that plant breeders might produce cultivars with quite different root systems for irrigated and unirrigated agriculture.

Leaf Adaptations. A number of morphological characteristics of shoots reduce water loss. These include changes in leaf orientation, such as rolling and drooping caused by wilting, which decrease the energy load. A thick coat of hairs or wax, often known as bloom, decrease the energy load, leaf temperature, and transpiration but also decrease photosynthesis (Ehleringer, 1980). Begg (1980) discussed morphological adaptations, including some interesting changes in leaf orientation with increasing water stress. Death and shedding of leaves reduce both the transpiring and photosynthetic surfaces. However, death of the lower, older leaves, which usually occurs first on stressed plants, probably has little effect because their rates of transpiration and photosynthesis are low compared to those of the younger, better-exposed upper leaves. Shedding of most or all of the leaves by some desert plants probably has some survival value, although the rate of water loss from sclerophyllous vegetation is low when the stomata are closed. However, death or shedding of leaves can be catastrophic for crop plants because yield is the chief consideration in agriculture. Leaf shedding was discussed by Kozlowski (1974, pp. 20–22). Among the puzzling observations is that some trees which lose their leaves at the beginning of the dry season often leaf out again before rains occur. This seemingly potentially dangerous behavior has not been explained.

Stomata. Reduction of water loss by stomatal closure is an effective means of postponing dehydration, especially where responsive stomata are associated with low cuticular transpiration, as in most xerophytes. Many xerophytes have high rates of transpiration in moist soil, but as the soil dries, the stomata close earlier in the morning and finally stay closed all day. Although this conserves water, it can reduce photosynthesis of sclerophyllous vegetation below the compensation point (Dunn *et al.*, 1976). Differences in stomatal control of transpiration are correlated with differences in drought tolerance of mesophytes such as sorghum cultivars (Teare *et al.*, 1973), races of monterey pine (Bennett and Rook, 1978), and poplar clones (Ceulemans *et al.*, 1978).

It was mentioned in Chapter 11 that the stomata of some kinds of plants are closed by decreasing atmospheric humidity even though the leaves remain turgid. This would be expected to conserve water by causing stomatal closure during the hottest, driest part of the day, and it is found in some desert plants, but not all. The stomata of species of *Atriplex, Larrea,* and *Tidestromia oblongifolia* from Death Valley do not show a response to humidity, whereas an *Atriplex* from the cool coastal region does. It appears that the stomata of *Opuntia compressa* are sensitive to atmospheric humidity (Conde and Kramer, 1975), but little is known about other species of cacti. Stomata of some kinds of trees are sensitive to atmospheric humidity (Davies and Kozlowski, 1974). Sheriff (1977b) surveyed the effects of change in atmospheric humidity on plants of a large number of species and concluded that the presence or absence of a response to humidity was not related to the normal habitat of the plants studied. This topic was reviewed by Osmond *et al.* (1980, pp. 277–280), and it seems that no general statement can be made concerning the adaptive value of sensitivity of guard cells to ambient humidity. There has been considerable discussion concerning what constitutes optimal stomatal behavior in terms of amount of carbon fixed per unit of water lost. Cowan and Farquhar (1977) proposed that stomata tend to adjust in a manner that keeps the ratio of photosynthesis to transpiration fairly constant over a wide range of stomatal opening. Farquhar *et al.* (1980) tested this hypothesis and found that it seems to hold for two species of plants. Ludlow (1980) also discussed the ecological and physiological importance of stomatal reactions to leaf water deficits and low humidity, and pointed out the need for more research.

Metabolic Adaptations. It was stated in Chapter 12 that water stress has important effects on the physiology and biochemistry of plants. For example, abscisic acid and proline accumulate in many kinds of plants when water stressed, and betaine accumulates in some. Such accumulation often is regarded as an adaptation to stress, but a correlation between accumulation and development of stress is not proof that the substance has any adaptive value in postponing stress or increasing stress tolerance. As stated earlier in this chapter, accumulation may be merely the incidental result of disturbance of normal metabolic pathways. For example, it is well established that proline accumulates because water stress (1) stimulates its synthesis from glutamate by loss of feedback inhibition, (2) decreases the rate of proline oxidation, and (3) decreases its incorporation into protein. Betaine (glycine betaine) is a quarternary ammonium compound that also accumulates during stress in some taxa such as the chenopods and grasses of the tribe *Hordeae,* but not in plants of other taxa (Jones, 1980; Wyn Jones and Storey in Paleg and Aspinall, 1981). Accumulation of betaine is associated with increased capacity for choline dehydrogenation, ethanolamine methylation, and serine decarboxylation. Thus betaine and proline accumulation occur because of disturbance of normal nitrogen metabolism and any beneficial effects, if such exist, are merely coincidental. Naylor (1972) and Bewley in Paleg and Aspinall (1981) discussed various effects of water stress on nitrogen metabolism.

As mentioned earlier, it is very difficult to prove that these compounds have important adaptive value. Abscisic acid seems to have a role with respect to stomatal closure and some investigators claim that it affects membrane permeability. However, there is no entirely satisfactory explanation of how proline or betaine might postpone dehydration or reduce injury from dehydration. The concentrations usually are too low to contribute significantly to lowering the osmotic potential of the vacuolar sap during osmotic adjustment. It has been suggested that they occur as osmotica in the cytoplasm, but this is uncertain. However, it seems that proline does play an important role in osmotic adjustment of microorganisms (Rains *et al.,* 1980).

According to Hanson and Hitz (1982) breeding experiments will provide the strongest evidence concerning the adaptive value of these compounds. Unless genetic variation can be found for a metabolic response to stress it cannot be used in selection and breeding programs to improve drought tolerance. Selection for high and low proline lines in cereals thus far has given equivocal results. In some experiments it appeared that selection for high proline really was selection for plants that developed water stress quickly, resulting in selection against drought tolerance. Hanson and Hitz (1982) stated that genetic and physiological studies are underway to determine the usefulness in breeding programs of ABA in wheat and pearl millet, betaine in barley and wheat, and proline in rice, but completion of such studies will require several years. Readers interested in further discussion of the problem of determining the adaptive value of various metabolic characters are referred to Hanson and Hitz (1982), Steward and Hanson (1980), and several chapters in Paleg and Aspinall (1981).

Such modifications in metabolism as crassulacean acid metabolism (CAM) and the C_4 carbon pathway are more clearly associated with postponement of dehydration. CAM refers to plants that have daytime closure of stomata and fix CO_2 as malic acid in darkness. This is decarboxylated during the day, releasing CO_2 that is refixed in carbohydrates by photosynthesis. Daytime closure of stomata combined with dark fixation of CO_2 greatly reduces water loss without an equivalent decrease in dry matter production. A number of succulents and other plants of dry habitats have CAM, and pineapple is the best example among cultivated plants. Its leaves are heavily cutinized and its stomata are at the bottom of deep furrows covered with hairs and do not open until late afternoon or evening. As a result it has one of the highest water use efficiencies among crop plants, using only 50 or 55 gm of water per gram of dry matter produced (Joshi *et al.*, 1965), compared with several hundred grams for most crop plants, as shown later in Table 13.3. Crassulacean acid metabolism is discussed by Kluge and Ting (1978) and by Osmond (1978). The daily pattern of transpiration of a C_3 and a CAM plant are shown in Fig. 13.1.

The C_4 carbon pathway often occurs in plants that are native to hot, dry habitats and its high water use efficiency may result in postponement of dehydration and production of more dry matter before water stress develops. According to Teare *et al.* (1973) the C_4 plant, sorghum, has better stomatal control of transpiration than soybean, thereby conserving soil water until late in the season.

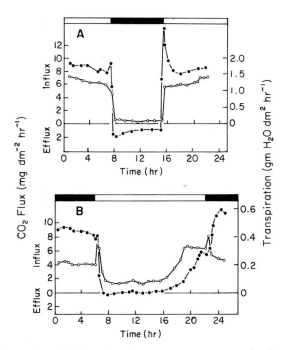

Fig. 13.1. Comparison of daily cycles of carbon dioxide exchange (●—●) and transpiration (○—○) of the C_3 plant, sunflower (A), and the CAM plant, *Agave americana* (B). Both carbon dioxide uptake and transpiration of sunflower ceased in darkness, and there was some efflux of carbon dioxide released by respiration. The situation was reversed in *Agave* with little transpiration and no carbon dioxide uptake during most of the light period. Note that the units for transpiration are over four times greater for sunflower than for *Agave* and the transpiration rate of sunflower is proportionately greater. (Adapted from Neales *et al.*, 1968.)

Such conservation is especially beneficial in regions where crops are subjected to late summer droughts. However, as Osmond *et al.* (1980, p. 355) point out, the C_4 carbon pathway confers no special advantage after plant water stress develops, and there is a wide range of drought tolerance in both C_3 and C_4 plants.

The possibility of improving drought tolerance by genetic manipulation is a challenge to modern genetics and plant breeding. It has been suggested that water use efficiency could be improved by introducing the C_4 pathway or CAM into C_3 plants, or by eliminating photorespiration. It was suggested by Csonkas *et al.* in Raper and Kramer (1983) that proline increases osmotic adjustment in bacteria; hence the dehydration tolerance of nitrogen fixation might be increased by introducing the plasmid for proline synthesis into the nitrogen-fixing Rhizobium bacteria.

Water Storage.   Most plants contain some water that can be used during periods when transpiration exceeds absorption. This is shown as a midday decrease in water content in Fig. 12.6. Stored water is a minor factor in long-term postponement of dehydration because the water loss on a hot, summer day often exceeds the total water content of mesophytes. According to Miller (1938, p. 412), in midsummer the daily transpiration of a Kansas corn plant may be twice its water content, or 4 kg versus 2 kg. Considerable water is stored in tree trunks and Waring and Running (1978) stated that this postpones water deficits in Douglas-fir forests, but this benefit is minimal because the water content of the tree trunks begins to decrease early in the growing season. However, if stored water enables photosynthesis to continue for an additional hour each day it would be beneficial.

In some plants water storage, combined with a low rate of transpiration, permits survival for many months without rain. Considerable water is stored in the enormous trunks of baobab trees (*Adansonia digitata*) which are sometimes several meters in diameter. Water storage is important in desert succulents such as Euphorbia and cacti which can survive for months or even a year or two without additional water. A large saguaro (*Carnegia gigantea*) may contain tons of water in the parenchyma cells of its stems. The loss by transpiration from desert succulents is negligible because of their thick cuticle and day time closure of stomata. Plants of this type also often have shallow, widely spreading root systems that absorb water when the surface soil is wetted by occasional showers.

In a few species water is stored in underground organs such as the greatly enlarged roots of *Welwitschia* and *Pachypodium bispinosum* of the South African desert. Some plants of humid and subhumid regions develop large roots or tubers in which water is stored, but the benefit of this characteristic is questionable.

This discussion of ways in which plants postpone dehydration emphasizes the fact that plants have become adapted to drought in many different ways. Goebel, a German plant morphologist, is said to have stated that the variety of structures in plants is much greater than the variety of environments in which they live. This certainly is true! Morphological adaptations to dry habitats are discussed in papers by Troll, and by Killian and Lemée in Volume 3 of Rhuland's Encyclopedia of Plant Physiology (1956).

Dehydration Tolerance

During prolonged droughts, the usefulness of the various mechanisms that contribute to postponement of dehydration is finally exhausted, and plants are subjected to severe dehydration that often results in irreversible injury and death. This section discusses differences in dehydration tolerance and some adaptations that seem to increase tolerance, including osmotic adjustment and recycling of CO_2.

Differences in Tolerance of Dehydration. The degree of water deficit that can be tolerated varies widely among species, ranging from -2.0 to -2.5 MPa in sunflower to far below -10.0 MPa in various xerophytes, and to the air-dry condition in a few plants. Dehydration of sunflower leaves to -1.5 MPa causes injury to about 10% of the cells, but dehydration to below -2.0 MPa causes so much injury to organelles and membranes that recovery is impossible (Fellows and Boyer, 1978). Giles *et al.* (1976) observed irreversible changes in cell structure of 25% of maize mesophyll cells at -1.8 MPa, but sorghum tolerated stresse of -3.7 MPa without serious injury. In contrast, some 60–70 species of ferns and seed plants and many species of algae, lichens, and mosses can be air-dried and will recover (Gaff, 1980).

The reasons for these differences are not fully understood, but they are certainly related to special properties of the protoplasm and protoplasmic membranes. Injury has been ascribed to mechanical rupture of protoplasm; degradation of cell membranes; protein denaturation, perhaps involving conversion of sulfhydryl to disulfide groups; and accelerated gene mutations. Relevant to the last point is Miller's (1965) observation that water-stressed pine seedlings were injured more by gamma radiation than well-watered seedlings. Probably in general terms, injury from dehydration can be attributed to physical changes in cell and organelle structure (Lee-Stadelmann and Stadelmann, 1976).

Although dehydration is usually accompanied by severe damage and disorganization of membranes and organelles, this does not always occur. The desiccation-tolerant moss *Tortula ruralis* appears to retain most of its structure and capacity for physiological activity, because it resumes normal respiration and protein synthesis about 30 min after rehydration and photosynthesis in 2 hr. In some resurrection plants there is severe disorganization of fine structure, and 1 or

2 days are required for recovery of physiological functions. The behavior of plants tolerant of severe dehydration is discussed in detail by Bewley (1979), by Gaff (1980) and by Poljakoff-Mayber in Paleg and Aspinall (1981). However, even after several decades of research, there is no adequate explanation of why the protoplasm of some kinds of plants can tolerate severe dehydration but most cannot.

Osmotic Adjustment.   This refers to a decrease in osmotic potential greater than can be explained by solute concentration during dehydration. For example, if the initial Ψ_s is -1.0 MPa and the volume of the vacuolar sap is reduced 10% by water loss, the new Ψ_s will be -1.11 MPa. However, accumulation of additional solutes might lower the Ψ_s to -1.3 or -1.5 MPa, permitting maintenance of turgor and turgor-dependent processes to a significantly lower water potential.

Thus, osmotic adjustment enables cell enlargement and growth to continue at water potentials that would otherwise be inhibitory. It also aids by keeping stomata open and the photosynthetic apparatus operating at lower leaf water potentials than if it did not occur. Osmotic adjustment usually develops in plants that are stressed slowly, although it has been reported to occur in response to midday water stress. It is not detectable in all kinds of plants or in all cultivars of a species. Turner and Jones (1980) list a dozen species in which it definitely occurs. It apparently occurs in some cultivars of soybean and sorghum, but not in others, and it has seldom been observed in woody species. It is also said to develop in roots and stem apices. Unfortunately, osmotic adjustment does not persist more than a few days after removal of stress, it can occur over only a limited range of water potential, and it does not fully maintain physiological processes. The subject was reviewed by Turner and Jones (1980), and Hellebust (1976) reviewed osmoregulation, chiefly in algae and fungi. Osmotic adjustment is also discussed in a book edited by Rains *et al.* (1980) and in Paleg and Aspinall (1981).

The solutes involved in osmotic adjustment vary but consist of inorganic ions, carbohydrates, and organic acids. In the tropical grasses studied by Ford and Wilson (1981), Na^+, K^+, and Cl^- ions accounted for most of the decrease in osmotic potential, and carbohydrates and organic acids were of minor importance. The importance of ion accumulation in the osmotic adjustment of plants to saline substrates was mentioned in Chapter 9 and is discussed in Rains *et al.* (1980). In stressed soybean seedlings, osmotic adjustment occurred in the hypocotyls by means of organic compounds exported from the cotyledons, but not used in growth (Meyer and Boyer, 1981). Osmotic adjustment occurred in both the elongating and nonelongating regions of leaves of stressed wheat seedlings (Munns and Weir, 1981). In high light intensities, osmotic adjustment of 0.12–0.34 MPa was caused largely by increased concentration of sugar supplied by photosynthesis, rather than from reserves. There is some evidence that increased concentration of CO_2 in the atmosphere increases osmotic adjustment in wheat, probably by increasing the carbohydrate supply (Sionit *et al.*, 1980a). Accumulation of amino acids probably occurs because of reduced use in growth. Accumulation of proline and betaine was discussed in the section Metabolic Adaptations. Munns and Weir (1981) suggested that use of organic substances for osmotic adjustment probably reduces growth. However, the substances used in osmotic adjustment become available for other processes when water stress ceases, and it seems probable that the net metabolic cost is not excessive.

Recycling of CO_2. According to Osmond *et al.* (1980), another physiological adaptation that may increase tolerance of water stress is recycling of carbon dioxide, because it may reduce injury from photoinhibition. When shade leaves are exposed to full sun, there is often inhibition of photosynthesis and bleaching of chlorophyll, sometimes termed solarization. It is also reported that water stress increases photoinhibition in the sun plant *Nerium oleander* (Björkman *et al.*, 1981). Osmond *et al.* (1980, pp. 312–316) suggest that when stomatal closure in bright sun reduces the intercellular carbon dioxide concentration, there may be a surplus of photochemical energy that inactivates the photosynthetic machinery. However, a sink for this energy may be provided by CO_2 produced by photorespiration in C_3 plants, by recycling of CO_2 between mesophyll and bundle sheath cells in C_4 plants, or by induction of CAM in some plants. This would decrease the degree of photoinhibition and related injury.

WATER USE EFFICIENCY

Water use efficiency refers in general terms to the amount of water used per unit of plant material produced, but it has been applied in various ways and the concept has gone under various names. Originally, it was termed water requirement by pioneer workers, such as Briggs and Shantz and their associates, who measured the water requirement of about 150 kinds of plants at Akron, in eastern Colorado, during the first quarter of this century. They found water requirements ranging from 216 for millet to 1131 g of water used per gram of dry matter produced for the native weed *Franseria* (Miller, 1938, pp. 496–502).

In those days, it was supposed that a low water requirement would be correlated with drought tolerance, but this proved untrue. It became evident that the term water requirement is a misnomer because there is no specific amount of water required to produce a unit of dry matter. The water used varied from 250 to 400 for maize and from 660 to 1000 for alfalfa over a period of 7 years at Akron, Colorado, depending on weather conditions (see Table 13.3). The term transpira-

TABLE 13.3 Water Requirement for the Years 1911–1917 at Akron, Colorado, and the Evaporation from a Free Water Surface[a]

Plant	1911	1912	1913	1914	1915	1916	1917
Alfalfa	1068	657	834	890	695	1047	822
Oats, Burt	639	449	617	615	445	809	636
Barley, Hannchen	527	443	513	501	404	664	522
Wheat, Kubanka	468	394	496	518	405	639	471
Corn, N.W. Dent	368	280	399	368	253	495	346
Millet, Kursk	287	187	286	295	202	367	284
Sorghum, Red Amber	298	239	298	284	303	296	272
Evaporation, April 1 to September 1 (mm)	1239	957	1092	1061	848	1196	1084

[a] From Miller (1938).

tion ratio replaced water requirement and, in turn, was later replaced by water use efficiency.

The term water use efficiency (WUE) is used in different ways by agronomists and physiologists. Agronomists usually define it in terms of the units of water used per unit of dry matter produced, often using total water lost by both evaporation and transpiration (Teare *et al.*, 1973).

$$\text{WUE} = \frac{\text{Water used in evapotranspiration}}{\text{Dry matter or crop yield}} \qquad (13.1)$$

Physiologists are more likely to discuss it in terms of photosynthesis, expressed as milligrams of CO_2 per gram of water or even as moles of CO_2 per mole of water (Fischer and Turner, 1978). In simple terms, at the level of gas exchange for single leaves:

$$\text{WUE} = \frac{\text{Net } CO_2 \text{ uptake in mg or gm}}{H_2O \text{ loss in gm or kg}} \qquad (13.2)$$

At the Physiological Level

To take into account the effects of CO_2 and water vapor gradients between leaf and air, stomatal opening, boundary layer resistance, and mesophyll resistance requires more complicated equations. Different treatments are used by Fischer and Turner (1978), Nobel (1980), and Osmond *et al.* (1980). Also, Nobel and Osmond use conductance, but Fischer and Turner use resistance, as we do in this book.

There are two important differences between the diffusion of CO_2 and that of water vapor into and out of leaves. First, the diffusivity of water vapor is 1.56 times that of CO_2, or conversely, the diffusivity of CO_2 is 0.64 times that of water vapor. The difference results from the fact that the speeds of diffusion are inversely proportional to the square roots of the molecular weights of the gases. The resistance of leaves is usually measured with porometers in terms of water vapor and must be corrected for the difference in diffusivity to be applicable to CO_2. Second, the resistance (r_m) to diffusion of CO_2 through the water phase between the cell surface and the reaction sites in mesophyll cells is significant, but that for evaporation of water is usually negligible and can be neglected. In practice, r_m for CO_2, treated as the difference between $r_a + r_s$ and the total resistance to CO_2 uptake, includes biochemical limitations on CO_2 fixation as well as physical limitations on diffusion. The equations for the two processes follow.

$$\text{Photosynthesis} = \frac{(\text{Conc}^{int}_{CO_2} - \text{Conc}^{ext}_{CO_2}) \times \text{diffusivity}}{r_a + r_s + r_m}$$

$$= \frac{\Delta CO_2 \times 0.64 D_{H_2O}}{\text{resistances}} \tag{13.3}$$

$$\text{Transpiration} = \frac{(\text{Conc}^{int}_{H_2O} - \text{Conc}^{ext}_{H_2O}) \times \text{diffusivity}}{r_a + r_s}$$

$$= \frac{\Delta H_2O \times D_{H_2O}}{\text{resistances}} \tag{13.4}$$

The two equations can be combined into the following equation:

$$\text{WUE} = \frac{0.64 \,\Delta CO_2}{\Delta H_2O} \frac{(r_a + r_s)}{(r_a + r_s + r_m)} \tag{13.5}$$

If r_m is large relative to $r_a + r_s$, as proposed by Gaastra (1959), partial closure of stomata should reduce water loss more than it reduces the rate of transpiration. This assumption underlies the use of antitranspirants, a topic to be discussed later. However, in practice the rate of photosynthesis often decreases at the same rate as transpiration when stomata close gradually (Fig. 12.13), suggesting that r_m may not be much larger than $r_a + r_s$. Also, there is uncertainty concerning the concentration of CO_2 in the intercellular spaces of leaves. It is sometimes arbitrarily set at zero (e.g., Fischer and Turner, 1978, p. 291; Nobel, 1980, p. 45) but such a low internal concentration seems improbable. In fact, some physiologists think it must be in excess of 200 ppm in order for the carboxylation reaction to proceed at the observed rates, at least in C_3 plants with their relatively high compensation points (Wong *et al.*, 1979). The internal concentration will materially affect the size of the difference in CO_2 concentration between the ambient air and the cell surfaces.

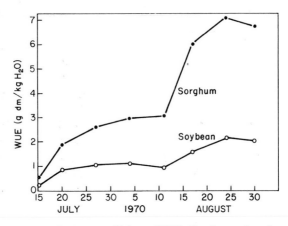

Fig. 13.2.   Differences in water use efficiency (WUE) of soybean and sorghum. (From Teare *et al.*, 1973.)

Nobel (1980) attempted to separate the effects of leaf anatomy on mesophyll resistance from the effects of changes in the photosynthetic machinery by introducing the ratio of mesophyll cell wall surface to leaf surface (A_{mes}/A) into the diffusivity term. Osmond *et al.* (1980) use the term intrinsic water use efficiency, P/C_s to compare species, i.e., the rate of photosynthesis (P) for any given stomatal conductance (C_s). High P/C_s values are particularly important where water deficits cause partial closure of stomata. Osmond *et al.* (1980, pp. 348–355) summarize considerable data indicating that under favorable environmental conditions C_4 plants have P/C_s ratios two and a half to three times those of C_3 plants. This is similar to the difference in dry matter production reported on a stand basis, given in the next section. However, Osmond *et al.* (1980, pp. 354–355) state that the differences between C_3 and C_4 plants tend to disappear in cool weather, an observation also made by Caldwell *et al.* (1977). This raises questions concerning the value of instantaneous measurements of WUE in predicting plant behavior in the field, because success of plants depends on rates of processes over the entire growing season.

Of Crops and Plant Stands

Agronomists and ecologists tend to consider WUE in terms of dry matter produced by stands of plants per unit of water used. Yield may be expressed as economic yield, such as grain or soybean seed, as forage in the case of grasses, or as dry weight of the entire plant, including roots. The plant parts used depend on the objectives of the experiments, but if the roots or other vegetative organs are neglected, the WUE will be lower than if the entire plant is used as a basis for computation. The other question is whether to use only the water lost by transpiration or the combined losses from evaporation and transpiration (ET). Use of ET results in more variability in WUE because E is affected by leaf cover and frequency of soil wetting, independently of T (Tanner, 1981), but it is more realistic with respect to agriculture.

An example of a field experiment is the study of sorghum and soybean by Teare *et al.* (1973), in which plants were grown in the field and root development and leaf area index were monitored along with evaporation and transpiration. WUE of sorghum was three times that of soybean in terms of both total dry matter and seed yield, supporting the statement made earlier that C_4 plants are two and a half to three times as efficient as C_3 plants. Sorghum had better stomatal control of transpiration than soybean and did not exhaust the soil water as rapidly. As a result, the net assimilation rate of sorghum was high in August, resulting in high WUE in the late summer (Fig. 13.2). Rawson *et al.* (1978) also reported that sorghum has a higher WUE than soybeans. Tanner (1981) made a study of the transpiration efficiency of potato and checked field data against an equation used by earlier workers:

$$\frac{\text{Yield}}{\text{Transpiration}} = \frac{k}{\Delta_e}$$

k being a constant that varies with the species and Δ_e the vapor pressure difference between leaf and air. This equation indicates that WUE should be greater in humid regions with a low Δ_e than in arid regions with a high Δ_e. Tanner's paper is a good introduction to the study of WUE in terms of dry matter production, and Fischer and Turner (1978, pp. 295–297) present considerable data on this kind of study. There are also some interesting papers on WUE in Monteith and Webb (1981).

Increasing Water Use Efficiency

There are two approaches to improvement of WUE by crop plants, improved management which chiefly affects the partitioning of water loss between evaporation and transpiration by increasing the fraction used by plants, and plant breeding which may provide plants that yield better under stress.

Plant Breeding. Fischer (1981) states that there are two general approaches to breeding for improvement of WUE: the "black box" and the "ideotype" system. The black box approach involves testing large numbers of cultivars for yield under stress and then looking for differences in structure or processes that might explain the differences in yield. The ideotype approach is based on an understanding of the water regime, of how water deficits affect plant growth and yield, and of what plant characteristics contribute to efficient use of water.

In the past, plant breeders have simply sought for drought tolerance, as evidenced by high yield under water stress, but drought tolerance is too broad a term to be an effective basis for a breeding program. It ought to be possible to specify what morphological and physiological characteristics are likely to improve yield under various conditions. For example, what is the relative importance of morphological characteristics such as deep roots and thick cuticle compared to physiological characteristics such as responsive stomata, C_4 carbon metabolism, osmotic adjustment, and protoplasmic tolerance of dehydration? Much of the material in this book provides the kind of background information needed for such a breeding program. Nevertheless, it remains difficult to identify those physiological characteristics likely to improve yield in specific situations. As desirable characteristics are identified, plant breeders will need simple physiological or biochemical tests that can be used to screen large populations. For example, at one time it was thought that development of proline might be a biochemical indicator of stress tolerance, but it seems to be chiefly an indicator of stress. Some of the problems encountered in using stomatal behavior as a character in breeding for drought tolerance are discussed by Jones, in Mussell and Staples (1979). In the same book, Blum discussed the problem of selecting characteristics to use in breeding for drought tolerance in sorghum. Hall and Grantz (1981) suggested that drought-tolerant cowpeas can be selected simply on the basis of early pod formation, and Boyer (1982) proposed that soybean yield can be related to low afternoon water stress. In regions where late summer droughts are common, there is increasing interest in cultivars suitable for early planting and early maturity that will truly avoid drought. Tolerance of dehydration is uncommon in crop plants and not generally useful because severely stressed crops are seldom profitable. However, the *latente* germplasm described by Castleberry (1983) seems to confer some degree of dehydration tolerance on corn, and at least some sorghum cultivars seem to have dehydration tolerance. Several aspects of the problem of selection for drought tolerance are discussed by various scientists in monographs edited by Eastin *et al.* (1969), Turner and Kramer (1980), Raper and Kramer (1983), Christiansen (1982), and Taylor *et al.* (1983); there is also a paper by Fischer (1981) on this subject. The paper by Nelson in Raper and Kramer (1983) is a summary of problems in breeding for stress tolerance. Eventually, genetic engineering and cell and tissue culture techniques should make important contributions to the improvement of water stress, and tissue culture has already proven useful in screening for salt tolerance (Rains *et al.*, 1980.)

Yield and Stress Tolerance. There is some question whether cultivars that yield well under stress are also relatively high yielding in the absence of stress. Orians and Solbrig (1977) claimed that the close association between loss of water and entrance of CO_2 through stomata and the morphological and physiological traits affecting these gas exchanges prevent plants capable of high rates of photosynthesis in moist soil from being able to maintain high rates in dry soil. In contrast, van den Driesche *et al.* (1971) reported that the biomass production of brigalow (*Acacia harpophylla*), a xeromorphic tree of semiarid regions in Australia, equals that of tropical rain forest trees. Bunce (1981) made extensive measurements of photosynthesis and stomatal and mesophyll conductances of drought-tolerant and drought-intolerant individuals of five species of plants: dandelion, beet, buckwheat, soybean, and sunflower. He found that plants with the highest rates of photosynthesis in moist soil were least able to maintain high rates of photosynthesis or to grow well in dry soil. The differences among plants at high leaf water potentials were caused by differences in stomatal conductance, but differences at low leaf water potentials were caused by differences in mesophyll conductance. Bunce cited several other examples of negative correlations between maximum rate of photosynthesis and drought tolerance. Fischer (1981) observed in a study of wheat cultivars that increase in yield was positively correlated with susceptibility to drought injury, but found no clear reason for this negative relationship. However, Reitz (1974) stated that hard winter wheats fall in three categories: those yielding relatively well only in unfavorable environments, those yielding well only in favorable environments, and those yielding relatively well in all environments, i.e., having a high degree of stability. Blum (1983) also seems to consider it possible to have high yields in both the presence and the absence of stress. However, considerable experimental data support the frequent existence of a negative correlation between high rates of photosynthesis, high stomatal conductance, and drought tolerance. Unfortunately, no adequate explanation is available and further investigation is needed.

An increase in the rate of photosynthesis per unit of leaf surface would increase WUE, and Nobel (1980) suggested that an increase in ratio of mesophyll cell surface to leaf surface might increase the WUE by increasing photosynthesis more than it increases transpiration. There has also been extensive interest in increasing the efficiency of photosynthesis by decreasing photorespiration (Zelitch, 1975) or by transferring the C_4 carbon pathway to C_3 plants by genetic engineering (Bassham, 1977; Radmer and Kok, 1977), but this will be difficult. It is also probable that the increasing CO_2 concentration of the atmosphere will increase WUE by increasing the rate of photosynthesis relative to transpiration. Numerous experiments have shown that an increased concentration of CO_2 results in an increase in dry matter production and yield, if other factors are not limiting (Kramer, 1981). Experiments of Sionit *et al.* (1980a) indicated that the yield of water-stressed wheat in air containing 1000 ppm of CO_2 was equal to that of unstressed wheat in air containing 350 ppm. The only well-known crop plant with a very high WUE is pineapple, and its efficiency results from the fact that it has CAM and its stomata are largely closed during the day (Ekern, 1965; Joshi *et al.,* 1965). However, its total dry matter production is not very high.

Management Practices. Management practices that reduce injury from drought, such as early planting and irrigation, are nearly as old as agriculture. However, their effectiveness is being increased by more detailed knowledge of when and how plants are damaged by water stress and by better weather predictions. For example, it is known that water stress at seed filling seriously reduces soybean yields and that hot, dry weather at silking reduces corn yields. If the probable dates of such weather perturbations are well established, farmers can choose varieties and planting times that minimize the probability of injury from drought. Success in this kind of management requires cooperative research among agronomists, soil scientists, physiologists, plant breeders, and crop climatologists to provide the necessary information.

Another approach is to use crop management practices that result in maximum storage of water in the soil and optimum use by crop plants (Bolton, 1981). These depend on local conditions but often involve fallowing, early planting, weed control, and rapid coverage of the soil by a crop canopy to reduce water loss by evaporation. The general objectives are to store as much water in the soil as possible and to use cultural methods that result in the minimum loss by evaporation and the maximum utilization by crop plants.

At present, it seems likely that WUE can be increased more by adjustment of cultural methods and improving the timing of plant development than by increasing the photosynthetic potential. Natural selection has operated for a long time to develop an optimal balance between plant water use and dry matter production, and this balance is difficult to change. Unfortunately, high rates of transpiration usually accompany high yields because conditions favorable for photosynthesis are favorable for transpiration. Management practices to minimize injury from stress are discussed in Monteith and Webb (1981) and in Raper and Kramer (1983). Irrigation is discussed in Chapter 4.

Antitranspirants

A special case of management is the use of antitranspirants. The usefulness of coatings that would reduce water loss by recently transplanted plants seems to have been recognized as early as the time of Theophrastus, 300 BC (Gale and Hagan, 1966). Reduction of water loss after transplanting permits plants to maintain turgor until root systems are reestablished, and reduction during droughts would enable plants to survive with minimal injury. Also, a reduction of transpiration on watersheds would increase the amount of runoff available for other purposes. There are two approaches to the reduction of transpiration: application of substances causing closure of stomata and application of films of material that cover the entire leaf surface and reduce water loss.

Unfortunately, treatments that decrease the escape of water vapor from leaves also decrease the entrance of CO_2. The use of antitranspirants is based on the assumption, mentioned earlier, that an increase in resistance at the leaf surface will decrease transpiration more than it will decrease CO_2 uptake. This is because the mesophyll resistance (r_m) to water loss is negligible but that for the entrance of CO_2 is assumed to be as high as or higher than the combined boundary layer and stomatal resistances ($r_a + r_s$). Thus, an increase of 50% in ($r_a + r_s$) would cut transpiration by 50%, but if $r_m = r_a + r_s$ such an increase would only decrease photosynthesis by 25%. Slatyer and Bierhuizen (1964) reported that phenylmercuric acetate actually did reduce transpiration of cotton more than it reduced photosynthesis. However, that compound is toxic to plants, and at present there seems to be no satisfactory compound to produce partial closure of stomata. There has been some interest in possible use of ABA or related compounds to reduce transpiration, and Davies and Kozlowski (1975) used ABA successfully on tree seedlings. However, ABA also reduces photosynthesis, and Davies *et al.* (1979) seem pessimistic about its usefulness.

Films such as emulsions of latex, polyvinyl waxes, polyethylene, and higher alcohols such as hexadecanol have been applied with varying results. Their effectiveness seems to depend on the species, stage of development, and atmospheric conditions at the time of application. Films are of limited usefulness on growing plants because repeated applications are required to cover new leaf surface. Although they should be useful on transplanted seedlings, the results have been inconsistent. In forest tree planting, much of the injury seems to occur long after planting, when the soil dries out. Davies and Kozlowski (1975) discussed the problems encountered in the use of antitranspirants on tree seedlings. Films have been used with some success on fruit trees to increase the size of ripening fruit and even to improve keeping qualities after picking (Uriu *et al.*, 1975). Solarova *et al.* (1981) reviewed the literature and tested various films and substances that cause stomatal closure. They concluded that antitranspirants have only limited practical usefulness.

There has been some concern that reduction of transpiration might cause injurious increases in leaf temperature, but this seems unlikely. Gale and Hagan (1966) reviewed the available data and concluded that if the rate of transpiration were halved, the leaf temperature would be increased by only 4°C. It has also been suggested that a decrease in transpiration might reduce the uptake and translocation of mineral nutrients, but this seems unlikely.

Overall, it seems that a reduction in transpiration would have several advantages and no disadvantages if it could be produced without reducing photosynthesis. Unfortunately, this seems impossible because no film-forming material exists that is much more permeable to CO_2 than to water vapor. Thus, although the concept is very attractive, the goal seems unattainable. The use of antitranspirants was reviewed by Gale and Hagan (1966), by Poljakoff-Mayber and Gale (1972), and by Solarova *et al.* (1981).

Energy Cost of Adaptations

In recent years there has been considerable interest in the energy cost of various adaptations. For example, the large root systems and extensive replacement of roots reported in Chapter 6 use a major fraction of the total photosynthate produced. It has also been suggested that root growth may compete with other organs, such as seeds, for photosynthate (see Chapter 6). However, in general, fruits and seeds are much stronger sinks than roots, and root growth is often greatly reduced during seed and fruit production (see Fig. 6.3,Chapter 6). Osmotic adjustment, thick cuticle, and responsive stomata should use relatively little energy compared to root growth. Although some scientists are much concerned about the energy cost of adaptations, the writer is not. Most plants have a greater potential rate of photosynthesis than is used, and in at least some instances, if sink size is increased in comparison to source size, the rate of photosynthesis will also increase. The effect of source–sink size on rate of photosynthesis is discussed by Peet and Kramer (1980), who cite some literature on this interesting problem.

SUMMARY

The term drought is used in this chapter to describe a meteorological event, a period without rain long enough to deplete soil moisture and injure plants. The length of the period without rain that must elapse to constitute a drought depends on the kind of plant, the water-storage capacity of the soil in the root zone, and the atmospheric conditions affecting the rates of evaporation and transpiration. Droughts are permanent in arid regions, seasonal in areas with well-defined wet and dry seasons, or random in many temperate, humid areas. Plants such as desert annuals and some plants of Mediterranean climates avoid drought by flowering and producing seed before the soil dries. Other plants show varying degrees of tolerance, either because they possess adaptations such as deep roots or good control of transpiration that postpone dehydration or because of adaptations that increase tolerance of dehydration, such as osmotic adjustment. A few seed plants and ferns possess protoplasm that survives air drying.

There is strong interest in the possibility of increasing drought tolerance and water use efficiency by plant breeding and better management of crops. At this time, it seems that selection and breeding for characters that increase water absorption or decrease water loss will be more productive than breeding for biochemical and physiological characteristics that are supposedly associated with dehydration tolerance. At present, it is uncertain whether tolerance of water stress is compatible with high yields in the absence of stress. Eventually, plant breeders may produce combinations of characteristics that will increase tolerance of dehydration and water use efficiency without loss of yield, but this will require much more research.

SUPPLEMENTARY READING

Bolton, F. E. (1981). Optimizing the use of water and nitrogen through soil and crop management. *Plant Soil* **58**, 231–247.

Christiansen, M. N., ed. (1982). "Breeding Plants for Less Favorable Environments." Wiley, New York.

Fischer, R. A., and Turner, N. C. (1978). Plant productivity in the arid and semiarid zones. *Annu. Rev. Plant Physiol.* **29**, 277–317.

Levitt, J. (1980). "Responses of Plants to Environmental Stresses," 2nd ed., Vol. 2. Academic Press, New York.

Monteith, J., and Webb, C., eds. (1981). Soil water and nitrogen in Mediterranean type environments. *Plant Soil* **58**, 1–434.

Mussell, H., and Staples, R. C., eds. (1979). "Stress Physiology in Crop Plants." Wiley, New York.

Osmond, C. B., Björkman, O., and Anderson, D. J. (1980). "Physiological Processes in Plant Ecology." Springer-Verlag, Berlin and New York.

Paleg, L. G., and Aspinall, D., eds. (1981). "The Physiology and Biochemistry of Drought Resistance in Plants." Academic Press, New York.

Poljakoff-Mayber, A., and Gale, J. (1972). Physiological basis and practical problems of reducing transpiration. *In* "Water Deficits and Plant Growth" (T. T. Kozlowski, ed.), pp. 277–306. Academic Press, New York.

Rains, D. W., Valentine, R. C., and Hollaender, A., eds. (1980). "Genetic Engineering of Osmoregulation." Plenum. New York.

Raper, C. D., Jr., and Kramer, P. J., eds. (1983). "Crop Reactions to Water and Temperature Stresses in Humid, Temperate Climates." Westview Press, Boulder, Colorado.

Taylor, H. W., Jordan, W. R., and Sinclair, T. B., eds. (1983). "Limitations to Efficient Water Use in Crop Production." Am. Soc. Agron., Madison, Wisconsin.

Turner, N. C., and Begg, J. E. (1981). Plant-water relations and adaptation to stress. *Plant Soil* **58**, 97–131.

Turner, N. C., and Kramer, P. J., eds. (1980). "Adaptation of Plants to Water and High Temperature Stress." Wiley, New York.

Bibliography

Abell, C. A., and Hursh, C. R. (1931). Positive gas and water pressures in oaks. *Science* **73**, 449.

Abu-Shakra, S., Akhtar, M., and Bray, D. W. (1969). Influence of irrigation interval and plant density on alfalfa seed production. *Agron. J.* **61**, 569–571.

Acevedo, E., Hsiao, T. C., and Henderson, D. W. (1971). Immediate and subsequent growth responses of maize leaves to changes in water status. *Plant Physiol.* **48**, 631–636.

Ackerson, R. C. (1980). Stomatal response of cotton to water stress and abscisic acid as affected by water stress history. *Plant Physiol.* **65**, 455–459.

Ackerson, R. C., and Krieg, D. R. (1977). Stomatal and nonstomatal regulation of water use in cotton, corn, and sorghum. *Plant Physiol.* **60**, 850–853.

Ackley, W. B. (1954). Seasonal and diurnal changes in the water contents and water deficits of Bartlett pear leaves. *Plant Physiol.* **29**, 445–448.

Adams, S., Strain, B. R., and Adams, M. S. (1970). Water-repellent soils, fire, and annual plant cover in a desert scrub community of southeastern California. Ecology **51**, 696–700.

Adams, J. A., Bingham, F. T., Kaufmann, M. R., Hoffman, G. J., and Yermanos, D. A. (1978). Responses of stomata and water, osmotic, and turgor potentials of jojoba to water and salt stress. *Agron, J.* **70**, 381–387.

Addoms, R. M. (1937). Nutritional studies on loblolly pine. *Plant Physiol.* **12**, 199–205.

Adeoye, K. B., and Rawlins, S. L. (1981). A split-root technique for measuring root water potential. *Plant Physiol.* **68**, 44–47.

Adjahossou, F., and Vieira da Silva, J. (1978). Teneur en glucides solubles et en amidon et resistance à la sécheresse chez le palmier à huile. *Oleagineux* **33**, 599–604.

Aitchison, G. D., Butler, P. F., and Gurr, C. G. (1951). Techniques associated with the use of gypsum block soil moistures meters. *Aust. J. Appl. Sci.* **2**, 56–75.

Alberte, R. S., Fiscus, E. L., and Naylor, A. W. (1975). The effects of water stress on the development of the photosynthetic apparatus in greening leaves. *Plant Physiol.* **55**, 317–321.

Alberte, R. S., Thornber, J. P., and Fiscus, E. L. (1977). Water stress effects on the content and organization of chlorophyll in mesophyll and bundle sheath chloroplasts of maize. *Plant Physiol.* **59**, 351–353.

Albertson, F. W., and Weaver, J. E. (1945). Injury and death or recovery of trees in prairie climate. *Ecol. Monogr.* **15**, 393–433.

Alberty, R. A., and Daniels, F. (1979). "Physical Chemistry," 5th ed. Wiley, New York.

Albrecht, W. A. (1940). Calcium-potassium-phosphorus relation as a possible factor in ecological array of plants. *J. Am. Soc. Agron.* **32**, 411–418.

Aldrich, W. W., Work, R. A., and Lewis, M. R. (1935). Pear root concentration in relation to soil-moisture extraction in heavy clay soil. *J. Agric. Res. (Washington, D.C.)* **50**, 975–988.

Allen, R. M. (1955). Foliage treatments improve survival of longleaf pine plantings. *J. For.* **53**, 724–727.

Alvim, P. de T. (1960). Moisture stress as a requirement for flowering of coffee. *Science* **132**, 354.

Alvim, P. de T. (1961). Stomatal opening as a practical indicator of water stress in plants. *Proc. Int. Bot. Congr. 9th, 1959,* Vol. 2, p. 5.

Alvim, P. de T. (1965). A new type of porometer for measuring stomatal opening and its use in irrigation studies. *Arid Zone Res.* **25,** 325–329.

Alvim, P. de T., and Havis, J. R. (1954). An improved infiltration series for studying stomatal opening as illustrated with coffee. *Plant Physiol.* **29,** 97–98.

Anderson, D. B., and Kerr, T. (1943). A note on the growth behavior of cotton bolls. *Plant Physiol.* **18,** 261–269.

Anderson, W. P., and House, C. R. (1967). A correlation between structure and function in the root of *Zea mays. J. Exp. Bot.* **18,** 544–555.

Andrews, F. C. (1976). Colligative properties of water. *Science* **194,** 567–571.

Andrews, R. E., and Newman, E. I. (1968). The influence of root pruning on the growth and transpiration of wheat under different soil moisture conditions. *New Phytol.* **67,** 617–630.

Arcichovsky, V., and Ossipov, A. (1931). Die Saugkraft der baumartigen Pflanzen der zentralasiastischen wüsten nebst Transpirationsmessungen am Saxaul (*Arthrophytum haloxylon* Litw.). *Planta* **14,** 552–565.

Arisz, W. H. (1956). Significance of the symplasm theory for transport in the root. *Protoplasma* **46,** 5–62.

Armstrong, W. (1968). Oxygen diffusion from the roots of woody species. *Physiol. Plant.* **21,** 539–543.

Armstrong, W. (1969). Biosphere oxidation in rice: an analysis of intervarietal differences in oxygen flux from the roots. *Physiol. Plant.* **22,** 296–303.

Armstrong, W. (1978). Root aeration in the wetland condition. *In* "Plant Life in Anaerobic Environments" (D. D. Hook and R. M. M. Crawford, eds.), pp. 269–297. Ann Arbor Sci. Press, Ann Arbor, Michigan.

Armstrong, W. (1979). Aeration in higher plants. *Adv. Bot. Res.* **7,** 225–332.

Armstrong, W., and Read, D. J. (1972). Some observations on oxygen transport in conifer seedlings. *New Phytol.* **71,** 55–62.

Army, T. J., and Kozlowski, T. T. (1951). Availability of soil moisture for active absorption in drying soil. *Plant Physiol.* **26,** 353–362.

Arndt, C. H. (1945). Temperature-growth relations of the roots and hypocotyls of cotton seedlings. *Plant Physiol.* **20,** 200–220.

Arntzen, C. J., Haugh, M. F., and Bobick, S. (1973). Induction of stomatal closure by *Helminthosporium maydis* pathotoxin. *Plant Physiol.* **52,** 569–574.

Ashton, F. M. (1956). Measurement of soil moisture. *Plant Physiol.* **31,** 266–274.

Askenasy, E. (1895). Ueber das saftsteigen. *Bot. Zentralbl.* **62,** 237–238.

Aspinall, D.(1980). Role of abscisic acid and other hormones in adaptation to water stress. *In* "Adaptation of Plants to Water and High Temperature Stress" (N. C. Turner and P. J. Kramer, eds.), pp. 154–172. Wiley, New York.

Aston, M. J. (1976). Variations of stomatal diffusion resistance with ambient humidity in sunflower (*Helianthus annuus*). *Aust. J. Plant Physiol.* **3,** 489–502.

Avery, G. S., Jr. (1933). Structure and development of the tobacco leaf. *Am. J. Bot.* **20,** 565–592.

Aylor, D. E., Parlange, J.-Y., and Krikorian, A. D. (1973). Stomatal mechanics. *Am. J. Bot.* **60,** 163–171.

Ayres, P. G. (1978). Water relations of diseased plants. *In* "Water Deficits and Plant Growth" (T. T. Kozlowski, ed.), Vol. 5, pp. 1–60. Academic Press, New York.

Baker, D. N., Hesketh, J. D., and Duncan, W. G. (1972). Simulation of growth and yield in cotton. I. Gross photosynthesis, respiration, and growth. *Crop Sci.* **12,** 431–435.

Baker, K. F., and Snyder, W. C., eds. (1965). "Ecology of Soil-Borne Plant Pathogens, Prelude to Biological Control." Univ. of California Press, Berkeley.

Bakke, A. L., and Noecker, N. L. (1933). The relation of moisture to respiration and heating in stored oats. *Iowa, Agric. Exp. Stn., Res. Bull.* **165.**

Bange, G. G. J. (1953). On the quantitative explanation of stomatal transpiration. *Acta Bot. Neerl.* **2,** 255–297.

Barber, D. A., Ebert, M., and Evans, N. T. S. (1962). Movement of ^{15}O through barley and rice plants. *J. Exp. Bot.* **13,** 397–403.

Barley, K. P. (1962). The effects of mechanical stress on the growth of roots. *J. Exp. Bot.* **13,** 95–110.

Barlow, E. W. R., Munns, R. E., and Brady, C. J. (1980). Drought responses of apical meristems. *In* "Adaptation of Plants to Water and High Temperature Stress" (N. C. Turner and P. J. Kramer, eds.), pp. 191–205. Wiley, New York.

Barlow, E. W. R., Jenka, B., and Vallance, R. J. (1981). The role of turgor maintenance in leaf expansion. *Abstr. Int. Bot. Congr., 13th, 1981,* p. 75.

Barney, C. W. (1951). Effects of soil temperature and light intensity on root growth of loblolly pine seedlings. *Plant Physiol.* **26,** 146–163.

Barrs, H. D. (1968a). Determination of water deficits in plant tissues. *In* "Water Deficits and Plant Growth" (T. T. Kozlowski, ed.), Vol. 1, pp. 235–368. Academic Press, New York.

Barrs, H. D. (1968b). Effect of cyclic variations in gas exchange under constant environmental conditions on the ratio of transpiration to net photosynthesis. *Physiol. Plant.* **21,** 918–929.

Barrs, H. D. (1971). Cyclic variations in stomatal aperture, transpiration, and leaf water potential under constant environmental conditions. *Annu. Rev. Plant Physiol.* **22,** 223–236.

Barrs, H. D., and Klepper, B. (1968). Cyclic variations in plant properties under constant environmental conditions. *Physiol. Plant.* **21,** 711–730.

Barrs, H. D., and Weatherley, P. E. (1962). A reexamination of the relative turgidity technique for estimating water deficits in leaves. *Aust. J. Biol. Sci.* **15,** 413–428.

Basler, E., Todd, G. W., and Meyer, R. E. (1961). Effects of moisture stress in absorption, translocation and distribution of 2,4-dichlorophenoxyacetic acid in bean plants. *Plant Physiol.* **36,** 573–576.

Bassett, J. R. (1964). Tree growth as affected by soil moisture availability. *Soil Sci. Soc. Am. Proc.* **28,** 436–438.

Bassham, J. A. (1977). Increasing crop production through more controlled photosynthesis. *Science* **197,** 630–638.

Bates, C. G. (1924). Relative resistance of tree seedlings to excessive heat. *U.S., Dep. Agric., Bull.* **1263.**

Batjer, L. P., Magness, J. P., and Regeimbal, L. O. (1939). The effect of root temperature on growth and nitrogen intake of apple trees. *Proc. Am. Soc. Hortic. Sci.* **37,** 11–18.

Batra, M. W., Edwards, K. L., and Scott, T. K. (1975). Auxin transport in roots: Its characteristics and relationship to growth. *In* "The Development and Function of Roots" (J. G. Torrey and D. T. Clarkson, eds.), pp. 299–325. Academic Press, New York.

Baughn, J. W., and Tanner, C. B. (1976). Leaf water potential: Comparison of pressure chamber and in situ hygrometer on five herbaceous species. *Crop. Sci.* **16,** 181–184.

Baver, L. D. (1948). "Soil Physics," 2nd ed. Wiley, New York.

Baver, L. D., Gardner, W. H., and Gardner, W. R. (1972). "Soil Physics," 4th ed. Wiley, New York.

Bayliss, W. M. (1924). "Principles of General Physiology," 4th ed., Chapter 8, Longmans, Green, New York.

Beament, J. W. L. (1965). The active transport of water: Evidence, models and mechanisms. *Symp. Soc. Exp. Biol.* **19,** 273–298.

Beardsell, M. F., Jarvis, P. G., and Davidson, B. (1972). A null-balance diffusion porometer suitable for use with leaves of many shapes. *J. Appl. Ecol.* **9,** 677–690.

Beasley, R. S. (1976). Contribution of subsurface flow from the upper slopes of forested watersheds to channel flow. *Soil Sci. Soc. Am. Proc.* **40**(6), 955–957.

Beckman, C. A. (1964). Host responses to vascular infection. *Annu. Rev. Phytopathol.* **2**, 231–252.

Beevers, H. (1979). Microbodies in higher plants. *Annu. Rev. Plant Physiol.* **30**, 159–193.

Begg, J. E. (1980). Morphological adaptations of leaves to water stress. *In* "Adaptation of Plants to Water and High Temperature Stress" (N. C. Turner and P. J. Kramer, eds.), pp. 33–42. Wiley, New York.

Begg, J. E., and Turner, N. C. (1976). Crop water deficits. *Adv. Agron.* **28**, 161–217.

Bennet-Clark, T. A. (1959). Water relations of cells. *In* "Plant Physiology" (F. C. Steward, ed.), vol. 2, pp. 105–191. Academic Press, New York.

Bennet-Clark, T. A., Greenwood, A. D., and Barker, J. W. (1936). Water relations and osmotic pressures of plant cells. *New Phytol.* **35**, 277–291.

Bennett, K. J., and Rook, D. A. (1978). Stomatal and mesophyll resistances in two clones of *Pinus radiata* D. Don known to differ in transpiration and survival rate. *Aust. J. Plant Physiol.* **5**, 231–238.

Berkowitz, G. A., and Hopper, N. W. (1980). A method of increasing the accuracy of diffusive resistance porometer calibrations. *Ann. Bot. (London)* [N.S.] **45**, 723–727.

Bernal, J. D. (1965). The structure of water and its biological implications. *Symp. Soc. Exp. Biol.* **19**, 17–32.

Bernstein, L., Gardner, W. R., and Richards, L. A. (1959). Is there a vapor gap around roots? *Science* **129**, 1750, 1753.

Berry, L. J. (1949). The influence of oxygen on the respiratory rate in different segments of onion roots. *J. Cell Comp. Physiol.* **33**, 41–66.

Bewley, J. D. (1979). Physiological aspects of desiccation tolerance. *Annu. Rev. Plant Physiol.* **30**, 195–238.

Bhardwaj, R., and Singhal, G. S. (1981). Effect of water stress on photochemical activity of chloroplasts during greening of etiolated barley seedlings. *Plant & Cell Physiol.* **22**, 155–162.

Bialoglowski, J. (1936). Effect of extent and temperature of roots on transpiration of rooted lemon cuttings. *Proc. Am. Soc. Hortic. Sci.* **34**, 96–102.

Biddulph, O., Nakayama, F. S., and Cory, R. (1961). Transpiration and ascension of calcium. *Plant Physiol.* **36**, 429–436.

Billings, W. D., and Mooney, H. A. (1968). The ecology of arctic and alpine plants. *Biol. Rev. Cambridge Philos. Soc.* **43**, 481–529.

Bingham, E. C., and Jackson, R. F. (1918). Standard substances for the calibration of viscometers. *Bur. Stand. (U.S.), Bull.* **14**, 59–86.

Biswell, H. H. (1935). Effect of the environment upon the root habits of certain deciduous forest trees. *Bot. Gaz. (Chicago)* **96**, 676–708.

Björkman, E. (1942). Über die Bedingungen der Mykorrhizabildung bei Kiefer und Fichte. *Symb. Bot. Ups.* **6**(2), 1–190.

Björkman, O., Downton, W. J. S., and Mooney, H. A. (1980). Response and adaptation to water stress in *Nerium oleander*. *Year Book—Carnegie Inst. Washington* **79**, 150–157.

Björkman, O., Powles, S. B., Fork, D. C., and Öquist, G. (1981). Interaction between high irradiance and water stress on photosynthetic reactions. *Year Book—Carnegie Inst. Washington* **80**, 57–59.

Black, C. A. (1968). "Soil-Plant Relationships." Wiley, New York.

Blake, G. R., Allred, E. R., van Bavel, C. H. M., and Whisler, F. D. (1960). Agricultural drought and moisture excesses in Minnesota. *Minn., Agric. Exp. Stn., Tech. Bull.* **235**.

Blinks, L. R., and Airth, R. L. (1951). The role of electroosmosis in living cells. *Science* **113**, 474–475.

Blizzard, W. E., and Boyer, J. S. (1980). Comparative resistance of the soil and the plant to water transport. *Plant Physiol.* **66**, 809–814.

Bloodworth, M. E., and Page, J. B. (1957). Use of thermistors for the measurement of soil moisture and temperature. *Soil Sci. Soc. Am. Proc.* **21**, 11–15.

Bloodworth, M. E., Page, J. B., and Cowley, W. R. (1956). Some applications of the thermoelectric method for measuring water flow rates in plants. *Agron. J.* **48**, 222–228.

Blum, A. (1979). Genetic improvement of drought resistance in crop plants: A case for sorghum. *In* "Stress Physiology in Crop Plants" (H. Mussell and R. C. Staples, eds.), pp. 429–445. Wiley, New York.

Blum, A. (1983). Breeding programs for improving crop resistance to water stress. *In* "Crop Reaction to Water and Temperature Stress in Humid, Temperate Climates" (C. D. Raper, Jr., and P. J. Kramer, eds.), pp. 263–275. Westview Press, Boulder, Colorado.

Bodman, G. B., and Colman, E. A. (1944). Moisture and energy conditions during downward entry of water into soils. *Soil Sci. Soc. Am. Proc.* **8**, 116–122.

Boehm, J. (1893). Capillarität und Saftsteigen. *Ber. Dtsch. Bot. Ges.* **11**, 203–212.

Böhm, W. (1979). "Methods of Studying Root Systems." Springer-Verlag, Berlin and New York.

Böhning, R. H., and Lusanandana, B. (1952). A comparative study of gradual and abrupt changes in root temperature on water absorption. *Plant Physiol.* **27**, 475–488.

Bollard, E. G. (1960). Transport in the xylem. *Annu. Rev. Plant Physiol.* **11**, 141–166.

Bolton, E. F., and Erickson, A. E. (1970). Ethanol concentration in tomato plants during soil flooding. *Agron. J.* **62**, 220–224.

Bolton, F. E. (1981). Optimizing the use of water and nitrogen through soil and crop management. *Plant & Soil* **58**, 231–247.

Bonner, J. (1959). Water transport. *Science* **129**, 447–450.

Bonner, J., and Varner, J. E., eds. (1976). "Plant Biochemistry," 3rd ed. Academic Press, New York.

Bordovsky, D. G., Jordan, W. R., Hiler, E. A., and Howell, T. A. (1974). Choice of irrigation timing indicator for narrow row cotton. *Agron. J.* **66**, 88–91.

Bormann, F. H. (1957). Moisture transfer between plants through intertwined root systems. *Plant Physiol.* **32**, 48–55.

Bormann, F. H., and Graham, B. F., Jr. (1959). The occurrence of natural root grafting in eastern white pine, *Pinus strobus* L., and its ecological implications. *Ecology* **40**, 677–691.

Bormann, F. H., and Graham, B. F., Jr. (1960). Translocation of silvicides through root grafts. *J. For.* **58**, 402–403.

Bose, J. C. (1923). "Physiology of the Ascent of Sap." Longmans, Green, New York.

Bose, J. C. (1927). "Plant Autographs and Their Revelations." Macmillan, New York.

Bouyoucos, G. T. (1949). Nylon electrical resistance unit for continuous measurement of soil moisture in the field. *Soil Sci.* **67**, 319–330.

Bouyoucos, G. J. (1951). Effect of fertilizers on the plaster of paris electrical resistance method of measuring soil moisture in the field. *J. Am. Soc. Agron.* **43**, 508–511

Bouyoucos, G. J. (1953a). An improved type of soil hydrometer. *Soil Sci.* **76**, 377–378.

Bouyoucos, G. J. (1953b). More durable plaster of paris moisture blocks. *Soil Sci.* **76**, 447–451.

Bouyoucos, G. J. (1954). New type electrode for plaster of paris moisture blocks. *Soil Sci.* **78**, 339–342.

Bouyoucos, G. J., and Mick, A. H. (1940). An electrical resistance method for the continuous measurement of soil moisture under field conditions. *Mich. Agric. Exp. Stn., Tech. Bull.* **172**.

Bouyoucos, G. J., and Mick, A. H. (1948). A comparison of electric resistance units for making a continuous measurement of soil moisture under field conditions. *Plant Physiol.* **23**, 532–543.

Bowen, G. D., and Theodorou, C. (1967). Studies of phosphate uptake by mycorrhizas. *Proc. IUFRO Congr., 14th, 1967,* Vol. 5, pp. 116–138.

Bowling, D. J. F. (1973). Measurement of a gradient of oxygen partial pressure across the intact root. *Planta* **111**, 323–328.

Boyce, S. G. (1954). The salt spray community. *Ecol. Monogr.* **24**, 29–67.

Boyer, J. S. (1965). Effects of osmotic water stress on metabolic rates of cotton plants with open stomata. *Plant Physiol.* **40**, 229–234.

Boyer, J. S. (1967). Matric potentials of leaves. *Plant Physiol.* **42**, 213–217.

Boyer, J. S. (1968). Relationship of water potential to growth of leaves. *Plant Physiol.* **43**, 1056–1062.

Boyer, J. S. (1969). Measurement of the water status of plants. *Annu. Rev. Plant Physiol.* **20**, 351–364.

Boyer, J. S. (1970a). Leaf enlargement and metabolic rates in corn, soybean, and sunflower at various leaf water potentials. *Plant Physiol.* **46**, 233–235.

Boyer, J. S. (1970b). Differing sensitivity of photosynthesis to low leaf water potentials in corn and soybean. *Plant Physiol.* **46**, 236–239.

Boyer, J. S. (1971). Nonstomatal inhibition of photosynthesis in sunflower at low leaf water potentials and high light intensities. *Plant Physiol.* **48**, 532–536.

Boyer, J. S. (1974). Water transport in plants: Mechanism of apparent changes in resistance during absorption. *Planta* **117**, 187–207.

Boyer, J. S. (1976). Water deficits and photosynthesis. *In* "Water Deficits and Plant Growth" (T. T. Kozlowski, ed.), Vol. 4, pp. 154–190. Academic Press, New York.

Boyer, J. S. (1983). Environmental stress and crop yields. *In* "Crop Reactions to Water and Temperature Stresses in Humid, Temperate Climates" (C. D. Raper, Jr. and P. J. Kramer, eds.), pp. 3–7. Westview Press, Boulder, Colorado.

Boyer, J. S., and Knipling, E. B. (1965). Isopiestic technique for measuring leaf water potentials with a thermocouple psychrometer. *Proc. Natl. Acad. Sci. U.S.A.* **54**, 1044–1051.

Boyer, J. S., and Potter, J. R. (1973). Chloroplast response to low leaf water potentials. I. Role of turgor. *Plant Physiol.* **51**, 989–992.

Boyer, J. S., Johnson, R. R., and Saupe, S. G. (1980). Afternoon water deficits and grain yields in old and new soybean cultivars. *Agron. J.* **72**, 981–986.

Boyko, H. (1966). "Salinity and Aridity: New Approaches to Old Problems." Junk, The Hague.

Bradford, K. J., and Yang, S. F. (1980). Xylem transport of 1-aminocyclopropane-1-carboxylic acid, an ethylene precursor, in waterlogged tomato plants. *Plant Physiol.* **65**, 322–326.

Bradshaw, A. D. (1965). Evolutionary significance of phenotypic plasticity in plants. *Adv. Genet.* **13**, 115–155.

Brady, N. C. (1974). "The Nature and Properties of Soils," 8th ed. Macmillan, New York.

Brauner, L. (1930). Uber polare Permeabilität. *Ber. Dtsch. Bot. Ges.* **49**, 109–118.

Brauner, L. (1956). Die Permeabilität der Zellwand. *Encycl. Plant Physiol.* **2**, 337–357.

Brauner, L., and Hasman, M. (1946). Untersuchungen über die anomale Komponente des osmotischen Potentials lebender Pflanzenzellen. *Rev. Fac. Sci. Univ. Istanbul, Ser. B* **11**, 1–37.

Bray, J. R. (1963). Root production and the estimation of net productivity. *Can. J. Bot.* **41**, 65–72.

Breazeale, J. F., and Crider, F. J. (1934). Plant association and survival, and the build-up of moisture in semi-arid soils. *Ariz. Agric. Exp. Stn., Tech. Bull.* **53**.

Briggs, G. E. (1967). "Movement of Water in Plants." Davis, Philadelphia, Pennsylvania.

Briggs, G. E., and Robertson, R. N. (1954). Apparent free space. *Annu. Rev. Plant Physiol.* **8**, 11–30.

Briggs, L. J. (1949). A new method of measuring the limiting negative pressure in liquids. *Science* **109**, 440.

Briggs, L. J., and Shantz, H. L. (1911). A wax seal method for determining the lower limit of available soil moisture. *Bot. Gaz. (Chicago)* **51**, 210–219.

Briggs, L. J., and Shantz, H. L. (1912). The relative wilting coefficients for different plants. *Bot. Gaz. (Chicago)* **53,** 229–235.

Bristow, K. L., Van Zyl, W. H., and DeJager, J. M. (1981). Measurement of leaf water potential using the J14 press. *J. Exp. Bot.* **32,** 851–854.

Brix, H. (1962). The effect of water stress on the rates of photosynthesis and respiration in tomato plants and loblolly pine seedlings. *Physiol. Plant.* **15,** 10–20.

Brix, H. (1966). Errors in measurement of leaf water potential of some woody plants with the Schardakow dye method. *Can., For. Branch, Dep. Publ.* **1164.**

Brouwer, R. (1953). Water absorption by the roots of *Vicia faba* at various transpiration strengths. I. II. *Proc. K. Ned. Akad. Wet. Ser. C* **56,** 106–115, 129–136.

Brouwer, R. (1954). Changes in water conductivity artificially obtained. *Proc. K. Ned. Akad. Wet., Ser. C* **57,** 68–80.

Brouwer, R. (1964). Responses of bean plants to root temperatures. I. Root temperatures and growth in the vegetative stage. *Jaarb. I.B.S.* (Wageningen) **22,** 11–22.

Brouwer, R. (1965). Ion absorption and transport in plants. *Annu. Rev. Plant Physiol.* **16,** 241–266.

Brouwer, R., and Hoogland, A. (1964). Responses of bean plants to root temperatures. II. Anatomical aspects. *Jaarb. I.B.S.* (Wageningen) **23,** 23–31.

Brouwer, R., Gasparikova, O., Kolek, J., and Loughman, B. G., eds. (1982). "Structure and Function of Plant Roots." Junk, The Hague.

Brown, E. M. (1939). Some effects of temperature on the growth and chemical composition of certain pasture grasses. *Res. Bull.—Mo., Agric. Exp. Stn.* **299.**

Brown, H. T., and Escombe, F. (1900). Static diffusion of gases and liquids in relation to the assimilation of carbon and translocation of plants. *Philos. Trans. R. Soc. London, Ser. B* **193,** 223–291.

Brown, R. W., and Van Haveren, B. P., eds. (1972). "Psychrometry in Water Relations Research." Utah Agric. Exp. Stn., Utah State University, Logan.

Brown, W. V. (1952). The relation of soil moisture to cleistogamy in *Stipa leucotricha. Bot. Gaz. (Chicago)* **113,** 438–444.

Brown, W. V., and Pratt, G. A. (1965). Stomatal inactivity in grasses. *Southwest. Nat.* **10**(1), 48–56.

Broyer, T. C. (1947). The movement of materials into plants. I. Osmosis and the movement of water into plants. *Bot. Rev.* **13,** 1–58.

Broyer, T. C. (1951). Experiments on imbibition and other factors concerned in the water relations of plant tissues. *Am. J. Bot.* **38,** 485–495.

Buckingham, E. A. (1907). Studies of the movement of soil moisture. *U.S. Dep. Agric., Bull.* **38.**

Bucks, D. A., Erie, L. J., and French, O. F. (1974). Quantity and frequency of trickle and furrow irrigation for efficient cabbage production. *Agron. J.* **66,** 53–57.

Bunce, J. A. (1977). Leaf elongation in relation to leaf water potential in soybean. *J. Exp. Bot.* **28,** 156–161.

Bunce, J. A. (1978a). Effects of shoot environment on apparent root resistance to water flow in whole soybean and cotton plants. *J. Exp. Bot.* **29,** 595–601.

Bunce, J. A. (1978b). Effects of water stress on leaf expansion, net photosynthesis, and vegetative growth of soybeans and cotton. *Can. J. Bot.* **56,** 1492–1498.

Bunce, J. A. (1981). Relationships between maximum photosynthetic rates and photosynthetic tolerance of low leaf water potentials. *Can. J. Bot.* **59,** 769–774.

Bunce, J. A. (1982). Effects of water stress on photosynthesis in relation to diurnal accumulation of carbohydrate in source leaves. *Can. J. Bot.* **60,** 195–200.

Burgerstein, A. (1920). "Die Transpiration der Pflanzen," vol. II. Fischer, Jena.

Burke, M. J., Bryant, R. C., and Weiser, C. J. (1974). Nuclear magnetic resonance of water in cold acclimating red osier dogwood stem. *Plant Physiol.* **54,** 392–398.

Burke, M. J., Gusta, L. V., Quamme, H. A., Weiser, C. J., and Li, P. H. (1976). Freezing and injury in plants. *Annu. Rev. Plant Physiol.* **27**, 507–528.

Burrows, F. J., and Milthorpe, F. L. (1976). Stomatal conductance in the control of gas exchange. *In* "Water Deficits and Plant Growth" (T. T. Kozlowski, ed.), Vol. 4, pp. 103–152. Academic Press, New York.

Burrows, W. J., and Carr, D. J. (1969). Effects of flooding the root system of sunflower plants on the cytokinin content of the xylem sap. *Physiol. Plant.* **22**, 1105–1112.

Burström, H. (1959). Growth and formation of intercellularies in root meristems. *Physiol. Plant.* **12**, 371–385.

Burström, H. (1965). The physiology of plant roots. *In* "Ecology of Soil-borne Plant Pathogens" (K. F. Baker and W. C. Snyder, eds.), pp. 154–159. Univ. of California Press, Berkeley.

Burström, H. (1971). Wishful thinking of turgor. *Nature (London)* **234**, 488.

Burton, G. W., DeVane, E. H., and Carter, R. L. (1954). Root penetration, distribution and activity in southern grasses measured by yields, drought symptoms, and p^{32} uptake. *Agron. J.* **46**, 229–233.

Bushnell, J. (1941). Exploratory tests of subsoil treatments inducing deeper rooting of potatoes on Wooster silt loam. *Agron. J.* **33**, 823–828.

Bushnell, W. R., and Rowell, J. B. (1968). Premature death of adult rusted wheat plants in relation to carbon dioxide evolution by root systems. *Phytopathology* **58**, 651–658.

Buswell, A. M., and Rodebush, W. H. (1956). Water. *Sci. Am.* **194**(4), 77–89.

Butler, G. W. (1953). Ion uptake by young wheat plants. II. The "apparent free space" of wheat roots. *Physiol. Plant.* **6**, 617–635.

Buttery, B. R., and Boatman, S. G. (1966). Manometric measurement of turgor pressures in laticiferous phloem tissues. *J. Exp. Bot.* **17**, 283–296.

Byott, G. S., and Sheriff, D. W. (1976). Water movement into and through *Tradescantia virginiana* L. leaves. II. Liquid flow pathways and evaporative sites. *J. Exp. Bot.* **27**, 634–639.

Caldwell, M. M. (1970). Plant gas exchange at high wind speeds. *Plant Physiol.* **46**, 535–537.

Caldwell, M. M. (1976). Root extension and water absorption. *In* "Water and Plant Life" (O. L. Lange, L. Kappen, and E.-D. Schulz, eds.), pp. 63–85. Springer-Verlag, Berlin.

Caldwell, M. M., White, R. S., Moore, R. T., and Camp, L. B. (1977). Carbon balance, productivity and water use of cold-winter desert shrub communities dominated by C$_3$ and C$_4$ species. *Oecologia* **29**, 275–300.

Camacho-B, S. E., Hall, A. E., and Kaufmann, M. R. (1974). Efficiency and regulation of water transport in some woody and herbaceous species. *Plant Physiol.* **54**, 169–172.

Cameron, S. H. (1941). The influence of soil temperature on the rate of transpiration of young orange trees. *Proc. Am. Soc. Hortic. Sci.* **38**, 75–79.

Campbell, G. S., and Campbell, M. D. (1974). Evaluation of a thermocouple hygrometer for measuring leaf water potential *in situ*. *Agron. J.* **66**, 24–27.

Campbell, G. S., Papendick, R. I., Rabie, E., and Shayo-Ngowi, A. J. (1979). A comparison of osmotic potential, elastic modulus, and apoplastic water in leaves of dryland winter wheat. *Agron. J.* **71**, 31–36.

Campbell, R. B., Bower, C. A., and Richards, L. A. (1949). Change of electrical conductivity with temperature and the relation of osmotic pressure to electrical conductivity and ion concentration for soil extracts. *Soil Sci. Soc. Am. Proc.* **13**, 66–69.

Cannell, R. Q., Gales, K., Snaydon, R. W., and Suhail, B. A. (1979). Effects of short-term waterlogging on the growth and yield of peas. *Ann. Appl. Biol.* **93**, 327–335.

Carlquist, S. (1975). "Ecological Strategies of Xylem Evolution." Univ. of California Press, Berkeley.

Carson, E. W., ed. (1974). "The Plant Root and Its Environment." University Press of Virginia, Charlottesville.

Carter, J. C. (1945). Wetwood of elms. *Circ.—Ill. Nat. Hist. Surv.* **23**, 401–448.

Cassell, D. K. (1983). Effects of soil characteristics and tillage practices on water storage and its availability to plant roots. *In* "Crop Reactions to Water and Temperature Stresses in Humid, Temperate Climates" (C. D. Raper, Jr. and P. J. Kramer, eds.), pp. 167–186. Westview Press, Boulder, Colorado.

Castleberry, R. M. (1983). Breeding programs for stress tolerance in corn. *In* "Crop Reactions to Water and Temperature Stresses in Humid, Temperate Climates" (C. D. Raper, Jr. and P. J. Kramer, eds.), pp. 277–287, Westview Press, Boulder, Colorado.

Čatsky, J. (1965). Leaf-disc method for determining water saturation deficit. *Arid Zone Res.* **25**, 353–360.

Caughey, M. G. (1945). Water relations of pocasin or bog shrubs. *Plant Physiol.* **20**, 671–689.

Cermak, J., and Kucera, J. (1981). The compensation of natural temperature gradient at the measuring point during the sap flow rate determination in trees. *Biol. Plant.* **23**, 469–471.

Ceulemans, R., Impens, I., Lemeur, R., Moermans, R., and Samsuddin, Z. (1978). II. Comparative study of transpiration regulation during water stress situations in four different poplar clones. *Oecol. Plant.* **13**, 139–146.

Chambers, T. C., Ritchie, I. M., and Booth, M. A. (1976). Chemical models for plant wax morphogenesis. *New Phytol.* **77**, 43–49.

Chaney, W. R. (1981). Sources of water. *In* "Water Deficits and Plant Growth" (T. T. Kozlowski, ed.), Vol. 6, pp. 1–47. Academic Press, New York.

Chaney, W. R., and Kozlowski, T. T. (1977). Patterns of water movement in intact and excised stems of *Fraxinus americana and Acer saccharum* seedlings. *Ann. Bot. (London)* [N.S.] **41**, 1093–1100.

Chang, H. T., and Loomis, W. E. (1945). Effect of carbon dioxide on absorption of water and nutrients by roots. *Plant Physiol.* **20**, 221–232.

Chapman, H. D., and Parker, E. R. (1942). Weekly absorption of nitrate by young bearing orange trees growing out of doors in solution cultures. *Plant Physiol.* **17**, 366–376.

Chatterton, N. J., Hanna, W. W., Powell, J. B., and Lee, D. R. (1975). Photosynthesis and transpiration of bloom and bloomless sorghum. *Can. J. Plant Sci.* **55**, 641–643.

Cheung, Y. N. S., Tyree, M. T., and Dainty, J. (1975). Water relations parameters on single leaves obtained in a pressure bomb and some ecological interpretations. *Can. J. Bot.* **53**, 1342–1346.

Chibnall, A. C. (1939). "Protein Metabolism in the Plant," pp. 265–266. Yale Univ. Press, New Haven, Connecticut.

Christensen, N. L., Burchell, R. B., Liggett, A., and Simms, E. L. (1981). The structure and development of pocosin vegetation. *In* "Pocosin Wetlands" (C. J. Richardson, ed.), pp. 43–61. Hutchinson Ross Publ. Co., Stroudsburg, Pennsylvania.

Christiansen, M. N., ed. (1982). "Breeding Plants for Less Favorable Environments." Wiley, New York.

Chrispeels, M. J. (1976). Biosynthesis, intracellular transport, and secretion of extracellular macromolecules. *Annu. Rev. Plant Physiol.* **27**, 19–38.

Chung, H.-H., and Kramer, P. J. (1975). Absorption of water and ^{32}P through suberized and unsuberized roots of loblolly pine. *Can. J. For. Res.* **5**, 229–235.

Clark, J., and Gibbs, R. D. (1957). Studies in tree physiology. IV. Further investigations of seasonal changes in moisture content of certain Canadian forest trees. *Can. J. Bot.* **35**, 219–253.

Clark, W. S. (1874). The circulation of sap in plants. *Mass. State Board Agric. Annu. Rep.* **21**, 159–204.

Clark, W. S. (1875). Observations upon the phenomena of plant life. *Mass. State Board Agric. Annu. Rep.* **22**, 204–312.

Clarkson, D. T. (1976). The influence of temperature on the exudation of xylem sap from detached root systems of rye (*Secale cereale*) and barley (*Hordeum vulgare*). *Planta* **132**, 297–304.

Clarkson, D. T., and Robards, A. W. (1975). The endodermis, its structural development and physiological role. *In* "The Development and Function of Roots" (J. G. Torrey and D. T. Clarkson, eds.), pp. 415–436. Academic Press, New York.

Clarkson, D. T., Robards, A. W., and Sanderson, J. (1971). The tertiary endodermis in barley roots: Fine structure in relation to radial transport of ions and water. *Planta* **96,** 292–305.

Clarkson, D. T., Mercer, E. R., Johnson, M. G., and Mattam, D. (1975). The uptake of nitrogen (ammonium and nitrate) by different segments of the roots of intact barley plants. *Annu. Rep.—Agric. Res. Counc., Letcombe Lab.* pp. 10–13.

Clarkson, D. T., Robards, A. W., Sanderson, J., and Peterson, C. A. (1978). Permeability studies on epidermal-hypodermal sleeves isolated from roots of *Allium cepa* (onion). *Can. J. Bot.* **56,** 1526–1532.

Clausen, J. J., and Kozlowski, T. T. (1965). Use of the relative turgidity technique for measurement of water stresses in gymnosperm leaves. *Can. J. Bot.* **43,** 305–316.

Clawson, K. L., and Blad, B. L. (1982). Infrared thermometry for scheduling irrigation of corn. *Agron. J.* **74,** 311–316.

Cleland, R. (1971). Cell wall extension. *Annu. Rev. Plant Physiol.* **22,** 197–223.

Clements, F. E. (1921). Aeration and air content. *Carnegie Inst. Washington Publ.* **315.**

Clements, F. E., and Martin, E. V. (1934). Effect of soil temperature on transpiration in *Helianthus annuus. Plant Physiol.* **9,** 619–630.

Clements, H. F. (1934). Significance of transpiration. *Plant Physiol.* **9,** 165–171.

Clements, H. F., and Kubota, T. (1942). Internal moisture relations of sugar cane—the selection of a moisture index. *Hawaii. Plant. Rec.* **46,** 17–36.

Cloutier, Y., and Siminovitch, D. (1982). Correlation between cold- and drought-induced frost hardiness in winter wheat and rye varieties. *Plant Physiol.* **69,** 256–258.

Coile, T. S. (1936). Soil samplers. *Soil Sci.* **42,** 139–142.

Coile, T. S. (1937). Distribution of forest tree roots in North Carolina Piedmont soils. *J. For.* **35,** 247–257.

Cole, F. D., and Decker, J. P. (1973). Relation of transpiration to atmospheric vapor pressure. *J. Ariz. Acad. Sci.* **8,** 74–75.

Cole, P. J., and Alston, A. M. (1974). Effect of transient dehydration on absorption of chloride by wheat roots. *Plant Soil* **40,** 243–247.

Colman, E. A., and Hendrix, T. M. (1949). The fiberglas electrical soil-moisture instrument. *Soil Sci.* **67,** 425–438.

Colton, C. E., and Einhellig, F. E. (1980). Allelopathic mechanisms of velvetleaf (*Abutilon theophrasti,* Medic., Malvaceae) on soybean. *Am. J. Bot.* **67,** 1407–1413.

Conde, L. F., and Kramer, P. J. (1975). The effect of vapor pressure deficit on diffusion resistance in *Opuntia compressa. Can. J. Bot.* **53,** 2923–2926.

Connor, D. J., Legge, N. J., and Turner, N. C. (1977). Water relations of mountain ash (*Eucalyptus regnans* F. Muell.) forests. *Aust. J. Plant Physiol.* **4,** 753–762.

Cooper, J. P. (1963). Species and population differences in climatic response. *In* "Environmental Control of Plant Growth" (L. T. Evans, ed.), pp. 381–403. Academic Press, New York.

Corey, A. T., and Kemper, W. D. (1961). Concept of total potential in water and its limitations. *Soil Sci.* **91,** 299–302; **92,** 281–283.

Corey, A. T., Slatyer, R. O., and Kemper, W. D. (1967). Comparative terminologies for water in the soil-plant-atmosphere system. *In* "Irrigation of Agricultural Lands" (R. M. Hagan, H. R. Haise, and T. W. Edminster, eds.), pp. 427–445. Am. Soc. Agron., Madison, Wisconsin.

Cormack, R. G. H. (1944). The effect of environmental factors on the development of root hairs in *Phleum pratense* and *Sporobolus cryptandrus. Am. J. Bot.* **31,** 443–449.

Cormack, R. G. H. (1945). Cell elongation and the development of root hairs in tomato roots. *Am. J. Bot.* **32,** 490–496.

Couchat, P., and Lasceve, G. (1980). Tritiated water vapour exchange method for the evaluation of whole plant diffusion resistance. *J. Exp. Bot.* **31,** 1217–1222.

Couchat, P., Moutonnet, P., Houelle, M., and Pickard, D. (1980). *In situ* study of corn seedling roots and shoot growth by neutron radiography. *Agron. J.* **72,** 321–324.

Cowan, I. R. (1965). Transport of water in the soil-plant-atmosphere system. *J. Appl. Ecol.* **2,** 221–239.

Cowan, I. R. (1972). Oscillations in stomatal conductance and plant functioning associated with stomatal conductance. I. Observations and a model. *Planta* **106,** 185–219.

Cowan, I. R. (1977). Stomatal behavior and environment. *Adv. Bot. Res.* **4,** 117–228.

Cowan, I. R. (1982). Regulation of water use in relation to carbon gain in higher plants. *Encycl. Plant Physiol., New Ser.* **12B,** 589–613. Springer-Verlag, Berlin and New York.

Cowan, I. R., and Farquhar, G. D. (1977). Stomatal function in relation to leaf metabolism and environment. *Symp. Soc. Exp. Biol.* **31,** 471–505.

Cowan, I. R., and Milthorpe, F. L. (1967). Resistance to water transport in plants—a misconception misconceived. *Nature (London)* **213,** 740–741.

Cowan, I. R., and Milthorpe, F. L. (1968). Plant factors influencing the water status of plant tissues. *In* "Water Deficits and Plant Growth" (T. T. Kozlowski, ed.), Vol. 1, pp. 137–193. Academic Press, New York.

Crafts, A. S. (1936). Further studies on exudation in cucurbits. *Plant Physiol.* **11,** 63–79.

Crafts, A. S., and Broyer, T. C. (1938). Migration of salts and water into xylem of the roots of higher plants. *Am. J. Bot.* **25,** 529–535.

Crafts, A. S., Currier, H. B., and Stocking, C. R. (1949). "Water in the Physiology of the Plant." Chronica Botanica, Waltham, Massachuetts.

Crawford, R. M. M. (1967). Alcohol dehydrogenase activity in relation to flooding tolerance in roots. *J. Exp. Bot.* **18,** 458–464.

Crawford, R. M. M. (1976). Tolerance of anoxia and the regulation of glycolysis in tree roots. *In* "Tree Physiology and Yield Improvement" (M. G. R. Cannell and F. T. Last, eds.), pp. 387–401. Academic Press, New York.

Crider, F. J. (1933). Selective absorption of ions not confined to young roots. *Science* **78,** 169.

Curtis, L. C. (1944). The exudation of glutamine from lawn grass. *Plant Physiol.* **19,** 1–5.

Curtis, O. F. (1937). Vapor pressure gradients, water distribution in fruits, and so-called infra-red injury. *Am. J. Bot.* **24,** 705–710.

Curtis, O. F., and Clark, D. G. (1950). "An Introduction to Plant Physiology." McGraw-Hill, New York.

Cutler, J. M., and Rains, D. W. (1978). Effects of water stress and hardening on the internal water relations and osmotic constituents of cotton leaves. *Physiol. Plant.* **42,** 261–268.

Cutler, J. M., Rains, D. W., and Loomis, R. S. (1977). On the importance of cell size in the water relations of plants. *Physiol. Plant.* **40,** 256–259.

Cutler, J. M., Shahan, K. W., and Steponkus, P. L. (1980). Alterations of the internal water relations of rice in response to drought hardening. *Crop Sci.* **20,** 307–310.

Dacey, J. W. H. (1980). Internal winds in water lilies: An adaptation for life in anaerobic sediments. *Science* **210,** 1017–1019.

Dainty, J. (1963). Water relations of plant cells. *Adv. Bot. Res.* **1,** 279–326.

Dainty, J. (1965). Osmotic flow. *Symp. Soc. Exp. Biol.* **19,** 75–85.

Dainty, J. (1976). Water relations of plant cells. *Encycl. Plant Physiol., New Ser.* **2,** Part A, 12–35.

Dainty, J., and Ginsburg, B. Z. (1964). The permeability of the protoplasts of *Chara australis* and *Nitella translucens* to methanol, ethanol, and isopropanol. *Biochim. Biophys. Acta* **79,** 122–128.

Dainty, J., and Hope, A. B. (1959). The water permeability of cells of *Chara australis* R. Br. *Aust. J. Biol. Sci.* **12,** 136–146.

Dalton, F. N., Raats, P. A. C., and Gardner, W. R. (1975). Simultaneous uptake of water and solutes by plants. *Agron. J.* **67,** 334–339.

Damptey, H. B., and Aspinall, D. (1976). Water deficit and inflorescence development in *Zea mays. Ann. Bot. (London)* [N.S.] **40,** 23–35.

Darwin, C. R. (1881). "The Formation of Vegetable Mould through the Action of Worms." Murray, London.

Darwin, F., and Pertz, D. F. M. (1911). On a new method of estimating the aperture of stomata. *Proc. R. Soc. London, Ser. B* **84,** 136–154.

Daum, C. R. (1967). A method for determining water transport in trees. *Ecology* **48,** 425–431.

Davidson, O. W. (1945). Salts in old greenhouse soils stunt flowers and vegetables. *Florists' Rev.* **95,** 17–19.

Davies, W. J. (1977). Stomatal responses to water stress and light in plants grown in controlled environments and in the field. *Crop Sci.* **17,** 735–740.

Davies, W. J., and Kozlowski, T. T. (1974). Stomatal responses of five woody angiosperms to light intensity and humidity. *Can. J. Bot.* **52,** 1525–1534.

Davies, W. J., and Kozlowski, T. T. (1975). Effect of applied abscisic acid and silicone on water relations and photosynthesis of woody plants. *Can. J. For.* **5,** 90–96.

Davies, W. J., Kozlowski, T. T., and Pereira, J. (1974). Effect of wind on transpiration and stomatal aperture of woody plants. *Bull.—R. Soc. N. Z.* **12,** 433–438.

Davies, W. J., Mansfield, T. A., and Wellburn, A. R. (1979). A role for abscisic acid in drought endurance and drought avoidance. *In* "Plant Growth Substances" (F. Skoog, ed.), pp. 242–253. Springer-Verlag, Berlin and New York.

Davis, R. E., Rosseau, D. L., and Board, R. D. (1971). "Polywater": Evidence from electron spectroscopy for chemical analysis (ESCA) of a complex salt mixture. *Science* **171,** 167–171.

Davis, R. M., and Lingle, J. C. (1961). Basis of shoot response to root temperature in tomato. *Plant Physiol.* **36,** 153–161.

Day, P. R., Bolt, G. H., and Anderson, D. M. (1967). Nature of soil water. *In* "Irrigation of Agricultural Lands" (R. M. Hagan, H. R. Haise, and T. W. Edminster, eds.), pp. 193–208. Am. Soc. Agron., Madison, Wisconsin.

de Candolle, A. P. (1832). "Physiologie Vegetale." Bechet jeune, Paris.

Decker, J. P., and Skau, C. M. (1964). Simultaneous studies of transpiration rate and sap velocity in trees. *Plant Physiol.* **39,** 213–215.

Decker, J. P., Gaylor, W. G., and Cole, F. D. (1962). Measuring transpiration of undisturbed tamarisk shrubs. *Plant Physiol.* **37,** 393–397.

Decker, W. L. (1983). Probability of drought for humid and subhumid regions. *In* "Crop Reactions to Water and Temperature Stresses in Humid, Temperate Climates" (C. D. Raper Jr. and P. J. Kramer, eds.), pp 11–19 Westview Press, Boulder, Colorado.

DeMichele, D. W., and Sharpe, P. J. H. (1973). An analysis of the mechanics of guard cell motion. *J. Theor. Biol.* **41,** 77–96.

Denmead, O. T., and Shaw, R. H. (1962). Availability of soil water to plants as affected by soil moisture content and meteorological conditions. *Agron. J.* **54,** 385–390.

Denny, F. E. (1917). Permeability of certain membranes to water. *Bot. Gaz. (Chicago)* **63,** 373–397.

DeRoo, H. C. (1961). Deep tillage and root growth. *Bull.—Conn. Agric. Exp. Stn., New Haven* **644.**

Desai, M. C. (1937). Effect of certain nutrient deficiencies on stomatal behavior. *Plant Physiol.* **12,** 253–283.

de Stigter, H. C. M. (1969). Growth relations between individual fruits and between fruits and roots in cucumber. *Neth. J. Agric. Sci.* **17,** 234–240.

de Vries, D. A., and Peck, A. J. (1958). On the cylindrical probe method of measuring thermal conductivity with special reference to soils I. Extension of theory and discussion of probe characteristics. *Aust. J. Phys.* **11,** 255–271.

Dimond, A. E. (1955). Pathogenesis in the wilt diseases. *Annu. Rev. Plant Physiol.* **6**, 329–350.

Dimond, A. E. (1966). Pressure and flow relations in vascular bundles of the tomato plants. *Plant Physiol.* **41**, 119–131.

Dimond, A. E. (1967). Physiology of wilt diseases. *In* "The Dynamic Role of Molecular Constituents in Plant-Parasite Interaction" (C. J. Mirocha and I. Uritanic, eds.), pp. 100–120. Am. Phytopath. Soc., St. Paul, Minnesota.

Dittmer, H. J. (1937). A quantitative study of the roots and root hairs of a winter rye plant (*Secale cereale*). *Am. J. Bot.* **24**, 417–420.

Dixon, H. H. (1914). "Transpiration and the Ascent of Sap in Plants." Macmillan, New York.

Dixon, H. H., and Joly, J. (1895). The path of the transpiration current. *Ann. Bot. (London)* **9**, 416–419.

Dobbs, R. C., and Scott, D. R. M. (1971). Distribution of diurnal fluctuations in stem circumference of Douglas-fir. *Can. J. For. Res.* **1**, 80–83.

Doley, D., and Leyton, L. (1968). Effects of growth regulating substances and water potential on the development of secondary xylem in *Fraxinus. New Phytol.* **67**, 579–594.

Donahue, R. L., Miller, R. W., and Shickluna, J. C. (1977). "Soils: An Introduction to Soils and Plant Growth," 4th ed. Prentice-Hall, Englewood Cliffs, New Jersey.

Doneen, L. D., and MacGillivray, J. H. (1946). Suggestions on irrigating commercial truck crops. *Lithoprint—Calif., Agric. Exp. Stn.*

Donnan, W. W., and Houston, C. E. (1967). Drainage related to irrigation management. *In* "Irrigation of Agricultural Lands" (R. M. Hagan, H. R. Haise, and T. W. Edminster, eds.), pp. 974–987. Am. Soc. Agron., Madison, Wisconsin.

Drew, M. C. (1977). Early effects of flooding on nitrogen deficiency and leaf chlorosis. *New Phytol.* **79**, 567–571.

Drew, M. C., and Siswora, E. J. (1979). The development of waterlogging damage in young barley plants in relation to plant nutrient status and changes in soil properties. *New Phytol.* **82**, 301–314.

Drew, M. C., Jackson, M. B., and Gifford, S. (1979). Ethylene-promoted adventitious rooting and development of cortical air spaces (aerenchyma) in roots may be an adaptive response to flooding in *Zea mays* L. *Planta* **147**, 83–88.

Drew, M. C., Chamel, A., Garrec, J. P., and Fourcy, A. (1980). Cortical air spaces (aerenchyma) in roots of corn subjected to oxygen stress. *Plant Physiol.* **65**, 506–511.

Drost-Hansen, W. (1965). Forms of water in biologic systems. *Ann. N. Y. Acad. Sci.* **125**, 249–272.

Drost-Hansen, W., and Clegg, J. S., eds. (1979). "Cell-Associated Water." Academic Press, New York.

Dube, P. A., Stevenson, K. R., Thurtell, G. W., and Neumann, H. H. (1975). Steady state resistance to water flow in corn under well watered conditions. *Can. J. Plant Sci.* **55**, 941–948.

Ducharme, E. P. (1977). Citrus tree decline and physiology of root-soil microbial interaction. *In* "Physiology of Root Microorganism Associations" (H. M. Vines, ed.), pp. 26–46. Am. Soc. Plant Physiol., Southern Sec., Atlanta, Georgia.

Duell, R. W., and Markus, D. K. (1977). Guttation deposits on turfgrass. *Agron. J.* **69**, 891–894.

Duke, S. H., Schrader, L. E., Henson, C. A., Servaites, J. C., Vogelsang, R. D., and Pendleton, J. W. (1979). Low root temperature effects on soybean nitrogen metabolism and photosynthesis. *Plant Physiol.* **63**, 956–962.

Dumbroff, E. B., and Peirson, D. R. (1971). Probable sites for passive movement of ions across the endodermis. *Can. J. Bot.* **49**, 35–38.

Duniway, J. M. (1977). Changes in resistance to water transport in safflower during the development of *Phytophthora* root rot. *Phytopathology* **67**, 331–337.

Dunkle, E. C., and Merkle, F. G. (1943). The conductivity of soil extracts in relation to germination and growth of certain plants. *Soil Sci. Soc. Am. Proc.* **8**, 185–188.

Dunn, E. L., Shropshire, F. M., Song, L. C., and Mooney, H. A. (1976). The water factor and convergent evolution in Mediterranean-type vegetation. *In* "Water and Plant Life" (O. L. Lange, L. Kappen, and E.-D. Schulze, eds.), pp. 492–505. Springer-Verlag, Berlin and New York.

Durbin, R. D. (1967). Obligate parasites: Effect on the movement of solutes and water. *In* "The Dynamic Role of Molecular Constituents in Plant-Parasite Reactions" (C. J. Mirocha and I. Uritani, eds.), pp. 80–99. Am. Phytopathol. Soc., St. Paul, Minnesota.

Durbin, R. D. (1978). Abiotic diseases induced by unfavorable water relations. *In* "Water Deficits and Plant Growth" (T. T. Kozlowski, ed.), Vol. 5, pp. 101–107. Academic Press, New York.

Duvdevani, S. (1953). Dew gradients in relation to climate, soil and topography. *Desert Res. Proc. Int. Symp., 1952* Spec. Publ. 2, pp. 136–152.

Duvdevani, S. (1957). Dew Research for arid agriculture. *Discovery* **18**, 330–334.

Dylla, A. S., Timmons, D. R., and Shull, H. (1980). Estimating water used by irrigated corn in west central Minnesota. *Soil Sci. Soc. Am. J.* **44**, 823–827.

Eames, A. J., and MacDaniels, L. H. (1947). "An Introduction to Plant Anatomy," 2nd ed. McGraw-Hill, New York.

Eastin, J. D., Haskins, F. A., Sullivan, C. Y., and van Bavel, C. H. M. eds. (1969). "Physiological Aspects of Plant Growth." Am. Soc. Agron. and Crop Sci. Soc. Am., Madison, Wisconsin.

Eaton, F. M. (1931). Root development as related to character of growth and fruitfulness of the cotton plant. *J. Agric. Res. (Washington, D.C.)* **43**, 875–883.

Eaton, F. M. (1941). Water uptake and root growth as influenced by inequalities in the concentration of the substrate. *Plant Physiol.* **16**, 545–564.

Eaton, F. M. (1942). Toxicity and accumulation of chloride and sulfate salts in plants. *J. Agric. Res. (Washington, D.C.)* **64**, 357–399.

Eaton, F. M. (1943). The osmotic and vitalistic interpretations of exudation. *Am. J. Bot.* **30**, 663–674.

Eaton, F. M., and Ergle, D. R. (1948). Carbohydrate accumulation in the cotton plant at low moisture levels. *Plant Physiol.* **23**, 169–187.

Edlefsen, N. E. (1941). Some thermodynamic aspects of the use of soil-moisture by plants. *Trans., Am. Geophys. Union* **22**, 917–940.

Edlefsen, N. E., and Bodman, G. B. (1941). Field measurements of water movement through a silt loam soil. *J. Am. Soc. Agron.* **33**, 713–731.

Edsall, J. T., and McKenzie, H. A. (1978). Water and proteins. I. The significance and structure of water: Its interaction with electrolytes and non-electrolytes. *Adv. Biophys.* **10**, 137–207.

Edwards, M., and Meidner, H. (1978). Stomatal responses to humidity and the water potentials of epidermal and mesophyll tissue. *J. Exp. Bot.* **29**, 771–780.

Ehleringer, J. (1980). Leaf morphology and reflectance in relation to water and temperature stress. *In* "Adaptation of Plants to Water and High Temperature Stress" (N. C. Turner and P. J. Kramer, eds.), pp. 295–308. Wiley, New York.

Eisenberg, D., and Kauzmann, W. (1969). "The Structure and Properties of Water." Oxford Univ. Press, London and New York.

Ekern, P. C. (1965). Evapotranspiration of pineapple in Hawaii. *Plant Physiol.* **40**, 736–739.

Elazari-Volcani, T. (1936). The influence of a partial interruption of the transpiration stream by root pruning and stem incisions on the turgor of citrus trees. *Palest. J. Bot. Hortic. Sci.* **1**, 94–96.

Elliott, L. F., Cochran, V. L., and Papendick, R. I. (1981). Wheat residue and nitrogen placement effects on wheat growth in the greenhouse. *Soil Sci.* **131**, 48–52.

Elrick, D. E., and Tanner, C. B. (1955). Influence of sample pretreatment on soil moisture retention. *Soil Sci. Soc. Am. Proc.* **19**, 279–282.

Elving, D. C., Kaufmann, M. R., and Hall, A. E. (1972). Interpreting leaf water potential measurements with a model of the soil-plant-atmosphere continuum. *Physiol. Plant.* **27**, 161–168.

Emerson, W. W., Bond, R. D., and Dexter, A. R., eds. (1978). "Modification of Soil Structure." Wiley, New York.

Emmert, E. M., and Ball, F. K. (1933). The effect of soil moisture on the availability of nitrate, phosphate, and potassium to the tomato plant. *Soil Sci.* **35,** 295–306.

England, C. B. (1965). Changes in fiber-glass soil moisture-electrical resistance elements in long-term installations. *Soil Sci. Soc. Am. Proc.* **29,** 229–231.

England, C. B., and Lesesne, E. H. (1962). Evapotranspiration research in Western North Carolina. *Agri. Eng.* **45,** 526–528.

Epstein, E. (1961). The essential role of calcium in selective cation transport by plant cells. *Plant Physiol.* **36,** 437–444.

Epstein, E., and Norlyn, J. D. (1977). Seawater-based crop production: A feasibility study. *Science* **197,** 249–251.

Erickson, A. E. (1965). Short-term oxygen deficiencies and plant responses. *In* "Conference on Drainage for Efficient Crop Production," pp. 11–12, 23. Am. Soc. Agric. Eng., St. Joseph, Michigan.

Erickson, L. C. (1946). Growth of tomato roots as influenced by oxygen in the nutrient solution. *Am. J. Bot.* **33,** 551–561.

Esau, K. (1941). Phloem anatomy of tobacco affected with curly top and mosaic. *Hilgardia* **13,** 437–470.

Esau, K. (1943). Vascular differentiation in the pear root. *Hilgardia* **15,** 299–324.

Esau, K. (1965). "Plant Anatomy" 2nd ed. Wiley, New York.

Esau, K. (1977). "Anatomy of Seed Plants," 2nd ed. Wiley, New York.

Etzler, F. M., and Drost-Hansen, W. (1979). A role for water in biological rate processes. *In* "Cell-Associated Water" (W. Drost-Hansen and J. S. Clegg, eds.), pp. 125–164. Academic Press, New York.

Evelyn, J. (1670). "Sylva." J. Martyn and J. Allestry, London.

Evenari, M. (1960). Plant physiology and arid zone research. *Arid Zone Res.* **18,** 175–195.

Fahn, A. (1979). "Secretory Tissues in Plants." Academic Press, New York.

Faiz, S. M. A., and Weatherley, P. E. (1978). Further investigations into the location and magnitude of the hydraulic resistances in the soil:plant system. *New Phytol.* **81,** 19–28.

Falk, M., and Kell, G. S. (1966). Thermal properties of water: Discontinuities questioned. *Science* **154,** 1013–1014.

Farmer, J. B. (1918). On the quantitative differences in the water conductivity of the wood in trees and shrubs. II. The deciduous plants. *Proc. R. Soc. London, Ser. B* **90,** 232–250.

Farquhar, G. D., and Sharkey, T. D. (1982). Stomatal conductance and photosynthesis. *Annu. Rev. Plant Physiol.* **33,** 317–345.

Farquhar, G. D., Schulze, E.-D., and Küppers (1980). Responses to humidity by stomata of *Nicotiana glauca and Corylus avellana* L. are consistent with the optimization of carbon dioxide uptake with respect to water loss. *Aust. J. Plant Physiol.* **7,** 315–327.

Fayle, D. C. F. (1968). "Radial Growth in Tree Roots," Tech. Rep. No. 9 Fac. For., University of Toronto, Toronto.

Federer, C. A. (1977). Leaf resistance and xylem potential differ among broad-leafed species. *For. Sci.* **23,** 411–419.

Feldman, L. J. (1975). Cytokinins and quiescent center activity in roots of *Zea. In* "The Development and Function of Roots" (J. G. Torrey and D. T. Clarkson eds.), pp. 55–72. Academic Press, New York.

Fellows, R. J., and Boyer, J. S. (1978). Altered ultrastructure of cells of sunflower leaves having low water potentials. *Protoplasma* **93,** 381–395.

Fensom, D. S. (1957). The bioelectric potentials of plants and their functional significance. *Can. J. Bot.* **35,** 573–582.

Fensom, D. S. (1958). The bioelectric potentials of plants and their functional significance. II. The patterns of bio-electric potential and exudation rate in excised sunflower roots and stems. *Can. J. Bot.* **36**, 367–383.

Fereres, E., Kiflas, P. M., Goldfein, R. E., Pruitt, W. O., and Hagan, R. M. (1981). Simplified but scientific irrigation scheduling. *Calif. Agric.* **36**(5&6), 19–21.

Ferguson, H., and Gardner, W. H. (1962). Water content measurement in soil columns by gamma ray absorption. *Soil Sci. Soc. Am. Proc.* **26**, 11–14.

Ferguson, I. B. (1973). Ion uptake and translocation by the root system of maize. *Annu. Rep.— Agric. Res. Counc. Letcombe Lab.* pp. 13–15.

Ferrell, G. T. (1978). Moisture stress threshold of susceptibility to fir engraver beetles in pole-size white fir. *For. Sci.* **24**, 85–92.

Ferrier, J. M., and Dainty, J. (1977). Water flow in *Beta vulgaris* storage tissue. *Plant Physiol.* **60**, 662–665.

Fischer, R. A. (1973). The effect of water stress at various stages of development on yield processes in wheat. *In* "Plant Response to Climatic Factors" (R. O. Slatyer, ed.), pp. 233–241. UNESCO, Paris.

Fischer, R. A. (1981). Optimizing the use of water and nitrogen through breeding of crops. *Plant Soil* **58**, 249–279.

Fischer, R. A., and Turner, N. C. (1978). Plant production in the arid and semiarid zones. *Annu. Rev. Plant Physiol.* **29**, 277–317.

Fiscus, E. L. (1972). In situ measurement of root-water potential. *Plant Physiol.* **50**, 191–193.

Fiscus, E. L. (1975). The interaction beteen osmotic- and pressure-induced water flow in plant roots. *Plant Physiol.* **55**, 917–922.

Fiscus, E. L. (1977). Determination of hydraulic and osmotic properties of soybean root systems. *Plant Physiol.* **59**, 1013–1020.

Fiscus, E. L. (1981a). Analysis of the components of area growth of bean root systems. *Crop Sci.* **21**, 909–913.

Fiscus, E. L. (1981b). Effects of abscisic acid on the hydraulic conductance of and the total ion transport through *Phaseolus* root systems. *Plant Physiol.* **68**, 169–174.

Fiscus, E. L. (1983). Water transport and balance within the plant: resistance to water flow in roots. *In* "Limitations to Efficient Water Use in Crop Production" (H. W. Taylor, W. R. Jordan, and T. B. Sinclair, eds.), in press. Am. Soc. Agron., Madison, Wisconsin.

Fiscus, E. L., and Kramer, P. J. (1970). Radial movement of oxygen in plants. *Plant Physiol.* **45**, 667–669.

Fiscus, E. L., and Kramer, P. J. (1975). General model for osmotic and pressure-induced flow in plant roots. *Proc. Natl. Acad. Sci. U.S.A.* **72**(8), 3114–3118.

Fiscus, E. L., and Markhart, A. H. (1979). Relationship between root system transport properties and plant size in *Phaseolus*. *Plant Physiol.* **64**, 770–773.

Fiscus, E. L., Parsons, L. R., and Alberte, R. S. (1973). Phyllotaxy and water relations in tobacco. *Planta* **112**, 285–292.

Fitting, H. (1911). Die Wasserversorgung und die osmotischen Druckverhältnisse der Wüsten-pflanzen. *Z. Bot.* **3**, 209–275.

Flowers, T. J., Troke, P. F., and Yeo, A. R. (1977). The mechanism of salt tolerance in halophytes. *Annu. Rev. Plant Physiol.* **28**, 89–121.

Ford, C. W., and Wilson, J. R. (1981). Changes in levels of solutes during osmotic adjustment to water stress in four tropical pasture species. *Aust. J. Plant Physiol.* **8**, 77–91.

Francis, C. M., Devitt, A. C., and Steele, P. (1974). Influence of flooding on the alcohol de-hydrogenase activity of roots of *Trifolium subterraneum* L. *Aust. J. Plant Physiol.* **1**, 9–13.

Franco, C. M., and Magalhaes, A. C. (1965). Techniques for the measurement of transpiration of individual plants. *Arid Zone Res.* **25**, 211–224.

Franks, F., ed. (1975). "Water: A Comprehensive Treatise." Plenum, New York.

Franks, F. (1981). "Polywater" MIT Press, Cambridge, Massachusetts.

Fraser, D. A., and Mawson, C. A. (1953). Movement of radio-active isotopes in yellow birch and white pine as detected with a portable scintillation counter. *Can. J. Bot.* **31**, 324–333.

Frey-Wyssling, A., and Mühlethaler, K. (1965). "Ultrastructural Plant Cytology." Am. Elsevier, New York.

Friesner, R. C. (1940). An observation on the effectiveness of root pressure in the ascent of sap. *Butler Univ. Bot. Stud.* **4**, 226–227.

Fritschen, L. J., and Doraiswamy, P. (1973). Dew: An addition to the hydrologic balance of Douglas fir. *Water Resour. Res.* **9**, 891–894.

Fritschen, L. J., Cox, L., and Kinerson, R. (1973). A 28-meter Douglas-fir in a weighing lysimeter. *For. Sci.* **19**, 256–261.

Fritts, H. C. (1976). "Tree Rings and Climate." Academic Press, New York.

Fry, K. E., and Walker, R. B. (1967). A pressure-infiltration method for estimating stomatal opening in conifers. *Ecology* **48**, 155–157.

Furkova, N. S. (1944). Growth reactions in plants under excessive watering. *C. R. Dokl. Acad. Sci. URSS* **42**, 87–90.

Furr, J. R., and Aldrich, W. W. (1943). Oxygen and carbon dioxide changes in the soil atmosphere of an irrigated date garden on calcareous very fine sandy loam soil. *Proc. Am. Soc. Hortic. Sci.* **42**, 46–52.

Furr, J. R., and Reeve, J. O. (1945). The range of soil-moisture percentages through which plants undergo permanent wilting in some soils from semiarid irrigated areas. *J. Agric. Res. (Washington, D.C.)* **71**, 149–170.

Furr, J. R., and Taylor, C. A. (1939). Growth of lemon fruits in relation to moisture content of the soil. *U.S., Dep. Agric., Tech. Bull.* **640.**

Gaastra, P. (1959). Photosynthesis of crop plants as influenced by light, carbon dioxide, temperature, and stomatal diffusion resistances. *Meded. Landbouwhogesch. Wageningen* **59**, 1–68.

Gaff, D. F. (1980). Protoplasmic tolerance of extreme water stress. *In* "Adaptation of Plants to Water and High Temperature Stress" (N. C. Turner and P. J. Kramer, eds.), pp. 207–230. Wiley, New York.

Gaff, D. F., and Carr, D. J. (1961). The quantity of water in the cell wall and its significance. *Aust. J. Biol. Sci.* **14**, 299–311.

Gale, J., and Hagan, R. M. (1966). Plant antitranspirants. *Annu. Rev. Plant Physiol.* **17**, 269–282.

Gardner, W. R. (1958). Some steady state solutions of the unsaturated moisture flow equation with application to evaporation from a water table. *Soil Sci.* **85**, 228–232.

Gardner, W. R. (1960). Dynamic aspects of water availability to plants. *Soil Sci.* **89**, 63–73.

Gardner, W. R. (1964). Relation of root distribution to water uptake and availability. *Agron. J.* **56**, 41–45.

Gardner, W. R., and Ehlig, C. F. (1962). Impedance to water movement in soil and plant. *Science* **138**, 522–523.

Gardner, W. R., and Ehlig, C. F. (1963). The influence of soil water on transpiration by plants. *JGR, J. Geophys. Res.* **68**, 5719–5724.

Gardner, W. R., and Ehlig, C. F. (1965). Physical aspects of the internal water relations of plant leaves. *Plant Physiol.* **40**, 705–710.

Gardner, W. R., and Fireman, M. (1958). Laboratory studies of evaporation from soil columns in the presence of a water table. *Soil Sci.* **85**, 244–249.

Gardner, W. R., and Nieman, R. H. (1964). Lower limit of water availability to plants. *Science* **143**, 1460–1462.

Gardner, W. R., Hillel, D., and Benyamini, Y. (1970). Post irrigation movement of soil water. I.

Redistribution. II. Simultaneous redistribution and evaporation. *Water Resour. Res.* **6**, 851–860, 1148–1153.

Gates, C. T. (1955). The response of the young tomato plant to a brief period of water shortage. I. The whole plant and its principal parts. *Aust. J. Biol. Sci.* **8**, 196–214.

Gates, C. T. (1968). Water deficits and growth of herbaceous plants. *In* "Water Deficits and Plant Growth" (T. T. Kozlowski, ed.), Vol. 2, pp. 135–190. Academic Press, New York.

Gates, C. T., and Bonner, J. (1959). IV. Effects of water stress on the ribonucleic acid metabolism of tomato leaves. *Plant Physiol.* **34**, 49–55.

Gates, D. M. (1968). Transpiration and leaf temperature. *Annu. Rev. Plant Physiol.* **19**, 211–238.

Gates, D. M. (1976). Energy exchange and transpiration. *In* "Water and Plant Life" (O. L. Lange, L. Kappen, and E.-D. Schulze, eds.), pp. 137–147. Springer-Verlag, Berlin and New York.

Gates, D. M. (1980). "Biophysical Ecology." Springer-Verlag, Berlin and New York.

Gates, D. M., Tibbals, E. C., and Kreith, F. (1965). Radiation and convection for ponderosa pine. *Am. J. Bot.* **52**, 66–71.

Geisler, G. (1963). Morphogenetic influence of (CO_2 + $-HCO_3$) on roots. *Plant Physiol.* **38**, 77–80.

Genkel, P. A. (1980). Linked and convergent resistance in plants. *Sov. Plant Physiol. (Engl. Transl.)* **26**(5, Pt. 1), 746–754.

Gerdemann, J. W. (1975). Vesicular-arbuscular mycorrhizae. *In* "The Development and Function of Roots" (J. G. Torrey and D. T. Clarkson, eds.), pp. 575–591. Academic Press, New York.

Geurten, I. (1950). Untersuchungen über den Gaswechsel von Baumrinden. *Forstwiss. Centralbl.* **69**, 704–743.

Gibbs, R. O. (1935). Studies of wood. II. The water content of certain Canadian trees, and changes in the water-gas system during seasoning and flotation. *Can. J. Res.* **12**, 727–760.

Giles, K. L., Cohen, D., and Beardsell, M. F. (1976). Effects of water stress on the ultrastructure of leaf cells of *Sorghum bicolor*. *Plant Physiol.* **57**, 11–14.

Gindel, I. (1973). "A New Ecophysiological Approach to Forest-Water Relationsbips in Arid Climates." Junk, The Hague.

Ginsburg, H. (1971). Model for iso-osmotic flow in plant roots. *J. Theor. Biol.* **32**, 147–158.

Ginsburg, H., and Ginzburg, B. Z. (1970). Radial water and solute flows in the roots of *Zea mays*. *J. Exp. Bot.* **21**, 580–592.

Girton, R. E. (1927). The growth of citrus seedlings as influenced by environmental factors. *Univ. Calif., Berkeley, Publ. Agric. Sci.* **5**, 83–117.

Glinka, Z., and Reinhold, L. (1964). Reversible changes in the hydraulic permeability of plant cell membranes. *Plant Physiol.* **39**, 1043–1050.

Glock, W. S., and Agerter, S. R. (1962). Rainfall and tree growth. *In* "Tree Growth" (T. T. Kozlowski, ed.), pp. 23–56. Ronald Press, New York.

González-Bernáldez, F., López-Sáez, J. F., and Garciá-Ferrero, G. (1968). Effect of osmotic pressure on root growth, cell cycle, and cell elongation. *Protoplasma* **65**, 255–262.

Goodwin, R. H., and Stepka, W. (1945). Growth and differentiation in the root tip of *Phleum pratense*. *Am. J. Bot.* **32**, 36–46.

Gortner, R. A. (1938). "Outlines of Biochemistry" 2nd ed. Wiley, New York.

Govindjee, Downton, W. J. S., Fork, D. C., and Armond, P. A. (1980). Chlorophyll a fluorescence transient as an indicator of water potential of leaves. *Year Book—Carnegie Inst. Washington* **79**, 191–193.

Grable, A. R., and Danielson, R. E. (1965). Influence of CO_2 on growth of corn and soybean seedlings. *Soil Sci. Soc. Am. Proc.* **29**, 233–238.

Gračanin, M. (1963). Über Unterscheide in der Transpiration von Blattspreite und Stamm. *Phyton* **10**, 216–224.

Gradmann, H. (1928). Untersuchungen über die Wasserverhältnisse des Bodens als Grundlage des Pflanzenwachstums. *Jahrb. Wiss. Bot.* **69**, 1–100.

Green, P. B., Erickson, R. O., and Buggy, J. (1971). Metabolic and physical control of cell elongation rate. *In vivo* studies in *Nitella*. *Plant Physiol.* **47**, 423–430.

Greenidge, K. N. H. (1952). An approach to the study of vessel length in hardwood species. *Am. J. Bot.* **39**, 570–574.

Greenidge, K. N. H. (1955). II. Experimental studies of fracture of stretched water columns in transpiring trees. *Am. J. Bot.* **42**, 28–37.

Greenidge, K. N. H. (1957). Ascent of sap. *Annu. Rev. Plant Physiol.* **8**, 237–256.

Greenidge, K. N. H. (1958). A note on the rates of upward travel of moisture in trees under differing experimental conditions. *Can. J. Bot.* **36**, 357–361.

Greenway, H. (1962). Plant responses to saline substrates. I. Growth and ion uptake of several varieties of *Hordeum* during and after sodium chloride treatment. *Aust. J. Biol. Sci.* **15**, 16–38.

Greenway, H., and Munns, R. (1980). Mechanisms of salt tolerance in nonhalophytes. *Annu. Rev. Plant Physiol.* **31**, 149–190.

Greenwood, D. J. (1971). Studies on the distribution of oxygen around the roots of mustard seedlings (*Sinapis alba* L.). *New Phytol.* **70**, 97–101.

Gregory, F. G., and Pearse, H. L. (1934). The resistance porometer and its application to the study of stomatal movement. *Proc. R. Soc. London, Ser. B* **114**, 477–493.

Grier, C. C., Vogt, K. A., Keyes, M. R., and Edmonds, R. L. (1981). Biomass distribution and above- and belowground production in young and mature *Abies amabilis* zone ecosystems of the Washington Cascades. *Can. J. For. Res.* **11**, 155–167.

Gries, G. A. (1943). Juglone (5-hydroxy-1,4-naphthoquinone)-a promising fungicide. *Phytopathology* **33**, 1112.

Grieve, B. J., and Went, F. W. (1965). An electric hygrometer apparatus for measuring water vapour loss from plants in the field. *Arid Zone Res.* **25**, 247–256.

Groenewegen, H., and Mills, J. A. (1960). Uptake of mannitol into the shoots of intact barley plants. *Aust. J. Biol. Sci.* **13**, 1–4.

Grossenbacher, K. A. (1939). Autonomic cycle of rate of exudation of plants. *Am. J. Bot.* **26**, 107–109.

Gunning, B. E. S., and Robards, A. W. (1976). Plasmodesmata and symplastic transport. *In* "Transport and Transfer Processes in Plants" (I. F. Wardlaw and J. B. Passioura, eds.), pp. 15–41. Academic Press, New York.

Gur, A., Bravdo, B., and Misrahi, Y. (1972). Physiological responses of apple trees to supraoptimal root temperature. *Physiol. Plant.* **27**, 130–138.

Gurr, C. G. (1962). Use of gamma rays in measuring water content and permeability in unsaturated columns of soil. *Soil Sci.* **94**, 224–229.

Gurr, C. G., Marshall, T. J., and Hutton, J. T. (1952). Movement of water in soil due to a temperature gradient. *Soil Sci.* **74**, 335–345.

Gutknecht, J., and Bisson, M. A. (1977). Ion transport and osmotic regulation in giant algal cells. *In* "Water Relations and Membrane Transport in Animals and Plants" (A. M. Jungrels, T. Hodges, A. M. Kleinzeller, and S. G. Schulz, eds.), pp. 3–14. Academic Press, New York.

Haas, A. R. C. (1936). Growth and water losses in citrus affected by soil temperature. *Calif. Citrog.* **21**, 467, 469.

Haas, A. R. C. (1948). Effect of the rootstock on the composition of citrus trees and fruit. *Plant Physiol.* **23**, 309–330.

Hack, H. R. B. (1980). The uptake and release of water vapour by the foam seal of a diffusion porometer as a source of bias. *Plant, Cell Environ.* **3**, 53–57.

Hacskaylo, E., and Palmer, J. G. (1957). Effects of several biocides on growth of seedling pines and incidence of mycorrhizae in field plots. *Plant Dis. Rep.* **41**, 354–358.

Hagan, P. M. (1949). Autonomic diurnal cycles in the water relations of nonexuding detopped root systems. *Plant Physiol.* **24**, 441–454.

Hagan, R. M. (1956). Factors affecting soil moisture-plant growth relations. *Int. Hort. Congr., Rep., 14th, 1955* pp. 82–98.

Hagan, R. M., Vaadia, Y., and Russell, M. B. (1959). Interpretation of plant responses to soil moisture regimes. *Adv. Agron.* **11,** 77–98.

Hagan, R. M., Haise, H. R., and Edminster, T. W., eds. (1967). "Irrigation of Agricultural Lands." Am. Soc. Agron., Madison, Wisconsin.

Haise, H. R., and Hagan, R. M. (1967). Soil, plant, and evaporative measurements as criteria for scheduling irrigation. *In* "Irrigation of Agricultural Lands" (R. M. Hagan, H. R. Haise, and T. W. Edminster, eds.), pp. 577–604. Am. Soc. Agron., Madison, Wisconsin.

Haise, H. R., and Kelley, O. J. (1946). Relation of moisture tension to heat transfer and electrical resistance in plaster of paris blocks. *Soil Sci.* **61,** 411–422.

Hales, S. (1727). "Vegetable Staticks." W. & J. Inneys and T. Woodward, London.

Halevy, A. (1956). Orange leaf transpiration under orchard conditions. IV. A contribution to the methodology of transpiration measurements in citrus leaves. *Bull. Res. Counc. Isr. Sect. D* **5,** 155–164.

Hall, A. E., and Grantz, D. A. (1981). Drought resistance of cowpea improved by selecting for early appearance of mature pods. *Crop. Sci.* **21,** 461–464.

Hall, A. E., and Kaufmann, M. R. (1975). Stomatal response to environment with *Sesamum indicum* L. *Plant Physiol.* **55,** 455–459.

Hall, A. E., and Schulze, E.-D. (1980). Stomatal response to environment and a possible interrelation between stomatal effects on transpiration and CO_2 assimilation. *Plant, Cell Environ.* **3,** 467–474.

Hall, A. E., Schulze, E.-D., and Lange, O. L. (1976). Current perspectives of steady-state stomatal responses to environment. *In* "Water and Plant Life" (O. L. Lange, L. Kappen, and E.-D. Schulze, eds.), pp. 169–188. Springer-Verlag, Berlin and New York.

Hall, A. E., Cannell, G. H., and Lawton, H. W., eds. (1979). "Agriculture in Semi-Arid Environments." Springer-Verlag, Berlin and New York.

Hall, N. S., Chandler, W. F., van Bavel, C. H. M., Reid, P. H., and Anderson, J. H. (1953). A tracer technique to measure growth and activity of plant root systems. *N. C., Agric. Exp. Stn., Tech. Bull.* **101.**

Hall, T. F., and Smith, G. E. (1955). Effects of flooding on woody plants, West Sandy dewatering project, Kentucky Reservoir. *J. For.* **53,** 281–285.

Hammel, H. T. (1967). Freezing of xylem sap without cavitation. *Plant Physiol.* **42,** 55–66.

Hammel, H. T. (1976). Colligative properties of a solution. *Science* **192,** 748–756.

Hammel, H. T., and Scholander, P. F. (1976). "Osmosis and Tensile Solvent." Springer-Verlag, Berlin and New York.

Hansen, C. (1926). The water-retaining power of the soil. *J. Ecol.* **14,** 111–119.

Hanson, A. D., and Hitz, W. D. (1982). Metabolic responses of mesophytes to plant water deficits. *Annu. Rev. Plant Physiol.* **33,** 163–203.

Hanson, J. B., and Biddulph, O. (1953). The diurnal variation in the translocation of minerals across bean roots. *Plant Physiol.* **28,** 356–370.

Harley, J. L. (1956). The mycorrhiza of forest trees. *Endeavour* **15,** 43–48.

Harley, J. L. (1969). "The Biology of Mycorrhiza." Leonard Hill, London.

Harley, J. L., and Russell, R. S., eds. (1979). "The Soil-Root Interface." Academic Press, New York.

Harms, W. R., and McGregor, W. H. D. (1962). A method for measuring the water balance of pine needles. *Ecology* **43,** 531–532.

Harris, J. A. (1934). "The Physico-chemical Properties of Plant Saps in Relation to Phytogeography." Univ. of Minnesota Press, Minneapolis.

Harris, J. A., and Harrisson, G. (1930). Osmotic concentration and water relations in the mistletoes. *Ecology* **11,** 687–700.

Harris, J. A., and Lawrence, J. V. (1916). On the osmotic pressure of the tissue fluids of Jamaican Loranthaceae parasitic on various hosts. *Am. J. Bot.* **3,** 438–455.

Harris, W. F., Kinerson, R. S., Jr., and Edwards, N. T. (1977). Comparison of belowground biomass of natural deciduous forests and loblolly pine plantations. *Range Sci. Dep. Sci. Ser. (Colo. State Univ.)* **26,** 29–37.

Hartt, C. E. (1967). Effect of moisture supply upon translocation and storage of ^{14}C in sugarcane. *Plant Physiol.* **42,** 338–346.

Hatch, A. B. (1937). The physical basis of mycotrophy in *Pinus. Black Rock For. Bull.* **6.**

Haynes, J. L., and Robbins, W. R. (1948). Calcium and boron as essential factors in the root environment. *J. Am. Soc. Agron.* **40,** 795–803.

Hayward, H. E., and Blair, W. M. (1942). Some responses of Valencia orange seedlings to varying concentrations of chloride and hydrogen ions. *Am. J. Bot.* **29,** 148–155.

Hayward, H. E., and Long, E. M. (1942). The anatomy of the seedling and roots of the Valencia orange. *U.S., Dep. Agric., Tech. Bull.* **786.**

Hayward, H. E., and Spurr, W. B. (1943). Effects of osmotic concentration of substrate on the entry of water into corn roots. *Bot. Gaz. (Chicago)* **105,** 152–164.

Hayward, H. E., and Spurr, W. (1944). Effects of isoosmotic concentrations of inorganic and organic substrates on entry of water into corn roots. *Bot. Gaz. (Chicago)* **106,** 131–139.

Hayward, H. E., Blair, W. M., and Skaling, P. E. (1942). Device for measuring entry of water into roots. *Bog. Gaz. (Chicago)* **104,** 152–160.

Hayward, H. E., Long, E. M., and Uhvits, R. (1946). Effect of chloride and sulfate salts on the growth and development of the Elberta peach on Shalil and Lovell rootstocks. *U.S., Dep. Agric., Tech. Bull.* **922.**

Head, G. C. (1964). A study of "exudation" from the root hairs of apple roots by time-lapse cinephotomicrography. *Ann. Bot. (London)* [N.S.] **28,** 495–498.

Head, G. C. (1965). Studies of diurnal change in cherry root growth and nutational movements of apple root tips by time-lapse cinematography. *Ann. Bot. (London)* [N.S.] **29,** 219–224.

Head, G. C. (1967). Effects of seasonal changes in shoot growth on the amount of unsuberized root on apple and plum trees. *J. Hortic. Sci.* **42,** 169–180.

Heap, A. J., and Newman, E. I. (1980). The influence of vesicular-arbuscular mycorrhizae on phosphorus transfer between plants. *New Phytol.* **85,** 173–179.

Heath, O. V. S. (1941). Experimental studies of the relation between carbon assimilation and stomatal movement. II. The use of the resistance porometer in estimating stomatal aperture and diffusive resistance. *Ann. Bot. (London)* [N.S.] **5,** 455–500.

Heath, O. V. S. (1959). The water relations of stomatal cells and the mechanisms of stomatal movement. *In* "Plant Physiology" (F. C. Steward, ed.), Vol. 2, pp. 193–250. Academic Press, New York.

Hellebust, J. A. (1976). Osmoregulation. *Annu. Rev. Plant Physiol.* **27,** 485–505.

Hellkvist, J., Richards, G. P., and Jarvis, P. G. (1974). Vertical gradients of water potential and tissue water relations in Sitka spruce trees measured with the pressure chamber. *J. Appl. Ecol.* **11,** 637–667.

Hellmers, H. (1963). Some temperature and light effects in the growth of Jeffrey pine seedlings. *For. Sci.* **9,** 189–201.

Hellmers, H., Horton, J. S., Juhren, G., and O'Keefe, J. (1955). Root systems of some chaparral plants in southern California. *Ecology* **36,** 667–678.

Henckel, P. A. (1964). Physiology of plants under drought. *Annu. Rev. Plant Physiol.* **15,** 363–386.

Henderson, L. (1934). Relation between root respiration and absorption. *Plant Physiol.* **9,** 283–300.

Henderson, L. J. (1913). "The Fitness of the Environment." Macmillan, New York.

Henry, C. (1978). Learning to irrigate—with 5000 tensiometers. *Irrig. Age* **13**(3), 6–8.

Herkelrath, W. N., Miller, E. E., and Gardner, W. R. (1977a). Water uptake by plants. I. Divided root experiments. *Soil Sci. Soc. Am. J.* **41,** 1033–1038.

Herkelrath, W. N., Miller, E. E., and Gardner, W. R. (1977b). Water uptake by plants. II. The root contact model. *Soil Sci. Soc. Am. J.* **41**, 1039–1043.

Hewlett, J. D. (1961). Soil moisture as a source of base flow from steep mountain watersheds. *U.S., For. Serv., Southeast. For. Exp. Stn., Stn. Pap.* **132**.

Hewlett, J. D., and Douglass, J. E. (1961). A method for calculating error of soil moisture volumes in gravimetric sampling. *For. Sci.* **7**, 265–272.

Hewlett, J. D., and Kramer, P. J. (1963). The measurement of water deficits in broadleaf plants. *Protoplasma* **57**, 381–391.

Hewlett, J. D., Douglass, J. E., and Clutter, J. L. (1964). Instrumental and soil moisture variance using the neutron-scattering method. *Soil Sci.* **97**, 19–24.

Hiler, E. A., and Clark, R. N. (1971). Stress day index to characterize effects of water stress on crop yields. *Trans. ASAE* **14**(4), 757–761.

Hiler, E. A., van Bavel, C. H. M., Hossain, M. M., and Jordan, W. R. (1972). Sensitivity of southern peas to plant water deficit at three growth stages. *Agron. J.* **64**, 60–64.

Hillel, D. (1980a). "Fundamentals of Soil Physics." Academic Press, New York.

Hillel, D. (1980b). "Applications of Soil Physics." Academic Press, New York.

Hillel, D., and Rawitz, E. (1972). Soil water conservation. *In* "Water Deficits and Plant Growth" (T. T. Kozlowski, ed.), Vol. 3, pp. 307–338. Academic Press, New York.

Hoad, G. V. (1975). Effects of osmotic stress on abscisic acid levels in xylem sap of sunflower (*Helianthus annuus* L.). *Planta* **124**, 25–29.

Hodgson, R. H. (1954). A study of the physiology of mycorrhizal roots on *Pinus taeda* L. M. A. Thesis, Duke University, Durham, North Carolina.

Hoffman, G. J., Rawlins, S. L., Garber, M. J., and Cullen, E. M. (1971). Water relations and growth of cotton as influenced by salinity and relative humidity. *Agron. J.* **63**, 822–826.

Holch, A. E. (1931). Development of roots and shoots of certain deciduous tree seedlings in different forest sites. *Ecology* **12**, 259–298.

Hollaender, A., ed. (1956). "Radiation Biology," Vol. 3. McGraw-Hill, New York.

Hollaender, A., Aller, J. C., Epstein, E., San Pietro, A., and Zaborsky, O. R., eds. (1979). "The Biosaline Concept." Plenum, New York.

Holmes, J. W. (1960). Water balance and the water-table in deep sandy soils of the upper south-east, South Australia. *Aust. J. Agric. Res.* **11**, 970–988.

Holmes, J. W., and Jenkinson, A. F. (1959). Techniques for using the neutron moisture meter. *J. Agric. Eng. Res.* **4**, 100–109.

Holmes, J. W., Taylor, S. A., and Richards, S. J. (1967). Measurement of soil water. *In* "Irrigation of Agricultural Lands" (R. M. Hagan, H. R. Haise, and T. W. Edminster, eds.), pp. 275–303. Am. Soc. Agron., Madison, Wisconsin.

Holmgren, P., Jarvis, P. G., and Jarvis, M. S. (1965). Resistances to carbon dioxide and water vapour transfer in leaves of different plant species. *Physiol. Plant.* **18**, 557–573.

Hölzl, J. (1955). Über Streuung der Transpirationswerte bei verschiedenen Blättern einer Pflanze und bei artgleichen Pflanzen eines Bestandes. *Sitzungs ber.—Oesterr. Akad. Wiss., Math-Natur-wiss. Kl., Abt. 1* **164**(9), 659–721.

Hook, D. D., and Brown, C. L. (1973). Root adaptations and relative flood tolerance of five hardwood species. *For. Sci.* **19**, 225–229.

Hook, D. D., and Crawford, R. M. M., eds. (1978). "Plant Life in Anaerobic Environments." Ann Arbor Sci. Press, Ann Arbor, Michigan.

Hook, D. D., Brown, C. L., and Kormanik, P. P. (1972). Inductive flood tolerance in swamp tupelo (*Nyssa sylvatica* var. *biflora* (Walt.) Sorg.). *J. Exp. Bot.* **22**, 78–89.

Hoover, M. D. (1944). Effect of removal of forest vegetation upon water yield. *Trans. Am. Geophys. Union* **25**, 969–977.

Hoover, M. D. (1949). Hydrologic characteristics of South Carolina Piedmont forest soils. *Soil Sci. Soc. Am. Proc.* **14**, 353–358.

Hough, W. A., Woods, F. W., and McCormack, M. L. (1965). Root extension of individual trees in surface soils of a natural longleaf pine-turkey oak stand. *For. Sci.* **11**, 223–242.

House, C. R., and Findlay, N. (1966). Water transport in isolated maize roots. *J. Exp. Bot.* **17**, 344–354.

Howard, A. (1925). The effect of grass on trees. *Proc. R. Soc. London, Ser. B* **97**, 284–321.

Hsiao, T. C. (1973). Plant responses to water stress. *Annu. Rev. Plant Physiol.* **24**, 519–570.

Hsiao, T. C., Acevedo, E., Fereres, E., and Henderson, D. W. (1976). Water stress, growth, and osmotic adjustment. *Philos. Trans. R. Soc. London, Ser. B* **273**, 479–500.

Huber, B. (1923). Transpiration in verschiedener Stammhöhe. I. *Sequoia gigantea. Z. Bot.* **15**, 465–501.

Huber, B. (1924). Die Beurteilung des Wasserhaushaltes der Pflanze. *Jahrb. Wiss. Bot.* **64**, 1–120.

Huber, B. (1928). Weitere quantitative Untersuchungen über das Wasserleitungssystem der Pflanzen. *Jahrb. Wiss. Bot.* **67**, 877–959.

Huber, B. (1932). Beobachtung und Messung pflanzlichen Saftströme. *Ber. Dtsch. Bot. Ges.* **50**, 89–109.

Huber, B. (1935). Die physiologische Bedeutung der Ring- und Zerstreutporigkeit. *Ber. Dtsch. Bot. Ges.* **53**, 711–719.

Huber, B. (1956). Die Gefässleitung. *Encycl. Plant Physiol.* **3**, 541–583.

Huber, B., and Schmidt, E. (1936). Weitere thermoelektrische Untersuchingen über den Transpirationsstrom der Bäume. *Tharandter Forstl. Jahrb.* **87**, 369–412.

Huber, B., and Schmidt, E. (1937). Eine Kompensationsmethode zur thermoelektrischen Messung Langsamer Saftstrome. *Ber. Dtsch. Bot. Ges.* **50**, 514–529.

Huck, M. G., Hageman, R. H., and Hanson, J. B. (1962). Diurnal variation in root respiration. *Plant Physiol.* **37**, 371–375.

Huck, M. G., Klepper, B., and Taylor, H. M. (1970). Diurnal variations in root diameter. *Plant Physiol.* **45**, 529–530.

Hudson, J. P. (1960). Relations between root and shoot growth in tomatoes. *Sci. Hortic. (Canterbury, Engl.)* **14**, 49–54.

Hunt, P. G., Campbell, R. B., Sojka, R. E., and Parsons, J. E. (1981). Flooding-induced soil and plant ethylene accumulation and water status response of field-grown tobacco. *Plant Soil* **59**, 427–439.

Hunter, A. S., and Kelley, O. J. (1946). A new technique for studying the absorption of moisture and nutrients from soil by plant roots. *Soil Sci.* **62**, 441–450.

Hurd, E. A. (1974). Phenotype and drought tolerance in wheat. *Agric. Meteorol.* **14**, 39–55.

Hurd, E. A. (1976). Plant breeding for drought resistance. *In* "Water Deficits and Plant Growth" (T. T. Kozlowski, ed.), Vol. 4, pp. 317–353. Academic Press, New York.

Hygen, G. (1953). Studies in plant transpiration. II. *Physiol. Plant.* **6**, 106–133.

Hylmö, B. (1953). Transpiration and ion absorption. *Physiol. Plant.* **6**, 333–405.

Idso, S. B., Jackson, R. D., and Reginato, R. J. (1975). Detection of soil moisture by remote surveillance. *Am. Sci.* **63**(5), 549–557.

Idso, S. B., Jackson, R. D., and Reginato, R. J. (1977). Remote sensing of crop yields. *Science* **196**, 19–25.

Idso, S. D., Reginato, R. J., Hatfield, J. L., Walker, G. K., Jackson, R. D., and Pinter, P. J., Jr. (1980). A generalization of the stress-degree-day concept of yield prediction to accommodate a diversity of crops. *Agric. Meteorol.* **21**, 205–211.

Iljin, W. S. (1957). Drought resistance in plants and physiological processes. *Annu. Rev. Plant Physiol.* **8**, 257–274.

Incoll, L. D., Long, S. P., and Ashmore, M. R. (1977). SI units in publications in plant science. *Current Advances in Plant Science* **9**, 331–343.

Ingelsten, B., and Hylmö, B. (1961). Apparent free space and surface film determined by a centrifugation method. *Physiol. Plant.* **14**, 157–170.

Ishihara, K., and Hirasawa, T. (1978). Relationship between leaf and xylem water potentials in rice plants. *Plant Cell Physiol.* **19,** 1289–1294.

Israel, D. W., Giddens, J. E., and Powell, W. W. (1973). The toxicity of peach tree roots. *Plant Soil* **39,** 103–112.

Itai, C., and Vaadia, Y. (1965). Kinetin-like activity in root exudate of water-stressed sunflower plants. *Physiol. Plant.* **18,** 941–944.

Ivanov, V. B. (1980). Specificity of spatial and time organization of root cell growth in connection with functions of the root. *Sov. Plant Physiol. (Engl. Transl.)* **26**(5, Pt. 1), 720–728.

Jackson, M. B., and Barlow, P. W. (1981). Root geotropism and the role of growth regulators from the cap: A reexamination. *Plant Cell Environ.* **4,** 107–123.

Jackson, M. B., and Campbell, D. J. (1976). Waterlogging and petiole epinasty in tomato: The role of ethylene and low oxygen. *New Phytol.* **76,** 21–29.

Jackson, M. B., Herman, B., and Goodenough, A. (1982). An examination of the importance of ethanol in causing injury to flooded plants. *Plant, Cell & Environ.* **5,** 163–172.

Jackson, M. B., Gales, K., and Campbell, D. J. (1978). Effect of waterlogged soil conditions on the production of ethylene and on water relationships in tomato plants. *J. Exp. Bot.* **29,** 183–193.

Jackson, W. T. (1955). The role of adventitious roots in recovery of shoots following flooding of the original root systems. *Am. J. Bot.* **42,** 816–819.

Jackson, W. T. (1956). The relative importance of factors causing injury to shoots of flooded tomato plants. *Am. J. Bot.* **43,** 637–639.

Jackson, W. T. (1962). Use of carbowaxes (polyethylene glycols) as osmotic agents. *Plant Physiol.* **37,** 513–519.

James, E. (1945). Effect of certain cultural practices on moisture conservation on a piedmont soil. *J. Am. Soc. Agron.* **37,** 945–952.

Jamison, V. C. (1946). The penetration of irrigation and rain water into sandy soils of central Florida. *Soil Sci. Soc. Am. Proc.* **10,** 25–29.

Jäntti, A., and Kramer, P. J. (1957). Regrowth of pastures in relation to soil moisture and defoliation. *Proc. Int. Grassl. Congr., 7th, 1956* pp. 1–12.

Jarvis, P. G. (1980). Stomatal response to water stress in conifers. *In* "Adaptation of Plants to Water and High Temperature Stress" (N. C. Turner and P. J. Kramer, eds.), pp. 105–122. Wiley, New York.

Jarvis, P. G., and Mansfield, T. A., eds. (1981). "Stomatal Physiology." Cambridge Univ. Press, London and New York.

Jarvis, P. G., and Slatyer, R. O. (1966). A controlled environment chamber for studies of gas exchange by each surface of a leaf. *CSIRO Div. Land Res., Tech. Pap,* **29.**

Jarvis, P. G., and Slatyer, R. O. (1970). The role of the mesophyll cell wall in leaf transpiration. *Planta* **90,** 303–322.

Jarvis, P. G., Rose, C. W., and Begg, J. E. (1967). An experimental and theoretical comparison of viscous and diffusive resistances to gas flow through amphistomatous leaves. *Agric. Meteorol.* **4,** 103–117.

Jeffree, C. E., Johnson, R. P. C., and Jarvis, P. G. (1971). Epicuticular wax in the stomatal antechamber of Sitka spruce and its effects on the diffusion of water vapour and carbon dioxide. *Planta* **98,** 1–10.

Jeje, A. A., and Zimmerman, M. H. (1979). Resistance to water flow in xylem vessels. *J. Exp. Bot.* **30,** 817–827.

Jemison, G. M. (1944). The effect of basal wounding by forest fires on the diameter growth of some southern Appalachian hardwoods. *Duke Univ. Sch. For. Bull.* **9.**

Jenny, H., and Grossenbacher, K. (1963). Root-soil boundary zones as seen in the electron microscope. *Soil Sci. Soc. Am. Proc.* **27,** 273–277.

Jensen, M. E., and Haise, H. R. (1963). Estimating evapotranspiration from solar radiation. *J. Irrig. Drain. Div., Am. Soc. Civ. Eng.* **89,** 15–41.

Jensen, R. D., Taylor, S. A., and Wiebe, H. H. (1961). Negative transport and resistance to water flow through plants. *Plant Physiol.* **36**, 633–638.

Johnson, H. B. (1975). Plant pubescence: An ecological perspective. *Bot. Rev.* **41**, 233–258.

Johnson, L. P. V. (1945). Physiological studies on sap flow in the sugar maple, *Acer saccharum* Marsh. *Can. J. Res., Sect. C* **23**, 192–197.

Jones, C. A., and Carabaly, A. (1980). Estimation of leaf water potential in tropical grasses with the Campbell-Brewster hydraulic press. *Trop. Agric. (Trinidad)* **57**, 305–307.

Jones, C. G., Edson, A. W., and Morse, W. J. (1903). The maple sap flow. *Bull.—Vt., Agric. Exp. Stn.* **103**.

Jones, H. G., and Higgs, K. H. (1980). Resistance to water loss from the mesophyll cell surface in plant leaves. *J. Exp. Bot.* **31**, 545–553.

Jones, R. G. W. (1980). An assessment of quaternary ammonium and related compounds as osmotic effectors in crop plants. *In* "Genetic Engineering of Osmoregulation" (D. W. Rains, R. C. Valentine, and A. Hollaender, eds.), pp. 155–170. Plenum, New York.

Jordan, W. R., and Miller, F. R. (1980). Genetic variability in sorghum root systems: Implications for drought tolerance. *In* "Adaptation of Plants to Water and High Temperature Stress" (N. C. Turner and P. J. Kramer, eds.), pp. 383–399. Wiley, New York.

Jordan, W. R., and Ritchie, J. T. (1971). Influence of soil water stress on evaporation, root absorption, and internal water status of cotton. *Plant Physiol.* **48**, 783–788.

Joshi, M. C., Boyer, J. S., and Kramer, P. J. (1965). Growth, carbon dioxide exchange, transpiration, and transpiration ratio of pineapple. *Bot. Gaz. (Chicago)* **126**, 174–179.

Juniper, B. E. (1976). Geotropism. *Annu. Rev. Plant Physiol.* **27**, 385–406.

Kalela, E. (1954). Mantysiemenpuiden japuustojen juuroisuhteista (On root relations of pine seed trees). *Acta For. Fenn.* **61**, 1–17.

Kamiya, N., and Tazawa, M. (1956). Studies of water permeability of a single plant cell by means of transcellular osmosis. *Protoplasma* **46**, 394–422.

Kamiya, N., Tazawa, M., and Takata, T. (1963). The relation of turgor pressure to cell volume in *Nitella* with special reference to mechanical properties of the cell wall. *Protoplasma* **57**, 501–521.

Kanemasu, E. T., Thurtell, G. W., and Tanner, C. B. (1969). Design, calibration and field use of a stomatal diffusion porometer. *Plant Physiol.* **44**, 881–885.

Karas, I., and McCully, M. E. (1973). Further study of the histology of lateral root development in *Zea mays. Protoplasma* **77**, 243–269.

Karsten, K. S. (1939). Root activity and the oxygen requirement in relation to soil fertility. *Am. J. Bot.* **26**, 855–860.

Kaufmann, M. R. (1968a). Water relations of pine seedlings in relation to root and shoot growth. Plant Physiol. 43, 281–288.

Kaufmann, M. R. (1968b). Evaluation of the pressure chamber technique for estimating plant water potential of forest tree species. *For. Sci.* **14**, 369–374.

Kaufmann, M. R. (1972). Water deficits and reproductive growth. *In* "Water Deficits and Plant Growth" (T. T. Kozlowski, ed.), Vol. 3, pp. 91–124. Academic Press, New York.

Kaufmann, M. R. (1975). Leaf water stress in Engelmann spruce: Influence of the root and shoot environments. *Plant Physiol.* **56**, 841–844.

Kaufmann, M. R. (1976a). Water transport through plants—current perspectives. *In* "Transport and Transfer Processes in Plants" (I. Wardlaw and J. Passioura, eds.), pp. 313–317. Academic Press, New York.

Kaufmann, M. R. (1976b), Stomatal response of Engelmann spruce to humidity, light, and water stress. *Plant Physiol.* **57**, 898–901.

Kaufmann, M. R. (1981). Automatic determination of conductance, transpiration, and environmental conditions in forest trees. *For. Sci.* **27**, 817–827.

Kaufmann, M. R. (1982a). Leaf conductance as a function of photosynthetic proton flux density and absolute humidity difference from leaf to air. *Plant Physiol.* **69**, 1018–1022.

Kaufmann, M. R. (1982b). Evaluation of season, temperature and water stress effects on stomata using a leaf conductance model. *Plant Physiol.* **69**, 1023–1026.

Kaufmann, M. R., and Eckard, A. N. (1977). A portable instrument for rapidly measuring conductance and transpiration of conifers and other species. *For. Sci.* **23**, 227–23.

Kaufmann, M. R., and Hall, A. E. (1974). Plant water balance—its relationship to atmospheric and edaphic conditions. *Agric. Meteorol.* **14**, 85–98.

Kaufmann, M. R., and Kramer, P. J. (1967). Phloem water relations and translocation. *Plant Physiol.* **42**, 191–194.

Kaufmann, M. R., and Troendle, C. A. (1981). The relationship of leaf area and foliage biomass to sapwood conducting area in four subalpine forest tree species. *For. Sci.* **27**, 477–482.

Kaul, O. N., and Kramer, P. J. (1965). Comparative drought resistance of two woody species. *Indian For.* **91**, 462–469.

Kavanau, J. L. (1964). "Water and Solute-Water Interactions." Holden-Day, San Francisco, California.

Kawase, M. (1974). Role of ethylene in induction of flooding damage in sunflower. *Physiol. Plant.* **31**, 29–38.

Kawase, M. (1976). Ethylene accumulation in flooded plants. *Physiol. Plant.* **36**, 236–241.

Kawase, M. (1979). Role of cellulase in aerenchyma development in sunflower. *Am. J. Bot.* **66**, 183–190.

Keck, R. W., and Boyer, J. S. (1974). Chloroplast response to low leaf water potentials. III. Differing inhibition of electron transport and photophosphorylation. *Plant Physiol.* **53**, 474–479.

Keeton, W. T. (1980). "Biological Science," 3rd ed. Norton, New York.

Kelley, O. J. (1944). A rapid method of calibrating various instruments for measuring soil moisture in situ. *Soil Sci.* **58**, 433–440.

Kelley, O. J., Hunter, A. S., Haise, H. R., and Hobbs, C. H. (1946). A comparison of methods of measuring soil moisture under field conditions. *J. Am. Soc. Agron.* **38**, 759–784.

Kemper, W. D., Robinson, C. W., and Golus, H. M. (1961). Growth rates of barley and corn as affected by changes in soil moisture stress. *Soil Sci.* **91**, 332–338.

Kende, H. (1965). Kinetinlike factors in the root exudate of sunflowers. *Proc. Natl. Acad. Sci. U.S.A.* **53**, 1302–1307.

Killian, C., and Lemée, G. (1956). Les xerophytes: Leur economie d'eau. *Encycl. Plant Physiol.* **3**, 787–824.

Kirkham, M. B., Gardner, W. R., and Gerloff, G. C. (1972). Regulation of cell division and cell enlargement by turgor pressure. *Plant Physiol.* **49**, 961–962.

Kiyosawa, K., and Tazawa, M. (1977). Hydraulic conductivity of tonoplast-free Chara Cells. *J. Membrane Biol.* **37**, 157–166.

Kleinendorst, A. (1975). An explosion of leaf growth after stress conditions. *Neth. J. Agric. Sci.* **23**, 139–144.

Kleinendorst, A., and Brouwer, R. (1972). The effect of local cooling on growth and water content of plants. *Neth. J. Agric. Sci.* **20**, 203–217.

Klepper, B., and Kaufmann, M. R. (1966). Removal of salt from xylem sap by leaves and stems of guttating plants. *Plant Physiol.* **41**, 1743–1747.

Kluge, M., and Ting, I. P. (1978). "Crassulacean Acid Metabolism." Springer-Verlag, Berlin and New York.

Knapp, R., Linskens, H. F., Lieth, H., and Wolf, R. (1952). Untersuchungen über die Bodenfeuchtigkeit in verschiedenen Pflanzengesellschaften nach neuren Methoden. *Ber. Dtsch. Bot. Ges.* **65**, 113–132.

Knight, D. H., Fahey, T. J., Running, S. W., Harrison, A. T., and Wallace, L. L. (1981). Transpiration from 100-year old lodgepole pine forests estimated with whole tree potometers. *Ecology* **62,** 717–726.

Knipling, E. B. (1967a). Measurement of leaf water potential by the dye method. *Ecology* **48,** 1038–1041.

Knipling, E. B. (1967b). Effect of leaf aging on water deficit-water potential relationship of dogwood leaves growing in two environments. *Physiol. Plant.* **20,** 65–72.

Knipling, E. B., and Kramer, P. J. (1967). Comparison of the dye method with the thermocouple psychrometer for measuring leaf water potentials. *Plant Physiol.* **42,** 1315–1320.

Knoerr, K. R. (1967). Contrasts in energy balances between individual leaves and vegetated surfaces. *In* "International Symposium on Forest Hydrology" (W. E. Sopper and H. W. Lull, eds.), pp. 391–401. Pergamon, Oxford.

Koeppe, D. E., Miller, R. J., and Bell, D. T. (1973). Drought-affected mitochondrial processes as related to tissue and whole plant responses. *Agron. J.* **65,** 566–569.

Kolata, G. B. (1979). Water structure and ion binding: a role in cell physiology? *Science* **192,** 1220–1222.

Kolattukudy, P. E. (1981). Structure, biosynthesis, and biodegradation of cutin and suberin. *Annu. Rev. Plant Physiol.* **32,** 539–567.

Kolek, J., ed. (1974). "Structure and Function of Primary Root Tissues." Slovak Acad. Sci., Bratislava.

Koller, D., and Samish, Y. (1964). A null-point compensating system for simultaneous and continuous measurement of net photosynthesis and transpiration by controlled gas-stream analysis. *Bot. Gaz. (Chicago)* **125,** 81–88.

Korstian, C. F. (1924). Density of cell sap in relation to environmental conditions in the Wasatch Mountains of Utah. *J. Agric. Res. (Washington, D.C.)* **28,** 845–909.

Korstian, C. F., and Fetherolf, N. J. (1921). Control of stem girdle of spruce transplants caused by excessive heat. *Phytopathology* **11,** 485–490.

Kŏvda, V. A. (1980). "Land Aridization and Drought Control." Westview Press, Boulder, Colorado.

Kŏvda, V. A., van den Berg, C., and Hagan, R. M., eds. (1967). "Irrigation and Drainage of Arid Lands in Relation to Salinity and Alkalinity." FAO-UNESCO, Paris.

Kozlowski, T. T. (1943). Transpiration rates of some forest tree species during the dormant season. *Plant Physiol.* **18,** 252–260.

Kozlowski, T. T. (1949). Light and water in relation to growth and competition of Piedmont forest trees. *Ecol. Monogr.* **19,** 207–231.

Kozlowski, T. T. (1961). The movement of water in trees. *For. Sci.* **7,** 177–192.

Kozlowski, T. T., ed. (1968–1981). "Water Deficits and Plant Growth," Vol. I–VI. Academic Press, New York.

Kozlowski, T. T. (1971). "Growth and Development of Trees," 2 vols. Academic Press, New York.

Kozlowski, T. T., ed. (1972). "Water Deficits and Plant Growth," Vol. 3. Academic Press, New York.

Kozlowski, T. T., ed. (1974). "Shedding of Plant Parts." Academic Press, New York.

Kozlowski, T. T., ed. (1976). "Water Deficits and Plant Growth," Vol. 4. Academic Press, New York.

Kozlowski, T. T., ed. (1978). "Water Deficits and Plant Growth," Vol. 5. Academic Press, New York.

Kozlowski, T. T., ed. (1981). "Water Deficits and Plant Growth," Vol. 6. Academic Press, New York.

Kozlowski, T. T. (1982a). Water supply and tree growth. I. Water deficits. *For. Abstr.* **43,** 57–95.

Kozlowski, T. T. (1982b). Water supply and tree growth. II. Flooding. *For. Abstr.* **43**, 145–161.

Kozlowski, T. T., and Cooley, J. C. (1961). Root grafting in northern Wisconsin. *J. For.* **59**, 105–107.

Kozlowski, T. T., and Scholtes, W. H. (1948). Growth of roots and root hairs of pine and hardwood seedlings in the Piedmont. *J. For.* **46**, 750–754.

Kozlowski, T. T., and Winget, C. H. (1963). Patterns of water movement in forest trees. *Bot. Gaz. (Chicago)* **124**, 301–311.

Kozlowski, T. T., Winget, C. H., and Torrie, J. H. (1962). Daily radial growth of oak in relation to maximum and minimum temperature. *Bot. Gaz. (Chicago)* **124**, 9–17.

Kozlowski, T. T., Hughes, J. F., and Leyton, L. (1966). Patterns of water movement in dormant gymnosperm seedlings. *Biorheology* **3**, 77–85.

Kozlowski, T. T., Hughes, J. F., and Leyton, L. (1967). Dye movement in gymnosperms in relation to tracheid alignment. *Forestry* **40**, 209–227.

Kramer, P. J. (1932). The absorption of water by root systems of plants. *Am. J. Bot.* **19**, 148–164.

Kramer, P. J. (1933). The intake of water through dead root systems and its relation to the problem of absorption by transpiring plants. *Am. J. Bot.* **20**, 481–492.

Kramer, P. J. (1934). Effects of soil temperature on the absorption of water by plants. *Science* **79**, 371–372.

Kramer, P. J. (1937). The relation between rate of transpiration and rate of absorption of water in plants. *Am. J. Bot.* **24**, 10–15.

Kramer, P. J. (1938). Root resistance as the cause of the absorption lag. *Am. J. Bot.* **25**, 110–113.

Kramer, P. J. (1939). The forces concerned in the intake of water by transpiring plants. *Am. J. Bot.* **26**, 784–791.

Kramer, P. J. (1940a). Causes of decreased absorption of water in poorly aerated media. *Am. J. Bot.* **27**, 216–220.

Kramer, P. J. (1940b). Root resistance as a cause of decreased water absorption by plants at low temperatures. *Plant Physiol.* **15**, 63–79.

Kramer, P. J. (1940c). Sap pressure and exudation. *Am. J. Bot.* **27**, 929–931.

Kramer, P. J. (1942). Species differences with respect to water absorption at low soil temperatures. *Am. J. Bot.* **29**, 828–832.

Kramer, P. J. (1949). ''Plant and Soil Water Relationships.'' McGraw-Hill, New York.

Kramer, P. J. (1950). Effects of wilting on the subsequent intake of water by plants. *Am. J. Bot.* **37**, 280–284.

Kramer, P. J. (1951). Causes of injury to plants resulting from flooding of the soil. *Plant Physiol.* **26**, 722–736.

Kramer, P. J. (1955a). Water content and water turnover in plant cells. *Encycl. Plant Physiol.* **1**, 196–223.

Kramer, P. J. (1955b). Bound water. *Encycl. Plant Physiol.* **1**, 223–242.

Kramer, P. J. (1955c). Physical chemistry of the vacuoles. *Encycl. Plant Physiol.* **1**, 649–660.

Kramer, P. J. (1956). The uptake of water by plant cells. *Encycl. Plant Physiol.* **2**, 316–336.

Kramer, P. J. (1957). Outer space in plants. *Science* **125**, 633–635.

Kramer, P. J. (1969). ''Plant and Soil Water Relationships: A Modern Synthesis.'' McGraw-Hill, New York.

Kramer, P. J. (1981). Carbon dioxide concentration, photosynthesis, and dry matter production. *BioScience* **31**, 29–33.

Kramer, P. J. (1982). Water and plant productivity or yield. *In* ''Handbook of Agricultural Productivity'' (M. Rechcigl, Jr., ed.), Vol. 1, pp. 41–47. CRC Press, Boca Raton, Florida.

Kramer, P. J., and Bullock, H. C. (1966). Seasonal variations in the proportions of suberized and unsuberized roots of trees in relation to the absorption of water. *Am. J. Bot.* **53**, 200–204.

Kramer, P. J., and Clark, W. S. (1947). A comparison of photosynthesis in individual pine needles and entire seedlings at various light intensities. *Plant Physiol.* **22**, 51–57.

Kramer, P. J., and Jackson, W. T. (1954). Causes of injury to flooded tobacco plants. *Plant Physiol.* **29**, 241–245.

Kramer, P. J., and Kozlowski, T. T. (1979). "Physiology of Woody Plants." Academic Press, New York.

Kramer, P. J., and Wilbur, K. M. (1949). Absorption of radioactive phosphorus by mycorrhizal roots of pine. *Science* **110**, 8–9.

Kreeb, K. (1965). Die ökologische Bedeutung der Bodenversalzung. *Angew. Bot.* **34**, 1–15.

Kreeb, K. (1967). Entgegnung an R. O. Slatyer. *Z. Pflanzenphysiol.* **56**, 95–97.

Kriedemann, P. E., and Barrs, H. D. (1982). Photosynthetic adaptation to water stress and implications for drought resistance. *In* "Crop Reactions to Water and Temperature Stresses in Humid, Temperate Climates" (C. D. Raper, Jr. and P. J. Kramer, eds.), pp. 201–230. Westview Press, Boulder, Colorado.

Kuiper, P. J. C. (1964). Water uptake of higher plants as affected by root temperature. *Meded. Landbouwhogesch. Wageningen* **63**, 1–11.

Kuiper, P. J. C. (1975). Role of lipids in water and ion transport. *In* "Recent Advances in the Chemistry and Biochemistry of Plant Lipids" (T. Galliard and E. I. Mercer, eds.), pp. 359–386. Academic Press, New York.

Kummerow, J. (1980). Adaptation of roots in water-stressed native vegetation. *In* "Adaptation of Plants to Water and High Temperature Stress" (N. C. Turner and P. J. Kramer, eds.), pp. 57–73. Wiley, New York.

Kuntz, I. D., and Kauzmann, W. (1974). Hydration of proteins and polypeptides. *Adv. Protein Chem.* **28**, 239–345.

Kuntz, J. E., and Riker, A. J. (1955). The use of radioactive isotopes to ascertain the role of root grafting in the translocation of water, nutrients, and disease-inducing organisms. *Proc. Int. Conf. Peaceful Uses At. Energy, 1st, 1955* Vol. 12, pp. 144–148.

Kurtzman, R. H., Jr. (1966). Xylem sap flow as affected by metabolic inhibitors and girdling. *Plant Physiol.* **41**, 641–646.

Ladefoged, K. (1960). A method for measuring the water consumption of larger intact trees. *Physiol. Plant.* **13**, 648–658.

Ladefoged, K. (1963). Transpiration of forest trees. *Physiol. Plant.* **16**, 378–414.

Lagerwerff, J. V., and Eagle, H. E. (1961). Osmotic and specific effects of excess salts on beans. *Plant Physiol.* **36**, 472–477.

Laing, H. E. (1940). The composition of the internal atmosphere of *Nuphar advenum* and other water plants. *Am. J. Bot.* **27**, 861–868.

Lamb, C. A. (1936). Tensile strength, extensibility, and other characteristics of wheat roots in relation to winter injury. *Ohio, Agric. Exp. Stn., Bull.* **568.**

Lamport, D. T. A. (1970). Cell wall metabolism. *Annu. Rev. Plant Physiol.* **21**, 235–270.

Land, S. B., Jr. (1974). Depth effects and genetic influences on injury caused by artificial sea water floods to loblolly and slash pine seedlings. *Can. J. For. Res.* **4**, 179–185.

Lange, O. L., Lösch, R., Schulze, E.-D., and Kappen, L. (1971). Responses of stomata to changes in humidity. *Planta* **100**, 76–86.

Lange, O. L., Kappen, L., and Schulze, E.-D., eds. (1976). "Water and Plant Life." Springer-Verlag, Berlin and New York.

Larson, P. R. (1964). Some indirect effects of environment on wood formation. *In* "The Formation of Wood in Forest Trees" (M. H. Zimmermann, ed.), pp. 345–365. Academic Press, New York.

Larson, P. R., and Isebrands, J. G. (1978). Functional significance of the nodal constricted zone in *Populus deltoides* Bartr. *Can. J. Bot.* **56**, 801–804.

LaRue, C. D. (1930). The water supply of the epidermis of leaves. *Pap. Mich. Acad. Sci. Art. Lett.* **13**, 131–139.

LaRue, C. D. (1952). Root grafting in tropical trees. *Science* **115**, 296.

Lassoie, J. P., Fetcher, N., and Salo, D. J. (1977). Stomatal infiltration pressures versus diffusion porometer measurements of needle resistance in Douglas-fir and lodgepole pine foliage. *Can. J. For. Res.* **7,** 192–196.

Lawlor, D. W. (1970). Absorption of polyethylene glycols by plants and their effects on plant growth. *New Phytol.* **69,** 501–513.

Lawlor, D. W. (1979). Effects of water and heat stress on carbon metabolism of plants with C_3 and C_4 photosynthesis. *In* "Stress Physiology in Crop Plants," (H. Mussell and R. C. Staples, eds.), pp. 303–326. Wiley, New York.

Lawlor, D. W., and Milford, G. F. J. (1975). The control of water and carbon dioxide flux in water-stressed sugar beet. *J. Exp. Bot.* **26,** 657–665.

Lebedeff, A. F. (1928). The movement of ground and soil waters. *Proc. Int. Congr. Soil Sci. 1st, 1927* Vol. 1, pp. 459–494.

Lee-Stadelmann, O. Y., and Stadelmann, E. J. (1976). Cell permeability and water stress. *In* "Water and Plant Life" (O. L. Lange, L. Kappen, and E.-D. Schulze, eds.), pp. 268–280. Springer-Verlag, Berlin and New York.

Lemon, E., Stewart, D. W., and Shawcroft, R. W. (1971). The sun's work in a cornfield. *Science* **174,** 371–378.

Leonard, O. A., and Pinckard, J. A. (1946). Effects of various oxygen and carbon dioxide concentrations on cotton root development. *Plant Physiol.* **21,** 18–36.

Leopold, A. C., Musgrove, M. E., and Williams, K. M. (1981). Solute leakage resulting from leaf desiccation. *Plant Physiol.* **68,** 1222–1225.

Leshem, B. (1965). The annual activity of intermediary roots of the Aleppo pine. *For. Sci.* **11,** 291–298.

Letey, J. Jr., and Stolzy, L. H. (1964). Measurement of oxygen diffusion rates with the platinum microelectrode. I. Theory and equipment. *Hilgardia* **35,** 545–554.

Letey, J., Jr., Stolzy, L. H., and Kemper, W. D. (1967). Soil aeration. *In* "Irrigation of Agricultural Lands" (R. M. Hagan, H. R. Haise, and T. W. Edminster, eds.), pp. 941–949. Am. Soc. Agron., Madison, Wisconsin.

Levitt, J. (1957). The significance of "apparent free space" (A.F.S.) in ion absorption. *Physiol. Plant.* **10,** 882–888.

Levitt, J. (1966). Resistance to water transport in plants—a misconception. *Nature (London)* **212,** 527.

Levitt, J. (1972). "Responses of Plants to Environmental Stresses." Academic Press, New York.

Levitt, J. (1980). "Responses of Plants to Environmental Stress," 2nd ed., Vol. 2. Academic Press, New York.

Levitt, J., Scarth, G. W., and Gibbs, R. D. (1936). Water permeability of isolated protoplasts in relation to volume change. *Protoplasma* **26,** 237–248.

Levy, Y. (1980). Effect of evaporative demand on water relations of *Citrus limon. Ann. Bot. (London)* [N.S.] **46,** 695–700.

Levy, Y., and Kaufmann, M. R. (1976). Cycling of leaf conductance in citrus exposed to natural and controlled environments. *Can. J. Bot.* **54,** 2215–2218.

Lewis, F. J. (1945). Physical condition of the surface of the mesophyll cell walls of the leaf. *Nature (London)* **156,** 407–490.

Lewis, G. N., and Randall, M. (1961). "Thermodynamics" (revised by K. S. Pitzer and L. Brewer), 2nd ed. McGraw-Hill, New York.

Liming, F. G. (1934). A preliminary study of the lengths of the open vessels in the branches of the American elm. *Ohio J. Sci.* **34,** 415–419.

Ling, G. N. (1969). A new model for the living cell. A summary of the theory and recent experimental evidence in its support. *Int. Rev. Cytol.* **26,** 1–61.

Livingston, B. E. (1918). Porous clay cones for the auto-irrigation of potted plants. *Plant World* **21,** 202–208.

Livingston, B. E., and Brown, W. H. (1912). Relation of the daily march of transpiration to variations in the water content of foliage leaves. *Bot. Gaz. (Chicago)* **53**, 309–330.

Livne, A., and Vaadia, Y. (1972). Water deficits and hormone relations. *In* "Water Deficits and Plant Growth" (T. T. Kozlowski, ed.), Vol. 3, pp. 255–275. Academic Press, New York.

Lloyd, F. E. (1908). The physiology of stomata. *Carnegie Inst. Washington Publ.* **82.**

Lockard, R. G., and Schneider, G. W. (1981). Stock and scion growth relationships and the dwarfing mechanism in apple. *Hortic. Rev.* **3**, 315–375.

Lodhi, M. A. K., and Killingbeck, K. T. (1980). Allelopathic inhibition of nitrification and nitrifying bacteria in a ponderosa pine (*Pinus ponderosa* Dougl.) community. *Am. J. Bot.* **67**, 1423–1429.

Loehwing, W. F. (1934). Physiological aspects of the effect of continuous soil aeration on plant growth. *Plant Physiol.* **9**, 567–583.

Loehwing, W. F. (1937). Root interactions of plants. *Bot. Rev.* **3**, 195–239.

Loftfield, J. V. G. (1921). The behavior of stomata. *Carnegie Inst. Washington Publ.* **314.**

Long, E. M. (1943). The effect of salt addition to the substrate on intake of water and nutrients by roots of approach grafted tomato plants. *Am. J. Bot.* **30**, 594–601.

Longstreth, D. J., and Kramer, P. J. (1980). Water relations during flower induction and anthesis. *Bot. Gaz. (Chicago)* **141**, 69–72.

Loomis, R. S., Williams, W. A., and Hall, A. E. (1971). Agricultural productivity. *Annu. Rev. Plant Physiol.* **22**, 431–468.

Loomis, W. E. (1934). Daily growth of maize. *Am. J. Bot.* **21**, 1–6.

Loomis, W. E. (1935). The translocation of carbohydrates in maize. *Iowa State Coll. J. Sci.* **9**, 509–520.

Lopushinsky, W. (1964a). Effect of water movement on ion movement into the xylem of tomato roots. *Plant Physiol.* **39**, 494–501.

Lopushinsky, W. (1964b). Calcium transport in tomato roots. *Nature (London)* **201**, 518–519.

Lopushinsky, W. (1969). Stomatal closure in conifer seedlings in response to leaf moisture stress. *Bot. Gaz. (Chicago)* **130**, 258–263.

Lopushinsky, W. (1980). Occurrence of root pressure exudation in Pacific Northwest conifer seedlings. *For. Sci.* **26**, 275–279.

Lopushinsky, W., and Beebe, T. (1976). Relationship of shoot-root ratio to survival and growth of outplanted Douglas fir seedlings. *USDA For. Serv. Res. Note PNW* **PNW-274.**

Lorio, P. L., Jr., and Hodges, J. D. (1968). Oleoresin exudation pressure and relative water content of the inner bark as an indicator of moisture stress in loblolly pine. *For. Sci.* **14**, 392–405.

Loustalot, A. J. (1945). Influence of soil-moisture conditions on apparent photosynthesis and transpiration of pecan leaves. *J. Agric. Res. (Washington, D.C.)* **71**, 519–532.

Louwerse, W. (1980). Effects of CO_2 concentration and irradiance on the stomatal behavior of maize, barley and sunflower plants in the field. *Plant, Cell Environ.* **3**, 391–398.

Loveless, A. R. (1961). A nutritional interpretation of sclerophylly based on differences in the chemical composition of sclerophyllous and mesophytic leaves. *Ann. Bot. (London)* [N.S.] **25**, 168–184.

Loveys, B. R. (1977). The intracellular location of abscisic acid in stressed and non-stressed leaf tissue. *Physiol. Plant.* **40**, 6–10.

Loweneck, M. (1930). Untersuchungen über wurzelatmung. *Planta* **10**, 185–228.

Lowry, M. W., Huggins, W. C., and Forrest, L. A. (1936). The effect of soil treatment on the mineral composition of exuded maize sap at different stages of development. *Ga., Agric. Exp. Stn., Bull.* **193.**

Lucas, W. J., and Alexander, J. M. (1981). Influence of turgor pressure manipulation on plasmalemma transport of HCO_3^- and OH^- in *Chara corallina*. *Plant Physiol.* **68**, 553–559.

Ludlow, M. M. (1980). Adaptive significance of stomatal responses to water stress. *In* "Adaptation of Plants to Water and High Temperature Stress'' (N. C. Turner and P. J. Kramer, eds.), pp. 123–138. Wiley, New York.

Lundegårdh, H. (1954). The transport of water in wood. *Ark. Bot.* **3,** 89–119.

Lunt, O. R., Sciaroni, R. H., and Enomoto, W. (1963). Organic matter and wettability for greenhouse soils. *Calif. Agric.* **17**(4), 6.

Lüttge, U., and Laties, G. G. (1966). Dual mechanisms of ion absorption in relation to long distance transport in plants. *Plant Physiol.* **41,** 1531–1539.

Lutz, H. J. (1944). Determinations of certain physical properties of forest soils. I. Methods utilizing samples collected in metal cylinders. *Soil Sci.* **57,** 475–487.

Lyford, W. H., and Wilson, B. F. (1964). Development of the root system of *Acer rubrum* L. *Harv. For. Pap.* **10.**

Lyon, T. L., and Buckman, H. O. (1943). ''The Nature and Properties of Soils,'' 4th ed. Macmillan, New York.

Lyons, J. M. (1973). Chilling injury in plants. *Annu. Rev. Plant Physiol.* **24,** 445–466.

Lyons, J. M., Graham, D., and Raison, J. K., eds. (1979). ''Low Temperature Stress in Crop Plants: The Role of the Membrane.'' Academic Press, New York.

Lyr, H., and Hoffmann, G. (1967). Growth rates and growth periodicity of tree roots. *Int. Rev. For. Res.* **2,** 181–236.

McComb, A. L., and Loomis, W. E. (1944). Subclimax prairie. *Bull. Torrey Bot. Club* **71,** 46–76.

McCree, K. J., and Davis, S. D. (1974). Effect of water stress and temperature on leaf size and on size and number of epidermal cells in grain sorghum. *Crop Sci.* **14,** 751–755.

McDermott, J. J. (1945). The effect of moisture content of the soil upon the rate of exudation. *Am. J. Bot.* **32,** 570–574.

MacDougal, D. T. (1926). The hydrostatic system of trees. *Carnegie Inst. Washington Publ.* **373.**

Machlis, L. (1944). The respiratory gradient in barley roots. *Am. J. Bot.* **31,** 281–282.

McIlrath, W. J., and Abrol, Y. P. (1963). Dehydration of seeds in intact tomato fruits. *Science* **142,** 1681–1682.

McKee, W. H., Jr., and Shoulders, E. (1974). Slash pine biomass response to site preparation and soil properties. *Soil Sci. Soc. Am. Proc.* **38,** 144–148.

McQuilkin, W. E. (1935). Root development of pitch pine, with some comparative observations in shortleaf pine. *J. Agric. Res. (Washington, D.C.)* **51,** 983–1016.

MacRobbie, E. A. C. (1962). Ionic relations of *Nitella translucens. J. Gen. Physiol.* **45,** 861–878.

McWilliam, J. R. (1983). Physiological basis for chilling stress and the consequences for crop production. *In* ''Crop Reactions to Water and Temperature Stress in Humid, Temperate Climates'' (C. D. Raper, Jr. and P. J. Kramer, eds.), pp. 113–132. Grandview Press, Boulder, Colorado.

McWilliam, J. R., and Kramer, P. J. (1968). The nature of the perennial response to Mediterranean grasses. I. Water relations and summer survival in *Phalaris. Aust. J. Agric. Res.* **19,** 381–395.

McWilliam, J. R., Kramer, P. J., and Musser, R. L. (1982). Temperature-induced water stress in chilling-sensitive plants. *Aust. J. Plant Physiol.* **9,** 343–352.

Madgwick, H. A. I., and Ovington, J. D. (1959). The chemical composition of precipitation in adjacent forest and open plots. *Forestry* **32,** 14–22.

Maercker, U. (1965). Zur Kenntnis der Transpiration der Schliesszellen. *Protoplasma* **60,** 61–78.

Magistad, O. C., and Reitemeier, R. F. (1943). Soil solution concentrations at the wilting point and their correlation with plant growth. *Soil Sci.* **55,** 351–360.

Mansfield, T. A., Wellburn, A. R., and Moreira, T. J. S. (1978). The role of abscisic acid and farnesol in the alleviation of water stress. *Philos. Trans. R. Soc. London, Ser. B* **284,** 471–482.

Mark, W. R., and Reid, C. P. P. (1971). Lodgepole pine-dwarf mistletoe xylem water potentials. *For. Sci.* **17,** 470–471.

Markhart, A. H., III, Fiscus, E. L., Naylor, A. W., and Kramer, P. J. (1979). Effect of temperature on water and ion transport in soybean and broccoli systems. *Plant Physiol.* **64,** 83–87.

Markhart, A. H., III, Peet, M. M., Sionit, N., and Kramer, P. J. (1980). Low temperature acclimation of root fatty acid composition, leaf water potential, gas exchange and growth of soybean seedlings. *Plant, Cell Environ.* **3,** 435–441.

Markhart, A. H., III, Sionit, N., and Siedow, J. N. (1981). Cell wall water dilution: An explanation of apparent negative turgor potentials. *Can. J. Bot.* **59,** 1722–1725.

Marks, G. C., and Kozlowski, T. T., eds. (1973). "Ectomycorrhizae: Their Ecology and Physiology." Academic Press, New York.

Maronek, D. M., Hendrix, J. W., and Kiernan, J. (1981). Mycorrhizal fungi and their importance in horticultural crop production. *Hortic. Rev.* **3,** 172–213.

Marshall, D. C. (1958). Measurement of sap flow in conifers by heat transport. *Plant Physiol.* **21,** 95–101.

Marshall, J. K., ed. (1977). "The Belowground Ecosystem," Range Sci. Dep. Sci. Ser. 26. Colorado State Univ., Fort Collins.

Marshall, T. J. (1959). Relations between water and soil. *Tech. Commun. Common. Bur. Soils* **50.**

Martin, E. V. (1943). Studies of evaporation and transpiration under controlled conditions. *Carnegie Inst. Washington Publ.* **550.**

Martin, E. V., and Clements, F. E. (1935). Studies of the effect of artificial wind on growth and transpiration in *Helianthus annuus*. *Plant Physiol.* **10,** 613–636.

Marvin, J. W. (1958). The physiology of maple sap flow. *In* "The Physiology of Forest Trees" (K. V. Thimann, ed.), pp. 95–124. Ronald Press, New York.

Marx, D. H. (1969). The influence of ectotrophic mycorrhizal fungi on the resistance of pine roots to pathogenic infection. *Phytopathology* **59,** 411–417.

Marx, D. H., Hatch, A. B., and Mendicino, J. F. (1977). High soil fertility decreases sucrose content and susceptibility of loblolly pine roots to ectomycorrhizal infection by *Pisolithus tinctorius*. *Can. J. Bot.* **55,** 1569–1574.

Masse, W. B. (1981). Prehistoric irrigation systems in the Salt River Valley, Arizona. *Science* **214,** 408–415.

Matile, P. (1976). Vacuoles. *In* "Plant Biochemistry" (J. Bonner and J. E. Varner, eds.), 3rd ed. pp. 189–224. Academic Press, New York.

Matile, P. (1978). Biochemistry and function of vacuoles. *Annu. Rev. Plant Physiol.* **29,** 193–213.

Maugh, T. H. II (1978). Soviet science: a wonder water from Kazakstan, *Science* **202,** 414.

Maximov, N. A. (1929). "The Plant in Relation to Water" (English tran. by R. H. Yapp) Allen & Unwin, London.

May, L. H., and Milthorpe, F. L. (1962). Drought resistance of crop plants. *Field Crop Abstr.* **15,** 171–179.

May, L. H., Milthorpe, E. J., and Milthorpe, F. L. (1962). Pre-sowing hardening of plants to drought. *Field Crop Abstr.* **15,** 93–98.

Mayoral, M. L., Atsmon, D. A., Shimshi, D., and Gromet-Elhanan, Z. (1981). Effect of water stress on enzyme activities in wheat and related wild species: Carboxylase activity, electron transport and photophosphorylation in isolated chloroplasts. *Aust. J. Plant Physiol.* **8,** 385–393.

Mederski, H. J. (1961). Determination of internal water stress of plants by beta ray gauging. *Soil Sci.* **92,** 143–146.

Mederski, H. J., and Alles, W. (1968). Beta gauging leaf water status: Influence of changing leaf characteristics. *Plant Physiol.* **43,** 470–472.

Meek, B. D., and Stolzy, L. H. (1978). Short-term flooding. *In* "Plant Life in Anaerobic Environments" (D. D. Hook and R. M. M. Crawford, eds.), pp. 351–373. Ann Arbor Sci. Press, Ann Arbor, Michigan.

Mees, G. C., and Weatherley, P. E. (1957). The mechanisms of water absorption by roots. I. II. *Proc. R. Soc. London, Ser. B* **147**, 367–380, 381–391.

Meidner, H. (1975). Water supply, evaporation, and vapour diffusion in leaves. *J. Exp. Bot.* **26**, 666–673.

Meidner, H. (1976). Vapour loss through stomatal pores with the mesophyll tissue excluded. *J. Exp. Bot.* **27**, 172–174.

Melin, E. (1953). Physiology of mycorrhizal relations in plants. *Annu. Rev. Plant Physiol.* **4**, 325–346.

Melin, E., Nilsson, H., and Hacskaylo, H. E. (1958). Translocation of cations to seedlings of *Pinus virginiana* through mycorrhizal mycelium. *Bot. Gaz. (Chicago)* **119**, 243–246.

Mengel, D. B., and Barber, S. A. (1974). Development and distribution of the corn root system under field conditions. *Agron. J.* **66**, 341–344.

Merkle, F. G., and Dunkle, E. C. (1944). Soluble salt content of greenhouse soils as a diagnostic aid. *J. Am. Soc. Agron.* **36**, 10–19.

Merwin, H. E., and Lyon, H. (1909). Sap pressure in the birch stem. *Bot. Gaz. (Chicago)* **48**, 442–458.

Meyer, B. S. (1931). Effects of mineral salts upon the transpiration and water requirement of the cotton plant. *Am. J. Bot.* **18**, 79–93.

Meyer, B. S. (1938). The water relations of plant cells. *Bot. Rev.* **4**, 531–547.

Meyer, B. S. (1945). A critical evaluation of the terminology of diffusion phenomena. *Plant Physiol.* **20**, 142–164.

Meyer, B. S., and Anderson, D. B. (1939). "Plant Physiology." D. Van Nostrand, New York.

Meyer, B. S., Anderson, D. B., Bohning, R. H., and Fratianne, D. G. (1973). "Introduction to Plant Physiology." Van Nostrand-Reinhold, Princeton, New Jersey.

Meyer, R. F., and Boyer, J. S. (1981). Osmoregulation, solute distribution, and growth in soybean seedlings having low water potentials. *Planta* **151**, 482–489.

Meyer, W. S., and Ritchie, J. T. (1980). Water status of cotton as related to taproot length. *Agron. J.* **72**, 577–580.

Michaelis, P. (1934). Ökologische Studien an der alpinen Baum-grenze. IV. Zur Kenntnis des winterlichen Wasserhaushaltes. *Jahrb. Wiss. Bot.* **80**, 169–247.

Milborrow, B. V. (1974). Biosynthesis of abscisic acid by a cell-free system. *Phytochemistry* **13**, 131–136.

Milburn, J. A. (1973). Cavitation in *Ricinus* by acoustic detection: Induction in excised leaves by various factors. *Planta* **110**, 253–265.

Milburn, J. A., and Johnson, R. P. C. (1966). The conduction of sap. II. Detection of vibrations produced by sap cavitation in *Ricinus* xylem. *Planta* **69**, 43–52.

Milburn, J. A., and Zimmermann, M. H. (1977). Preliminary studies on sapflow in *Cocos nucifera* L. II. Phloem transport. *New Phytol.* **79**, 543–558.

Millar, A. A., Gardner, W. R., and Goltz, S. M. (1971). Internal water stress and water transport in seed onion plants. *Agron. J.* **63**, 779–784.

Millar, B. D. (1966). Relative turgidity of leaves: Temperature effects in measurement. *Science* **154**, 512–513.

Miller, D. E., and Aarstad, J. S. (1976). Yields and sugar content of sugarbeets as affected by high frequency irrigation. *Agron. J.* **68**, 231–234.

Miller, D. E., and Hang, A. N. (1980). Deficit, high frequency irrigation of sugar beets with the line source technique. *Soil Sci. Soc. Am. J.* **44**, 1295–1298.

Miller, D. R., Vavrina, C. A., and Christensen, T. W. (1980). Measurement of sap flow and transpiration in ring-porous oaks using a heat pulse velocity technique. *For. Sci.* **26**, 485–494.

Miller, E. C. (1938). "Plant physiology," 2nd ed. McGraw-Hill, New York.

Miller, L. N. (1965). Changes in radiosensitivity of pine seedlings subjected to water stress during chronic gamma irradiation. *Health Phys.* **11,** 1653–1662.

Miller, L. N. (1972). Matric potentials in plants: Means of estimation and eco-physiological significance. *In* "Psychrometry in Water Relations Research" (R. W. Brown and B. P. van Haveren, eds.), pp. 211–217. Utah Agric. Exp. Stn., Utah State University, Logan.

Miller, R. J., and Beard, B. H. (1967). Effects of irrigation management on chemical composition of soybeans in the San Joaquin Valley. *Calif. Agric.* **21**(10), 8–10.

Miller, S. A., and Mazurak, A. P. (1958). Relationships of particle and pore sizes to the growth of sunflowers. *Soil Sci. Soc. Am. Proc.* **22,** 275–278.

Milthorpe, L. L., and Spencer, E. (1957). Experimental studies of the factors controlling transpiration. *J. Exp. Bot.* **8,** 413–437.

Minshall, W. H. (1964). Effect of nitrogen-containing nutrients on the exudation from detopped tomato plants. *Nature (London)* **202,** 925–926.

Minshall, W. H. (1968). Effects of nitrogenous materials on translocation and stump exudation in root systems of tomato. *Can. J. Bot.* **46,** 363–376.

Mitchell, H. L., Finn, R. F., and Rosendahl, R. O. (1937). The relation between mycorrhizae and the growth and nutrient absorption of coniferous seedlings in nursery beds. *Black Rock For. Pap.* **11,** 57–73.

Mohanty, P., and Boyer, J. S. (1976). Chloroplast response to low leaf water potential. IV. Quantum yield is reduced. *Plant Physiol.* **57,** 704–709.

Moinat, A. D. (1943). An auto-irrigator for growing plants in the laboratory. *Plant Physiol.* **18,** 280–287.

Molisch, H. (1912). Das Offen-und Geschlossensein des Spaltöffnungen, veranschaulicht durch eine neue Methode (infiltrationsmethode). *Z. Bot.* **4,** 106–122.

Molz, F. J., and Boyer, J. S. (1978). Growth-induced water potentials in plant cells and tissues. *Plant Physiol.* **62,** 423–429.

Molz, F. J., and Ferrier, J. M. (1982). Mathematical treatment of water movement in plant cells and tissue: a review. *Plant, Cell Environ.* **5,** 191–206.

Molz, F. J., and Ikenberry, E. (1974). Water transport through plant cells and cell walls: Theoretical development. *Soil Sci. Soc. Am. J.* **38,** 699–704.

Monteith, J. L. (1963). Dew: Facts and fallacies. *In* "The Water Relations of Plants" (A. J. Rutter and F. H. Whitehead, eds.), pp. 37–56. Wiley, New York.

Monteith, J. L., and Owen, P. C. (1958). A thermocouple method for measuring relative humidity in the range 95–100%. *J. Sci. Instrum.* **35,** 443–446.

Monteith, J. L., and Webb, C., eds. (1981). Soil Water and Nitrogen in Mediterranean-type Environments. *Plant Soil* **58,** 1–434.

Mooney, H. A. (1969). Dark respiration of related evergreen and deciduous mediterranean plants during induced drought. *Torrey Bot. Club Bull.* **96,** 550–555.

Mooney, H. A., Björkman, O., and Collatz, G. J. (1977). Photosynthetic acclimation to temperature and water stress in the desert shrub *Larrea divaricata. Year Book—Carnegie Inst. Washington* **76,** 328–335.

Moreland, D. E. (1950). A study of translocation of radioactive phosphorous in loblolly pine (*Pinus taeda* L.). *J. Elisha Mitchell Sci. Soc.* **66,** 175–181.

Morrow, P. A., and Slatyer, R. O. (1971). Leaf resistance measurements: Precautions in calibration and use. *Agric. Meteorol.* **8,** 223–233.

Moss, D. N. (1963). The effect of environment on gas exchange of leaves. *Bull.—Conn., Agric. Exp. Stn., New Haven* **664.**

Moss, G. I., and Downey, L. A. (1971). Influence of drought stress on female gametophyte development in corn (*Zea mays* L.) and subsequent grain yield. *Crop Sci.* **11,** 368–372.

Mothes, K. (1932). Ernährung, Structur, und Transpiration. Ein Beitrag zur kausalen Analyse der Xeromorphosen. *Biol. Zentralbl.* **52,** 193–233.

Mozhaeva, L. V., and Pil'shchikova, N. (1980). The motive force behind bleeding in plants. *Sov. Plant Physiol. (Engl. Transl.)* **26**(5), 802–807.

Mudd, J. B., and Kozlowski, T. T., eds. (1975). "Responses of Plants to Air Pollution." Academic Press, New York.

Muller, C. H. (1969). Allelopathy as a factor in ecological process. *Vegetatio* **18**, 348–357.

Müller-Stoll, W. R. (1947). Der Einfluss der Ernährung auf die Xeromorphie der Hochmoorpflanzen. *Planta* **35**, 225–251.

Münch, E. (1930). "Die Stoffbewegung in der Pflanze." Fischer, Jena.

Munns, R., and Weir, R. (1981). Contribution of sugars to osmotic adjustment in elongating and expanded zones of wheat leaves during moderate water deficits at two light levels. *Aust. J. Plant Physiol.* **8**, 93–105.

Mussell, H., and Staples, R. C., eds. (1979). "Stress Physiology in Crop Plants." Wiley, New York.

Musser, R. L., Kramer, P. J., Naylor A. W., and Thomas, S. A. (1982). Physiological effects of root versus shoot chilling of soybean. *Plant Physiol.* **69** (supplement to No. 4), 121.

Mustafa, M. A., and Letey, J. (1970). Factors influencing effectiveness of two surfactants on water-repellent soils. *Calif. Agric.* **24**(6), 12–13.

Myers, G. M. P. (1951). The water permeability of unplasmolryzed tissues. J. Exp. Bot. 2, 129–144.

Nagashi, G., Thomson, W. W., and Leonard, R. T. (1974). The casparian strip as a barrier to the movement of lanthanum in corn roots. *Science* **183**, 670–671.

Nakayama, F. S., and Ehrler, W. L. (1964). Beta ray gauging technique for measuring leaf water content changes and moisture status of plants. *Plant Physiol.* **39**, 95–98.

Nangju, D. (1980). Soybean response to indigenous *Rhizobia* as influenced by cultivar origin. *Agron. J.* **72**, 403–406.

Naylor, A. W. (1972). Water deficits and nitrogen metabolism. *In* "Water Deficits and Plant Growth" (T. T. Kozlowski, ed.), Vol. 3, pp. 241–254. Academic Press, New York.

Neales, T. F., and Davies, J. A. (1966). The effect of photoperiod duration upon the respiratory activity of the roots of wheat seedlings. *Aust. J. Biol. Sci.* **19**, 471–480.

Neales, T. F., Patterson, A. A., and Hartney, V. J. (1968). Physiological adaptation to drought in the carbon assimilation and water loss of xerophytes. *Nature* **219**, 469–472.

Nelson, O. E. (1983). Genetics and plant breeding in relation to stress tolerance. *In* "Crop Reactions to Water and Temperature Stresses in Humid, Temperate Climates" (C. D. Raper, Jr. and P. J. Kramer, eds.), pp. 351–357. Westview Press, Boulder, Colorado.

Néméthy, G., and Scheraga, H. A. (1962). Structure of water and hydrophobic bonding in proteins. I. A model for the thermodynamic properties of liquid water. *J. Chem. Phys.* **36**, 3382–3400.

Neumann, H. H., and Thurtell, G. W. (1972). A Peltier cooled thermocouple dewpoint hygrometer for *in situ* measurement of water potentials. *In* "Psychrometry in Water Relations Research" (R. W. Brown and B. P. Van Haveren, eds.), pp. 103–112. Utah Agric. Exp. Stn., Utah State University, Logan.

Neumann, H. H., Thurtell, G. W., and Stevenson, K. R. (1974). In situ measurements of leaf water potential and resistance to water flow in corn, soybean, and sunflower at several transpiration rates. *Can. J. Plant Sci.* **54**, 175–184.

Newman, E. I. (1966). Relationship between root growth of flax (*Linum usitatissimum*) and soil water potential. *New Phytol.* **65**, 273–283.

Newman, E. I. (1969). Resistance to water flow in soil and plant. I. Soil resistance in relation to amounts of roots: Theoretical estimates. *J. Appl. Ecol.* **6**, 1–12.

Newman, E. I. (1974). Root and soil water relations. *In* "The Plant Root and its Environment" (E. W. Carson, ed.), pp. 363–440. Univ. Press of Virginia, Charlottesville.

Newman, E. I. (1976). Water movement through root systems. *Philos. Trans. R. Soc. London, Ser. B* **273**, 463–478.

Ney, D., and Pilet, P. E. (1981). Nutation of growing and georeacting roots. *Plant, Cell Environ.* **4**, 339–343.

Nielsen, D. R., Jackson, R. D., Cory, J. W., and Evans, D. D., eds. (1972). "Soil Water." Am. Soc. Agron. and Soil Sci. Soc. Am., Madison, Wisconson.

Nightingale, G. T. (1935). Effects of temperature on growth, anatomy, and metabolism of apple and peach roots. *Bot. Gaz. (Chicago)* **96**, 581–639.

Nishiyama, I. (1975). A break in the Arrhenius plot of germination activity in rice seeds. *Plant Cell Physiol.* **16**, 535–536.

Nobel, P. S. (1974). "Introduction to Biophysical Plant Physiology." Freeman, San Francisco, California.

Nobel, P. S. (1980). Leaf anatomy and water use efficiency. *In* "Adaptation of Plants to Water and High Temperature Stress" (N. C. Turner and P. J. Kramer, eds.), pp. 43–55. Wiley, New York.

Norby, R. J., and Kozlowski, T. T. (1980). Allelopathic potential of ground cover species on *Pinus resinosa* seedlings. *Plant Soil* **57**, 363–374.

Norris, R. F., and Bukovac, M. J. (1968). Structure of the pear leaf cuticle with special reference to cuticular penetration. *Am. J. Bot.* **55**, 975–983.

North, C. P., and Wallace, A. (1955). Soil temperature and citrus. *Calif. Agric.* 9(11), 13.

Noy-Meir, I., and Ginzburg, B. Z. (1967). An analysis of the water potential isotherm in plant tissue. I. The theory. *Aust. J. Biol. Sci.* **20**, 695–721.

Noy-Meir, I., and Ginzburg, B. Z. (1969). An analysis of the water potential isotherm in plant tissue. II. Comparative studies on leaves of different types. *Aust. J. Biol. Sci.* **22**, 35–52.

Nutman, F. J. (1933). The root-system of *Coffea arabica*, Part II. The effect of some soil conditions in modifying the "normal" root-system. *Emp. J. Exp. Agric.* **1**, 285–296.

Nutman, F. J. (1934). The root-system of *Coffea arabica*. Part III. The spatial distribution of the absorbing area of the root. *Emp. J. Exp. Agric.* **2**, 293–302.

Nutman, F. J. (1941). Studies of the physiology of *Coffea arabcia*. III. Transpiration rates of whole trees in relation to natural environmental conditions. *Ann. Bot. (London)* [N.S.] **5**, 59–82.

Obloj, H., and Kacperska, A. (1981). Desiccation tolerance changes in winter rape leaves grown under different environmental conditions. *Biol. Plant.* **23**, 209–213.

Oertli, J. J. (1966). Active water transport in plants. *Physiol. Plant.* **19**, 809–817.

O'Leary, J. W. (1969). The effect of salinity on permeability of roots to water. *Isr. J. Bot.* **18**, 1–9.

O'Leary, J. W., and Knecht, G. W. (1971). The effect of relative humidity on growth, yield and water consumption of bean plants. *J. Am. Soc. Hortic. Sci.* **96**, 2563–265.

O'Leary, J. W., and Kramer, P. J. (1964). Root pressure in conifers. *Science* **145**, 284–285.

O'Leary, J. W., and Prisco, J. T. (1970). Response of osmotically stressed plants to growth regulators. *Adv. Front. Plant Sci.* **25**, 129–139.

Olszyk, D. M., and Tibbits, T. W. (1981). Stomatal response and leaf injury of *Pisum sativum* L. with SO_2 and O_3 exposures. I. Influence of pollutant level and leaf maturity. *Plant Physiol.* **67**, 539–544.

Oppenheimer, H. R. (1941). Root cushions, root stalagmites and similar structures. *Palest. J. Bot., Rehovot Ser.* **4**, 11–19.

Oppenheimer, H. R., and Elze, D. L. (1941). Irrigation of citrus trees according to physiological indicators. *Palest. J. Bot., Rehovot Ser.* **4**, 20–46.

Ordin, L., and Kramer, P. J. (1956). Permeability of *Vicia faba* root segments to water as measured by diffusion of deuterium hydroxide. *Plant Physiol.* **31**, 468–471.

Orians, G. H., and Solbrig, O. T. (1977). A cost-income model of leaves and roots with special reference to arid and semi-arid areas. *Am. Nat.* **111**, 677–690.

Osmond, C. B. (1978). Crassulacean acid metabolism: A curiosity in context. *Annu. Rev. Plant Physiol.* **29**, 379–414.

Osmond, C. B. (1980). Integration of photosynthetic carbon metabolism during stress. *In* "Genetic

Engineering of Osmoregulation'' (D. W. Rains, R. C. Valentine, and A. Hollaender, eds.), pp. 171–185. Plenum, New York.

Osmond, C. B., Björkman, O., and Anderson, D. J. (1980). ''Physiological Processes in Plant Ecology.'' Springer-Verlag, Berlin and New York.

Osmond, D. L., Wilson, R. F., and Raper, C. D., Jr. (1982). Fatty acid composition and nitrate uptake of soybean roots during acclimation to low temperature. *Plant Physiol.* **70**, 1689–1693.

O'Toole, J. C., and Maya, T. B. (1978). Genotypic variation in maintenance of leaf water potential in rice. *Crop Sci.* **18**, 873–876.

Owen, P. C. (1952). The relation of germination of wheat to water potential. *J. Exp. Bot.* **3**, 188–203.

Owen, P. C., and Watson, D. J. (1956). Effect on crop growth of rain after prolonged drought. *Nature (London)* **177**, 847.

Painter, L. I. (1966). Method of subjecting growing plants to a continuous soil moisture stress. *Agron. J.* **58**, 459–460.

Paleg, L. G., and Aspinall, D. eds. (1981). ''The Physiology and Biochemistry of Drought Resistance in Plants.'' Academic Press, New York.

Pallardy, S. G., and Kozlowski, T. T. (1980). Cuticle development in the stomatal region of *Populus* clones. *New Phytol.* **85**, 363–368.

Pallas, J. E., and Williams, G. G. (1962). Foliar absorption and translocation of p^{32} and 2,4-dichlorophenoxyaetic acid as affected by soil-moisture tension. *Bot. Gaz. (Chicago)* **123**, 175–180.

Pallas, J. E., Stansell, J. R., and Bruce, R. R. (1977). Peanut seed germination as related to soil water regime during pod development. *Agron. J.* **69**, 381–383.

Palzkill, D. A., and Tibbits, T. W. (1977). Evidence that root pressure flow is required for calcium transport to head leaves of cabbage. *Plant Physiol.* **60**, 854–856.

Parker, J. (1949). Effects of variation in the root-leaf ratio on transpiration rate. *Plant Physiol.* **24**, 739–743.

Parker, J. (1950). The effects of flooding on the transpiration and survival of some southeastern forest tree species. *Plant Physiol.* **25**, 453–460.

Parker, J. (1964). Autumn exudation from black birch. *Sci. Tree Top.* **2**(10), 9–11.

Parker, J. (1965). Physiological diseases of trees and shrubs. *Adv. Front. Plant Sci.* **12**, 97–248.

Parker, J. (1972). Protoplasmic resistance in water deficit. *In* ''Water Deficits and Plant Growth'' (T. T. Kozlowski, ed.), Vol. 3, pp. 125–176. Academic Press, New York.

Parkhurst, D. F., and Loucks, O. L. (1972). Optimal leaf size in relation to environment. *J. Ecol.* **60**, 505–537.

Parks, R. Q. (1951). Irrigation, agriculture and soil research in the United States. *Adv. Agron.* **3**, 323–344.

Parsons, L. R. (1978). Water relations, stomatal behavior, and root conductivity during acclimation to freezing temperatures. *Plant Physiol.* **62**, 64–70.

Parsons, L. R. (1982). Plant responses to water stress. *In* ''Breeding Plants for Less Favorable Environments'' (M. N. Christiansen, ed.), pp. 175–192. Wiley, New York.

Parsons, L. R., and Kramer, P. J. (1974). Diurnal cycling in root resistance to water movement. *Physiol. Plant.* **30**, 19–23.

Passioura, J. B. (1972). The effect of root geometry on the yield of wheat growing on stored water. *Aust. J. Agric. Res.* **23**, 745–52.

Passioura, J. B. (1973). Sense and nonsense in simulation. *J. Aust. Inst. Agric. Sci.* **39**, 181–183.

Passioura, J. B. (1980). The meaning of matric potential. *J. Exp. Bot.* **31**, 1161–1169.

Patric, J. H., and Lyford, W. H. (1980). Soil-water relations at the headwaters of a forest stream in central New England. *Harv. For. Pap.* **22**.

Patric, J. H., Douglass, J. E., and Hewlett, J. D. (1965). Soil water absorption by mountain and Piedmont forests. *Soil. Sci. Soc. Am. Proc.* **29**, 303–308.

Patterson, D. T. (1981). Effects of allelopathic chemicals on growth and physiological responses of soybean (*Glycine max*). *Weed Sci.* **29**, 53–59.

Patterson, D. T., Bunce, J. A., Alberte, R. S., and Van Volkenburgh, E. (1977). Photosynthesis in relation to leaf characteristics of cotton from controlled and field environments. *Plant Physiol.* **59**, 384–387.

Pavlychenko, T. K. (1937). Quantitative study of the entire root system of weed and crop plants under field conditions. *Ecology* **18**, 62–79.

Pearson, G. A. (1931). Forest types in the southwest as determined by climate and soil. *U.S., Dep. Agric., Tech. Bull.* **247**.

Peet, M. M., and Kramer, P. J. (1980). Effects of decreasing source/sink ratio in soybeans on photosynthesis, photorespiration, transpiration and yield. *Plant, Cell Environ.* **3**, 201–206.

Pereira, J. S., and Kozlowski, T. T. (1977). Influence of light intensity, temperature, and leaf area on stomatal aperture and water potential of woody plants. *Can. J. For. Res.* **7**, 145–153.

Persson, H. (1979). Fine root production, mortality, and decomposition in forest ecosystems. *Vegetatio* **41**, 101–109.

Peschel, G. (1976). The structure of water in the biological cell. *In* "Water and Plant Life" (O. L. Lange, L. Kappen, and E.-D. Schulze, eds.), pp. 6–58. Springer-Verlag, Berlin and New York.

Peters, D. B. (1957). Water uptake of corn roots as influenced by soil moisture content and soil moisture tension. *Soil Sci. Soc. Am. Proc.* **21**, 481–484.

Peters, D. B., and Russell, M. B. (1959). Relative water losses by evaporation and transpiration in field corn. *Soil Sci. Soc. Am. Proc.* **23**, 170–173.

Petty, J. A. (1978). Fluid flow through the vessels of birch wood. *J. Exp. Bot.* **29**, 1463–1469.

Pfeffer, W. F. P. (1877). "Osmotische Untersuchungen." Engelmann, Leipzig.

Philip, J. R. (1957). Evaporation, and moisture and heat fields in the soil. *J. Meteorol.* **14**, 354–366.

Philip, J. R. (1958). The osmotic cell, solute diffusibility, and the plant water economy. *Plant Physiol.* **33**, 264–271.

Philip, J. R., and D. A. de Vries. (1957). Moisture movement in porous materials under temperature gradients. *Trans. Am. Geophys. Union,* **38**, 222–232.

Phillips, I. D. (1964). The importance of an aerated root system in the regulation of growth levels in the shoot of *Helianthus annuus. Ann. Bot. (London)* [N.S.] **28**, 17–36.

Philpott, J. (1956). Blade tissue organization of foilage leaves of some Carolina shrub bog species are compared with their Appalachian Mountain affinities. *Bot. Gaz. (Chicago)* **118**, 88–105.

Pickard, W. F. (1973). A heat pulse method of measuring water flux in woody plant stems. *Math. Biosci.* **16**, 247–262.

Pierre, W. H., and Pohlman, G. G. (1934). The phosphorus concentration of the exuded sap of corn as a measure of the available phosphorus in the soil. *J. Am. Soc. Agron.* **25**, 160–171.

Pisek, A., and Berger, E. (1938). Kutikuläre Transpiration und Trockenresistenz isolierter Blätter und Sprosse. *Planta* **28**, 124–155.

Pisek, A., and Cartelliers, E. (1932). Zur Kenntnis des Wasserhaushaltes der Pflanzen. I. Sonnenpflanzen. *Jahrb. Wiss. Bot.* **75**, 195–251.

Pleasants, A. L. (1930). The effect of nitrate fertilizer on stomatal behavior. *J. Elisha Mitchell Sci. Soc.* **46**, 95–116.

Plymale, E. L., and Wylie, R. B. (1944). The major veins of mesomorphic leaves. *Am. J. Bot.* **31**, 99–106.

Pohlman, G. G. (1946). Effect of liming different soil layers on yield of alfalfa and on root development and nodulation. *Soil Sci.* **62**, 255–266.

Poljakoff-Mayber, A., and Gale, J. (1972). Physiological basis and practical problems of reducing transpiration. *In* "Water Deficits and Plant Growth" (T. T. Kozlowski, ed.), Vol. 3, pp. 277–306. Academic Press, New York.

Poljakoff-Mayber, A., and Gale, J., eds. (1975). "Plants in Saline Environments." Springer-Verlag, Berlin and New York.

Pollard, J. K., and Sproston, T. (1954). Inorganic constituents of sap exuded from the sapwood of *Acer saccharum*. *Plant Physiol.* **29**, 360–364.

Pospisilova, J., and Solorova, J., eds. (1977–1981). "Water in Plants Bibliography," Vols. 1–5. Junk, The Hague.

Post, K., and Seeley, J. G. (1943). Automatic watering of greenhouse crops. *Bull.—N.Y., Agric. Exp. Stn. (Ithaca)* **793.**

Postlethwait, S. N., and Rogers, B. (1958). Tracing the path of the transpiration stream in trees by the use of radioactive isotopes. *Am. J. Bot.* **45**, 753–757.

Powell, D. B. B. (1978). Regulation of plant water potential by membranes of the endodermis in young roots. *Plant, Cell Environ.* **1**, 69–76.

Preston, R. D. (1952). Movement of water in higher plants. *In* "Deformation and Flow in Biological Systems" (A. Frey-Wyssling, ed.), pp. 257–321. North-Holland Publ., Amsterdam.

Preston, R. D. (1961). Theoretical and practical implications of the stresses in the water-conducting system. "Recent Advances in Botany," pp. 1144–1149. Univ. of Toronto Press, Toronto.

Preston, R. D. (1974). "The Physical Biology of Plant Cell Walls." Chapman & Hall, London.

Preston, R. D. (1979). Polysaccharide conformation and cell wall function. *Annu. Rev. Plant Physiol.* **30**, 55–78.

Priestley, J. H. (1922). Further observations upon the mechanism of root pressure. *New Phytol.* **21**, 41–48.

Priestley, J. H. (1935). Radial growth and extension growth in the tree. *Forestry* **9**, 84–95.

Prihar, S. S., Khera, K. L., Shandhu, K. S., and Sandhu, B. S. (1976). Comparison of irrigation schedules based on pan evaporation and growth stages in winter wheat. *Agron. J.* **68**, 650–653.

Prikryl, Z., and Vancura, V. (1980). Root exudates of plants. VI. Wheat root exudation as dependent on growth, concentration gradient of exudates and the presence of bacteria. *Plant Soil* **57**, 69–83.

Proebsting, E. L. (1943). Root distribution of some deciduous fruit trees in a California orchard. *Proc. Am. Soc. Hortic. Sci.* **43**, 1–4.

Proebsting, E. L., and Gilmore, A. E. (1941). The relation of peach root toxicity to the re-establishing of peach orchards. *Proc. Am. Soc. Hortic. Sci.* **38**, 21–26.

Prokof'ev, A. A., Rybalova, B. A., and Zavadskaya, O. Y. (1981). Water exchange of ripening poppy fruits. *Sov. Plant Physiol. (Engl. Transl.)* **28**, 98–106.

Putnam, A. R., and Duke, W. B. (1978). Allelopathy in agroecosystems. *Annu. Rev. Phytopathol.* **16**, 431–451.

Queen, W. H. (1967). Radial movement of water and ^{32}P through suberized and unsuberized roots of grape. Ph.D. Dissertation, Duke University, Durham, North Carolina.

Raber, O. (1937). Water utilization by trees, with special reference to the economic forest species of the north temperate zone. *Misc. Publ.—U.S., Dep. Agric.* **257.**

Radin, J. W. (1982). Control of plant growth by nitrogen: differences between cereals and broadleaf species. *Plant, Cell Environ.* **5**, in press.

Radin, J. W., and Boyer, J. S. (1982). Control of leaf expansion by nitrogen nutrition in sunflower plants. *Plant Physiol.* **69**, 771–775.

Radin, J. W., and Parker, L. L. (1979). Water relations of cotton plants under nitrogen deficiency. II. Environmental interactions on stomata. *Plant Physiol.* **64**, 499–501.

Radler, F. (1965). Reduction of loss of moisture by the cuticle wax components of grapes. *Nature (London)* **207**, 1002–1003.

Radmer, R., and Kok, B. (1977). Photosynthesis: Limited yields, unlimited dreams. *BioScience* **27**, 599–605.

Radulovich, R. A., Phene, C. J., Davis, K. R., and Brownell, J. R. (1982). Comparison of water stress of cotton from measurements with the hydraulic press and the pressure chamber. *Agron. J.* **74**, 383–385.

Railton, I. D., and Reid, D. M. (1973). Effects of benzyladenine on the growth of waterlogged tomato plants. *Planta* **111**, 261–266.

Rains, D. W., Valentine, R. C., and Hollaender, A., eds. (1980). "Genetic Engineering of Osmoregulation." Plenum, New York.

Ramos, C., and Kaufmann, M. R. (1979). Hydraulic resistance of rough lemon roots. *Physiol. Plant.* **45**, 311–314.

Rand, R. H. (1977). Gaseous diffusion in the leaf interior. Trans. *ASAE* **20**, 701–704

Raney, F. C., and Mihara, Y. (1967). Water and soil temperature. *In* "Irrigation of Agricultural Lands" (R. M. Hagan, H. R. Haise, and T. W. Edminster, eds.), pp. 1024–1036. Am. Soc. Agron., Madison, Wisconsin.

Raney, F. C., and Vaadia, Y. (1965). Movement and distribution of THO in tissue water and vapor transpired by shoots of *Helianthus* and *Nicotina*. *Plant Physiol.* **40**, 383–388.

Raper, C. D., Jr., and Barber, S. A. (1970). Rooting systems of soybeans. I. Differences in root morphology among varieties. II. Physiological effectiveness as nutrient absorption surfaces. *Agron. J.* **62**, 581–584, 585–588.

Raper, C. D., Jr., and Kramer, P. J., eds. (1983). "Crop Reactions to Water and Temperature Stresses in Humid, Temperate Climates." Westview Press, Boulder, Colorado.

Raschke, K. (1975). Stomatal action. *Annu. Rev. Plant Physiol.* **26**, 309–340.

Raschke, K. (1976). How stomata resolve the dilemma of opposing priorities. *Philos. Trans. R. Soc. London, Ser. B* **273**, 551–560.

Raschke, K. (1979). Movements of Stomata. *Encycl. Plant Physiol. New Ser.* **7**, 383–441.

Rawlins, S. L. (1963). Resistance to water flow in the transpiration stream. *Bull.—Conn., Agric. Exp. Stn., New Haven* **664**, 69–85.

Rawlins, S. L. (1976). Measurement of water content and the state of water in soils. *In* "Water Deficits and Plant Growth" (T. T. Kozlowski, ed.), Vol. 4, pp. 1–55. Academic Press, New York.

Rawlins, S. L., and Raats, P. A. C. (1975). Prospects for high frequency irrigation. *Science* **188**, 604–610.

Rawson, H. M., Turner, N. C., and Begg, J. E. (1978). Agronomic and physiological responses of soybean and sorghum crops to water deficits. IV. Photosynthesis, transpiration and water use efficiency of leaves. *Aust. J. Plant Physiol.* **5**, 195–209.

Read, D. W. L., Fleck, S. V., and Pelton, W. L. (1962). Self-irrigating greenhouse pots. *Agron. J.* **54**, 467–470.

Reed, H. S., and MacDougal, D. T. (1937). Periodicity in the growth of the orange tree. *Growth* **1**, 371–373.

Reed, J. F. (1939). Root and shoot growth of shortleaf and loblolly pines in relation to certain environmental conditions. *Duke Univ. Sch. For. Bull.* **4**.

Reeve, R. C., and Fireman, M. (1967). Salt problems in relation to irrigation. *In* "Irrigation of Agricultural Lands" (R. M. Hagan, H. R. Haise, and T. W. Edminister, eds.), pp. 988–1008. Am. Soc. Agron., Madison, Wisconsin.

Reicosky, D. C., and Ritchie, J. T. (1976). Relative importance of soil resistance and plant resistance in root water absorption. *Soil Sci. Soc. Am. J.* **40**, 293–297.

Reid, D. M., and Crozier, A. (1971). Effects of waterlogging on the gibberellin content and growth of tomato plants. *J. Exp. Bot.* **22**, 39–48.

Reimann, E. G., Van Doren, C. A., and Stauffer, R. S. (1946). Soil moisture relationships during crop production. *Soil Sci. Soc. Am. Proc.* **10**, 41–46.

Reitz, L. P. (1974). Breeding for more efficient water use—is it real or a mirage? *Agric. Meteorol.* **14**, 3–11.

Renner, O. (1912). Versuche zur Mechanik der Wasserversorgung. 2. Über wurzeltätigkeit. *Ber. Dtsch. Bot. Ges.* **30**, 642–648.

Renner, O. (1929). Versuche zur Bestimmung des Filtrationswiderstandes der Wurzeln. *Jahrb. Wiss. Bot.* **70**, 805–838.

Repp, G. I., McAlister, D. R., and Wiebe, H. H. (1959). Salt resistance of protoplasm as a test for salt tolerance of agricultural plants. *Agron. J.* **51**, 314–314.

Rhoades, E. D. (1964). Inundation tolerance of grasses in flooded areas. *Trans. ASAE* **7**(2), 164, 165, 166, 169.

Rice, E. L. (1974). "Allelopathy." Academic Press, New York.

Rice, E. L. (1979). Allelopathy—an update. *Bot. Rev.* **45**, 15–109.

Richards, L. A. (1949). Methods of measuring soil moisture tension. *Soil Sci.* **68**, 95–112.

Richards, L. A., ed. (1954). "Diagnosis and Improvement of Saline and Alkaline Soils," U.S. Dep. Agric. Handb. 60. USDA, Washington, D.C.

Richards, L. A., and Campbell, R. B. (1950). The effect of salinity on the electircal resistance of gypsum, nylon, and fiberglass soil moisture measuring units. *U.S. Reg. Salinity Lab. Res. Rep.* **42**, 1–8.

Richards, L. A., and Loomis, W. E. (1942). Limitations of auto-irrigators for controlling soil moisture under growing plants. *Plant Physiol.* **17**, 223–235.

Richards, L. A., and Ogata, G. (1958). Thermocouple for vapor pressure measurements in biological and soil systems at high humidity. *Science* **128**, 1089–1090.

Richards, L. A., and Wadleigh, C. H. (1952). Soil water and plant growth. *In* "Soil Physical Conditions and Plant Growth" (B. T. Shaw, ed.), pp. 73–251. Academic Press, New York.

Richards, L. A., and Weaver, L. R. (1944). Moisture retention by some irrigated soils as related to soil-moisture tension. *J. Agric. Res. (Washington, D.C.)* **69**, 215–235.

Richards, S. J., and Marsh, A. W. (1961). Irrigation based on soil suction measurements. *Soil. Sci. Soc. Am. Proc.* **25**, 65–69.

Richards, S. J., Hagan R. M., and McCalla, T. M. (1952). Soil temperature and plant growth. *In* "Soil Physical Conditions and Plant Growth" (B. T. Shaw, ed.), pp. 303–480. Academic Press, New York.

Richards, S. J., Weeks, L. V., and Johnson, J. C. (1958). Effects of irrigation treatments and rates of nitrogen fertilization on young Hass avocado trees. *Proc. Am. Soc. Hortic. Sci.* **71**, 292–297.

Richter, C., and Marschner, H. (1973). Umtausch von Kalium in verschiedenei Wurzelzonen von Maiskeimpflanzen. *Z. Pflanzen-physiol.* **70**, 211–221.

Richter, H. (1973). Frictional potential losses and total water potential in plants: A reevaluation. *J. Exp. Bot.* **24**, 983–994.

Richter, H. (1976). The water status in the plant-experimental evidence. *In* "Water and Plant Life" (O. L. Lange, L. Kappen, and E.-D. Schulze, eds.), pp. 42–58. Springer-Verlag, Berlin and New York.

Rickman, R. W., Letey, J., and Stolzy, L. H. (1966). Plant responses to oxygen supply and physical resistance in the root environment. *Soil Sci. Soc. Am. Proc.* **30**, 304–307.

Ritchie, J. T., and Burnett, E. (1968). A precision weighing lysimeter for row crop water use studies. *Agron. J.* **60**, 545–549.

Ritchie, G. A., and Hinckley, T. M. (1975). The pressure chamber as an instrument for ecological research. *Adv. Ecol. Res.* **9**, 165–254.

Robards, A. W., ed. (1974). "Dynamic Aspects of Plant Ultrastructure." McGraw-Hill, New York.

Robards, A. W., and Clarkson, D. T. (1976). The role of plasmodesmata in the transport of water and nutrients across roots. *In* "Intercellular Communication in Plants: Studies on Plasmodesmata" (B. E. S. Gunning and A. W. Robards, eds.), pp. 181–202. Springer-Verlag, Berlin and New York.

Roberts, B. R. (1964). Effects of water stress on the translocation of photosynthetically assimilated carbon-14 in yellow poplar. *In* "The Formation of Wood in Forest Trees" (M. H. Zimmermann, ed.), pp. 273–288. Academic Press, New York.

Roberts, B. R. (1969). Light as a source of error in estimates of water potential by vapor equilibration. *Plant Physiol.* **44,** 937–938.

Roberts, F. L. (1948). A study of the absorbing surfaces of the roots of loblolly pine. M.A. Thesis, Duke University, Durham, North Carolina.

Roberts, J. (1977). The use of tree-cutting techniques in the study of water relations of mature *Pinus sylvestris* L. *J. Exp. Bot.* **28,** 751–767.

Roberts, S. W. (1978). A comparative study of the leaf water relationship in four forest tree species, using the pressure-volume analytical technique. Ph.D. Dissertation, Duke University, Durham, North Carolina.

Roberts, S. W., and Knoerr, K. R. (1977). Components of water potential estimated from xylem pressure measurements in five tree species. *Oecologia* **28,** 191–202.

Robinson, F. E. (1964). Required percent air space for normal growth of sugar cane. *Soil Sci.* **98,** 206–207.

Rogers, W. S. (1939). Apple root growth in relation to root-stock, soil, seasonal and climatic factors. *J. Pomol. Hortic. Sci.* **17,** 99–130.

Rokach, A. (1953). Water transfer from fruits to leaves in the Shamouti orange tree and related topics. *Palest. J. Bot., Rehovot Ser.* **8,** 146–151.

Romberger, J. A. (1963). Meristems, growth, and development in woody plants. *U.S., Dep. Agric., Tech. Bull.* **1293.**

Rook, D. A. (1973). Conditioning radiata pine seedlings to transplanting by restricted watering. *N. Z. J. For. Sci.* **3**(1), 54–69.

Rose, C. W., and Stern, W. R. (1965). The drainage component of the water balance equation. *Aust. J. Soil Res.* **3,** 95–100.

Rose, C. W., Stern, W. R., and Drummond, J. E. (1965). Determination of hydraulic conductivity as a function of depth and water content for soil *in situ. Aust. J. Soil Res.* **3,** 1–9.

Rosene, H. F. (1937). Distribution of the velocities of absorption of water in the onion root. *Plant Physiol.* **12,** 1–19.

Rovira, A. D., and Davey, C. B. (1974). Biology of the rhizosphere. *In* "The Plant Root and its Environment" (E. W. Carson, ed.), pp. 153–204. Univ. Press of Virginia, Charlottesville.

Ruben, S., Randall, M., and Hyde, J. L. (1941). Heavy oxygen (O^{18}) as a tracer in the study of photosynthesis. *J. Am. Chem. Soc.* **63,** 877–879.

Rudinsky, J. A., and Vité, J. P. (1959). Certain ecological and phylogenetic aspects of the pattern of water conduction in conifers. *For. Sci.* **5,** 259–266.

Rufelt, H. (1956). Influence of the root pressure on the transpiration of wheat plants. *Physiol. Plant.* **9,** 154–164.

Rufty, T. W., Jr., Volk, R. J., McClure, P. R., Israel, D. W., and Raper, C. D., Jr. (1982). Relative content of NO_3^- and reduced N in xylem exudates as an indicator of root reduction of concurrently absorbed $^{15}NO_3$. *Plant Physiol.* **69,** 166–170.

Ruhland, W., ed. (1956). "Encyclopedia of Plant Physiology," Vol. 2. Springer-Verlag, Berlin and New York.

Rundel, P. W. (1973). The relationship between basal fire scars and crown damage in giant sequoia. *Ecology* **54,** 210–213.

Running, S. W. (1980). Field estimates of root and xylem resistances in *Pinus contorta* using root excision. *J. Exp. Bot.* **31**, 555–569.

Running, S. W., Waring, R. H., and Rydell, R. A. (1975). Physiologicol control of water flux in conifers. A simulation computer model. *Oecologia* **18**, 1–16.

Rush, D. W., and Epstein, E. (1981). Comparative studies on the sodium, potassium and chloride relations of a wild halophyte and a domestic salt-sensitive tomato species. *Plant Physiol.* **68**, 1308–1313.

Rushin, J. W., and Anderson, J. E. (1981). An examination of the leaf quaking adaptation and stomatal distribution in *Populus tremuloides* Michx. *Plant Physiol.* **67**, 1264–1266.

Russell, E. W. (1973). "Soil Conditions and Plant Growth," 10th ed. Longmans, Green, New York.

Russell, M. B., and Woolley, J. T. (1961). Transport processes in the soil-plant system. *In* "Growth in Living Systems" (M. X. Zarrow, ed.), pp. 695–721. Basic Books, New York.

Russell, R. S. (1977). "Plant Root Systems." McGraw-Hill, New York.

Safir, G. R., Boyer, J. S., and Gerdemann, J. W. (1972). Nutrient status and mycorrhizal enhancement of water transport in soybean. *Plant Physiol.* **49**, 700–703.

Salim, M. H., and Todd, G. W. (1968). Seed soaking as a presowing, drought hardening treatment in wheat and barley seeds. *Agron. J.* **60**, 179–182.

Sammis, T. W. (1980). Comparison of sprinkler, trickle subsurface, and furrow irrigation methods for row crops. *Agron. J.* **72**, 701–704.

Sanchez-Diaz, M. F., and Kramer, P. J. (1971). Behavior of corn and sorghum under water stress and during recovery. *Plant Physiol.* **48**, 613–616.

Sanchez-Diaz, M. F., and Mooney, H. A. (1979). Resistance to water transfer in desert shrubs native to Death Valley, California. *Physiol. Plant.* **46**, 139–146.

Sands, R., Fiscus, E. L., and Reid, C. P. P. (1982). Hydraulic properties of pine and bean roots with varying degrees of suberization, vascular differentiation and mycorrhizal infection. *Aust. J. Plant Physiol.* **9**, 559–569.

Sauter, J. J. (1971). Physiology of sugar maple. *Annu. Rep.—Harvard For.* pp. 10–11.

Sayre, J. D. (1926). Physiology of the stomata of *Rumax patientia*. *Ohio J. Sci.* **26**, 233–267.

Schimper, A. F. W. (1903). "Plant Geography upon a Physiological Basis" (Engl. transl.) Oxford Univ. Press (Clarendon), London and New York.

Schnepf, E. (1974). Gland cells. *In* "Dynamic Aspects of Plant Ultrastructure" (A. W. Robards, ed.), pp. 331–357. McGraw-Hill, New York.

Schoeneweiss, D. F. (1978). Water stress as a predisposing factor in plant disease. *In* "Water Deficits and Plant Growth" (T. T. Kozlowski, ed.), Vol. 5, pp. 61–99. Academic Press, New York.

Scholander, P. F. (1958). The rise of sap in lianas. *In* "The Physiology of Forest Trees" (K. V. Thimann, ed.), pp. 3–17. Ronald Press, New York.

Scholander, P. F., Love, W. E., and Kanwisher, J. W. (1955). The rise of sap in tall grapevines. *Plant Physiol.* **30**, 93–104.

Scholander, P. F., Ruud, B., and Leivestad, H. (1957). The rise of sap in a tropical liana. *Plant Physiol.* **32**, 1–6.

Scholander, P. F., Hammel, H. T., Hemmingsen, E. A., and Bradstreet, E. D. (1964). Hydrostatic pressure and osmotic potential in leaves of mangroves and some other plants. *Proc. Natl. Acad. Sci. U.S.A.* **52**, 119–125.

Scholander, P. F., Hammel, H. T., Bradstreet, E. D., and Hemmingsen, E. A. (1965). Sap pressure in vascular plants. *Science* **148**, 339–346.

Schönherr, J. (1976). Water permeability of isolated cuticular membranes: The effect of cuticular waxes on diffusion of water. *Planta* **131**, 159–164.

Schönherr, J., and Ziegler, H. (1980). Water permeability of *Betula* periderm. *Planta* **147**, 345–354.

Schönherr, J., Eckl, K., and Gruler, H. (1979). Water permeability of plant cuticles: The effect of temperature on diffusion of water. *Planta* **147,** 21–26.

Schroeder, C. A., and Wieland, P. A. (1956). Diurnal fluctuations in size in various parts of the avocado tree and fruit. *Proc. Am. Soc. Hortic. Sci.* **68,** 253–258.

Schroeder, R. A. (1939). The effect of root temperature upon the absorption of water by the cucumber. *Res. Bull.—Mo., Agric. Exp. Stn.* **309,** 1–27.

Schultz, R. P. (1972). Intraspecific root grafting in slash pine. *Bot. Gaz. (Chicago)* **133,** 26–29.

Schulze, E.-D., Lange, O. L., Buschbom, U., Kappen, L., and Evenari, M. (1972). Stomatal responses to changes in humidity in plants growing in the desert. *Planta* **108,** 259–270.

Scofield, C. S. (1945). The water requirement of alfalfa. *U.S., Dep. Agric., Circ.* **735.**

Scott, F. M. (1950). Internal suberization of tissues. *Bot. Gaz. (Chicago)* **111,** 378–394.

Scott, F. M. (1963). Root hair zone of soil-grown plants. *Nature (London)* **199,** 1009–1010.

Scott, F. M. (1964). Lipid deposition in the intercellular space. *Nature (London)* **203,** 164–165.

Scott, F. M. (1965). The anatomy of plant roots. *In* "International Symposium on Factors Determining the Behavior of Plant Pathogens in Soil" (K. F. Baker and W. C. Snyder, eds.), pp. 145–153. Univ. of California Press, Berkeley.

Scott, F. M., Schroeder, M. R., and Turrell, F. M. (1948). Development, cell shape, suberization of internal surface, and abscission in the leaf of the Valencia orange, *Citrus sinensis. Bot. Gaz. (Chicago)* **109,** 381–411.

Scott, G. D. (1969). "Plant Symbiosis." St. Martin's Press, New York.

Scott, L. I., and Priestley, J. H. (1928). The root as an absorbing organ. I. A reconsideration of the entry of water and salts into the absorbing region. *New Phytol.* **27,** 125–141.

Sendak, P. E. (1978). Birch sap utilization in the Ukraine. *J. For.* **76,** 120–121.

Shah, C. B., and Loomis, R. S. (1965). Ribonucleic acid and protein metabolism in sugar beet during drought. *Physiol. Plant.* **18,** 240–254.

Shaner, D. L., and Boyer, J. S. (1976). Nitrate reductase activity in maize (*Zea mays* L.) leaves. II. Regulation at low leaf water potential. *Plant Physiol.* **58,** 505–509.

Shantz, H. L. (1925). Soil moisture in relation to the growth of plants. *J. Am. Soc. Agron.* **17,** 705–711.

Sharkey, T. D., and Raschke, K. (1980). Effects of phaseic acid and dihydrophaseic acid on stomata and the photosynthetic apparatus. *Plant Physiol.* **65,** 291–297.

Sharkey, T. D., and Raschke, K. (1981). Separation and measurement of direct and indirect effects of light on stomata. *Plant Physiol.* **68,** 33–40.

Sharp, R. E., and Davies, W. J. (1979). Solute regulation and growth by roots and shoots of water-stressed maize plants. *Planta* **147,** 43–49.

Shaw, B. T., ed. (1952). "Soil Physical Conditions and Plant Growth" Agron. II. Academic Press, New York.

Shawcroft, R. W., Lemon, E. R., Stewart, L. H., Jr., and Jensen, S. E. (1974). The soil-plant-atmosphere model and some of its predictions. *Agric. Meteorol.* **14,** 287–307.

Shayo-Ngowi, A., and Campbell, G. S. (1980). Measurement of matric potential in plant tissue with a hydraulic press. *Agron. J.* **72,** 567–568.

Sheriff, D. W. (1972). A new apparatus for the measurement of sap flux in small shoots with the magnetohydrodynamic technique. *J. Exp. Bot.* **23,** 1086–1095.

Sheriff, D. W. (1974). Magnetohydrodynamic sap flux meters: An instrument for laboratory use and the theory of calibralium. *J. Exp. Bot.* **25,** 675–683.

Sheriff, D. W. (1977a). Evaporation sites and distillation in leaves. *Ann. Bot. (London)* [N.S.] **41,** 1081–1082.

Sheriff, D. W. (1977b). The effect of humidity on water uptake by, and viscous flow resistance of excised leaves of a number of species: Physiological and anatomical observations. *J. Exp. Bot.* **28,** 1399–1407.

Shimshi, D. (1963). Effect of soil moisture and phenylmercuric acetate upon stomatal aperture, transpiration, and photosynthesis. *Plant Physiol.* **38**, 713–721.

Shiraishi, M., Hashimoto, Y., and Kuraishi, S. (1978). Cyclic variations of stomatal aperture observed under the scanning electron microscope. *Plant and Cell Physiol.* **19**, 637–645.

Shirk, H. G. (1942). Freezable water content and the oxygen respiration in wheat and rye grain at different stages of ripening. *Am. J. Bot.* **29**, 105–109.

Shirley, H. L. (1936). Lethal high temperatures for conifers and the cooling effect of transpiration. *J. Agric. Res. (Washington, D.C.)* **53**, 239–258.

Shive, J. B., Jr., and Brown, K. W. (1978). Quaking and gas exchange in leaves of cottonwood (*Populus deltoides,* Marsh). *Plant Physiol.* **61**, 331–333.

Shmueli, E. (1953). Irrigation studies in the Jordan Valley. I. Physiological activity of the banana in relation to soil moisture. *Bull. Res. Counc. Isr.* **3**, 228–247.

Shmueli, E. (1971). The contribution of research to the efficient use of water in Israel agriculture. *Z. Bewässerungswirtsch.* **6**(1), 38–58.

Shmueli, E., and Cohen, O. P. (1964). A critique of Walter's hydrature concept and of his evolution of water status measurements. *Isr. J. Bot.* **13**, 199–207.

Shoemaker, E. M., and Srivastava, L. M. (1973). The mechanics of stomatal opening in corn (*Zea mays* L.). *J. Theoretical Biol.* **42**, 219–225.

Shull, C. A. (1916). Measurement of the surface forces in soils. *Bot. Gaz. (Chicago)* **62**, 1–31.

Shull, C. A. (1930). Absorption of water and the forces involved. *J. Am. Soc. Agron.* **22**, 459–471.

Sierp, H., and Brewig, A. (1935). Quantitative Untersuchungen über die Wasserabsorptionzone der Wurzeln. *Jahrb. Wiss. Bot.* **82**, 99–122.

Sinclair, W. B., and Bartholomew, E. T. (1944). Effects of rootstock and environment on the composition of oranges and grapefruit. *Hilgardia* **16**, 125–176.

Sinclair, T. R., Spaeth, S. C., and Vendeland, J. S. (1981). Microclimatic limitations to crop yield. *In* "Breaking the Climate/Soil Barriers to Crop Yield" (M. H. Miller, D. M. Brown, and E. G. Beauchamps, eds.) Univ. of Guelph, Guelph, Ontario.

Singer, S. J. (1974). The molecular organization of membranes. *Annu. Rev. Biochem.* **43**, 805–833.

Sionit, N., and Kramer, P. J. (1977). Effect of water stress during different stages of growth of soybeans. *Agron. J.* **69**, 274–277.

Sionit, N., Hellmers, H., and Strain, B. R. (1980a). Growth and yield of wheat under CO_2 enrichment and water stress. *Crop Sci.* **20**, 687–690.

Sionit, N., Teare, I. D., and Kramer, P. J. (1980b). Effects of repeated application of water stress on water status and growth of wheat. *Physiol. Plant.* **50**, 11–15.

Skau, C. M., and Swanson, R. H. (1963). An improved heat pulse velocity meter as an indicator of sap speed and transpiration. *JGR, J. Geophys. Res.* **68**, 4743–4749.

Skene, K. G. M. (1967). Gibberellin-like substances in root exudate of *Vitis vinifera*. *Planta* **74**, 250–262.

Skene, K. G. M. (1975). Cytokinin production by roots as a factor in the control of plant growth. *In* "The Development and Function of Roots" (J. G. Torrey and D. T. Clarkson, eds.), pp. 365–396. Academic Press, New York.

Skene, K. G. M., and Antcliff, A. J. (1972). A comparative study of cytokinin levels in bleeding sap of *Vitis vinifera* (L.) and the two grapevine rootstocks, Salt Creek and 1613. *J. Exp. Bot.* **23**, 283–293.

Skidmore, E. L., and Stone, J. F. (1964). Physiological role in transpiration rate of the cotton plant. *Agron. J.* **56**, 405–410.

Slankis, V. (1973). Hormonal relationship in mycorrhizal development. *In* "Ectomycorrhizae" (G. C. Marks and T. T. Kozlowski, eds.), pp. 232–298. Academic Press, New York.

Slatyer, R. O. (1956). Absorption of water from atmospheres of different humidity and its transport through plants. *Aust. J. Biol. Sci.* **9**, 552–558.

Slatyer, R. O. (1957). The significance of the permanent wilting percentage in studies of plant and soil water relations. *Bot. Rev.* **23**, 585–636.

Slatyer, R. O. (1958). The measurement of diffusion pressure deficit in plants by a method of vapor-equilibration. *Aust. J. Biol. Sci.* **11**, 349–365.

Slatyer, R. O. (1960). Aspects of the tissue water relationships of an important arid zone species (*Acacia aneura* F. Muell.) in comparison with two mesophytes. *Bull. Res. Counc. Isr., Sect. D* **8**, 159–168.

Slatyer, R. O. (1961). Effects of several osmotic substrates on water relations of tomato. *Aust. J. Biol. Sci.* **14**, 519–540.

Slatyer, R. O. (1966). Some physical aspects of non-stomatal control of leaf transpiration. *Agric. Meteorol.* **3**, 281–292.

Slatyer, R. O. (1967). "Plant Water Relationships." Academic Press, New York.

Slatyer, R. O. (1969). Physiological significance of internal water relations to crop yield. *In* "Physiological Aspects of Crop Yield" (J. D. Eastin, F. A. Haskins, C. Y. Sullivan, and C. H. M. van Bavel, eds.), pp. 53–83. Am. Soc. Agron. and Crop Sci. Soc. Am., Madison, Wisconsin.

Slatyer, R. O., and Bierhuizen, J. F. (1964). A differential psychrometer for continuous measurements of transpiration. *Plant Physiol.* **39**, 1051–1056.

Slatyer, R. O., and Jarvis, P. G. (1966). A gaseous-diffusion porometer for continuous measurement of diffusive resistance of leaves. *Science* **151**, 574–576.

Slatyer, R. O., and Lake, J. V. (1966). Resistance to water transport in plants—whose misconception? *Nature (London)* **212**, 1585–1586.

Slatyer, R. O., and McIlroy, D. C. (1961). "Practical Micro-climatology." UNESCO, Paris.

Slatyer, R. O., and Morrow, P. A. (1977). Altitudinal variation in the photosynthetic characteristics of snow gum, *Eucalyptus pauciflora* Sieb. × Spreng. I. Seasonal changes under field conditions in the Snowy Mountains area of South-Eastern Australia. *Aust. J. Bot.* **25**, 1–20.

Slatyer, R. O., and Shmueli, E. (1967). Measurement of internal water status and transpiration. *In* "Irrigation of Agricultural Lands" (R. M. Hagan, H. R. Haise, and T. W. Edminster, eds.), pp. 337–353. Am. Soc. Agron., Madison, Wisconsin.

Slatyer, R. O., and Taylor, S. A. (1960). Terminology in plant and soil-water relations. *Nature (London)* **187**, 922.

Slavik, B. (1963). On the problem of the relationship between hydration of leaf tissue and intensity of photosynthesis and respiration. *In* "The Water Relations of Plants" (A. J. Rutter and F. H. Whitehead, eds.), pp. 225–234. Wiley, New York.

Slavik, B. (1974). "Methods of Studying Plant Water Relations." Springer-Verlag, Berlin and New York.

Slavikova, J. (1964). Horizontales Gradient der Saughkraft eines Wurzelastes und seine Zusammenhang mit dem Wassertransport in der Wurzel. *Acta Horti Bot. Pragensis* pp. 73–79.

Slavikova, J. (1967). Compensation of root suction force within a single root system. *Biol. Plant.* **9**, 20–27.

Smart, R. E., and Bingham, G. E. (1974). Rapid estimates of relative water content. *Plant Physiol.* **53**, 258–260.

Smith, H. W. (1962). The plasma membrane, with notes on the history of botany. *Circulation* **26**, 987–1012.

Smith, R. M., and Browning, D. R. (1946). Some suggested laboratory standards of subsoil permeability. *Soil Sci. Soc. Am. Proc.* **11**, 21–26.

Smith, W. K. (1978). Temperatures of desert plants: Another perspective on the adaptability of leaf size. *Science* **201**, 614–616.

Smith, W. O. (1943). Thermal transfer of moisture in soils. *Trans., Am. Geophys. Union* **24**, 511–523.

Sneed, R. E., and Patterson, R. P. (1983). In "Crop Reactions to Water and Temperature Stress in

Humid, Temperate Climates'' (C. D. Raper, Jr., and P. J. Kramer, eds.), pp. 187–199. Westview Press, Boulder, Colorado.

Sojka, R. E., and Stolzy, L. H. (1980). Soil-oxygen effects on stomatal response. *Soil Sci.* **130**, 350–358.

Solarova, J., Pospisilova, J., and Slavik, B. (1981). Gas exchange regulation by changing of epidermal conductance with antitranspirants. *Photosynthetica* **15**, 365–400.

Somers, G. F. (1979). Production of food plants in areas supplied with highly saline water: Problems and prospects. *In* "Stress Physiology in Crop Plants" (H. Mussell and R. C. Staples, eds.), pp. 107–125. Wiley, New York.

Soran, V., and Cosma, D. (1962). Effectul transpiratiei asupra activatatii absorbante a diferitelor regiuni ale sistemului radical. *Stud. Univ. Babes-Bolyai [Ser.] Biol. Fasc.* **1**, 75–87. (Summary in French.)

Spanner, D. C. (1951). The Peltier effect and its use in the measurement of suction pressure. *J. Exp. Bot.* **2**, 145–168.

Spanner, D. C. (1956). Energetics and mathematical treatment of diffusion. *Encycl. Plant Physiol.* **2**, 125–138.

Spanner, D. C. (1958). The translocation of sugar in sieve tubes. *J. Exp. Bot.* **9**, 332–342.

Spanner, D. C. (1964). "Introduction to Thermodynamics." Academic Press, New York.

Spanner, D. C. (1972). Plants, water, and some other topics. *In* "Psychrometry in Water Relations Research" (R. W. Brown and B. P. Van Haveren, eds.), pp. 29–39. Utah Agric. Exp. Stn., Utah State University, Logan.

Spoehr, H. A., and Milner, H. W. (1939). Starch dissolution and amylolytic activity of leaves. *Proc. Am. Philos. Soc.* **81**, 37–78.

Stadelmann, E. (1971). The protoplasmic basis for drought resistance. In "Food, Fiber, and the Arid Lands" (W. G. McGinnies, B. J. Goldman, and P. Paylore, eds.), pp. 337–352. Univ. Arizona Press, Tucson.

Stålfelt, M. G. (1932). Der stomatäre Regulator in der Pflanzlichen Transpiration. *Planta* **17**, 22–85.

Stålfelt, M. G. (1956). Morphologie und Anatomie des Blattes als Transpirationsorganen. *Encycl. Plant Physiol.* **3**, 324–341.

Staple, W. J., and Lehane, J. J. (1962). Variability in soil moisture sampling. *Can. J. Soil Sci.* **42**, 157–161.

Stark, N. M., and Jordan, C. F. (1978). Nutrient retention by the root mat of an Amazonian rain forest. *Ecology* **59**, 434–437.

Stebbins, G. L., and Shah, S. S. (1960). Developmental studies of cell differentiation in the epidermis of monocotyledons. II. Cytological features of stomatal development in the Gramineae. *Dev. Biol.* **2**, 477–500.

Steponkus, P. L. (1980). A unified concept of stress in plants. *In* "Genetic Engineering of Osmoregulation" (D. W. Rains, R. C. Valentine, and A. Hollaender, eds.), pp. 235–255. Plenum, New York.

Sterne, R. E., Kaufmann, M. R., and Zentmeyer, G. A. (1978). Effect of Phytophthora root rot on water relations of avocado: Interpretation with a water transport model. *Phytopathology* **68**, 595–602.

Stevens, C. L. (1931). Root growth of white pine (*Pinus strobus* L.) *Bull.—Yale Univ. Sch. For.* **32**, 1–62.

Stevens, C. L., and Eggert, R. L. (1945). Observations on the causes of flow of sap in red maple. *Plant Physiol.* **20**, 636–648.

Steward, F. C., Prévot, P., and Harrison, J. A. (1942). Absorption and accumulation of rubidium bromide by barley plants. Localization in the root of cation accumulation and of transfer to the shoot. *Plant Physiol.* **17**, 411–421.

Stewart, C. R., and Hanson, A. D. (1980). Proline accumulation as a metabolic response to water stress. *In* "Adaptation of Plants to Water and High Temperature Stress" (N. C. Turner and P. J. Kramer, eds.), pp. 173–189. Wiley, New York.

Stillinger, F. H. (1980). Water revisited. *Science* **209**, 451–457.

Stocker, O. (1929). Das Wasserdefizit von Gefässpflanzen in verschiedenen Klimazonen. *Planta* **7**, 382–387.

Stocker, O. (1960). Physiological and morphological changes in plants due to water deficiency. *Arid Zone Res.* **15**, 63–104.

Stocking, C. R. (1945). The calculation of tensions in *Cucurbita pepo*. *Am. J. Bot.* **32**, 126–134.

Stocking, C. R. (1956). Excretion by glandular organs. *Encycl. Plant Physiol.* Vol. **3**, 503–510.

Stolzy, L. H., Focht, D. D. and Fluehler, H. (1981). Indicators of soil aeration. *Flora* **171**, 136–265.

Stone, E. C. (1957). Dew as an ecological factor. II. The effect of artificial dew on the survival of *Pinus ponderosa* and associated species. *Ecology* **38**, 414–422.

Stone, E. C., and Fowells, H. A. (1955). The survival value of dew as determined under laboratory conditions. I. *Pinus ponderosa*. *For. Sci.* **1**, 183–188.

Stone, E. C., and Norberg, E. A. (1979). Root growth capacity: One key to bare-root survival. *Calif. Agric.* **33**(5), 14–15.

Stone, J. E., and Stone, E. L. (1975a). Water conduction in lateral roots of red pine. *For. Sci.* **21**, 53–60.

Stone, J. E., and Stone, E. L. (1975b). The communal root system of red pine: Water conduction through root grafts. *For. Sci.* **21**, 255–262.

Strasburger, E. (1891). Ueber den Bau and die Verrichtungen der Leitungsbahnen in der Pflanzen. *Histol. Beitr.* **3**, 849–877.

Strogonov, B. P. (1964). "Physiological Basis of Salt Tolerance of Plants" (In Russian, Engl. Transl. by A. Poljakoff-Mayber and A. M. Mayer). Oldbourne Press, London.

Strugger, S. (1943). Der Aufsteigende Saftstrom in der Pflanze. *Naturwissenschaften* **31**, 181–194.

Strugger, S. (1949). "Praktikum der Zell-und Gewebe-physiologie der Pflanzen," 2nd ed. Springer-Verlag, Berlin and New York.

Stuckey, I. H. (1941). Seasonal growth of grass roots. *Am. J. Bot.* **28**, 486–491.

Sung, F. J. M., and Krieg, D. R. (1979). Relative sensitivity of photosynthetic assimilation and translocation of [14] carbon to water stress. *Plant Physiol.* **64**, 852–856.

Swanson, C. A. (1943). Transpiration in American holly in relation to leaf structure. *Ohio J. Sci.* **43**, 43–46.

Swanson, R. H. (1966). Seasonal course of transpiration of lodgepole pine and Engelmann spruce. *In* "Forest Hydrology" (W. E. Sopper and H. W. Lull, eds.), pp. 419–433. Pergamon, Oxford.

Swanson, R. H., and Lee, R. (1966). Measurement of water movement from and through shrubs and trees. *J. For.* **64**, 187–190.

Takaoki, T. (1969). Measurement of osmotic quantities in higher plants. *J. Sci. Hiroshima Univ., Ser. B, Div. 2* **12**(2), 199–210.

Tal, M. (1966). Abnormal stomatal behavior in wilty mutants of tomato. *Plant Physiol.* **41**, 1387–1391.

Talboys, P. W. (1978). Disfunction of the water system. *In* "Plant Disease" (J. G. Horsfall and E. B. Cowling, eds.), Vol. 3, pp. 141–162. Academic Press, New York.

Tanford, C. (1963). The structure of water and aqueous solutions. *In* "Temperature—its Measurement and Control in Science and Industry" (C. M. Herzfeld, ed.), Vol. 3, pp. 123–129. D. Van Nostrand, New York.

Tanford, C. (1980). "The Hydrophobic Effect." Wiley, New York.

Tanner, C. B. (1967). Measurement of evapotranspiration. *In* "Irrigation of Agricultural Lands" (R.

M. Hagan, H. R. Haise, and T. W. Edminster, eds.), pp. 534–574. Am. Soc. Agron., Madison, Wisconsin.

Tanner, C. B. (1981). Transpiration efficiency of potato. *Agron. J.* **73,** 59–64.

Tanton, T. W., and Crowdy, S. H. (1972). Water pathways in higher plants. II. Water pathways in roots. *J. Exp. Bot.* **23,** 600–618.

Taylor, H. M. (1980). Modifying root systems of cotton and soybeans to increase water absorption. *In* "Adaptation of Plants to Water and High Temperature Stress" (N. C. Turner and P. J. Kramer, eds.), pp. 75–84. Wiley, New York.

Taylor, H. M., and Klepper, B. (1973). Rooting density and water extraction patterns for corn (Zea mays L.). *Agron. J.* **65,** 965–968.

Taylor, H. M., and Klepper, B. (1975). Water uptake by cotton root systems: An examination of assumptions in the single root model. *Soil Sci.* **120,** 57–67.

Taylor, H. M., and Terrell, E. E. (1982). Rooting pattern and plant productivity. *In* "Handbook of Agricultural Productivity" (M. Rechcigl, Jr., ed.), Vol. 1, pp. 185–200. CRC Press, Boca Raton, Florida.

Taylor, H. M., Jordan, W. R., and Sinclair, T. B., eds. (1983). "Limitations to Efficient Water Use in Crop Production." Am. Soc. Agron., Madison, Wisconsin.

Tazaki, T., Ishihara, K., and Usijima, T. (1980). Influence of water stress on the photosynthesis and productivity of plants in humid areas. *In* "Adaptation of Plants to Water and High Temperature Stress" (N. C. Turner and P. J. Kramer, eds.), pp. 309–321. Wiley, New York.

Teare, I. D., and Peet, M. M., eds. (1982). "Crop Water Relations." Wiley, New York.

Teare, I. D., Sionit, N., and Kramer, P. J. (1982). Changes in water status during water stress at various stages of development in wheat. *Physiol. Plant.* **55,** 296–300.

Teare, I. D., Kanemasu, E. T., Powers, W. L., and Jacobs, B. S. (1973). Water-use efficiency and its relation to crop canopy area, stomatal regulation and root distribution. *Agron. J.* **65,** 207–211.

Tepfer, M., and Taylor, I. E. P. (1981). The permeability of plant cell walls as measured by gel filtration chromatography. *Science* **213,** 761–763.

Thill, D. C., Schirman, R. D., and Appleby, A. P. (1979). Osmotic stability of mannitol and polyethylene glycol 20,000 solutions used as seed germination media. *Agron. J.* **71,** 105–108.

Thoday, D. (1918). On turgescence and the absorption of water by the cells of plants. *New Phytol.* **17,** 108–113.

Thomas, W. A. (1967). Dye and calcium ascent in dogwood trees. *Plant Physiol.* **42,** 1800–1802.

Thompson, L. M. (1952). "Soils and Soil Fertility." McGraw-Hill, New York.

Thornthwaite, C. W. (1948). An approach toward a rational classification of climate. *Geogr. Rev.* **38,** 55–94.

Thut, H. F. (1932). Demonstrating the lifting power of transpiration. *Am. J. Bot.* **19,** 358–364.

Thut, H. F., and Loomis, W. E. (1944). Relation of light to growth of plants. *Plant Physiol.* **19,** 117–130.

Tibbits, T. W. (1979). Humidity and plants. *BioScience* **29,** 358–368.

Tinker, P. B. (1976). Roots and water. Transport of water to plant roots in soil. *Philos. Trans. R. Soc. London, Ser. B* **273,** 445–461.

Tinoco, I., Sauer, K., and Wang, J. C. (1978). "Physical Chemistry: Principles and Applications in Biological Sciences." Prentice-Hall, Englewood, New Jersey.

Todd, G. W. (1972). Water deficits and enzymatic activity. *In* "Water Deficits and Plant Growth" (T. T. Kozlowski, ed.), Vol. 3, pp. 177–216. Academic Press, New York.

Tomar, V. S., and Ghildyal, B. P. (1975). Resistances to water transport in rice plants. *Agron. J.* **67,** 269–272.

Torrey, J. G., and Clarkson, D. T., (1975). eds. "The Development and Function of Roots." Academic Press, New York.

Toumey, J. W. (1929). Initial root habit in American trees and its bearing on regeneration. *Proc. Int. Bot. Congr., 4th, 1926* Vol. 1, pp. 713–728.

Tranquillini, W. (1969). Photosynthese und Transpiration einiger Holzarten bei verschieden starkem Wind. *Centralbl. Gesamte Forstwes.* **85**, 43–49.

Transeau, E. N. (1905). Forest centers of eastern North America. *Am. Nat.* **39**, 875–889.

Traube, M. (1867). Experimente zur Theorie der Zellbildung und Endosmose. *Arch. Anat. Physiol. Wiss. Med.* pp. 87–165.

Travis, A. J., and Mansfield, T. A. (1981). Light saturation of stomatal opening on the adaxial and abaxial epidermis of *Commelina communis*. *J. Exp. Bot.* **32**, 1169–1179.

Trewavas, A. (1981). How do plant growth substances work? *Plant, Cell Environ.* **4**, 203–228.

Triplett, E. W., Barnett, N. M., and Blevins, D. G. (1980). Organic acids and ionic balance in xylem exudate of wheat during nitrate or sulfate absorption. *Plant Physiol.* **65**, 610–613.

Tubbs, F. R.(1973). Research fields in the interaction of rootstock and scions in woody perennials. Parts I and II. *Hort. Abstr.* **43**, 247–253, 325–335.

Tukey, H. B., Jr., Mecklenburg, R. A., and Morgan, J. V. (1965). A mechanism for the leaching of metabolites from foliage. *Isot. and Radiat. in Soil-Plant Nutr. Stud., Proc. Symp., 1965* pp. 371–385. IAEA, Vienna.

Turner, L. M. (1936). Root growth of seedlings of *Pinus echinata* and *Pinus taeda*. *J. Agric. Res. (Washington, D.C.)* **53**, 145–149.

Turner, N. C. (1974). Stomatal behavior and water status of maize, sorghum, and tobacco under field conditions. II. At low soil water potential. *Plant Physiol.* **53**, 360–365.

Turner, N. C. (1981). Techniques and experimental approaches for the measurement of plant water stress. *Plant Soil* **58**, 339–366.

Turner, N. C., and Begg, J. E. (1973). Stomatal behavior and water status of maize, sorghum, and tobacco under field conditions. *Plant Physiol.* **51**, 31–36.

Turner, N. C., and Begg, J. E. (1978). Responses of pasture plants to water deficits. *In* "Plant Relations in Pastures" (J. R. Wilson, ed.), pp. 50–66. CSIRO, Melbourne.

Turner, N. C., and Begg, J. E. (1981). Plant-water relations and adaptations to stress. *Plant Soil* **58**, 97–131.

Turner, N. C., and Jones, M. M. (1980). Turgor maintenance by osmtic adjustment: A review and evaluation. *In* "Adaptation of Plants to Water and High Temperature Stress" (N. C. Turner and P. J. Kramer, eds.), pp. 87–103. Wiley, New York.

Turner, N. C., and Kramer, P. J., eds. (1980). "Adaptation of Plants to Water and High Temperature Stress." Wiley, New York.

Turrell, F. M. (1936). The area of the internal exposed surface of dicotyledon leaves. *Am. J. Bot.* **23**, 255–264.

Turrell, F. M. (1944). Correlation between internal surface and transpiration rate in mesomorphic and xeromorphic leaves grown under artificial light. *Bot. Gaz. (Chicago)* **105**, 413–425.

Tyree, M. T. (1969). The thermodynamics of short-distance translocation in plants. *J. Exp. Bot.* **20**, 341–349.

Tyree, M. T. (1970). The symplast concept: A general theory of symplastic transport according to the dynamics of irreversible processes. *J. Theor. Biol.* **26**, 181–214.

Tyree, M. T. (1976). Negative turgor pressure in plants: Fact or fallacy. *Can. J. Bot.* **54**, 2738–2746.

Tyree, M. T., and Hammel, H. T. (1972). The measurement of the turgor pressure and the water relations of plants by the pressure-bomb technique. *J. Exp. Bot.* **23**, 267–282.

Tyree, M. T., and Karamanos, A. J. (1980). Water stress as an ecological factor. *In* "Plants and Their Atmospheric Environment" (J. Grace, E. D. Ford, and P. G. Jarvis, eds.), pp. 237–261. Blackwell, Oxford.

Tyree, M. T., and Yianoulis, P. (1980). The site of water evaporation from sub-stomatal cavities, liquid path resistances and hydroactive stomatal closure. *Ann. Bot. (London)* [N.S.] **46**, 175–193.

Tyree, M. T., Cruiziat, P., Benis, M., LoGullo, M. A., and Salleo, S. (1981). The kinetics of rehydration of detached sunflower leaves from different initial water deficits. *Plant, Cell Environ.* **4**, 309–317.

Tyree, M. T., and Jarvis, P. G. (1982). Water in tissues and cells. *Encycl. Plant Physiol. New Ser.* **12B**, 35–78. Springer-Verlag, Berlin and New York.

Uhvits, R. (1946). Effect of osmotic pressure on water absorption and germination of alfalfa seeds. *Am. J. Bot.* **33**, 278–285.

Uriu, K. (1964). Effect of post-harvest soil moisture depletion on subsequent yield of apricots. *Proc. Am. Soc. Hortic. Sci.* **84**, 93–97.

Uriu, K., Davenport, D. C., and Hagan, R. M. (1975). Antitranspirant effects on fruit growth of "Manzanillo olive." *J. Am. Soc. Hortic. Sci.* **100**, 666–669.

Ursprung, A. (1915). Uber die Kohäsion des Wassers im Farnanulus. *Ber. Dtsch. Bot. Ges.* **33**, 153–162.

Ursprung, A. (1929). The osmotic quantities of the plant cell. *Proc. Int. Bot. Congr., 4th, 1926* Vol. 2, pp. 1081–1094.

Ursprung, A., and Blum, G. (1916). Zur Methode der Saugkraftmessung. *Ber. Dtsch. Bot. Ges.* **34**, 525–539.

Ussing, H. H. (1953). Transport through biological membranes. *Annu. Rev. Physiol.* **15**, 1–20.

Vaadia, Y. (1960). Autonomic diurnal fluctuations in rate of exudation and root pressure of decapitated sunflower plants. *Physiol. Plant.* **13**, 701–717.

Vaclavik, J. (1966). The maintaining of constant soil moisture levels (lower than maximum capillary capacity) in pot experiments. *Biol. Plant.* **8**, 80–85.

Valoras, N., Osborne, J. F., and Letey, J. (1974). Wetting agents for erosion control on burned watersheds. *Calif. Agric.* **28**(5), 12–13.

van Bavel, C. H. M. (1953). A drought criterion and its application in evaluating drought incidence and hazard. *Agron. J.* **45**, 167–172.

van Bavel, C. H. M., and Verlinden, F. J. (1956). Agricultural drought in North Carolina. *N. C., Agric. Exp. Stn., Tech. Bull.* **122**.

van Bavel, C. H. M., Fritschen, L. J., and Lewis, W. E. (1963). Transpiration by Sudangrass as an externally controlled process. *Science* **141**, 269–270.

van Bavel, C. H. M., Nakayama, F. S., and Ehrler, W. L. (1965). Measuring transpiration resistance of leaves. *Plant Physiol.* **40**, 535–540.

van den Driesche, R., Connor, D. J., and Tunstall, B. R. (1971). Photosynthetic response of brigalow to irradiance, temperature and water potential. *Photosynthetica* **5**, 210–217.

van den Honert, T. H. (1948). Water transport as a catenary process. *Discuss. Faraday Soc.* **3**, 146–153.

Van der Post, C. J. (1968). Simultaneous observations on root and top growth. *Acta Hortic.* **7**, 138–144.

van Eijk, M. (1939). Analyze der Wirkung des NaCl auf die Entwicklung, Sukkulenz und Transpiration bei *Salicornia herbacea*, sowie Untersuchungen über den Einfluss der Salzaufnahme auf die Wurzelatmung bei Aster Tripolium. *Recl. Trav. Bot. Neerl.* **36**, 559–657.

Van Fleet, D. S. (1961). Histochemistry and function of the endodermis. *Bot. Rev.* **27**, 166–220.

van Overbeek, J. (1942). Water uptake by excised root systems of the tomato due to non-osmotic forces. *Am. J. Bot.* **29**, 677–683.

van Raalte, M. H. (1940). On the oxygen supply of rice roots. *Ann. Jard. Bot. Buitenz.* **50**, 99–113.

van Schilfgaarde, J., ed. (1974). "Drainage for Agriculture." Am. Soc. Agron., Madison, Wisconsin.

Van Volkenburgh, E., and Davies, W. J. (1977). Leaf anatomy and water relations of plants grown in controlled environments and in the field. *Crop Sci.* **16**, 353–358.

Veihmeyer, F. J. (1927). Some factors affecting the irrigation requirements of deciduous orchards. *Hilgardia* **2**, 125–284.

Veihmeyer, F. J., and Hendrickson, A. H. (1938). Soil moisture as an indication of root distribution in deciduous orchards. *Plant Physiol.* **13**, 169–177.

Veihmeyer, F. J., and Hendrickson, A. H. (1950). Soil moisture in relation to plant growth. *Annu. Rev. Plant Physiol.* **1**, 285–304.

Veto, F. (1963). Mobilization of fluids in biological objects by means of temperature gradient. *Acta Physiol. Acad. Sci. Hung.* **24**, 119–128.

Vieira da Silva, J., Naylor, A. W., and Kramer, P. J. (1974). Some ultrastructural and enzymatic effects of water stress in cotton (*Gossypium hirsutum* L.) leaves. *Proc. Natl. Acad. Sci. U.S.A.* **71**, 3243–3247.

Viets, F. G., Jr. (1972). Water deficits and nutrient availability. *In* "Water Deficits and Plant Growth" (T. T. Kozlowski, ed.), Vol. 3, pp. 217–239. Academic Press, New York.

Vité, P. J. (1961). The influence of water supply on oleoresin exudation pressure and resistance to bark beetle attack in *Pinus ponderosa*. *Contrib. Boyce Thompson Inst.* **21**(2), 37–66.

Volk, G. M. (1947). Significance of moisture translocation from soil zones of low moisture tension to zones of high moisture tension by plant roots. *J. Am. Soc. Agron.* **39**, 93–106.

Vomocil, J. A. (1954). In situ measurement of soil bulk density. *Agric. Eng.* **35**, 651–654.

Vomocil, J. A., and Flocker, W. J. (1961). Effect of soil compaction on storage and movement of soil air and water. *Trans. ASAE* **4**, 242–245.

Wadleigh, C. H. (1946). The integrated soil moisture stress upon a root system in a large container of saline soil. *Soil Sci.* **61**, 225–238.

Wadleigh, C. H., and Ayers, A. D. (1945). Growth and biochemical composition of bean plants as conditioned by soil moisture tension and salt concentration. *Plant Physiol.* **20**, 106–132.

Wadleigh, C. H., Gauch, H. G., and Magistad, O. C. (1946). Growth and rubber accumulation in guayule as conditioned by soil salinity and irrigation regime. *U.S., Dep. Agric. Tech. Bull.* **925.**

Wadleigh, C. H., Gauch, H. G., and Strong, D. G. (1947). Root penetration and moisture extraction in saline soil by crop plants. *Soil Sci.* **63**, 341–349.

Waisel, Y. (1958). Dew absorption by plants of arid zones. *Bull. Res. Counc. Isr. Sect. D* **6**, 180–186.

Wallace, A., Soufi, S. M., and Hemaidan, N. (1966). Day-night differences in accumulation and translocation of ions by tobacco plants. *Plant Physiol.* **41**, 102–104.

Wallihan, E. F. (1946). Studies of the dielectric method of measuring soil moisture. *Soil Sci. Soc. Am. Proc.* **10**, 39–40.

Wallihan, E. F. (1964). Modification and use of an electric hygrometer for estimating relative stomatal apertures. *Plant Physiol.* **39**, 86–90.

Walter, H. (1931). Die hydratur der Pflanze and ihre physiologische ökologische Bedeutung,'' pp. 118–121. Fischer, Jena.

Walter, H. (1963). Zur Klärung des Spezifischen Wasserzustandes in Plasma und in der Zellwand bei höheren pflanze und seine Bestimmung. *Ber. Dtsch. Bot. Ges.* **76**, 40–71.

Walter, H. (1965). Klärung des spezifischen Wasserzustandes in Plasma. *Ber. Dtsch. Bot. Ges.* **78**, 104–114.

Walter, H. (1979). "Vegetation of the Earth," 2nd ed. Springer-Verlag, Berlin and New York.

Walton, D. C., Harrison, M. A., and Coté, P. (1976). The effects of water stress on abscisic-acid levels and metabolism in roots of *Phaseolus vulgaris* L. and other plants. *Planta* **131**, 141–144.

Wardlaw, I. F. (1968). The control and pattern of movement of carbohydrates in plants. *Bot. Rev.* **34**, 79–105.

Waring, R. H., and Cleary, B. D. (1967). Plant moisture stress: Evaluation by pressure bomb. *Science* **155**, 1248, 1253–1254.

Waring, R. H., and Running, S. W. (1978). Sapwood water storage: Its contribution to transportation and effect upon water conductance through the stems of old-growth Douglas-fir. *Plant, Cell Environ.* **1,** 131–140.

Watts, W. R. (1974). Leaf extension in *Zea mays.* III. Field measurement of leaf extension in response to temperature and leaf water potential. *J. Exp. Bot.* **25,** 1085–1096.

Weatherley, P. E. (1950). Studies in the water relations of the cotton plant I. The field measurements of water deficits in leaves. *New Phytol.* **49,** 81–97.

Weatherley, P. E. (1951). Studies in the water relations of the cotton plant. II. Diurnal and seasonal fluctuations and environmental factors. *New Phytol.* **50,** 36–51.

Weatherley, P. E. (1963). The pathway of water movement across the root cortex and leaf mesophyll of transpiring plants. *In* "The Water Relations of Plants" (A. J. Rutter, and F. H. Whitehead, eds.), pp. 85–100. Wiley, New York.

Weatherley, P. E. (1965). The state and movement of water in the leaf. *Symp. Soc. Exp. Biol.* **19,** 157–184.

Weatherley, P. E. (1970). Some aspects of water relations. *Adv. Bot. Res.* **3,** 171–206.

Weatherley, P. E., and Slatyer, R. O. (1957). Relationship between relative turgidity and diffusion pressure deficit in leaves. *Nature (London)* **179,** 1085–1086.

Weatherspoon, C. P. (1968). The significance of the mesophyll resistance in transpiration. Ph.D. Dissertation, Duke University, Durham, North Carolina.

Weaver, H. A., and Jamison, V. C. (1951). Limitations in the use of electrical resistance soil moisture units. *Agron. J.* **43,** 602–605.

Weaver, J. E. (1919). The ecological relations of roots. *Carnegie Inst. Washington Publ.* **286.**

Weaver, J. E. (1920). Root development in the grassland formation. *Carnegie Inst. Washington Publ.* **292.**

Weaver, J. E. (1925). Investigations on the root habits of plants. *Am. J. Bot.* **12,** 502–509.

Weaver, J. E. (1926). "Root Development of Field Crops." McGraw-Hill, New York.

Weaver, J. E., and Bruner, W. E. (1927). "Root Development of Vegetable Crops." McGraw-Hill, New York.

Weaver, J. E., and Clements, F. E. (1938). "Plant Ecology," 2nd ed. McGraw-Hill, New York.

Weaver, J. E., and Crist, J. W. (1922). Relation of hardpan to root penetration in the Great Plains. *Ecology* **3,** 237–249.

Weaver, J. E., and Darland, R. W. (1947). A method of measuring vigor of range grasses. *Ecology* **28,** 146–162.

Weaver, J. E., and Zink, E. (1946). Length of life of roots of ten species of perennial range and pasture grasses. *Plant Physiol.* **21,** 201–217.

Weaver, J. E., Jean, F. C., and Crist, J. W. (1922). Development and activities of roots of crop plants. *Carnegie Inst. Washington Publ.* **316.**

Weiland, R. T., and Stuttle, C. A. (1980). Concomitant determination of foliar nitrogen loss, net carbon uptake, and transpiration. *Plant Physiol.* **65,** 403–406.

Weisz, P. B., and Fuller, M. S. (1962). "The Science of Botany." McGraw-Hill, New York.

Wenger, K. F. (1955). Light and mycorrhiza development. *Ecology* **36,** 518–520.

Wenkert, W. (1980). Measurement of tissue osmotic pressure. *Plant Physiol.* **65,** 614–617.

Wenkert, W., Lemon, E. R., and Sinclair, T. R. (1978). Leaf elongation and turgor pressure in field-grown soybean. *Agron. J.* **70,** 761–764.

Went, F. W. (1938). Specific factors other than auxin affecting growth and root formation. *Plant Physiol.* **13,** 55–80.

Went, F. W. (1943). Effect of the root system on tomato stem growth. *Plant Physiol.* **18,** 51–65.

Went, F. W. (1975). Water vapor absorption in *Prosopis. In* "Physiological Adaptation to the Environment" (F. J. Vernberg, ed.), pp. 67–75. Intext Educational Publications, New York.

Went, F. W., and Stark, N. (1968). The biological and mechanical role of soil fungi. *Proc. Natl. Acad. Sci. U.S.A.* **60**, 497–504.

White, L. M., and Ross, W. H. (1939). Effect of various grades of fertilizers on the salt content of the soil solution. *J. Agric. Res. (Washington, D.C.)* **59**, 81–100.

White, P. R. (1938). "Root-pressure"—an unappreciated force in sap movement. *Am. J. Bot.* **25**, 223–227.

White, P. R., Schuker, E., Kern, J. R., and Fuller, F. H. (1958). Root-pressure in gymnosperms. *Science* **128**, 308–309.

Whiteman, P. C., and Koller, D. (1964). Saturation deficit of the mesophyll evaporating surfaces in a desert halophyte. *Science* **146**, 1320–1321.

Whitfield, C. J. (1932). Ecological aspects of transpiration. II. Pikes Peak and Santa Barbara regions: Edaphic and climatic aspects. *Bot. Gaz. (Chicago)* **94**, 183–196.

Whittington, W. J., ed. (1969). "Root Growth." Butterworth, London.

Wiebe, H. H. (1966). Matric potential of several plant tissues and biocolloids. *Plant Physiol.* **41**, 1439–1442.

Wiebe, H. H. (1978). The significance of plant vacuoles. *BioScience* **28**, 327–331.

Wiebe, H. H. (1981). Measuring water potential (activity) from free water to oven dryness. *Plant Physiol.* **68**, 1218–1221.

Wiebe, H. H., and Brown, R. W. (1979). Temperature gradient effects on in situ hygrometer measurements of soil water potential. II. Water movement. *Agron. J.* **71**, 397–401.

Wiebe, H. H., and Kramer, P. J. (1954). Translocation of radioactive isotopes from various regions of roots of barley seedlings. *Plant Physiol.* **29**, 342–348.

Wiebe, H. H., and Prosser, R. J. (1977). Influence of temperature gradients on leaf water potential. *Plant Physiol.* **59**, 256–258.

Wiebe, H. H., and Wirheim, S. E. (1962). The influence of internal moisture stress on translocation. *Radioisoto. Soil-Plant Nutr. Stud., Proc. Symp., 1962* pp. 279–288. IAEA, Vienna.

Wiebe, H. H., Brown, R. W., Daniel, T. W., and Campbell, E. (1970). Water potential measurement in trees. *Bio Science* **20**, 225–226.

Wiebe, H. H., Campbell, G. S., Gardner, W. H., Rawlins, S. L., Cary, J. W., and Brown, R. W. (1971). Measurement of plant and soil water status. *Bull.—Utah, Agric. Exp. Stn.* **484**.

Wieler, A. (1893). Das Bluten der Pflanzen. *Beitr. Biol. Pflanz.* **6**, 1–211.

Wiersma, J. V., and Bailey, T. B. (1975). Estimation of leaflet, trifoliate, and total leaf areas of soybeans. *Agron. J.* **67**, 26–30.

Wiggans, C. C. (1936). The effect of orchard plants on subsoil moisture. *Proc. Am. Soc. Hortic. Sci.* **33**, 103–107.

Wilcox, H. (1954). Primary organization of active and dormant roots of noble fir, *Abies procera*. *Am. J. Bot.* **41**, 812–821.

Wilcox, H. (1962). Growth studies of the root of incense cedar *Libocedrus decurrens*. II. Morphological features of the root system and growth behavior. *Am. J. Bot.* **49**, 237–245.

Will, G. M. (1966). Root growth and dry-matter production in a high-producing stand of *Pinus radiata*. *N. Z. For. Res. Notes* **44**.

Will, G. M., and Stone, E. L. (1967). Pumice soils as a medium for tree growth. 1. Moisture storage capacity. *N. Z. J. For.* **12**, 189–199.

Williams, G.C.(1966)."Adaptation and Natural Selection." Princeton Univ. Press, Princeton, New Jersey.

Williams, H. F. (1933). Absorption of water by the leaves of common mesophytes. *J. Elisha Mitchell Sci. Soc.* **48**, 83–100.

Williams, W. T. (1950). Studies in stomatal behavior. IV. The water-relations of the epidermis. *J. Exp. Bot.* **1**, 114–131.

Williams, W. T., and Barber, D. A. (1961). The functional significance of aerenchyma in plants. *Symp. Soc. Exp. Biol.* **15**, 132–144.

Williamson, C. E. (1950). Ethylene, a metabolic product of diseased or injured plants. *Phytopathology* **40**, 205–208.

Wilson, B. F. (1967). Root growth around barriers. *Bot. Gaz. (Chicago)* **128**, 79–82.

Wilson, C. C. (1948). The effect of some environmental factors on the movements of guard cells. *Plant Physiol.* **23**, 5–37.

Wilson, C. C., and Kramer, P. J. (1949). Relation between root respiration and absorption. *Plant Physiol.* **24**, 55–59.

Wilson, C. C., Boggess, W. R., and Kramer, P. J. (1953). Diurnal fluctuations in the moisture content of some herbaceous plants. *Am. J. Bot.* **40**, 97–100.

Wilson, J. M. (1983). Interaction of chilling and water stress. *In* "Crop Reactions to Water and Temperature Stresses in Humid, Temperature Climates" (C. D. Raper, Jr. and P. J. Kramer, eds.), pp. 133–147. Grandview Press, Boulder, Colorado.

Wilson, K. (1947). Water movement in submerged aquatic plants, with special reference to cut shoots of *Ranunculus fluitans*. *Ann. Bot. (London)* [N.S.] **11**, 91–122.

Wilson, K., and Honey, J. N. (1966). Root contraction in *Hyacinthus orientalis*. *Ann. Bot. (London)* [N.S.] **30**, 47–61.

Wind, G. P. (1955). Flow of water through plant roots. *Neth. J. Agric. Sci.* **3**, 259–264.

Winneberger, J. H. (1958). Transpiration as a requirement for growth of land plants. *Physiol. Plant.* **11**, 56–61.

Wolf, F. A. (1962). "Aromatic or Oriental Tobaccos." Duke Univ. Press, Durham, North Carolina.

Woodhams, D. H., and T. T. Kozlowski. (1954). Effects of soil moisture stress on carbohydrate development and growth in plants. *Am. J. Bot.* **41**, 316–320.

Woodroof, J. G., and N. C. Woodroof. (1934). Pecan root growth and development. *J. Agr. Res.* **49**, 511–530.

Woods, F. W. (1957). Factors limiting root penetration in deep sands of the southeastern Coastal Plain. *Ecology* **38**, 357–359.

Woods, F. W., and K. Brock. (1970). Interspecific transfer to inorganic materials by root systems of woody plants. *Ecology* **45**, 886–889.

Woods, T. E. (1980). Biological and chemical control of phosphorus cycling in a northern hardwood forest. Ph.D. Thesis. Yale University, New Haven, Connecticut.

Woolley, J. T. (1961). Mechanisms by which wind influences transpiration. *Plant Physiol.* **36**, 112–114.

Woolley, J. T. (1965). Radial exchange of labeled water in intact maize roots. *Plant Physiol.* **40**, 711–717.

Woolley, J. T. (1966). Drainage requirements of plants. Proc. Conf. on Drainage for Efficient Crop Production, pp. 2–5. Am. Soc. Agr. Eng., St. Joseph, Michigan.

Wright, J. L., Stevens, J., and Brown, M. J. (1981). Controlled cooling of onion umbels by periodic sprinkling. *Agron. J.* **73**, 481–490.

Wuenscher, J. E., and Kozlowski, T. T. (1971). The response of transpiration resistance to leaf temperature as a desiccation resistance mechanism in tree seedlings. *Physiol. Plant.* **24**, 254–259.

Wylie, R. B. (1938). Concerning the conductive capacity of the minor veins of foliage leaves. *Am. J. Bot.* **25**, 567–572.

Wylie, R. B. (1943). The role of the epidermis in foliar organization and its relations to the minor venation. *Am. J. Bot.* **30**, 273–280.

Wylie, R. B. (1952). The bundle sheath extension in leaves of dicotyledons. *Am. J. Bot.* **39**, 645–651.

Yadava, V. L., and Doud, S. L. (1980). The short life and replant problems of deciduous fruit trees. *Hortic. Rev.* **2**, 1–116.

Yarwood, C. E. (1978). Water and the infection process. *In* ''Water Deficits and Plant Growth'' (T. T. Kozlowski, ed.), Vol. 5, pp. 141–173. Academic Press, New York.

Yegappan, T. M., Paton, D. M., Gates, C. T., and Miller, W. J. (1980). Water stress in sunflower (*Helianthus annuus* L.) I. Effect on plant development. *Ann. Bot. (London)* [N.S.] **46**, 61–70.

Yelenosky, G. (1964). Tolerance of trees to deficiencies of soil aeration. *Int. Shade Tree Conf. Proc., 40th, 1964* pp. 127–147.

Young, K. K., and Dixon, J. D. (1966). Overestimation of water content at field capacity by use of sieved sample data. *Soil Sci.* **101**, 104–107.

Yu, G. H. (1966). A study of radial movement of salt and water in roots. Ph.D. Dissertation, Duke University, Durham, North Carolina.

Zahner, R. (1968). Water deficits and growth of trees. *In* ''Water Deficits and Plant Growth'' (T. T. Kozlowski, ed.), Vol. 2, pp. 191–254. Academic Press, New York.

Zak, B. (1964). Role of mycorrhizae in root disease. *Annu. Rev. Phytopathol.* **2**, 377–392.

Zeikus, J. G., and Ward, J. C. (1974). Methane formation in living trees: A microbial origin. *Science* **184**, 1181–1183.

Zelitch, I. (1961). Biochemical control of stomatal opening in leaves. *Proc. Natl. Acad. Sci. U.S.A.* **47**, 1423–1433.

Zelitch, I. (1969). Stomatal control. *Annu. Rev. Plant Physiol.* **20**, 329–350.

Zelitch, I. (1979). Improving the efficiency of photosynthesis. *Science* **188**, 626–633.

Zentmyer, G. A. (1966). Soil aeration and plant disease. *Proc. Cont. Drain. Efficient Crop Prod., 1966* pp. 15–16.

Zholkevich, V. N., Sinitsyna, Z. A., Peisakhzon, B. I., Abutalybov, V. F., and D'yachenko, I. V. (1980). On the nature of root pressure. *Sov. Plant Physiol. (Engl. Transl.)* **26**(5), 790–802.

Zimmermann, M. H. (1964a). Sap movement in trees. *Biorheology* **2**, 15–27.

Zimmermann, M. H. (1964b). Effect of low temperature on ascent of sap in trees. *Plant Physiol.* **39**, 568–572.

Zimmermann, M. H. (1973). The monocotyledons. Their evolution and comparative biology. IV. Transport problems in arborescent monocotyledons. *Q. Rev. Biol.* **48**, 314–321.

Zimmermann, M. H. (1978a). Hydraulic architecture of some diffuse porous trees. *Can. J. Bot.* **56**, 2286–2295.

Zimmermann, M. H. (1978b). Structural requirements for optimal water conduction in tree stems. *In* ''Tropical Trees as Living Systems'' (P. B. Tomlinson and M. H. Zimmerman, eds.), pp. 517–532. Cambridge Univ. Press, London and New York.

Zimmermann, M. H., and Brown, C. L. (1971). ''Trees: Structure and Function.'' Springer-Verlag, Berlin and New York.

Zimmermann, M. H., and McDonough, J. (1978). Dysfunction in the flow of food. *In* ''Plant Disease'' (J. G. Horsfall, and E. B. Cowling, eds.), Vol. 3, pp. 117–140. Academic Press, New York.

Zimmermann, M. H., and Milburn, J. A., eds. (1975). ''Transport in Plants 1: Phloem Transport.'' Springer-Verlag, Berlin and New York.

Zimmermann, M. H., and Tomlinson, P. B. (1974). Vascular patterns in palm stems: Variation of the *Rhapis* principle. *J. Arnold Arbor. Harv. Univ.* **55**, 402–424.

Zimmermann, U. (1978). Physics of turgor and osmoregulation. *Annu. Rev. Plant Physiol.* **29**, 121–148.

Zimmermann, U., and Dainty, J., eds. (1974). ''Membrane Transport in Plants.'' Springer-Verlag, Berlin and New York.

Zimmermann, U., and Steudle, E. (1975). The hydraulic conductivity and volumetric elastic modulus of cells and isolated cell walls of Nitella and Chara spp. pressure and volume effects. *Aust. J. Plant Physiol.* **2,** 1–12.

Zur, B. (1967). Osmotic control of the matric soil-water potential. II. Soil-plant system. *Soil Sci.* **103,** 301–38.

Zur, G., Boote, K. J., and Jones, J. W. (1981). Changes in internal water relations and osmotic properties of leaves in maturing soybean plants. *J. Exp. Bot.* **32,** 1181–1191.

Index